Integrated Water Management for Enhanced Water Quality and Reuse to Create a Sustainable Future

In Focus – Special Book Series

Integrated Water Management for Enhanced Water Quality and Reuse to Create a Sustainable Future

Editors

Eldon R. Rene, Veeriah Jegatheesan and Li Shu

IWA PUBLISHING

Published by

IWA Publishing
Unit 104–105, Export Building
1 Clove Crescent
London E14 2BA, UK
Telephone: +44 (0)20 7654 5500
Fax: +44 (0)20 7654 5555
Email: publications@iwap.co.uk
Web: www.iwapublishing.com

First published 2021

British Library Cataloguing in Publication Data
A CIP catalogue record for this book is available from the British Library

ISBN: 9781789062939 (paperback)

Contents

doi: 10.2166/ws.2021.198

Editorial: Integrated water management for enhanced water quality and reuse to create a sustainable future

Safe drinking water and sanitation are very important for the survival of human life. With the rapid proliferation of industries, growth in population and different forms of pollution, i.e. in water, air, soil and sediments, the living environment and the ecosystem is constantly polluted. In this context, integrating different water resources for enhanced water quality and reuse is important to solve the persisting problems and challenges in developing and the developed nations. Integrated water management offers environmental, economic and social benefits because it aims at maximizing the existing resources and prevents further depletion of the ecosystem.

In this special issue, a total of 22 papers were selected for publication from an open call for papers in association with the 13th International Conference on Challenges in Environmental Science and Engineering, CESE-2020. These papers covered the following aspects as illustrated in Figure 1: (i) the performance of different bioreactor configurations for wastewater treatment, (ii) rain water harvesting, (iii) water/wastewater reuse, (iv) the application of mathematical models, neural networks and remote sensing for water quality monitoring, treatment plant performance prediction and water resource management, and (v) the implementation of the water-energy-agriculture concept in practice.

Alsulaili & Refaie (2021) applied artificial neural networks (ANNs) for predicting the performance of wastewater treatment plants (WWTPs) in Kuwait using temperature, pH, conductivity, and BOD as the input parameters. According to the results, increasing the number of model inputs beyond three did not enhance the ANN performance, while the influent BOD and conductivity showed the highest effect on the effluent water quality. The results from sensitivity analysis showed that the influent COD had the highest impact on effluent BOD. The authors recommended ANN model as a reliable option for the intelligent control and monitoring of WWTPs in practice. In a study aimed at water quality monitoring for timely decision making in Namibia, Kapalanga *et al.* (2021) used reflectance values and field measured water quality data to develop regression analysis-based retrieval algorithms. The authors used turbidity, total suspended solids (TSS), nitrates, ammonia, total nitrogen (TN), total phosphorus (TP) and total algae counts as the major water quality parameters. It was observed that the turbidity levels exceeded the recommended limits for potable water, while the TN and TP values were within the acceptable values. According to the authors, an assessment and monitoring of water quality using the combination of remote sensing and in-situ measurements will help to provide a good estimate of the water quality. Furthermore, the authors had also used remote sensing as a framework for continuous monitoring of water quality in Olushandja Dam (Namibia). Another study also used ANNs for predicting the removal of heavy metals, i.e. Cu (II), Cd (II) and Pb (II) from synthetic wastewater in a rotating biological contactor (RBC) (Kiran *et al.* 2021). The authors used a feed-forward back propagation neural network to predict the removal of heavy metals and COD using 90 data sets obtained over a period of three months from the RBC. Through vigorous trial and error for training and testing, and by proper optimization of the internal network parameters, the authors ascertained a network architecture of 2-12-2 and hydraulic retention time (HRT) as the most sensitive input parameter for heavy metal and COD removal in the RBC.

Garin *et al.* (2021) conducted a survey in a French wine-growing region where the public authorities were considering irrigating fields and green spaces with treated wastewater. The authors conducted 845 face-to-face interviews with potential consumers and the following interesting results were envisaged in their study: (a) attitudes and intended behaviours were driven by disgust, environmental sensitivity, risk perception, and type of water reuse, (ii) working women more reluctant than men to wastewater reuse, (iii) the farmers rejected the idea of wastewater reuse, and (iv) with the help of proper communication, farmers could strengthen the social acceptability of wastewater reuse, but with a loss of their wine customers. In this work, the authors clearly presented the attitudes of the consumers buying their fruit (i.e. the grapes), vegetables and wines in this area towards the implementation of a wastewater reuse project. The authors also mentioned that the reluctant group, significantly less open to the media and made suspicious by health scandals, will be difficult to convince regarding the

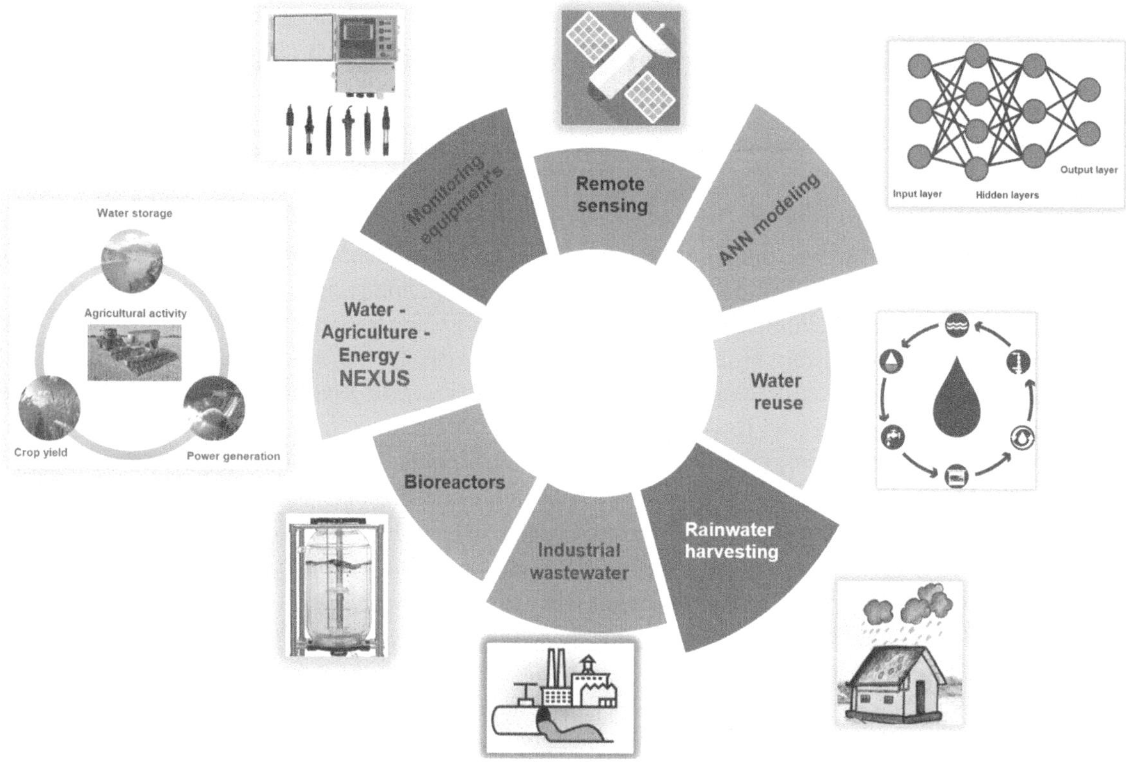

Figure 1 | The application of environmental concepts, technologies and modelling tools for environmental sustainability.

implementation of any wastewater reuse projects in the wine-growing region. In another questionnaire-based study, Vasudevan & Natarajan (2021) prepared structured questionnaires and conducted a survey to realise the awareness of people about improving the rainwater harvesting (RWH) infrastructure in Tamilnadu, India. The authors formulated 45 multiple-choice type questions for the survey under various categories such as demographic details (11%), details about source of water (11%), access to financial support systems (20%), awareness on water quality and treatment (22%), and technical know-how about the operation, troubleshooting and cleaning protocols (36%). The results from this study showed that a majority of the people use harvested water for domestic purposes either collected from rooftops or from the porch/garden, while some people prefer direct groundwater recharge. Furthermore, this study also highlighted an important concern from the people for improved water quality from RWH storage tanks with the help of a treatment step before direct consumption. Thus, the authors recommended that groundwater recharge has to be approached with caution to improve the priority-wise usage of the harvested rainwater and an integrated public-private partnership for RWH has to emerge in India with due consideration for eco-socio-economic variability.

Taylor *et al.* (2021a) determined the suitability of waste activated sludge (WAS) and algae produced from a brewery effluent treatment system as a fertiliser in agriculture. In this study, the change in soil characteristics and the growth of a crop fertilised with WAS or algae was compared with a conventional inorganic fertiliser. It was observed that, Swiss chard production was similar when fertilised with fertiliser, WAS or algae and both WAS and algae increased the soil's sodium absorption ratio. Furthermore, the soil pH, exchangeable acidity and electrical conductivity were similar between all fertiliser treatments, while the soil phosphorus and potassium concentrations were higher in the inorganic-fertilised soil compared with algae or WAS fertiliser soils, respectively. Besides, the authors also observed that the nitrogen applied to the soil from algae and WAS biomass leached out of the soil less than the nitrogen supplied by inorganic the fertilisers. In another study involving algal biomass, Blanco *et al.* (2021) studied the effect of using anaerobically digested gelatin industry wastewater as a culture medium for the cultivation of *Chlorella vulgaris* microalgae in bubble column photobioreactors (PBRs). The authors determined the kinetic parameters of microbial growth and substrate consumption, the biomass productivity, the efficiency of the removal of nitrogen compounds (nitrate and other N compounds) and the lipid productivity. The authors proved that the use of wastewater in bubble column photobioreactors (PBRs) provided a 58% increase in algal biomass productivity and ammoniacal nitrogen was preferred compared with nitrate during the PBR operation. Concerning lipid production, the authors reported a total lipid content of 0.13 g, i.e. a mass fraction of 12.45%, while the lipid productivity was 17.6 mg/L.d. The authors attributed the low lipid production to the availability of substrate as the limiting step for the growth algal biomass and production of fatty acids.

Taylor *et al.* (2021b) grew cabbage (*Brassica oleracea*) in recirculating hydroponic systems fed with post-anaerobically digested brewery effluent (BE), both with pH adjustment in the range of 6.5–7.0 or unaltered pH (8.0–8.5). The study is interesting from the viewpoint that brewery effluent is not an ideal candidate for its use in hydroponic crop production and there are only a few studies that have reported its land application for growing tomatoes, cabbage and lettuce. The authors were able to ascertain the effect of nutrient solution type and pH on the health, growth and leaf chemical composition of cabbage plants and elucidate the effect of pH on the removal of nutrients and other elements by cabbage plants grown in a hydroponic production system. The hydroponic systems fed with post-primary facultative pond brewery adjusted to pH 6.5 decreased the dissolved inorganic nitrogen, ammonium, phosphate and COD concentrations from 54.8, 36.9, 27.3 and 216 mg/L to 14.9, 0.56, 18.6 and 105 mg/L, respectively. Santos *et al.* (2021) studied the effect of decreasing the hydraulic retention time (HRT) on the stability and efficiency of a novel hybrid anaerobic biofilm baffled reactor (HABBR) treating simulated fat- and salt-rich dairy wastewater. The authors used the treated water for agriculture. During the long-term investigation of 328 days, hydraulic detention times of 72, 24, and 12 h were tested and organic matter removal >90% and biogas in the range of 41–64% was produced in the HABBR. From an application view point, the following recommendations were made by the authors: (i) agricultural reuse of the anaerobic-treated effluent could guarantee yields of the lettuce crops, (ii) 50% dose of dairy anaerobically treated effluent resulted in the best conditions for residue utilization in agriculture, and (iii) a reduction in 50% of the nitrogenous mineral fertilizer can be achieved.

Ozay *et al.* (2021) conducted a series of well-planned experiments to investigate the treatability of plasticizer production industry wastewater using nanofiltration (NF, NF270 and NF90) and reverse osmosis (RO, BW30) membranes. For these two systems, the authors tested the effect of operating pressure, pH of the wastewater, and sequential treatment option on the permeate flux, COD, phthalate, and micropollutant removal efficiencies. Furthermore, the effect of transmembrane pressure and pH of wastewater were tested on the flux and permeate quality. The following important results were reported: (i) the NF270 (NF system) had the highest permeate flux value due to its larger pore size compared with NF90 and BW30 membranes and (ii) COD, phthalate and micropollutant removal efficiency increased with the adjustment of operating conditions for NF90 and NF90 + BW30 membranes. Another study also focused on solving an important industrial wastewater issue arising from the electroplating sectors. Rajasimman *et al.* (2021) tested the removal of zinc ions from wastewater using a novel green emulsion liquid membrane technology. This liquid membrane was prepared using waste cooking oil and a surfactant (SPAN 80), with sulfuric acid as the internal phase. The authors used Box-Behnken Design (BBD) to optimize the operating variables for zinc removal, namely, the external pH, surfactant concentration, internal phase concentration, zinc concentration, external phase to emulsion volume ratio and carrier concentration. To achieve >97% zinc removal, the following optimal conditions were reported: pH - 3.8, surfactant concentration 4% (*v/v*), internal phase concentration – 1.61 N, zinc concentration – 742 mg/L, external phase to emulsion volume ratio - 0.94 and carrier concentration – 8.9%. Sivapragasam *et al.* (2021) tested real-time monitoring as an adaptive strategy towards green treatment of textile effluent using biosorbent prepared from *Acalypha indica*. The authors conducted batch experiments with synthetic and real-time dye effluents under the following optimal conditions: pH = 3.0, dosage of the biosorbent = 1.0 g/L and time = 1 h and achieved the highest adsorption capacity of 6 mg/g and 2 mg/g, respectively. For real-time monitoring, the authors used LabVIEW for the measurement of pH and temperature of the effluent and recommended that there is enormous scope in optimizing the entire process-flow of textile wastewater treatment in order to withdraw, treat and reuse the effluents in a continuous manner. The authors also showed substantial reduction in COD (74%), TDS (20%) and turbidity (53%) due to the application of adsorption technology.

Kan & Duan (2021) highlighted the importance of non-point source (NPS) pollution and suggested a resource-based approach for controlling agricultural NPS pollution using biological agents. According to the authors, in this novel approach, the domestic sewage disposal was carried out under powered or non-powered conditions (i.e. with or without electricity supply), and the domestic organic waste, sewage sludge, and livestock manure were economically utilized to produce biological agents to replace the harmful chemical fertilizers and pesticides that contribute to NPS pollution. The proposed approach was applied in Xichuan County and Xixia County since 2010 and the results have successfully demonstrated its value with regard to NPS pollution control. From a longevity view point of this approach, the following recommendations were given by the authors: (i) to improve water quality in a sustainable manner, the environmental protection infrastructure in the townships and villages should be improved, and (ii) a corporate and market-oriented operation of rural pollution control should be implemented by involving local stakeholders and the farming communities. Another similar study dealt with the removal of fly ash using a biocatalyst, i.e. *Penicillium chrysogenum,* and ascertained its toxicity effects (Taştan 2021). The author compared two different removal mechanisms, namely, bio-removal and bio-sorption. The results of this study were promising because the highest bio-removal yield was 100% at a fly ash concentration of 0.5%, while the bio-sorption yield was 95% after 24 h of contact time. The author also recommended that fungi are very effective microorganisms for selectively removing heavy metals (i.e. by bio-sorption) from fly ash and reducing their toxicity effects on the environment.

In a study that applied adsorption and biodegradation for removing natural organic matters (NOM), Gong *et al.* (2021) used mesoporous activated carbon MCGL-4 for simultaneous enhancement of adsorption and bio-degradation by multistage depth-activation (MDA) in a pilot-scale bio-enhanced activated carbon (BEAC) system. In that study, in order to determine the synergistic effect between bio-degradation and adsorption, rapid small-scale column tests system (RSSCTs) tests were performed. The adsorption capacities for humic-like organics were 67,725 mg DOC/(kg carbon) and the removal due to biodegradation was 31,674 mg DOC/(kg carbon), respectively. The authors ascertained that the mesopores played an important role for the adsorption of humic-like organics. Jin-xi & Singh (2021) evaluated the performance of a modified flocculant (PEI-T) for the removal of turbidity, heavy metal (Hg^{2+}) and the activator EDC.HCl. The novel PEI-T was prepared under the following conditions: room temperature, nPEI: nTGA = 1:2, PEI concentration = 4%, time t = 12 h, catalyst EDC = 6 mL, and initial pH = 2.5. In this study, the authors deduced the removal mechanism as follows: (i) complexation reaction with Hg^{2+} in the form of coordination bonds and covalent bonds to form a stable insoluble heavy metal complex to precipitate PEIT-Hg, (ii) redox reaction, and (iii) the presence of Na^{2+}, Mg^{2+}, and Ca^{2+} in wastewater promoted the removal of heavy metals by PEI-T. Nandi *et al.* (2021) studied the biodegradation of benzyl butyl phthalate (BBP) and dibutyl phthalate (DBP) by *Arthrobacter* sp. via micellar solubilization in a surfactant-aided system. In systems assisted by the action of surfactants, the bioavailability of hydrophobic organic pollutants is increased either by decreasing the interfacial tension between the aqueous and non-aqueous phases or by the solubilization of hydrophobic organic compounds in the core of the micelles formed by surface-active agents. The authors used four chemical surfactants, e.g. Tween 20, Tween 80, Triton X-100 and sodium dodecyl sulphate (SDS), and assessed their capability to solubilize BBP and DBP in aqueous solutions. Based on a series of batch tests, Tween 80 was found to be highly effective for enhancing the bioavailability of both BBP and DBP and their biodegradation by *Arthrobacter* sp. Zamora *et al.* (2021) compared coriander (*Coriandrum sativum* L.), cv. Verdão yields under fertigation via drip irrigation with a continuous application and in pulse type applications, with 40, 60, 80, 100 and 120% of the crop evapotranspiration (ETc) being recovered. The authors deduced that the frequency of watering and fertilization affected the crop yield, whereas pulse type drip irrigation saves 40% water compared to continuous irrigation. From an application view point, pulsed fertigation was recommended by the authors for coriander crop because it provides higher shoot and root values of fresh and dry mass as well as plant height and yield values, with evapotranspiration replacement of $<100\%$.

Ramasamy *et al.* (2021) highlighted the need for an improved and controlled Smart Energy Distribution Network (SEDN) in industries and used Python-based simulations of weighted least squares (WLS) and principal component analysis (PCA) based Data Reconciliation (DR) techniques on selected flow streams of SEDN. Four important results were reported in that study: (i) the DR technique showed good ability to treat random errors present in flow sensor data used in water distribution networks, (ii) the performance of DR techniques for small and large scale networks using Python-based simulation, (iii) RMSE is the best metric for evaluation because it is very sensitive to small changes in the measurement and estimates, and (iv) when the difference in magnitudes between process variables increases, the PCA-DR combination performs less accurately compared to the WLS-DR. Another study highlighted the importance of applying the most popular Soil Conservation Service Curve Number (SCS-CN) method for calculating storm depth from a rainfall event (Upreti & Ojha 2021). In order to reduce the error in predicting runoff calculations during an unexpected rise in the SCS-CN, the authors considered pre-storm rainfall as an event-based runoff model, and evaluated the relative significance of antecedent precipitation (P5) on the calculated runoff amount. According to the results of this model-based study, the maximum potential retention value (S, mm) changed due to the antecedent 5 days' rainfall values and the authors were able to test the performance of this model using rainfall-runoff events of 114 watersheds located in the USA.

Chen & Wang (2021) applied an inexact multi-stage interval-parameter partial information programming model (IMIPM) for urban water resources planning and management under uncertainties. The authors combined the following optimization techniques: two-stage stochastic programming (TSP), interval-parameter programming (IPP), linear partial information theory (LPI) and multistage stochastic programming (MSP) into one framework and applied it for a real case study in Harbin where the manager had to allocate water from multi-water sources to multi-water users during multi-planning time periods. The developed framework was used to verify the rationality and practicality of this approach by managers in Harbin and helped them to propose multiple decision-making alternatives. Torres *et al.* (2021) formulated a method for classifying the interrelation between sectoral regulatory laws and the 'water-energy-agriculture nexus' concept in Brazil. In that study, the authors identified the level of interrelation between sectoral regulatory laws and the main characteristics present in the nexus concept and determined the main strengths and weaknesses of sectoral regulatory laws upon the nexus concept. The main aspect used from multicriteria analysis was the classification phase of alternatives according to selected multiple criteria and three different steps were used by the authors: I: Definition of the criteria – Main characteristics of the composition of the nexus concept, II: Definition of alternatives – regulatory legislation, and III: Evaluation and classification of the interrelation between criteria and alternatives. The authors recommended that the Brazilian energy sector needs to be tightly integrated with water and agriculture legislations, especially in matters concerning biofuels and biodiesel policies that have a strong association with the growing agricultural sectors.

The Editors of this special issue are confident that the papers published in this issue are from a wide array of multidisciplinary fields, both from the industrial and academic community, and they will certainly attract broad readership from scholars and practising engineers from the environmental, modelling, biochemical and agricultural sectors. Our special thanks to Ms Emma Buckingham, Editorial Operations Manager, IWA Publishing and the entire production team at IWA Publishing for their valuable support in bringing out this issue successfully. We also thank all the reviewers for providing critical reviews, comments and feedback on manuscripts that were submitted for this thematic special issue.

Guest Editors
Eldon R. Rene
IHE Delft Institute for Water Education, Delft, The Netherlands

Veeriah Jegatheesan
RMIT University, Melbourne, Australia

Li Shu
Edith Cowan University, Joondalup, Australia

REFERENCES

Alsulaili, A. & Refaie, A. 2021 Artificial neural network modeling approach for the prediction of five-day biological oxygen demand and wastewater treatment plant performance. *Water Supply* **21** (5), 1861–1877.

Blanco, G. C., Stablein, M. J. & Tommaso, G. 2021 Cultivation of *Chlorella vulgaris* in anaerobically digested gelatin industry wastewater. *Water Supply* **21** (5), 1953–1965.

Chen, H. & Wang, Z. 2021 An inexact multi-stage interval-parameter partial information programming model for water resources management under uncertainties. *Water Supply* **21** (5), 2139–2157.

Garin, P., Montginoul, M. & Noury, B. 2021 Waste water reuse in France – social perception of an unfamiliar practice. *Water Supply* **21** (5), 1913–1926.

Gong, X.-J., Dong, Y.-Q. & Li, W.-G. 2021 Mesoporous structure regulation of activated carbons and effects on the synergistic efficacy of synchronous adsorption and bio-degradation in biological activated carbon process. *Water Supply* **21** (5), 2058–2074.

Jin-xi, W. & Singh, R. P. 2021 Experimental study on the performance of modified flocculant (PEI-T) for removal of turbidity and heavy metals from aqueous solution. *Water Supply* **21** (5), 2075–2083.

Kan, D. & Duan, P. 2021 Resource-based agricultural non-point source pollution control using biological agents. *Water Supply* **21** (5), 2035–2046.

Kapalanga, T. S., Hoko, Z., Gumindoga, W. & Chikwiramakomo, L. 2021 Remote-sensing-based algorithms for water quality monitoring in Olushandja Dam, north-central Namibia. *Water Supply* **21** (5), 1878–1894.

Kiran, M. G., Das, R., Behera, S. K., Pakshirajan, K. & Das, G. 2021 Modelling a rotating biological contactor treating heavy metal contaminated wastewater using artificial neural network. *Water Supply* **21** (5), 1895–1912.

Nandi, M., Paul, T., Kanaujiya, D. K., Baskaran, D., Pakshirajan, K. & Pugazhenthi, G. 2021 Biodegradation of benzyl butyl phthalate and dibutyl phthalate by *Arthrobacter* sp. via micellar solubilization in a surfactant-aided system. *Water Supply* **21** (5), 2084–2098.

Ozay, Y., Canli, O., Unal, B. O., Keskinler, B. & Dizge, N. 2021 Investigation of plasticizer production industry wastewater treatability using pressure-driven membrane process. *Water Supply* **21** (5), 1994–2007.

Rajasimman, M., Rajamohan, N. & Sujatha, S. 2021 Recovery of zinc from electroplating wastewater using Green emulsion liquid membrane. *Water Supply* **21** (5), 2008–2018.

Ramasamy, J., Devanathan, S. & Jayaraman, D. 2021 Comparative analysis of select techniques and metrics for data reconciliation in smart energy distribution network. *Water Supply* **21** (5), 2109–2121.

Santos, K. A., Gomes, T. M., Rossi, F., Kushida, M. M., Bianchi, V. L. D., Ribeiro, R., Alves, M. S. M. & Tommaso, G. 2021 Water reuse: dairy effluent treated by a hybrid anaerobic biofilm baffled reactor and its application in lettuce irrigation. *Water Supply* **21** (5), 1980–1993.

Sivapragasam, C., Janani, V. A., Andappan, A., Archana, B., Vasudevan, M. & Natarajan, N. 2021 Real-time monitoring as an adaptive strategy towards Green treatment of textile effluent using biosorbent from *Acalypha indica*. *Water Supply* **21** (5), 2019–2034.

Taştan, B. E. 2021 Effective removal of fly ash by *Penicillium chrysogenum* and determination of direct fly ash toxicity with *Daphnia magna*. *Water Supply* **21** (5), 2047–2057.

Taylor, R. P., Jones, C. L. & Laubscher, R. K. 2021a Agricultural fertiliser from brewery effluent – the recovery of nutrients from the biomass of activated sludge and high rate algal pond treatment systems. *Water Supply* **21** (5), 1939–1952.

Taylor, R. P., Jones, C. L. W. & Laing, M. 2021b Effect of pH on nutrient removal and crop production of hydroponic systems treating brewery effluent. *Water Supply* **21** (5), 1966–1979.

Torres, C. J. F., de Lima, C. H. P., Fontes, A. S., Ribeiro, D. V., Moreira, Í. T. A. & Medeiros, Y. D. P. 2021 A method for classifying interrelation between sectoral regulatory laws and the 'water-energy-agriculture nexus concept' in Brazil. *Water Supply* **21** (5), 2158–2172.

Upreti, P. & Ojha, C. S. P. 2021 Comparison of antecedent precipitation based rainfall-runoff models. *Water Supply* **21** (5), 2122–2138.

Vasudevan, M. & Natarajan, N. 2021 An investigation on the adaptability of residential rainwater harvesting system in Tamil Nadu – technoeconomic considerations and the way forward. *Water Supply* 21 (5), 1927–1938.

Zamora, V. R. O., da Silva, M. M., Júnior, J. A. S., da Silva, G. F., Menezes, D. & de Almeida, C. D. G. C. 2021 Assessing the productivity of coriander under different irrigation depths and fertilizers applied with continuous and pulsed drip systems. *Water Supply* **21** (5), 2099–2108.

Artificial neural network modeling approach for the prediction of five-day biological oxygen demand and wastewater treatment plant performance

Abdalrahman Alsulaili IWA and Abdelrahman Refaie

ABSTRACT

The measurement of the wastewater BOD_5 level requires five days, and the use of a prediction model to estimate BOD_5 saves time and enables the adoption of an online control system. This study investigates the application of artificial neural networks (ANNs) in predicting the influent BOD_5 concentration and the performance of WWTPs. The WWTP performance was defined in terms of the COD, BOD, and TSS concentrations in the effluent. Sensitivity analysis was performed to identify the best-performing ANN network structure and configuration. The results showed that the ANN model developed to predict the BOD concentration performed the best among the three outputs. The top-performing ANN models yielded R^2 values of 0.752, 0.612, and 0.631 for the prediction of the BOD, COD, and TSS concentrations, respectively. The optimal performing models were obtained (three inputs – one output), which indicated that the influent temperature and conductivity greatly affect the WWTP performance as inputs in all models. The developed prediction model for the influent BOD_5 concentration attained a high accuracy, i.e., $R^2 = 0.754$, which implies that the model is viable as a soft sensor for online control and management systems for WWTPs. Overall, the ANN model provides a simple approach for the prediction of the complex processes of WWTPs.

Key words | artificial neural networks, modeling, sensitivity analysis, wastewater plant, wastewater treatment

Abdalrahman Alsulaili IWA (corresponding author)
Abdelrahman Refaie
Civil Engineering Department, College of Engineering and Petroleum,
Kuwait University,
P.O. Box-5969, 13060, Safat,
Kuwait
E-mail: *a.alsulaili@ku.edu.kw*

HIGHLIGHTS

- ANN model provides an assessment tool for WWTP design and performance.
- Increasing the number of model inputs beyond three inputs was not beneficial.
- Influent BOD and conductivity have the highest effect on the WWTP effluent.
- COD input parameter had the highest impact on BOD_5 prediction model.
- BOD_5 soft-sensor development is viable using ANN model.

doi: 10.2166/ws.2020.199

GRAPHICAL ABSTRACT

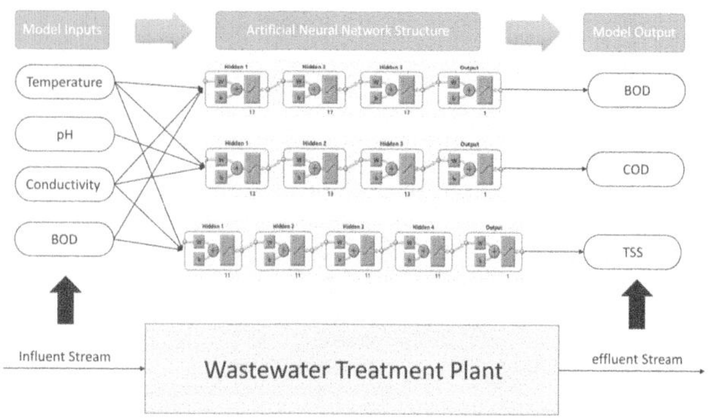

INTRODUCTION

Rapid urban development leads to an increase in wastewater flow rates, which requires a high pace of advancement in treatment methods. Wastewater treatment plant (WWTP) effluent quality is a key factor in environmental and health concerns (Hamed *et al.* 2004). Thus, it is important to be able to model and simulate WWTPs to identify optimal conditions and to avoid future operation failures. However, the characteristics and variables involved in WWTPs are numerous and exhibit a high level of complexity, which leads to difficulties in modeling through linear regression models (Hamed *et al.* 2004). Moreover, physical and chemical simulations through software can be performed to simulate WWTPs. Ferrer *et al.* (2008) applied Design and Simulation of Activated Sludge Systems (DESASS) software to model and optimize WWTP performance in the steady or transient state. Other software includes anaerobic digestion models (Batstone *et al.* 2002) and the Activated Sludge Model (ASM) (Henze *et al.* 1987, 1995, 1999). However, these software programs require a large amount of input data regarding the process at each stage of the operation.

Data-driven simulation and modeling is an alternative approach in the modeling of complex systems (Elmolla *et al.* 2010). Some of these methods include fuzzy logic and artificial neural networks (ANNs). Data-driven modeling relies on advancements in the computing field and artificial intelligence (AI). ANNs can be applied to predict the performance of WWTPs. Due to their high prediction accuracy, ANNs were developed to model the water treatment process (Manu & Thalla 2017; Newhart *et al.* 2019; Alver & Kazan 2020). ANN modeling greatly relies on the quality of historical data. Poor historical data quality could lead to poor ANN modeling performance. However, ANN modeling requires a relatively small amount of data to provide acceptable prediction results. The use of neural network models to simulate wastewater plants provides a framework for plant operation monitoring (Mjalli *et al.* 2007). This monitoring framework leads to the minimization of the cost of operation and the determination of the quality of environmental stability. Mjalli *et al.* (2007) collected data over a span of one year (measurements were conducted every five days). It was determined that the outputs of the secondary treatment effluent (STE) were the inputs of their model. The overall data set was divided at a ratio of 4:2:1 into training, validation, and testing data sets. It should be noted that Mjalli *et al.* (2007) employed a feed-forward neural network (FFNN), which was developed using MATLAB software. In all single-input networks, the network structure consists of three layers (input, hidden, and output layers). In multi-input networks, the network structure consists of four layers (two hidden layers). The optimal network structure was found to be 1–40–1 with the COD as the input variable ($R^2 = 0.987$ and MSE = 0.021). The training output function was based on the mean square error (MSE) between the network prediction and observed values. Nasr *et al.* (2012) applied an ANN to predict the performance of wastewater treatment plants in

Alexandria. The plant performance was studied in terms of the total suspended solids (TSS), biological oxygen demand (BOD), and chemical oxygen demand (COD) over a one-year period. The gathered data were classified into four groups, where each group represents three months of the year. The feedforward network was developed and trained using back propagation. The predicted values were highly correlated with the measured values ($R^2 = 0.90317$).

Guo *et al.* (2015) applied machine learning through a feedforward ANN and support vector machine (SVM) to predict the total nitrogen concentration in the effluent of a WWTP in Ulsan, Korea. The inputs of the models were the total nitrogen, total phosphorus, and TSS of the influent flow. Based on the coefficient of determination (R^2), Nash–Sutcliffe efficiency (NSE) and relative efficiency (d_{rel}) criteria, the results showed that both models were effective in prediction ($R^2 = 0.55$, NSE $= 0.56$, and $d_{rel} = 0.8$). Even though the SVM model attained a higher prediction efficiency, the ANN model consisting of three layers (input, hidden, and output layers) was more effective in correlating the input values to the T-N concentrations. Nezhad *et al.* (2016) employed the MATLAB ANN toolbox to apply machine learning techniques to predict the effluent quality index (EQI) for wastewater treatment plants in Tehran. A feedforward back-propagated neural network was developed that consisted of three hidden layers. The inputs of the model included BOD, TDS, TSS, FC, PO_4, NH_4, and pH. The results indicated that the optimal network structure was 8–7–1, which led to a high EQI prediction efficiency ($R = 0.96$, and MSE $= 0.1$).

Moreover, Abba & Elkiran (2017) predicted the COD of the effluent of wastewater treatment plants using a feedforward neural network (FFNN). The COD of the effluent is considered a major parameter to assess the performance of WWTPs. An FFNN was applied to predict the COD, BOD, pH, T-P, T-N, TSS, and conductivity of the effluent at WWTP inlets. Several structures and input combinations were considered; however, the FFNN model with an 8–8–1 network structure, utilizing all four input parameters, yielded good performance and the highest accuracy in effluent COD prediction ($R^2 = 0.7$, and RMSE $= 0.0108$).

The use of neural network models to simulate wastewater plants provides a framework for plant operation monitoring (Mjalli *et al.* 2007). This monitoring framework enables minimizing the cost of operation and determining the quality of environmental stability. Moreover, previous studies did not present the method used for determining the ANN configuration. Thus, this study provides a simple approach to identify the optimal ANN configuration for WWTP plant performance prediction.

Moreover, several regression models and artificial neural networks were developed to predict the five-day biological oxygen demand (BOD_5) (Kasem *et al.* 2018; Baki *et al.* 2019; Najafzadeh & Ghaemi 2019). The importance of BOD_5 modeling stems from the extensive laboratory procedures performed to measure the BOD_5 concentration, as these tests require approximately five days. A regression model based on the relation between the characteristics of wastewater and BOD_5 was developed and proved to attain a high accuracy (R^2 up to 0.7966) (Baki *et al.* 2019). Furthermore, Kasem *et al.* (2018) developed a software sensor to monitor the BOD_5 concentration in the Sefidrood River in Iran using a feedforward ANN as a function of the dissolved oxygen level. The developed ANN performance indicated was suitable and a high R^2 value was generated (up to 0.89). In WWTPs, deep neural networks and genetic algorithms were also used to design a BOD_5 soft sensor (Qiu *et al.* 2016). The developed sensor was tested under three conditions, i.e., dry, rainy, and stormy weather conditions on the BSM1 simulation platform, and the results revealed good performance under extreme weather conditions. Although there have been a few studies that developed soft sensors for BOD monitoring, there is a lack of parameter analysis and its effect on predicting the inlet BOD concentration, which will be performed in this study.

Most of the literature utilized TSS, COD, and BOD to predict the performance of WWTPs using ANN-based models. These parameters are considered the controlling parameters of the effluent quality from WWTPs (Abba & Elkiran 2017). Furthermore, WWTPs commonly consist of a series of complex processes that cannot be modeled using simple regression techniques due to the large number of input parameters and data points required to capture the plant performance. Thus, the capacity of ANN modeling makes it a reliable solution in modeling the performance of wastewater treatment plants. The objective of this paper is to investigate the application of ANNs in WWTP modeling and to provide a simple algorithm to determine the optimal ANN configuration capturing the complex behavior of treatment plants. In contrast to most of the previous literature, this study aims to provide a simple determination method for the optimal configuration, which is defined in this study as a simple and general configuration that can be extrapolated, in future studies, to multiple treatment plants with similar processes. Additionally, the generated model could be utilized in the design of control systems and the monitoring of plant performance and water quality parameters. This study also aims to develop a viable ANN to act as a soft sensor of the influent BOD_5 in WWTPs, as there is a lack of this specific approach in the previous literature. The above soft sensor is expected to decrease the five-day period of the BOD_5 measurement to several hours, which will allow the use of online control systems.

METHODOLOGY AND MODEL DEVELOPMENT

Data collection and preparation

The data analyzed in this study were collected from the Kabd WWTP, which is located in Kuwait, over a seven-year period (2013 to 2019). A total of six wastewater parameters were adopted as model inputs, including the influent temperature, pH, conductivity, and TSS, COD, and BOD concentrations. These chosen input parameters sufficiently describe the WWTP performance (Mjalli *et al.* 2007; Nasr *et al.* 2012). These parameters were employed to predict the WWTP performance, which is represented by the effluent BOD, COD, and TSS concentrations. The influent temperature, pH, conductivity, TSS, and COD were used to predict the influent BOD_5 concentration. The total number of data points was 2,397 points; however, they included missing values. Thus, IBM SPSS Statistics 26 was applied to conduct missing value analysis along with descriptive analysis on the data set, as indicated in Table 1. The BOD_5 concentration had the most missing values among all the other parameters, which could be explained by worker negligence in performing the tedious five-day experimental procedure to estimate the BOD_5 value. Thus, a BOD_5 prediction model is a valuable approach to calculate these missing values. Listwise deletion of missing data points was conducted using a developed MATLAB code. This code located the missing values of any parameter and deleted the whole data row. After removing the data points with missing values for all of the parameters, the total number of data points was 1,032 points.

Further data refining was performed by excluding any extreme values that were not within the 1.5 interquartile ranges (IQRs) of (Q1 − 1.5*IQR) and (Q3 + 1.5*IQR). Moreover, data normalization was performed according to Equation (1):

$$Y_{\text{norm}} = \frac{Y - Y_{\text{min}}}{Y_{\text{max}} - Y_{\text{min}}} \tag{1}$$

Table 1 | Descriptive and missing value analysis results on the data set used in this study

Descriptive statistics						Missing		No. of extremes[a]	
	N	Minimum	Maximum	Mean	Standard deviation	Count	Percent	Low	High
Temperature (°C)	1,992	16.6	36.9	28.8	4.3	405	16.9	1	0
pH	1,973	6.48	10.12	7.0	0.2	424	17.7	4	31
Conductivity (μs/cm)	1,991	728	2,100	1,504.1	194.1	406	16.9	7	1
TSS (mg/L)	1,990	72	1,124	173.4	52.6	407	17.0	10	86
COD (mg/L)	1,989	196	1,570	613.9	143.8	408	17.0	16	126
BOD_5 (mg/L)	1,365	17	541	279.6	88.2	1,032	43.1	21	14
TSS effluent (eff.) (mg/L)	1,989	0	92	7.1	5.1	408	17.0	3	50
COD eff. (mg/L)	1,988	0	217	25.3	12.3	409	17.1	2	70
BOD_5 eff. (mg/L)	1,365	1	41	6.2	3.4	1,032	43.1	0	14

[a]Number of cases outside the range of (Q1 − 1.5*IQR, Q3 + 1.5*IQR).

where Y represents the values of each parameter studied and $Y_{\min}$ and $Y_{\max}$ are the minimum and maximum values, respectively, of the variable to be normalized. Furthermore, a correlation matrix was developed to ensure the independence of the input parameters. Data preprocessing and preparation were performed using MATLAB 2017b.

Artificial neural network

The idea of strong computing methods, known as neural networks, as an equivalent to the human brain, was developed in the late 1800s (Lippmann 1988). Artificial neural networks consist of artificial neurons in a connected layered structure to provide the desired output. The network is trained through the continuous addition of data. Then, generalization of the network is realized through the introduction of new unseen data. The principal advantages of ANNs lie in (a) their high learning speed and data processing and (b) their capacity to represent highly nonlinear systems. However, the major drawback is the black box aspect of ANNs (Lippmann 1988).

Figure 1 shows a common FFNN structure that consists of three main network layers. The first layer is called the input layer, which receives the input data and sends these data to the hidden layers. The hidden layers are the heart of the neural network and consist of multiple neurons. In a single network, there can be multiple hidden layers with numerous neurons. Usually, the number of hidden layers is proportional to the complexity of the system to be modeled (Dreyfus 2002). The working principle of ANN is basically the same for the many network types. The basic processing element, the neuron, receives input signals then processes them through an activation function and provides an output signal. Furthermore, the weight of each neuron and the transfer functions are responsible for passing signals from one layer to the next layer. The mathematical expression of the neural network working principle is given in Equation (2):

$$Y_i = f\left(\sum_{j=1}^{M} W_{ij}\, X_j + b_i\right) \tag{2}$$

where Y_i is the value of predicted output i, f is the activation function, W_{ij} is the weight assigned to each input j, M is the total number of inputs, and b_i is the bias for each output.

The most common type of activation function is the sigmoid function (Haykin 2009). Examples of the sigmoid function are the tan-sigmoid function and the log-sigmoid function, expressed by Equations (3) and (4), respectively

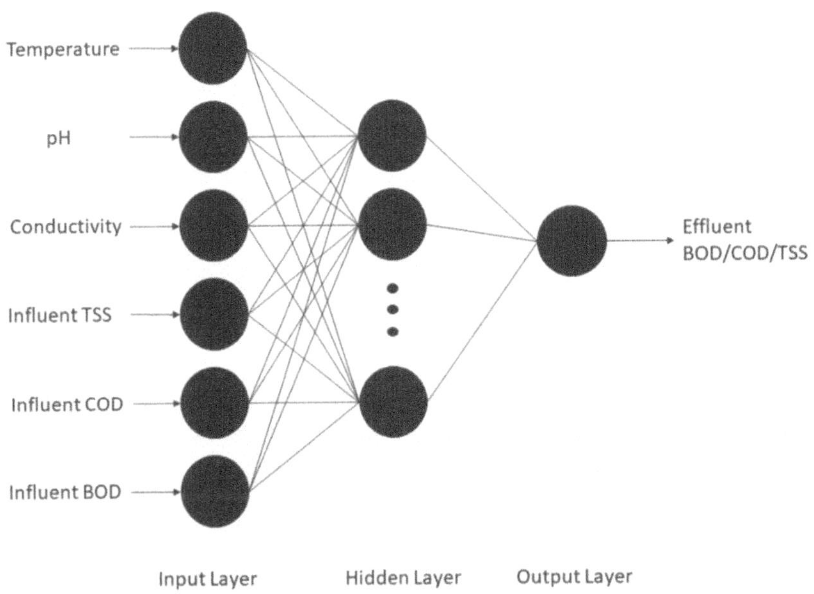

Figure 1 | Typical feedforward structure of an artificial neural network.

(Gangi Setti & Rao 2014):

$$f(x) = \frac{2}{1 + \exp(-2x)} - 1 \quad (3)$$

$$f(x) = \frac{1}{1 + \exp(-x)} \quad (4)$$

where x is the input of the activation function. In this study, the MATLAB 2017b neural network tool is applied to develop and train the feedforward neural network with the tan-sigmoid transfer function (Equation (3)) for all layers. The ANN is trained using one input matrix containing up to six inputs for each of the three outputs. Table 2 summarizes the combinations used in this study. The input data are pretreated before training for better network training performance.

At present, to determine the optimal number of hidden layers and neurons in each layer, several network structures should be tested, and the performance of each network should be evaluated. Although a small number of hidden layers and neurons could lead to a suitable generalization, it may also lead to an underfitted network. In contrast, a large number of hidden layers could lower the training error but could increase network overfitting (Geman *et al.* 1992). Various configurations of the network structure were tested, and the optimal structure for each output was determined through trial and error. The configuration parameters included the number of inputs, hidden layers, and neurons in the hidden layer.

Table 2 | Input combinations used in this study

Input	Parameter combination
input 1	Temperature
input 2	pH
input 3	Conductivity_inf
input 4	TSS_inf
input 5	COD_inf
input 6	BOD5_inf
input 7	TSS_inf + COD_inf + BOD5_inf
input 8	Conductivity_inf + COD_inf + BOD5_inf
input 9	Conductivity_inf + TSS_inf + BOD5_inf
input 10	Conductivity_inf + TSS_inf + COD_inf
input 11	pH_inf + COD_inf + BOD5_inf
input 12	pH_inf + TSS_inf + BOD5_inf
input 13	pH_inf + TSS_inf + COD_inf
input 14	pH_inf + Conductivity_inf + BOD5_inf
input 15	pH_inf + Conductivity_inf + COD_inf
input 16	pH_inf + Conductivity_inf + TSS_inf
input 17	Temp_inf + COD_inf + BOD5_inf
input 18	Temp_inf + TSS_inf + BOD5_inf
input 19	Temp_inf + TSS_inf + COD_inf
input 20	Temp_inf + Conductivity_inf + BOD5_inf
input 21	Temp_inf + Conductivity_inf + COD_inf
input 22	Temp_inf + Conductivity_inf + TSS_inf
input 23	Temp_inf + pH_inf + BOD5_inf
input 24	Temp_inf + pH_inf + COD_inf
input 25	Temp_inf + pH_inf + TSS_inf
input 26	Temp_inf + pH_inf + Conductivity_inf
input 27	Temp_inf + Conductivity + pH + BOD5_inf + COD_inf + TSS_inf

RESULTS AND DISCUSSION

Correlation matrix

The developed correlation matrix, summarized in Table 3, indicates the absence of any linear relationship among the parameters. However, this does not indicate the absence of any other types of relationships, as the small correlation coefficient indicates that conventional regression methods are unsuitable to predict such a complex system. Moreover, the input parameters lack any correlation among each other, which is required to train a reliable ANN. The presence of correlated input parameters biases the ANN towards the effect of these parameters.

ANN modeling and prediction

The development of the ANN model was realized using various structures to obtain the ANN with the optimal performance. First, each of the influent parameters was applied as input to predict each of the effluent parameters (one input to one output). Then, the number of inputs was gradually increased until all the inputs (six inputs) were applied to predict each output. The number of hidden

Table 3 | Correlation matrix of the parameters in this study

	Temperature	pH	Conductivity	TSS influent (inf.)	COD inf.	BOD5 inf.	TSS eff.	COD eff.	BOD5 eff.
Temp.	1.000								
pH	0.365	1.000							
Conductivity	0.422	0.190	1.000						
TSS	0.018	−0.019	0.012	1.000					
COD	0.009	0.010	−0.006	0.146	1.000				
BOD5	0.051	−0.235	−0.010	0.176	0.177	1.000			
TSS eff.	0.118	−0.146	0.207	0.182	0.007	0.189	1.000		
COD eff.	−0.056	−0.108	0.109	0.025	0.041	−0.017	0.353	1.000	
BOD5 eff.	0.197	−0.192	0.149	0.209	0.111	0.497	0.435	0.228	1.000

layers and the number of neurons were also scanned for the optimal network parameters. The data set was divided into training, validation, and test sets at a ratio of 70:15:15.

The key performance indicator considered in this study is the coefficient of determination (R^2) (Equation (5)) and the index of agreement (d) (Equation (6)) between the predicted values and the measured values for each ANN developed:

$$R^2 = \frac{\left[\sum_{i=1}^{N} (OBS_i - \overline{OBS})(P_i - \bar{P})\right]^2}{\sum_{i=1}^{N} (OBS_i - \overline{OBS})^2 \sum_{i=1}^{N} (P_i - \bar{P})^2} \tag{5}$$

$$d = 1 - \frac{\sum_{i=1}^{N} (OBS_i - P_i)^2}{\sum_{i=1}^{N} (|P_i - \overline{OBS}| + |OBS_i - \overline{OBS}|)^2} \tag{6}$$

where N is the total number of points, OBS is the measured value of the variable, P is the corresponding predicted value of the variable, and the bar indicates the mean value of the variable.

The results of developing a one input to one output ANN for the estimation of the effluent TSS, BOD and COD concentrations from the WWTP plant are summarized in Table 4. The low R^2 value indicates the complexity of the system, as it cannot be modeled using only one input parameter. In comparison, Mjalli *et al.* (2007) reported high R^2 values while using a one input and one output ANN structure. However, the high performance occurred due to the large number of neurons used in their model, which led to overfitting and reduced model generalization.

The best input performances in predicting the effluent BOD were obtained for the influent BOD, pH, and temperature. The top three COD predictors were the temperature, pH, and conductivity. The effluent TSS was best predicted by the temperature, influent BOD, and influent TSS. The effect of the initial pH on the BOD rate and wastewater treatment process has been examined in previous literature

Table 4 | One input and one output ANN model performance for BOD, COD, and TSS prediction

	Effluent BOD		Effluent COD		Effluent TSS	
Input parameter	ANN structure	R^2	ANN structure	R^2	ANN structure	R^2
Temperature	1–25–1	0.2082	1–25–1	0.1052	1–25–1	0.1636
pH	1–25–1	0.2228	1–24–1	0.1258	1–24–1	0.1495
Conductivity	1–24–1	0.1013	1–24–1	0.1406	1–25–1	0.1544
TSS	1–24–1	0.1310	1–24–1	0.0657	1–25–1	0.1562
COD	1–25–1	0.1547	1–24–1	0.1001	1–25–1	0.1482
BOD	1–25–1	0.3440	1–25–1	0.0998	1–25–1	0.1733

(Mukherjee *et al.* 1968; Liu *et al.* 2007). The removal efficiency of COD and TSS is known to increase with the temperature (Ahsan *et al.* 2005). The next step was to train three input and one output ANN models. It should be noted that every possible combination of three input parameters was considered with different numbers of hidden layers.

Effluent BOD prediction

Figure 2 shows the R^2 values for predicting the effluent BOD for each input combination using the various numbers of hidden layers. It is found that the input-20 combination (temperature–conductivity–inf. BOD) outperforms the other input combinations. The results also show that for most of the input combinations, the prediction accuracy peaks when using three hidden layers. This explains the underfitting that occurs with a small number of hidden layers and the overfitting that occurs when using more than three hidden layers.

The input-20 combination was trained for various network structures, and the optimal ANN structure ($R^2 = 0.752$ and $d = 0.928$) contained three hidden layers with 17 neurons for each layer (1–17–17–17–1), as shown in Figure 3. Figures 4 and 5 show the performance of the three-input ANN model as a parity plot between the measured and predicted values and as a comparison of the prediction and measured values along the data sequence, respectively. Although the R^2 value is deemed inadequate, the modeled values match the measured values closely except for some extent values. These extent values might be caused by the noisy data set.

The resulting model attains a lower performance than that of Mjalli *et al.* (2007); however, they adopted a much smaller data set (80 data points) than the data set used in this study (1,365 data points). In addition, Mjalli *et al.* employed the secondary treatment effluent as the input of their model, which is strongly correlated to the effluent of the plant. Thus, their neural network only predicts the tertiary treatment process, while in our study, the artificial

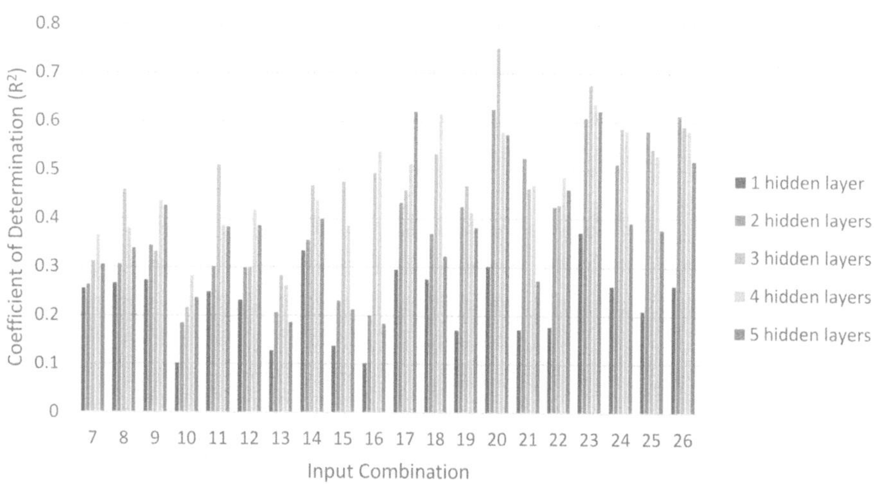

Figure 2 | BOD model prediction accuracy for all three-input combinations using various numbers of hidden layers.

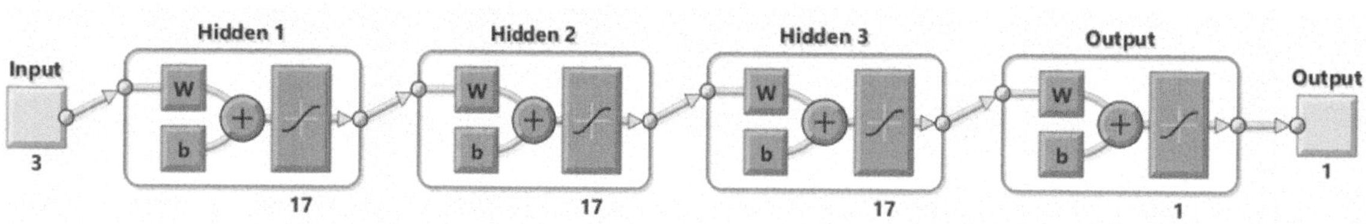

Figure 3 | Optimal ANN structure for BOD prediction.

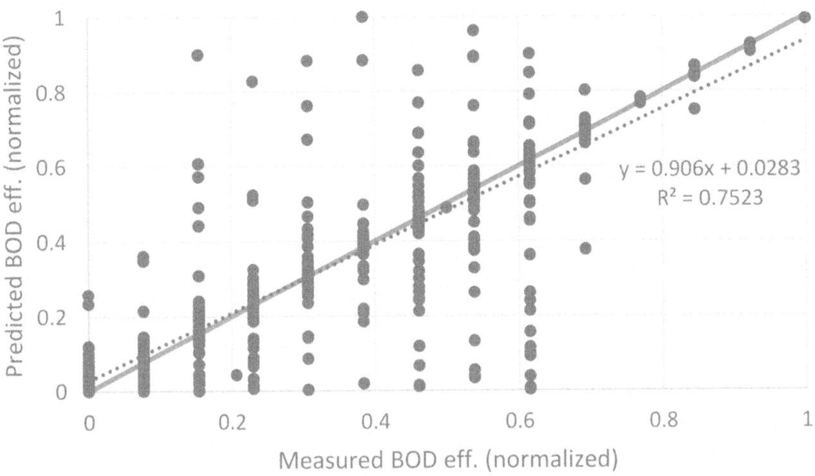

Figure 4 | Parity plot for the BOD prediction model.

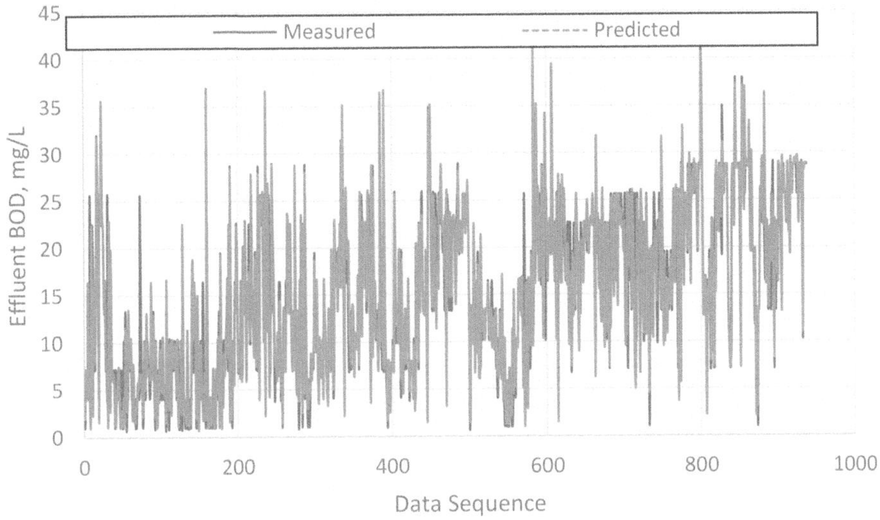

Figure 5 | BOD prediction values against measured values for all data sequences.

neural network is designed to predict the whole treatment process.

Effluent COD prediction

The accuracy results for each input combination are shown in Figure 6 using an ANN with different numbers of hidden layers. The input-26 combination (temperature–conductivity–pH) outperformed the other input combinations. It was also observed that the combinations containing temperature + conductivity, or temperature + pH (inputs 20 to 26) resulted in a higher prediction performance than that of the other combinations. The results also revealed performance spikes for the other combinations; however, these were not considered due to the lack of consistency.

The temperature–conductivity–pH combination was trained for various network structures, and the optimal ANN structure ($R^2 = 0.6115$ and $d = 0.877$) contained three hidden layers with 13 neurons for each layer (1–13–13–13–1), as shown in Figure 7. Figures 8 and 9 show the performance of the three-input ANN model as a parity

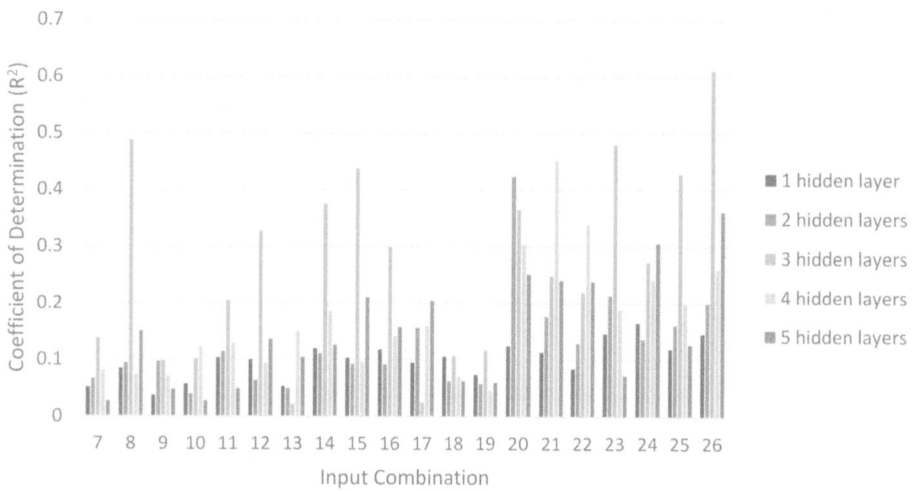

Figure 6 | COD model prediction accuracy for all three-input combinations using various numbers of hidden layers.

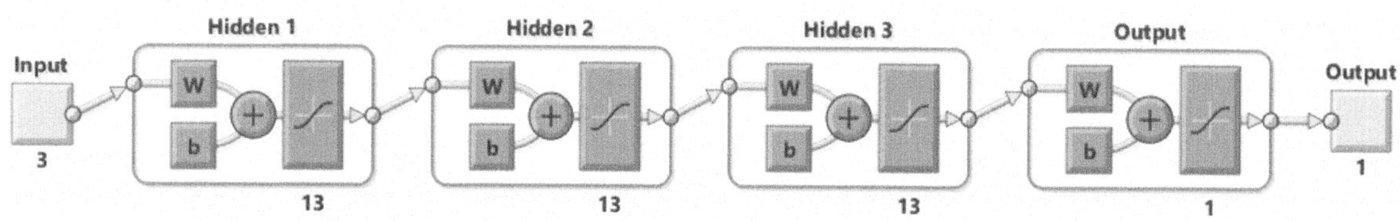

Figure 7 | Optimal ANN structure for COD prediction.

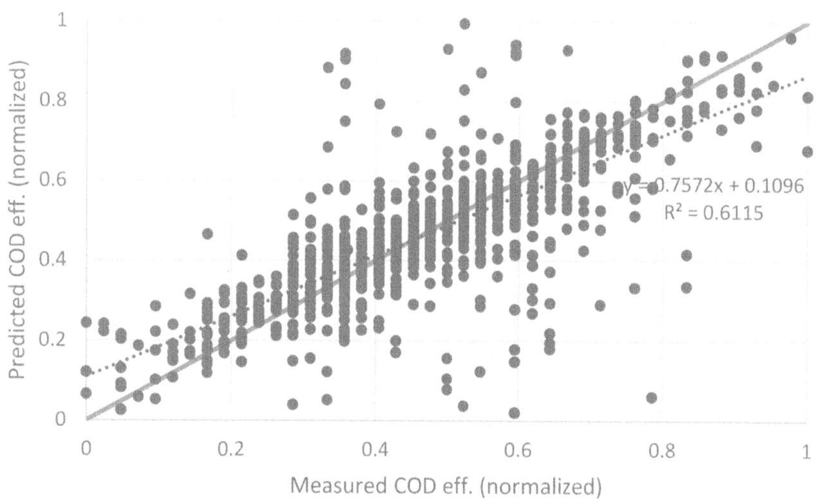

Figure 8 | Parity plot for the COD prediction model.

plot between the measured and predicted values and as a comparison of the prediction and measured values along the data sequence, respectively. Figure 9 indicates that the prediction accuracy is low at the extent values. Except for these extent values, the prediction values are nearly identical to the measured values.

Additionally, Mjalli *et al.* (2007) generated a better-performing ANN model with many fewer data points and

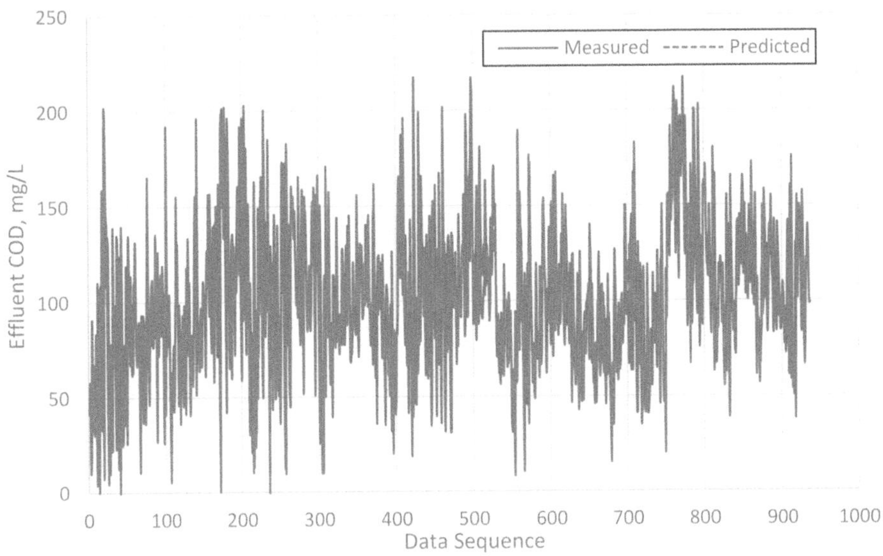

Figure 9 | COD prediction values against measured values for all data sequences.

different input parameters. Moreover, the performance of the three-input ANN ($R^2 = 0.6115$) is comparable to that of the ANN model developed with six inputs ($R^2 = 0.7034$) (Abba & Elkiran 2017). This shows the reliability of our procedure of determining a simple, accurate ANN prediction model.

Effluent TSS prediction

For the TSS ANN model, the input-20 combination (temperature–conductivity–BOD) yielded better results than the other combinations, as shown in Figure 10. Moreover, the input combinations containing the conductivity attained a higher prediction accuracy than the other combinations. In contrast to BOD and COD, optimum TSS modeling occurred with four hidden layers rather than with three hidden layers. This indicates the amount of complexity involved in TSS modeling over BOD and COD modeling.

The temperature–conductivity–BOD combination was trained for various network structures, and the optimal ANN structure ($R^2 = 0.6308$ and $d = 0.884$) consisted of four hidden layers with 11 neurons for each layer

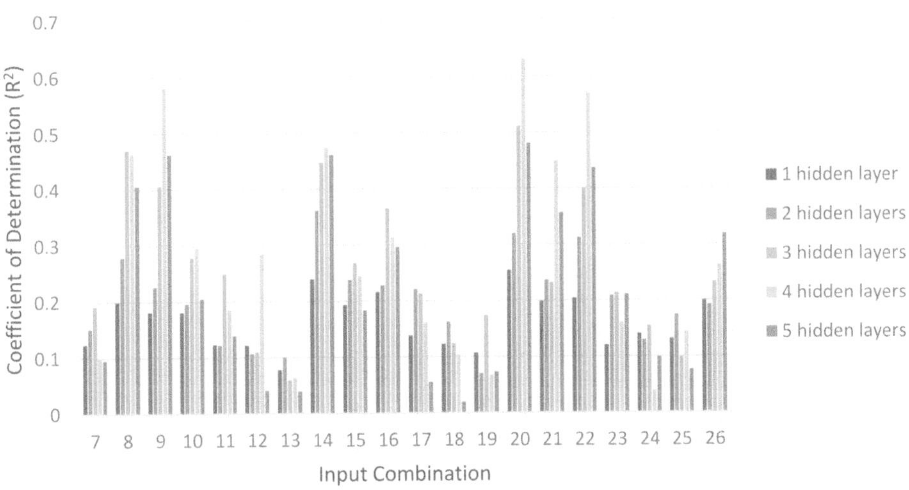

Figure 10 | TSS model prediction accuracy for all three-input combinations using various numbers of hidden layers.

(1–11–11–11–11–1), as shown in Figure 11. Figures 12 and 13 show the performance of the three-input ANN model as a parity plot between the measured and predicted values and as a comparison of the prediction and measured values along the data sequence, respectively. TSS was found to be the most challenging parameter to model. However, the developed ANN managed to predict TSS with a sufficient accuracy in regard to the measured values. As with

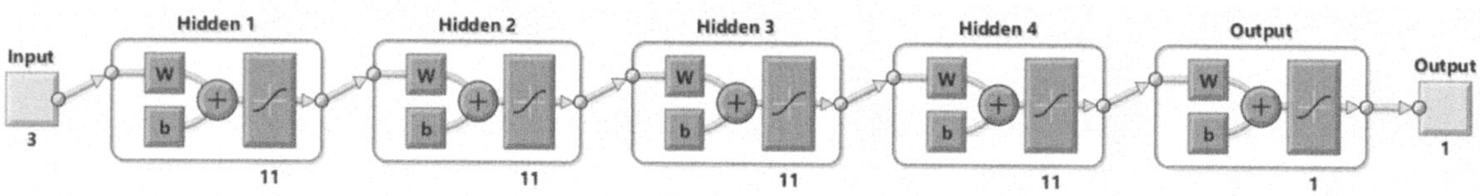

Figure 11 | Optimal ANN structure for TSS prediction.

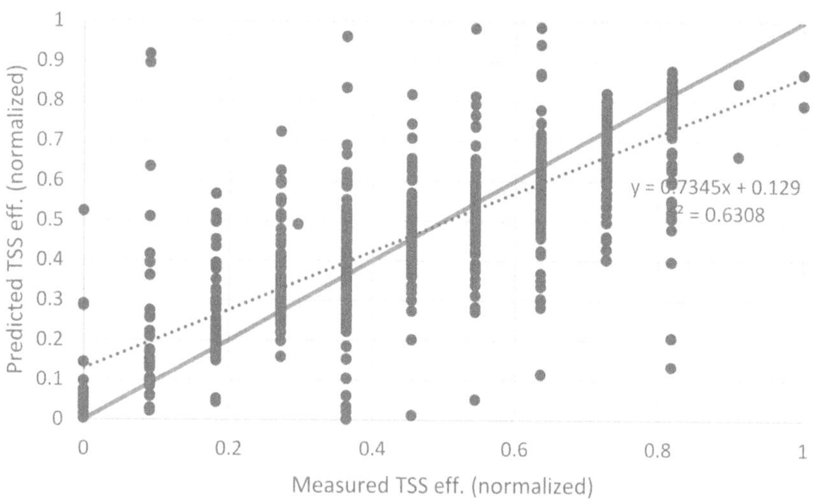

Figure 12 | Parity plot for the TSS prediction model.

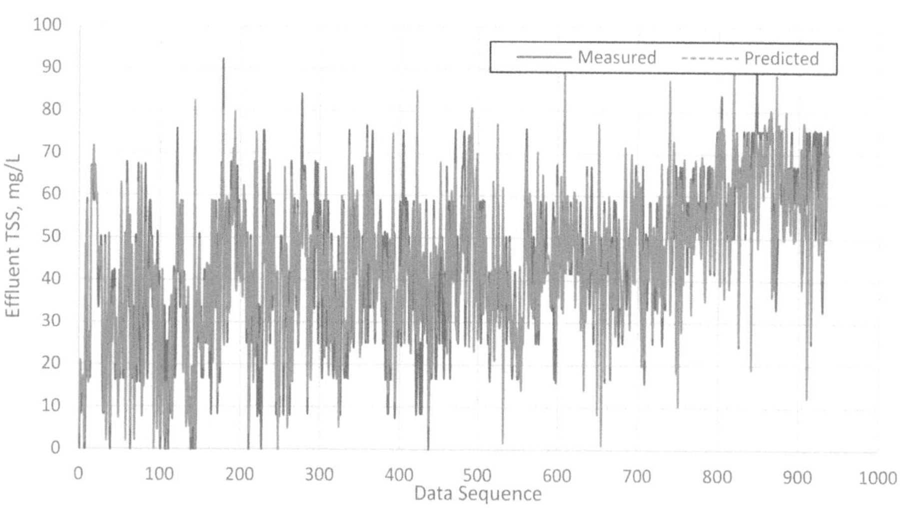

Figure 13 | TSS prediction values against measured values for all data sequences.

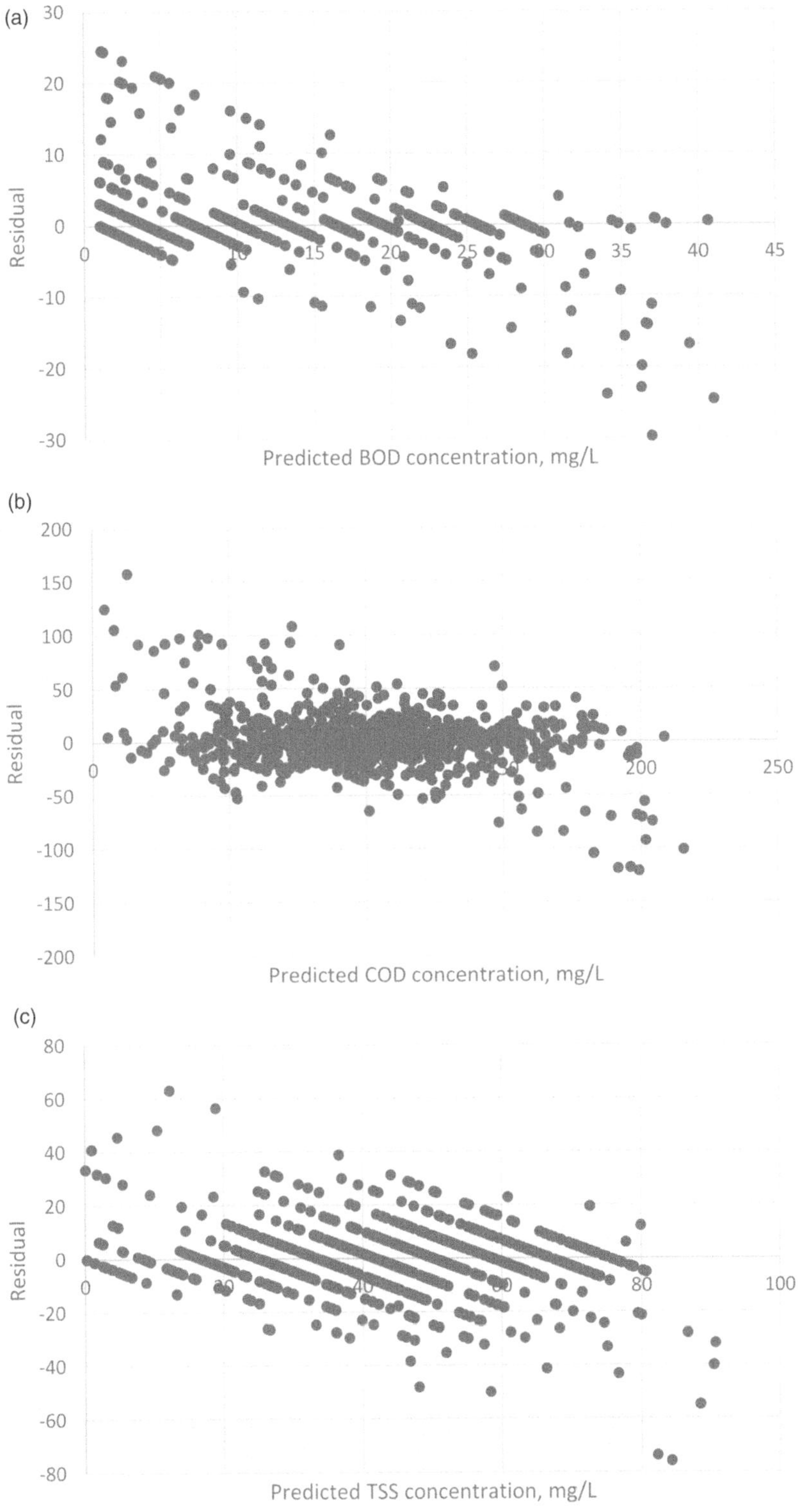

Figure 14 | Residual plots of (a) the effluent BOD, (b) effluent COD, and (c) effluent TSS ANN models.

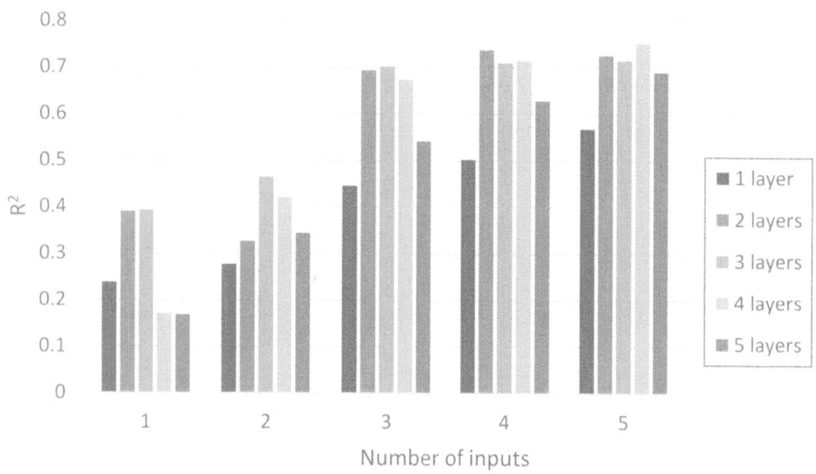

Figure 15 | Values of R^2 of the best of each ANN configuration for influent BOD_5 prediction.

the BOD and COD predictions, with a smaller data set and a less generalized model, Mjalli *et al.* (2007) developed an ANN model with a high performance.

Furthermore, six inputs of ANN were also trained for BOD, COD and TSS; however, they yielded a decreased performance ($R^2 = 0.649$, 1–15–15–15–1; $R^2 = 0.47$, 1–8–8–8–1; and $R^2 = 0.6974$, 1–11–11–11–1; respectively). It is evident here that increasing the number of inputs is not always a benefit (Zare *et al.* 2011; Hamada *et al.* 2018). For all outputs in this study, using six inputs to develop the ANN failed to provide a higher performance (Wei 2013). However, the R^2 values are comparable to those reported by Abba & Elkiran (2017) when predicting the effluent COD concentration.

To study the cause of the low value of the performance indicator for the models developed in this paper, residual plots were generated, which depict the goodness of fit of the created ANN models (Guo *et al.* 2015). Figure 14 shows the residual plots for each model. The low relationship between the residual and predicted data indicates that the low model performance is mainly due to data noise. Hence, the ANN model would yield a higher prediction accuracy using smoother data. Moreover, ANN models provide sufficient accuracy for the prediction of the effluent concentration.

Influent BOD_5 prediction

Several ANN configurations (i.e., various input combinations, layers, and neurons) were developed and tested for the prediction of the BOD_5 concentration in the influent wastewater treatment plant. Figure 15 shows the best prediction results of the test data set for each ANN configuration with respect to the coefficient of determination (R^2). The difference between underfitting and overfitting is clearly demonstrated in Figure 15, where the R^2 values are lower using either a small (one layer) or large (five layers) number of layers, and the maximum R^2 value occurs between these extremes.

Moreover, sensitivity analysis of the input parameters was conducted utilizing the ANN prediction accuracy under the various input combinations. Table 5 summarizes the best results obtained under each number of inputs and corresponding ANN configuration. The table indicates that the COD concentration imposes the highest impact on the ANN prediction accuracy, followed by the temperature

Table 5 | ANN configurations with the highest influent BOD_5 prediction accuracy for the different numbers of input parameters

	Input parameters	ANN configuration	R^2
ANN–V	Temperature–pH–Conductivity–TSS–COD	(1–9–9–9–9–1)	0.754
ANN–IV	Temperature–Conductivity–TSS–COD	(1–20–20–1)	0.737
ANN–III	Temperature–Conductivity–COD	(1–13–13–13–1)	0.702
ANN–II	Temperature–COD	(1–10–10–10–1)	0.464
ANN–I	COD	(1–19–1)	0.176

and conductivity. From the results, the order of significance of the input parameters is as follows:

COD > temperature > conductivity > TSS > pH

Furthermore, the introduction of the conductivity parameter greatly increased the prediction accuracy. Thus, for an ANN to sufficiently predict the influent BOD_5, at least the influent COD, temperature, and conductivity are required as input parameters. Thus, decreasing the time to estimate BOD_5 from five days to several hours is the time required to measure the COD concentration of any sample. From the different configurations listed in Table 4, ANN-V was adopted to predict BOD_5, and the results are shown in Figure 16. The results demonstrate the high accuracy of the ANN in predicting BOD_5 with the exception of a few data points. The spikes deviating from the measurement values occur due to noise in the training data set similar to the previously developed ANN.

The R^2 values from the literature and previous models designed to predict BOD_5 in WWTPs are listed in Table 6. The previous regression models, in the literature, attain a slightly higher performance than the ANN developed in this study. However, regression models tend to require a large amount of data and rely on a large number of input parameters. For instance, Tables S1 to S6 (Supplementary Material) list the results of linear regression for the data used in this study and using the input parameters of the optimal ANN developed. The results reveal low R^2 values for the linear regression model. Moreover, the soft sensor of Qiu *et al.* (2016) yields a comparable or better performance in predicting BOD_5 than the model in this study. However, the model developed in this study is notably simpler, as it does not require the use of a genetic algorithm.

CONCLUSION

The developed ANN models suitably predicted the WWTP performance, which was defined based on the effluent TSS, BOD, and COD concentrations in the treatment plant, with a high degree of reliability. A low model performance mainly occurred due to noise in the data used to develop the ANN.

The results demonstrated that preliminary data analysis and preparation are essential for ANN training. It is recommended to gradually increase the complexity (number of inputs, hidden layers, and neurons) of the network configuration until no further improvement is recorded. In this study, a three input and one output configuration was sufficient, and a further increase in inputs resulted in overfitting of the system. The results also indicated that increasing the number of inputs was not always beneficial. Another finding from the developed models is the importance of the input parameters in the wastewater treatment process. The parameter significance could be interpreted from the model input combination that yielded the highest prediction accuracy. For instance, the influent temperature and conductivity greatly affected the WWTP performance as they were used as inputs in all models. However, the influent BOD concentration was important in the treatment process regarding the effluent

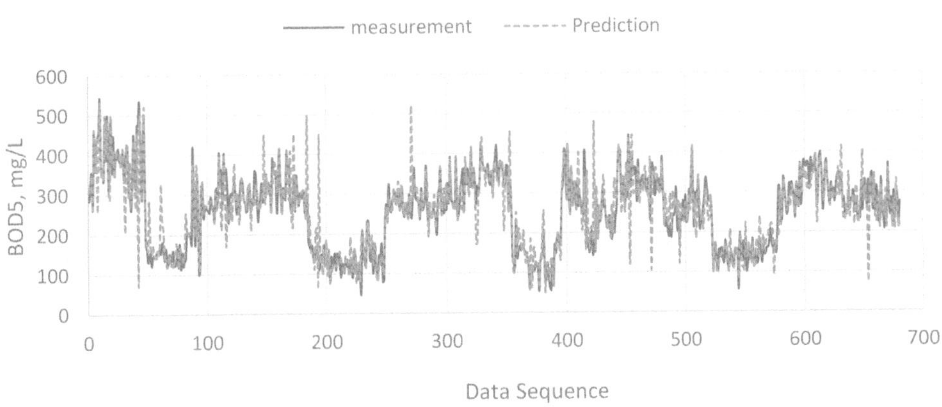

Figure 16 | Influent BOD prediction values against measured values for all data sequences using ANN–V.

Table 6 | The R^2 values in previous studies

References	Model type	R^2
Baki *et al.* (2019)	Regression models	0.7966
Ebrahimi *et al.* (2017)	Stepwise Multivariate Regression Analysis	0.82–0.83
Qiu *et al.* (2016)	FFNN and Genetic Algorithm	0.7–0.96

BOD and TSS concentrations, while the effluent COD concentration was indirectly dependent on the influent pH.

In this study, the developed prediction model for the influent BOD_5 concentration worked well and attained a high performance accuracy, namely, $R^2 = 0.754$. This result demonstrates the viability of using this model as a soft sensor for online control and management systems for WWTPs. A further recommendation is to implement a more robust control system entailing frequent measurements to further tune the model with data highly representative of real-time operation conditions.

FUNDING

This research did not receive a specific grant from any funding agencies in the public, commercial, or not-for-profit sectors.

DATA AVAILABILITY STATEMENT

Data cannot be made publicly available; readers should contact the corresponding author for details.

REFERENCES

Abba, S. I. & Elkiran, G. 2017 Effluent prediction of chemical oxygen demand from the wastewater treatment plant using artificial neural network application. *Procedia Computer Science* **120**, 156–163. https://doi.org/10.1016/j.procs.2017.11.223.

Ahsan, S., Rahman, M. A., Kaneco, S., Katsumata, H., Suzuki, T. & Ohta, K. 2005 Effect of temperature on wastewater treatment with natural and waste materials. *Clean Technologies and Environmental Policy* **7** (3), 198–202. https://doi.org/10.1007/s10098-005-0271-5.

Alver, A. & Kazan, Z. 2020 Prediction of full-scale filtration plant performance using artificial neural networks based on principal component analysis. *Separation and Purification Technology* **230**, 115868. https://doi.org/10.1016/j.seppur.2019.115868.

Baki, O. T., Aras, E., Akdemir, U. O. & Yilmaz, B. 2019 Biochemical oxygen demand prediction in wastewater treatment plant by using different regression analysis models. *Desalination and Water Treatment* **157**, 79–89.

Batstone, D. J., Keller, J., Angelidaki, I., Kalyuzhnyi, S. V., Pavlostathis, S. G., Rozzi, A., Sanders, W. T. M., Siegrist, H. & Vavilin, V. A. 2002 The IWA Anaerobic Digestion Model No 1 (ADM1). *Water Science and Technology* **45** (10), 65–73. https://doi.org/10.2166/wst.2002.0292.

Dreyfus, T. 2002 Advanced mathematical thinking processes. In: *Advanced Mathematical Thinking*. (D. Tall, ed.), Springer, Dordrecht, The Netherlands, pp. 25–41.

Ebrahimi, M., Gerber, E. L. & Rockaway, T. D. 2017 Temporal performance assessment of wastewater treatment plants by using multivariate statistical analysis. *Journal of Environmental Management* **193**, 234–246. https://doi.org/10.1016/j.jenvman.2017.02.027.

Elmolla, E. S., Chaudhuri, M. & Eltoukhy, M. M. 2010 The use of artificial neural network (ANN) for modeling of COD removal from antibiotic aqueous solution by the Fenton process. *Journal of Hazardous Materials* **179** (1–3), 127–134. https://doi.org/10.1016/j.jhazmat.2010.02.068.

Ferrer, J., Seco, A., Serralta, J., Ribes, J., Manga, J., Asensi, E., Morenilla, J. J. & Llavador, F. 2008 DESASS: A software tool for designing, simulating and optimising WWTPs. *Environmental Modelling & Software* **23** (1), 19–26. https://doi.org/10.1016/j.envsoft.2007.04.005.

Gangi Setti, S. & Rao, R. N. 2014 Artificial neural network approach for prediction of stress–strain curve of near β titanium alloy. *Rare Metals* **33** (3), 249–257. https://doi.org/10.1007/s12598-013-0182-2.

Geman, S., Bienenstock, E. & Doursat, R. 1992 Neural networks and the bias/variance dilemma. *Neural Computation* **4** (1), 1–58.

Guo, H., Jeong, K., Lim, J., Jo, J., Kim, Y. M., Park, J., Kim, J. H. & Cho, K. H. 2015 Prediction of effluent concentration in a wastewater treatment plant using machine learning models. *Journal of Environmental Sciences* **32**, 90–101. https://doi.org/10.1016/j.jes.2015.01.007.

Hamada, M., Adel Zaqoot, H. & Abu Jreiban, A. 2018 Application of artificial neural networks for the prediction of Gaza wastewater treatment plant performance – Gaza strip. *Journal of Applied Research in Water and Wastewater* **5** (1), 399–406. https://doi.org/10.22126/arww.2018.874.

Hamed, M. M., Khalafallah, M. G. & Hassanien, E. A. 2004 Prediction of wastewater treatment plant performance using artificial neural networks. *Environmental Modelling & Software* **19** (10), 919–928. https://doi.org/10.1016/j.envsoft.2003.10.005.

Haykin, S. S. 2009 *Neural Networks and Learning Machines*, 3rd edn. Prentice Hall, Upper Saddle River, NJ, USA.

Henze, M., Grady, C. P. L. Jr, Gujer, W., Marais, G. v. R. & Matsuo, T. 1987 *Activated Sludge Model No 1*, IAWPRC Scientific and Technical Report No 1. IAWPRC, London, UK.

Henze, M., Gujer, W., Mino, T., Matsuo, T., Wentzel, M. C. & Marais, G. v. R. 1995 *Activated Sludge Model No 2*, IAWQ Scientific and Technical Report. IAWQ, London, UK.

Henze, M., Gujer, W., Mino, T., Matsuo, T., Wentzel, M. C., Marais, G. v. R. & Van Loosdrecht, M. C. M. 1999 Activated Sludge Model No 2d, ASM2D. *Water Science & Technology* **39** (1), 165–182.

Kasem, R., ALabdeh, D., Noori, R. & Karbassi, A. 2018 A software sensor for in-situ monitoring of the 5-day biochemical oxygen demand. *Rudarsko-geološko-naftni Zbornik* **33** (1), 15–23.

Lippmann, R. P. 1988 An introduction to computing with neural nets. In: *Artificial Neural Networks: Theoretical Concepts* (V. Vemuri, ed.), IEEE Computer Society Press, Washington, DC, USA, pp. 36–54.

Liu, Y., Chen, Y. & Zhou, Q. 2007 Effect of initial pH control on enhanced biological phosphorus removal from wastewater containing acetic and propionic acids. *Chemosphere* **66** (1), 123–129. https://doi.org/10.1016/j.chemosphere.2006.05.004.

Manu, D. S. & Thalla, A. K. 2017 Artificial intelligence models for predicting the performance of biological wastewater treatment plant in the removal of Kjeldahl Nitrogen from wastewater. *Applied Water Science* **7** (7), 3783–3791. https://doi.org/10.1007/s13201-017-0526-4.

Mjalli, F. S., Al-Asheh, S. & Alfadala, H. E. 2007 Use of artificial neural network black-box modeling for the prediction of wastewater treatment plants performance. *Journal of Environmental Management* **83** (3), 329–338. https://doi.org/10.1016/j.jenvman.2006.03.004.

Mukherjee, S. K., Chatterji, A. K. & Saraswat, I. P. 1968 Effect of pH on the rate of BOD of wastewater. *Journal (Water Pollution Control Federation)* **40** (11), 1934–1939. JSTOR.

Najafzadeh, M. & Ghaemi, A. 2019 Prediction of the five-day biochemical oxygen demand and chemical oxygen demand in natural streams using machine learning methods. *Environmental Monitoring and Assessment* **191** (6), 380. https://doi.org/10.1007/s10661-019-7446-8.

Nasr, M. S., Moustafa, M. A. E., Seif, H. A. E. & El Kobrosy, G. 2012 Application of artificial neural network (ANN) for the prediction of EL-AGAMY wastewater treatment plant performance – Egypt. *Alexandria Engineering Journal* **51** (1), 37–43. https://doi.org/10.1016/j.aej.2012.07.005.

Newhart, K. B., Holloway, R. W., Hering, A. S. & Cath, T. Y. 2019 Data-driven performance analyses of wastewater treatment plants: a review. *Water Research* **157**, 498–513. https://doi.org/10.1016/j.watres.2019.03.030.

Nezhad, M. F., Mehrdadi, N., Torabian, A. & Behboudian, S. 2016 Artificial neural network modeling of the effluent quality index for municipal wastewater treatment plants using quality variables: south of Tehran waste water treatment plant. *Journal of Water Supply: Research and Technology – Aqua* **65** (1), 18–27. https://doi.org/10.2166/aqua.2015.030.

Qiu, Y., Liu, Y. & Huang, D. 2016 Date-driven soft-sensor design for biological wastewater treatment using deep neural networks and genetic algorithms. *Journal of Chemical Engineering of Japan* **49** (10), 925–936. https://doi.org/10.1252/jcej.16we016.

Wei, X. 2013 *Modeling and Optimization of Wastewater Treatment Process with a Data-Driven Approach.* PhD thesis, University of Iowa, Iowa City, IA, USA. https://doi.org/10.17077/etd.wwzj01nf.

Zare, A. H., Bayat, V. M. & Daneshkare, A. P. 2011 Forecasting nitrate concentration in groundwater using artificial neural network and linear regression models. *International Agrophysics* **25** (2), 187–192.

First received 9 June 2020; accepted in revised form 9 August 2020. Available online 24 August 2020

Remote-sensing-based algorithms for water quality monitoring in Olushandja Dam, north-central Namibia

Taimi S. Kapalanga, Zvikomborero Hoko, Webster Gumindoga and Loyd Chikwiramakomo

ABSTRACT

Frequent and continuous water quality monitoring of Olushandja Dam in Namibia is needed to inform timely decision making. This study was carried out from November 2014 to June 2015 with Landsat 8 reflectance values and field measured water quality data that were used to develop regression-analysis-based retrieval algorithms. Water quality parameters considered included turbidity, total suspended solids (TSS), nitrates, ammonia, total nitrogen (TN), total phosphorus (TP) and total algae counts. Results show that turbidity levels exceeded the recommended limits for raw water for potable water treatment while TN and TP values are within acceptable values. Turbidity, TN, and TP and total algae count showed a medium to strong positive linear relationship between Landsat predicted and measured water quality data while TSS showed a weak linear relationship. The regression coefficients between predicted and measured values were: turbidity ($R^2 = 0.767$); TN ($R^2 = 0.798$,); TP ($R^2 = 0.907$); TSS ($R^2 = 0.284$,) and total algae count ($R^2 = 0.851$). Prediction algorithms are generally the best fit to derive water quality parameters. Remote sensing is recommended for frequent and continuous monitoring of Olushandja Dam as it has the ability to provide rapid information on the spatio-temporal variability of surface water quality.

Key words | Landsat 8, reflectance values, regression algorithms, remote sensing, retrieval, water quality

Taimi S. Kapalanga
Zvikomborero Hoko
Webster Gumindoga (corresponding author)
Loyd Chikwiramakomo
Department of Civil Engineering,
University of Zimbabwe,
Box MP 167 Mt Pleasant, Harare,
Zimbabwe
E-mail: *wgumindoga@gmail.com*

HIGHLIGHTS

- Over past years, frequent and continuous water quality monitoring has been problematic in Namibia.
- A linear regression can now be used to develop algorithms for retrieving water quality data.
- Good prediction accuracy for turbidity, TN, TP and total algae count.
- More sampling points needed to further improve regression model accuracy.
- Remote sensing provides rapid information on water quality spatio-temporal variability.

doi: 10.2166/ws.2020.290

GRAPHICAL ABSTRACT

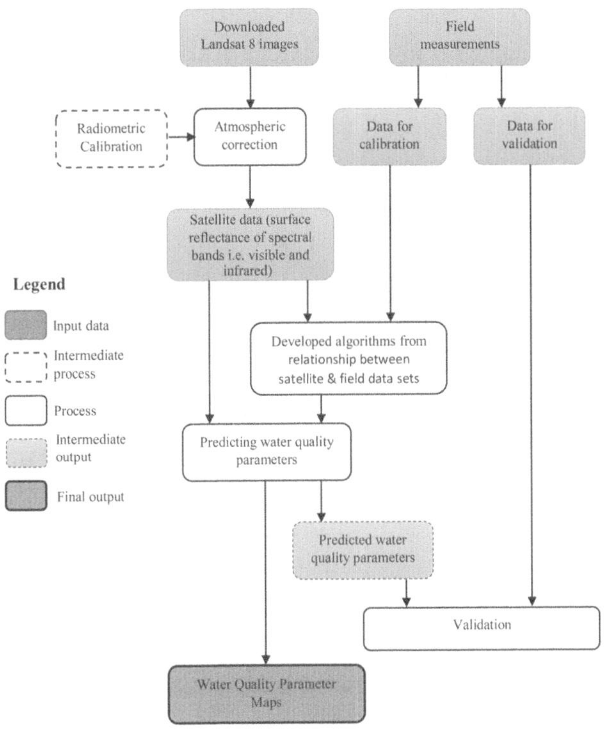

INTRODUCTION

Fresh water is a finite resource that is essential for human existence (Bhuyar *et al.* 2019a, 2019b; Bhuyar *et al.* 2020). Without freshwater of adequate quantity and quality, sustainable development will not be possible (UN-Water 2011; McMillan *et al.* 2017). Effective management of this scarce resource ensures that present generations' needs do not deprive future generations the same privileges of access in both quantity and quality as envisioned in the Sustainable Development Goals (SDGs) (Bain *et al.* 2020). Therefore, Integrated Water Resources Management (IWRM) is an essential approach as it emphasises effective management of water resources within the basin.

In the Southern Africa Region, pollution of both surface and ground water is on the increase, particularly from mining, agricultural and industrial activities. Man-made reservoirs such as dams and lakes are threatened by nutrient enrichment and heavy metals and their water quality is continuously degraded as a result (Oberholster & Ashton 2008; Nhapi 2009; Lehmann 2010). For example, the water quality in Von Bach and Swakoppoort Dam in the Central Area of Namibia has been reported to be poor due to treated and at times partially treated wastewater from the City of Windhoek and Okahandja town (Lehmann 2010; NamWater 2012).

In the north-central regions of Namibia, poor sanitation practices are notably high (IWRMPJVN 2010), with about 67% of the population having no access to improved sanitation facilities. North-central Namibia is more vulnerable to effects of climate change and variability. Combined effects of environmental degradation, social vulnerability to poverty and a changing climate will compromise water and sanitation provision (Angula & Kaundjua 2016). Flash floods are further predicted to impact overall sanitation and human health conditions. Therefore, during high rainfall events, cholera outbreak has been reported due to wash-away of water and wastewater aided by poor and inadequate sanitation facilities in the areas (UN 2011).

Shuuya & Hoko (2014) found that the quality of water in the Calueque–Oshakati canal is deteriorating from upstream to downstream due to human activities along the canal such as agriculture, settlements and mining. All these activities tend to affect the quantity and quality of water leading to an increase in the cost of purifying water for human consumption. IWRMPJVN (2010) noted that lack of data, continuous monitoring and poor data management exist in the Namibian water sector. Water quality monitoring in many southern African countries, including Namibia, has been noted to be hindered by lack of data, capacity and resources. Water quality monitoring in Namibia has been noted to be one of the key issues that requires urgent attention by researchers and water managers.

Few studies that are available for the Kunene River and Cuvelai Basins in the north-central regions of Namibia focus on ad-hoc monitoring. Studies have focused on the quality of water in the canals that bring water into the Olushandja Dam and to the treatment plants (Shuuya & Hoko 2014) and transport water from the dam (SDP10 2001). These studies mainly employ traditional in-situ methods of determining water quality. According to IWRMPJVN (2010), determining water quality using traditional techniques is costly and may be one of the major contributors to poor monitoring frameworks not only in Namibia but also in many developing countries in the world. Ritchie *et al.* (2003) stated that traditional methods for assessing and monitoring water quality are expensive and time-consuming. In addition, they do not give the spatial or temporal views of water quality needed for accurate assessment of water bodies, therefore there is a need for more robust techniques which have a spatial and a temporal dimension (Kallio 2000).

Thus, this study explored the applicability of remote sensing in combination with in-situ observed measurements to develop algorithms and prediction of selected water quality parameters at Olushandja Dam. Remote-sensing-based water quality assessment is an economical way to monitor water quality, since it allows routine monitoring of large areas in a short time and on a repetitive basis (Hellweger *et al.* 2004; Somvanshi *et al.* 2012). According to Ritchie *et al.* (2003), the use of remote sensing in water quality dates back to the 1970s. Namibia is one of the countries with the greatest number of sunshine days or cloud-free skies, which make it easy for optical remote sensing application in the visible and infrared regions of the electromagnetic spectrum. Information obtained from this study will help the institutions responsible for management of Olushandja Dam such as NamWater and Namibian community-based water management (CBWM; Kelbert 2016) to carry out continuous monitoring of the quality of water in an economical way for decision-making.

However the major constraints would be the lack of reliable retrieval algorithms, cost of satellite data and equipment for in-situ, on-site and laboratory measurements of water quality parameters (Bauer *et al.* 2007). For the retrieval of water quality parameters from different satellite sensors, a number of methods have been used which range from empirical, semi-empirical to analytical techniques (Schalles *et al.* 1998; Wang & Ma 2001; Dekker *et al.* 2002; Brando & Dekker 2003; Vignolo *et al.* 2006; Chen *et al.* 2007; He *et al.* 2008; Maillard & Pinheiro Santos 2008; Salama *et al.* 2009; Olet 2010; Chawira *et al.* 2013; Kibena *et al.* 2014). The advantage of satellite remote sensing techniques lies in their capacity to allow managers to detect and control the pollutants before they reach alarming levels. The water quality parameters that have been assessed include chlorophyll-*a*, suspended matter and turbidity as they are most likely to change the water colour (Schalles *et al.* 1998; Li 2009; Salama *et al.* 2009; Olet 2010; Kibena *et al.* 2014). In Belbok Spruit, South Africa, du Plessis *et al.* (2014) predicted water quality from landcover changes using partial least squares regression analysis. Yang & Jin (2010) predicted NO_3 using the spatial regression method for Lowa River in the United States and obtained an acceptable R^2 of 0.8. Most researchers in sub-Saharan Africa have also applied earth observation techniques in the retrieval of water quality parameters mainly for lakes and dams (Chawira *et al.* 2013; Kibena *et al.* 2014). There is still need to explore in greater depth the application of earth observation techniques for predicting water quality parameters, especially in developing countries where land use changes are pertinent (du Plessis *et al.* 2014; Chapra 2015).

A few studies have attempted to monitor and model nutrients (such as total nitrogen, phosphorus, nitrate etc.) and proven the ability of remote sensing for these predictions (Alparslan *et al.* 2007; He *et al.* 2008; Chen & Quan 2012). Most of the nutrient prediction models developed so far are based on statistical regression approaches. Predicting water quality characteristics from remote sensing requires ground-truthing and validation data (Schaeffer *et al.* 2013). Therefore, assessment and monitoring of water quality using the combination of remote sensing and in-situ measurements plays a significant role in providing reasonable and accurate optical constituents of water (Salama *et al.* 2009).

Remote-sensing-based water quality assessment and monitoring may use the same method of retrieval or predicting water quality parameters, but the available sensor types may differ (Tomppo *et al.* 2002; USGS 2013). These include Moderate Resolution Imaging Spectroradiometer (MODIS), Medium Resolution Imaging Spectrometer (MERIS), SPOT, and Landsat, which differ in spatial, temporal, and number of spectral bands and pixels. Landsat, used in this study, has a fair revisit time (16 days) and better resolution (30 m) than most medium-resolution sensors. In addition, Landsat imagery is available free of charge from an archive hosted by the USGS Earth Resources Observation and Science (EROS) Center. The Landsat mission represents the longest continuous satellite record of the Earth, beginning in 1972 with Landsat 1 and currently with the operation of Landsat 8 (USGS 2013). The different Landsat satellite mission images have been used in a number of water quality studies (Zuccari Fernandes Braga *et al.* 1993; Wang *et al.* 2004; Vignolo *et al.* 2006; He *et al.* 2008; Olet 2010; Somvanshi *et al.* 2012; Waxter 2014; Bonansea *et al.* 2015). The best correlations between remote sensing signal and in-situ observed water quality parameters have been found mainly in the visible (blue, green, red) and near-infrared spectral range.

The objectives of this study were: (i) to characterize the status of the quality of water in Olushandja Dam through in-situ observed measurements, (ii) to develop algorithms for predicting selected water quality parameters through satellite and in-situ observed measurements and (iii) to predict selected water quality parameters from remote sensing as a framework for continuous monitoring of water quality in Olushandja Dam.

STUDY AREA

Olushandja Dam is located in the upper western part of Omusati Region in the Olushandja Sub-Basin of the Cuvelai-Etosha Basin in north-central Namibia (Figure 1). The Namibian part of the Cuvelai Basin covers an area between the Okavango and Kunene Rivers and ends up in the Etosha Pan. The basin covers an area of approximately 250,053 km^2 from the Namibia–Angola border terminating in the Etasha Pan (Kolberg 2002). The Cuvelai lies within a relatively small depression along the western margins of the vast Kalahari Basin that covers much of south-central Africa. The basin is made up of shallow pans known as 'Iishana', which form a network of shallow ephemeral river systems.

The Cuvelai Basin in north-central Namibia comprises a unique system of seasonal wetlands and constitutes some of the areas having the highest human population densities in Namibia (50–100 persons/km^2). The region carries about 40% of Namibia's population of about 2.2 million people. The Cuvelai-Etosha Basin and Namibia as a whole are characterized by a semi-arid to arid climate and desert environment (Mendelsohn *et al.* 2000, 2013). The average temperatures vary between 20 °C and greater than 22 °C in most areas. This translates to potential evaporation that ranges between 2,600 and 3,200 mm per year, much higher than the yearly amount of rainfall (CuveWaters 2014). The geomorphology of the Cuvelai Basin is complex. Mendelsohn *et al.* (2016) has provided a comprehensive description of the Cuvelai Basin landscapes and how the characteristics of these landscapes have influenced human settlement patterns.

The Olushandja Dam lies between 1,100 and 1,150 metres above sea level (CuveWaters 2014). Olushandja Dam has a surface area of about 29.0 km^2 and capacity of around 42.331 Mm^3 when full (NamWater 2013). The dam is about 20 km long and varies between 200 m and 2,000 m in width (Mendelsohn *et al.* 2000). The dam was built in 1973 during the country's liberation struggle in the old Etaka Channel to act as a balancing reservoir that stores water during excess flows of the Kunene River. In addition, the dam was built to provide a strategic reserve in the event of supplies from Calueque Dam in Angola being interrupted. In 1990 the system was renovated and now raw surface water is transferred from Calueque Dam

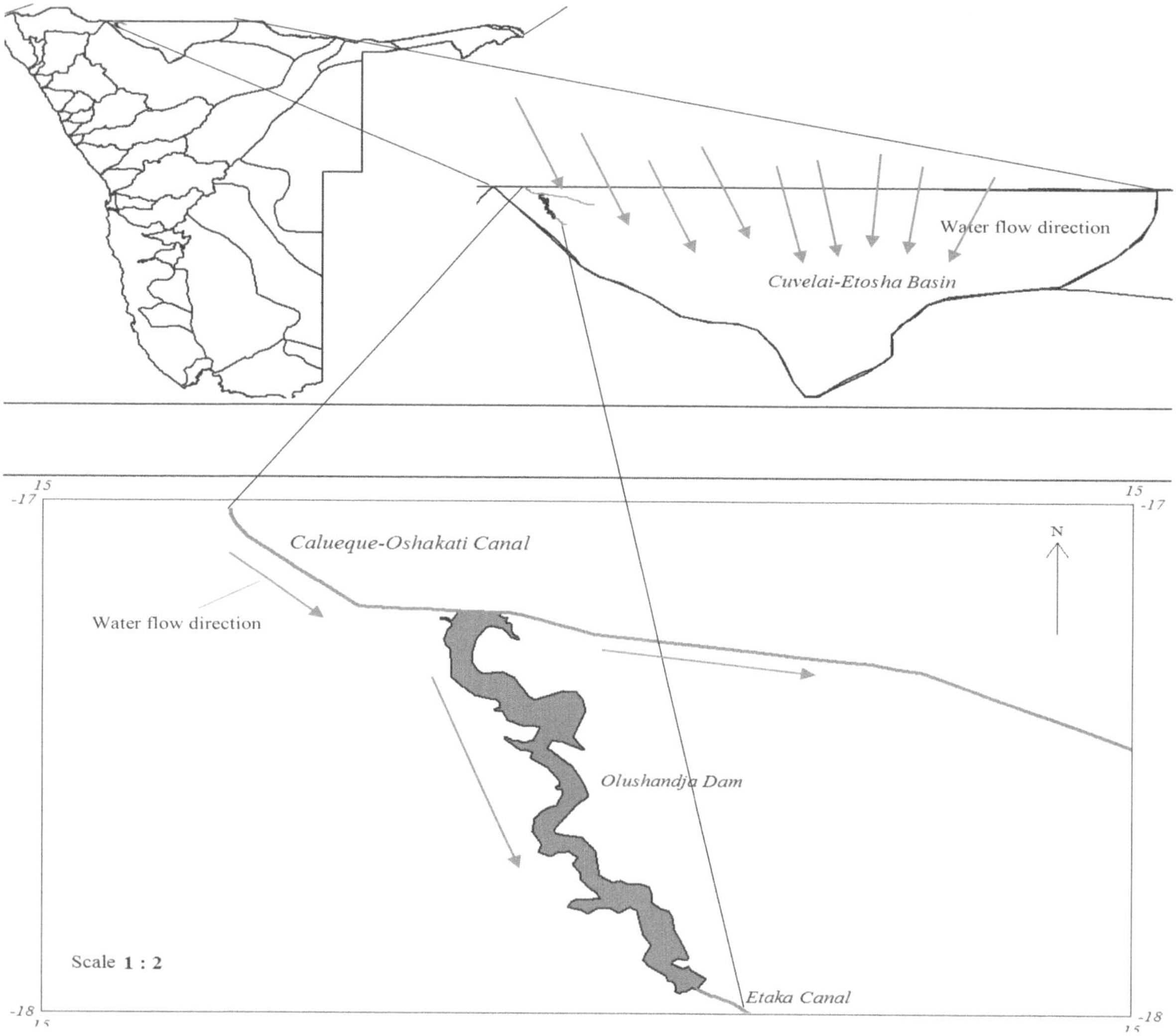

Figure 1 | The study area in relation to Cuvelai-Etosha Basin.

in Angola through a concrete open canal into the dam and also to four treatment plants in the region. In recent years, the dam has been held at 50% capacity and since its construction, more than 100 households have been built in the dam flood plain (Mendelsohn *et al.* 2000; NamWater 2013).

MATERIALS AND METHODS

Field data collection

Sampling points (Figure 2) in the dam were selected systematically to reflect on the impacts of major tributaries' landcover activities on the dam water quality. Parameters were also selected taking into account the relationship between pollution sources and different water quality parameters including their potential impacts on human, aquatic life and environmental health. In addition, the selection considers water quality parameters that are known to be successfully retrieved from satellite data. Turbidity, algae and nutrients are known to have been successfully retrieved from satellite data (Zuccari Fernandes Braga *et al.* 1993; Hellweger *et al.* 2004; Wang *et al.* 2004; He *et al.* 2008; Li 2009; Olet 2010; Irenosen *et al.* 2012; Chawira *et al.* 2013; Waxter 2014; Bonansea *et al.* 2015).

Representative water samples were collected from six systematically selected locations in the Olushandja Dam

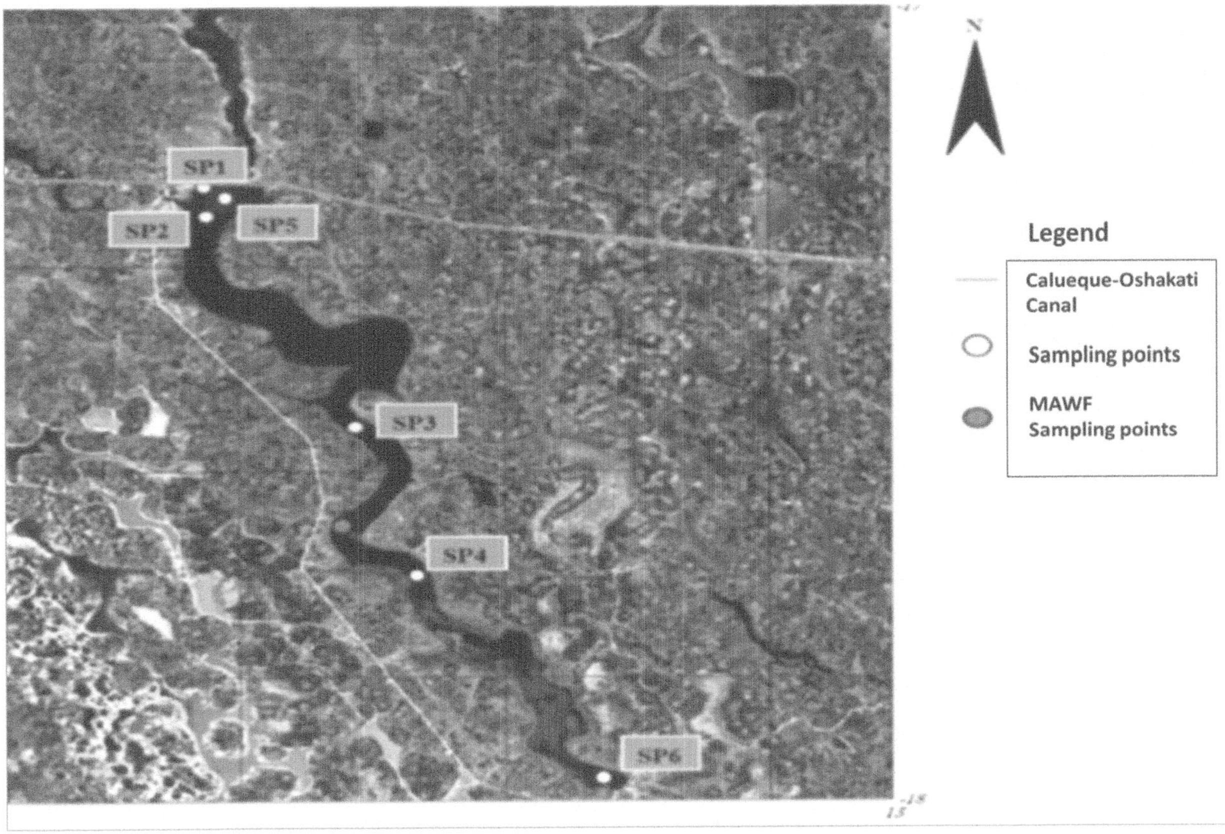

Figure 2 | Sampling point locations on a geometrically corrected Landsat 8 OLI image of 2015.

following the Landsat imagery capturing calendar obtained from the USGS (earthexplorer.usgs.gov). At each sampling point, grab water samples were taken at about 10 cm depth below the surface using three 500 ml bottles with screw-caps. The sampling campaigns were done twice a month between January and April 2015 from 08:30 am to 14:00 pm. On the same days that satellite images were taken, the in-situ observed measurements were done because a time difference between in-situ observed measurements and satellite measurements is an important aspect (Hellweger *et al.* 2004). The water samples were collected, preserved, transported and analysed according to the Standard Methods for Water and Wastewater described by Wait *et al.* (2020).

Satellite data retrieval

Landsat 8 images of Path 180, Row 072, which covers Olushandja Dam, were downloaded from the Landsat imagery archive hosted by the USGS Earth Resources Observation and Science (EROS) Centre using the USGS Globalization Visualization Viewer (Glovis) tool. Figure 3 displays a sample Landsat 8 image tile for 15 April 2015 that covers Olushandja Dam.

The Digital Elevation Model (DEM) map covering the study area was obtained from USGS Earth Explorer (http://www.earthexplorer.usgs.gov) and four ground control points (GCP) were established. The dam outline and canal route were obtained by digitizing a Google Earth satellite image and converting to a shape file (.shp) format in Quantum Geographical Information System (GIS or QGIS) software. All files were then imported into ILWIS and were assigned to the same coordinate system as that of the study area.

The Landsat 8 imagery (cloud-free or with cloud cover of $<10\%$) over the area was selected for analysis. Radiometric calibration to convert digital number (DN) values to physical units, at sensor spectral radiance

Figure 3 | Landsat 8 image tile of path 180, row 072, showing the location of Olushandja Dam.

(watts/(m^2/srad/μm)) and then to top-of-atmosphere (TOA) reflectance was done in GIS software, Environment for Visualizing Images (ENVI 5.0.3) using the same equations described by Chander *et al.* (2009) and USGS (2013). In the ENVI program, a band math expression was applied to rescale the floating reflectance value to surface reflectance of between 0 and 1.

After the radiometric calibration and atmospheric correction, pre-processed bands with reflectance were imported in the Integrated Land and Water Information System (ILWIS) GIS environment. The image raster maps were geometrically corrected by using the nearest-neighbour resampling technique and assigning the correct georeference as that of the study area. This study used pixel reflectance values for the four visible and near-infrared bands (2, 3, 4 and 5) of the Landsat 8 images for each selected date based on the GPS locations of every sampling point. This was done in ILWIS through the 'map cross' operation that matched up the sampling point raster map with a raster map of each processed band (after image re-sampling). Reflectance values of each band on every image at specific sampling points were determined.

Development of algorithms

The algorithms were then developed through a simple linear regression analysis to determine the best-fit model. The surface reflectances of each band and water quality parameter values at a specific sampling point were regressed. Reflectance values of all bands were selected to be independent of each water quality variable. Independent variables that form a relationship with reflectance by showing a high coefficient of determination (R^2) were selected based on the literature (Alparslan *et al.* 2007; He *et al.* 2008; El-Saadi *et al.* 2014). The R^2 compares estimated and actual *y*-values, and ranges in value from 0 to 1. A value of 1 denotes a strong correlation in the sample, meaning there is no difference between the estimated value and the actual value. At the other extreme, if the coefficient of determination is 0, the regression equation is not helpful in predicting a *y*-value (Alparslan *et al.* 2007).

Application of the algorithms and validation

The developed algorithms were applied to resampled multispectral reflectance image bands. Interpolation was performed on the six sampling locations to come up with a spatial variation map of each parameter. The inverse distance weighted (IDW) interpolation (Parida *et al.* 2017) technique was used for interpolating water quality data at respective point locations to grid points for comparison with satellite retrieved estimates.

Furthermore, the predicted values for each parameter were extracted from every parameter satellite raster map based on GPS location. The predicted values were validated against field measured data to test whether the satellite-based predictions were able to estimate water quality parameters. This was done by simple regression analysis using field data (for 25 January and 10 February 2015) that were randomly selected from the main dataset and not used in the development of the algorithms to reduce errors. To assess the detailed spatial distribution of water quality in Olushandja Dam, quantitative attribute maps were produced by applying algorithms to the original Landsat datasets.

RESULTS

Field measured data

The mean field measured water quality parameter data acquired from four sampling campaigns (26 Feb–15 April 2015) were calculated at each of the six points. The field data that correspond to the same dates as the satellite data (Table 1) are eventually used for model development.

Remote sensed data

The spectral band reflectance values at each sampling point on the selected four images corresponding to field data (26 February and 15 April 2015) are given in Table 2. From the table, low reflectance values (<0.1) on the water body can be noticed and this could be due to the fact that the water is relatively clean and this makes it have low reflective spectral radiation (He *et al.* 2008).

Table 1 | Average values of measured data for four sampling campaigns at six sampling points in Olushandja Dam

Sampling point	Algae counts (cells/ml)	Turbidity (NTU)	NH_3 (mg/L)	TN (mg/L)	TPU (mg/L)	TSS (mg/L)
1	380	18.70	0.10	0.90	0.19	11.75
2	590	9.88	0.02	0.93	0.04	4.50
3	2,374	5.97	0.02	0.95	0.04	6.13
4	684	19.74	0.33	1.42	0.06	22.88
5	864	10.77	0.02	1.03	0.05	6.33
6	3,028	54.97	0.06	3.78	0.36	27.20

Note: NH_3 = ammonia, TN = total nitrogen, TP = total phosphorus and TSS = total suspended solids.

Table 2 | Average spectral band reflectance for four sampling campaigns at six sampling points in Olushandja Dam

Sampling point	Band2	Band3	Band4	Band5
1	0.011	0.042	0.035	0.095
2	0.016	0.047	0.026	0.020
3	0.059	0.057	0.057	0.057
4	0.004	0.012	0.012	0.008
5	0.010	0.050	0.027	0.015
6	0.000	0.004	0.024	0.004

Developed algorithms

Field measured data (Table 1) were regressed with reflectance values (given in Table 2) using a simple linear regression method and algorithms were formulated and are given in Table 3. Looking at the R^2 (coefficient of determination) and *F*-values (which approach 1 for good performance), the results of regression analysis can be interpreted as follows; all the derived regression models have good regressive correlation for all parameters (i.e. total algae, turbidity, TN, TP, TSS and NH_3). Total algae, turbidity, TN, TP and TSS have R^2 values of 0.999, 0.986, 0.987, 0.980, and 0.988 respectively. The R^2 values show that there is a high accuracy in predicting all the water quality parameters for the whole dam.

Table 3 | Derived retrieval algorithms for each parameter through regression analyses

Water quality variable	(R^2)	Derived algorithms
Algae	0.999	$= -54.7081 - 26{,}766.6(\lambda 2) - 42{,}687.5(\lambda 3) + 137{,}000.1(\lambda 4) - 23{,}819.7(\lambda 5)$
Turbidity	0.986	$= 15.31856 - 956.806(\lambda 2) - 747.376(\lambda 3) + 1{,}742.455(\lambda 4) + 165.173(\lambda 5)$
TN	0.987	$= 1.047532 - 54.928(\lambda 2) - 46.2947(\lambda 3) + 120.8943(\lambda 4) - 19.223(\lambda 5)$
TP	0.98	$= -0.00309 - 8.78139(\lambda 2) - 4.99958(\lambda 3) + 15.31713(\lambda 4) - 0.3916(\lambda 5)$
TSS	0.988	$= 27.08987 + 10.80036(\lambda 2) - 507.708(\lambda 3) + 95.37331(\lambda 4) + 27.8424(\lambda 5)$

Note: TN = total nitrogen, TP = total phosphorus and TSS = total suspended solids; λ = pixel reflectance values at bands 2–5.

The R^2 values for all water quality parameters were high and close to 1, which shows that there is a significant relationship between satellite data and in-situ observed measurements.

The *F*-value in one-way analysis of variance (ANOVA) aided in assessing whether the variance between the means of Landsat 8 derived water quality values and field measured water quality relate. Calculated *F*-values (not displayed) show that all algorithms exceeded the 95% level of confidence. Li (2009) also found algorithms for turbidity in Shakespeare Bay to exceed the 90% confidence level. Furthermore, NH_3 showed no significant correlation with satellite data and had low *F*-value (less than 4.53), thus it was excluded in the prediction.

Prediction of water quality parameter from satellite data

The study demonstrated that all four bands of Landsat 8 data contribute to the derived water quality parameters. The equations in Table 3 were applied to multispectral Landsat datasets to predict water quality parameters over the entire dam and at each sampling point. Five water quality variables in Olushandja Dam (turbidity, total suspended solids, total nitrogen, total phosphorus and total algae counts) were derived from Landsat 8 imagery on 26 February, 14 and 30 March as well as 15 April 2015 (four sampling campaigns). This was done after applying the developed algorithms to the original bands with surface reflectance and summary statistics (Table 4), and the spatial distribution (Figure 4(a)–4(e)) of the predicted five water quality parameters are presented.

Table 4 | Summary statistics of predicted water qualities over the entire Olushandja Dam

	Concentrations				
Statistics	Algae (cells/ml)	Turbidity (NTU)	TN (mg/L)	TP (mg/L)	TSS (mg/L)
Minimum	0	4.7	0.00	0.00	0.0
Maximum	21,204	356.0	13.20	2.60	42.0
Average	1,650	83.2	1.39	0.43	8.8
Standard deviation	3,143	63.2	2.18	0.40	16.3
NamWater	–	5	–	–	–
Canadian	–	50	–	–	–
UN/ECE	–	–	< 300	< 10	–

Spatial variation of predicted water quality parameters

The resultant quantitative maps for the five water quality variables (total algae content, turbidity, and concentration of total nitrogen, total phosphorus and total suspended solids) are presented in Figure 4(a)–4(e). The resultant maps clearly show that different water quality parameters yield several spatially distributed patterns over the reservoir over time. Some parameters show very low predicted values, especially in areas with dense vegetation in the inner and along the edge of the dam closer to the land and some show the opposite. This could be due to the reflectance ability of specific bands, especially band 5 (near-infrared) that reflects more highly on vegetation and interaction of water and land than on water (USGS 2013).

The resultant maps demonstrated that total algae contents and concentrations TN and TSS are highly variable over the dam while turbidity and TP have less spatial variance. Turbidity is found to be above the recommended

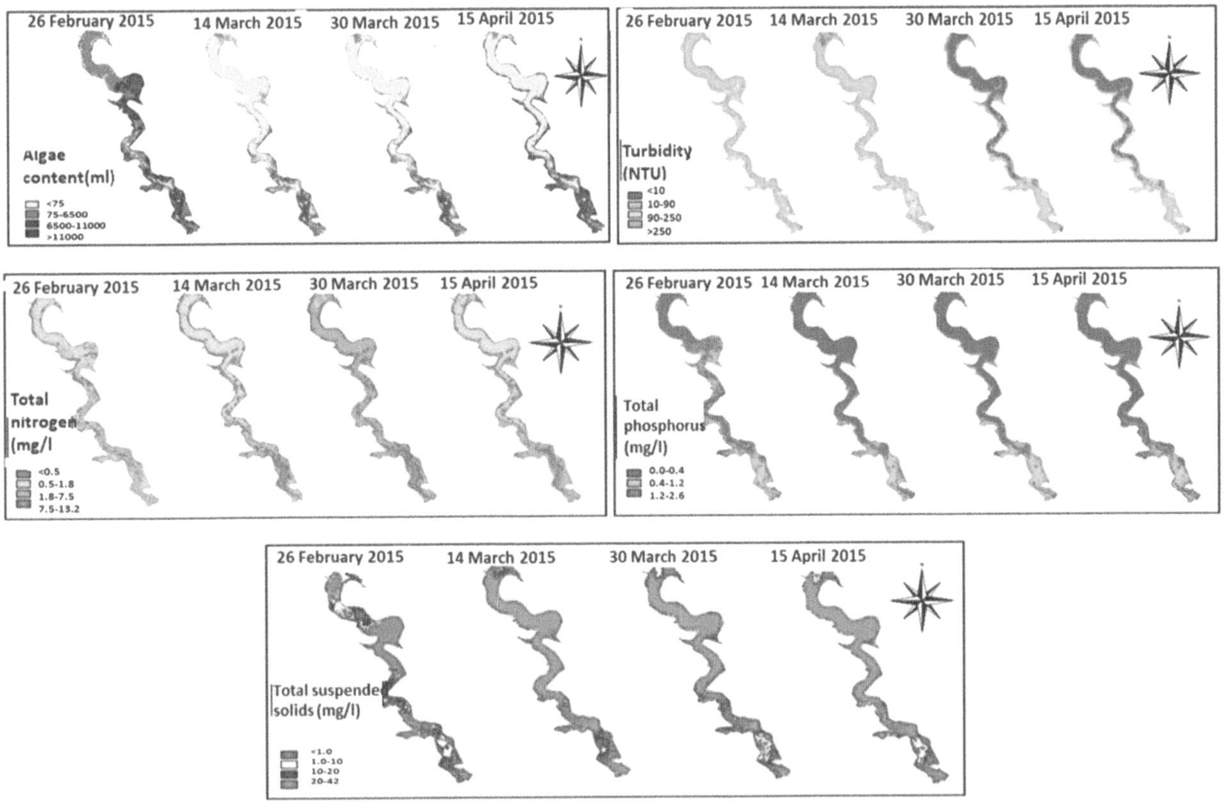

Figure 4 | Spatial variation of predicted water quality parameters.

limits of drinking water standards by NamWater as well as for recreational purposes by Canadian guidelines (50 NTU). Other studies (He *et al.* 2008; Li 2009; Lehmann 2010) found high turbidity levels in different water bodies. Total nitrogen and total phosphorus values are found to be within the UNECE (1994) standards for maintaining aquatic life, which are <300 mg/L.

Validation of algorithms

The relationship between Landsat predicted values and in-situ observed measurements of water quality parameters at each of the six sampling points for 15 April 2015 are presented in Figure 5(a)–5(e). Turbidity, TN, and TP and total algae count showed medium to strong positive linear relationship between Landsat predicted and measured values while TSS showed a weak linear relationship. The regression coefficients between predicted and measured values were: turbidity ($R^2 = 0.767$); TN ($R^2 = 0.798$); TP ($R^2 = 0.907$); TSS ($R^2 = 0.284$) and total algae count ($R^2 = 0.851$). Total suspended solids showed a lower coefficient than all other parameters. Looking at the R^2 values, it can be concluded that the prediction models are best fitted to derive the four water quality parameters other than TSS.

Figure 5(a)–5(e) confirm the strength of the regression models by showing correlation with high coefficient of determination (R^2) between predicted values and measured data. It was found that TP has higher R^2 followed by total algae counts, TN and turbidity respectively. TSS had a low correlation coefficient.

Methodological framework for monitoring water quality

Based on the result of this study, Landsat data has proven its ability to derive water quality parameters, data that are comparable to the in-situ observed measurements. Therefore, a proposed framework for predicting water qualities for Olushandja Dam is presented in Figure 6.

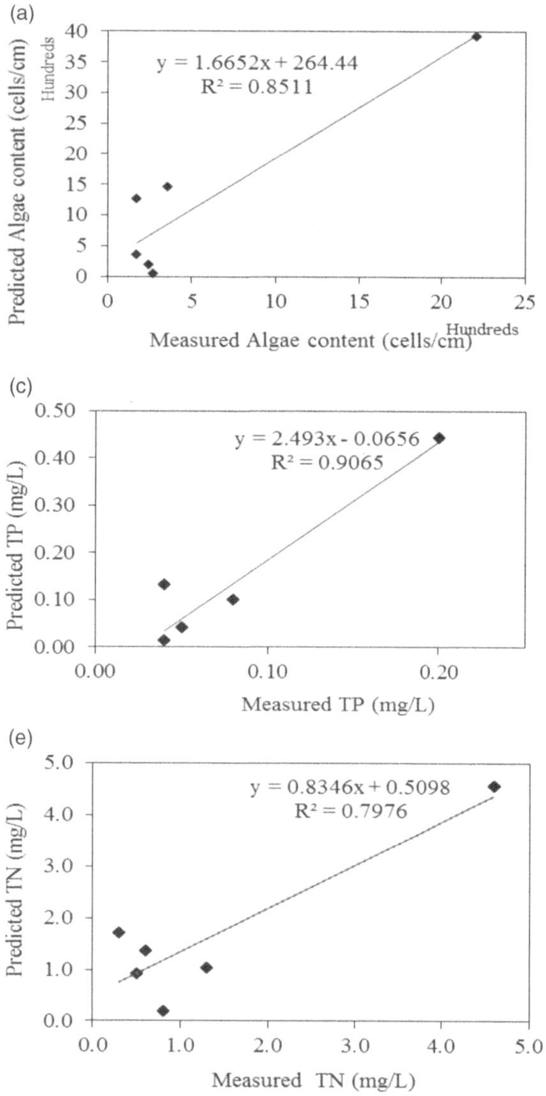

Figure 5 | (a–e): Relationship between Landsat predicted versus field measured data: (a) total algae content, (b) turbidity level, (c) total nitrogen (TN), (d) total phosphorus (TP) and (e) total suspended solids (TSS).

The framework will help responsible institutions such as NamWater and the Department of Water Affairs under the Directorate of Resources Management in the Ministry of Agriculture, Water and Forestry and interested members of the community to implement a continuous monitoring of water quality for improved decision-making. The framework will also be useful to the Ministry of Fisheries as there are fishery resources, the Ministry of Health and Social Services, students, community members, water resources managers, planners, developers and any other interested parties. The application of this framework would require among others: personnel with some knowledge of GIS and remote sensing; and financial resources (for transportation and laboratory costs).

DISCUSSION

Summary

This research found that regression analysis-based retrieval algorithms are ideal for water quality retrieval in Olushandja

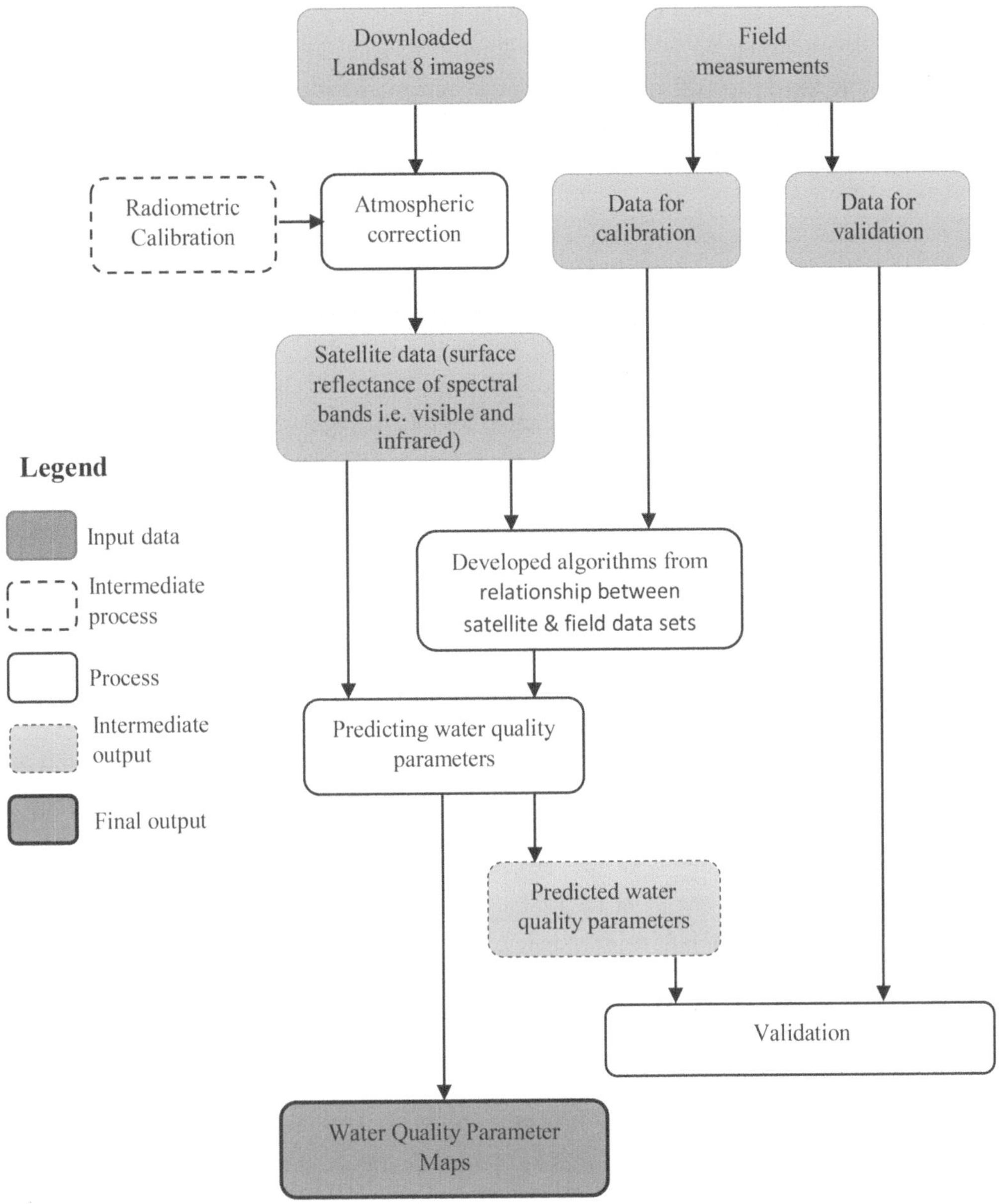

Figure 6 | A proposed framework for predicting water quality parameters based on remote sensing.

Dam, north-central Namibia. Obtaining the spatial variation of water quality parameters allows decision makers to manage the critical resource in near-real-time. Remote sensing is therefore recommended for frequent and continuous monitoring of Olushandja Dam as it has the ability to provide the information on the spatio-temporal variability of surface water quality. Detailed discussion and future applications of the present work are described in the following sections.

Regression-analysis-based algorithms for water quality retrieval

In-situ-based water quality retrieval (Bhuyar *et al.* 2019a, 2019b) brings limitation in comprehensive water quality assessment. This study used Landsat 8, 30 m spatial resolution imagery reflectance values and in-situ observed data to develop regression-analysis-based retrieval algorithms from November 2014 to June 2015. The study found that the majority of water quality parameter regression algorithms had high correlation coefficients (R^2) between 0.75 and 0.99. Therefore, the study concludes that the developed regression algorithms are the best fit to predict water quality parameters from satellite data. The study further applied the developed algorithms to derive water quality parameters at selected points in the dam. Correlation analysis was performed between Landsat 8 predicted and field measured data (Masocha *et al.* 2018). Alparslan *et al.* (2007) used Landsat ETM pixel reflectance at Ömerli Dam, which is a vital potable water resource of Istanbul City, Turkey and also found suspended solid matter and total phosphate to have high R^2 of 0.9999 and 0.9906, respectively. On the other hand, Reza (2008) showed that there is a relationship between the level of suspended solids and MODIS radiance or reflectance in the seawater region around Penang Island of northern Malaysia. Lim *et al.* (2008) revealed that the total suspended solids (TSS) algorithm developed in the same island (Penang) from optical properties is a promising TSS model for high-accuracy TSS mapping using satellite data. Furthermore, He *et al.* (2008) found a fairly high correlation between Landsat TM imagery DN values with water quality variables (algae, turbidity, TN and TP) at the Guanting Reservoir in Beijing, China. All the above studies have concluded that water quality can be successfully derived through remote sensing data. Our study shows that there is generally a good correlation for all the parameters except total suspended solids, suggesting that the algorithms can be used to retrieve and predict water quality data.

Spatial variation of retrieved water quality parameters

The spatial distribution of water quality parameters retrieved from satellite data demonstrate similar patterns as the in-situ observed measurements based on a statistical comparison (R^2 of >0.75 except for total suspended solids). Hafeez *et al.* (2019) found similar results with turbidity amongst other parameters, in a comparison of satellite retrieval of reflectance data with in situ reflectance data for a study in Hong Kong. The results of this study show that there is substantial spatial variation in the concentration of water quality parameters both longitudinally and latitudinally due to several factors. These include rainfall runoff processes (Bonansea *et al.* 2015), nutrient accumulation due to decomposition of plants and animals as well as low or no flow especially at the end part of the dam. In addition, it could be due to human activities such as fishing, abstraction of water and swimming, which would encourage mixing of water constituents. All these factors affect water quality.

In addition to the six sampling points, the Landsat predicted data at specific sampling points were also compared with water quality standards for the different uses. Turbidity was found to be above the NamWater (1998) guidelines of 5 NTU but below the Canadian (2012) guidelines for recreational water quality (50 NTU). On the other hand, TN and TP are below ECE (1994) standards for maintenance of aquatic life, which are <300 mg/L and <10 mg/L respectively. Water quality parameters outside recommended ranges are likely to cause complications in drinking water treatment as well as for human and aquatic life.

Suitability of medium-resolution satellite images for water quality monitoring

This study proves that medium resolution images such as those of Landsat are able to predict water quality parameters based on high regression coefficients (>0.75), as observed in studies by Concha & Schott (2014), Chen & Quan (2012), He *et al.* (2008) and Hellweger *et al.* (2004). The study also showed that Landsat 8 has capabilities in modelling the water quality of relatively clean water bodies that have low spectral radiation just like Landsat TM as recommended by Peterson *et al.* (2020), Pu *et al.* (2019) and He *et al.* (2008). Li (2009), however, used different Landsat missions (Landsat 5) in the Chesapeake Bay, USA, for water quality monitoring. However, the in-situ observed measurements remain critical in the process of monitoring water quality using remote sensing techniques (Ritchie *et al.* 2003). Namibia has cloud-free sky most times of the year, which makes remote sensed water quality monitoring possible.

Future application of the present work

The future practical application of the present work includes the improved capacity to predict water quality status for data-scarce environments such as north-central Namibia where there is a deterioration of water quality. Such efforts potentially contribute towards improved environmental and watershed management in an integrated water resources management approach. Based on the strong relationship between field-based water quality results and satellite-based water quality retrieval, our results also help in rapid water quality appraisals of other reservoirs in Namibia. In addition, the developed algorithms can be used to assess anthropogenic land-use activities in pollution of reservoirs. Multiple linear regression analyses of land-use and land-cover variable combinations can further be developed and used to develop equations for estimating water quality parameters. Example applications of such work can also be found in many global water resources studies (Christian *et al.* 2000; Goodman *et al.* 2020; Kim *et al.* 2020; Pasika & Gandla 2020; Topp *et al.* 2020). A water quality prediction framework for Olushandja Dam, north-central Namibia, will continue to use freely available earth observation data to contribute towards improved management of pollution. Future work should also focus on developing a mobile-based and a web-based application for rapid retrieval and communication of water quality information.

CONCLUSIONS AND RECOMMENDATIONS

The capacity to predict water quality status for data-scarce environments can potentially contribute towards improved water quality management. The study demonstrated that a linear regression statistical approach can be used to develop algorithms for retrieving water quality data from satellite imagery as the water quality values retrieved from remote sensing techniques had a high correlation with field measured data ($R^2 > 0.98$). The algorithms developed have a good prediction accuracy for turbidity, total nitrogen, total phosphorus and total algae count as the linear relationship between Landsat predicted and measured values had R^2 of 0.77–0.91 while prediction for total suspended solids showed a weak linear relationship ($R^2 = 0.28$). The developed algorithms can be used to predict water quality variables for the whole Olushandja Dam but require more sampling points to further improve the accuracy of the regression models. We recommend further development of a GIS web-based application for rapid retrieval and communication of water quality information by water managers, water users and stakeholders in the Cuvelai Basin in north-central regions of Namibia.

ACKNOWLEDGEMENTS

This paper presents part of the research results of an MSc study by Taimi Kapalanga at the University of Zimbabwe (Department of Civil Engineering) under a WaterNet Fellowship under the supervision team of Eng. Zvikomborero Hoko and Mr Webster Gumindoga. The authors would like to thank NamWater and NamLab for laboratory equipment and water quality analysis. Mr Lloyd Chikwiramakomo improved the maps in this manuscript.

DATA AVAILABILITY STATEMENT

All relevant data are included in the paper or its Supplementary Information.

REFERENCES

Alparslan, E., Aydöner, C., Tufekci, V. & Tüfekci, H. 2007 Water quality assessment at Ömerli Dam using remote sensing techniques. *Environmental Monitoring and Assessment* **135**, 391–398.

Angula, M. N. & Kaundjua, M. B. 2016 The changing climate and human vulnerability in north-central Namibia. *Jàmbá: Journal of Disaster Risk Studies* **8** (2), 200. http://dx.doi.org/10.4102/jamba.v8i2.20.

Bain, R., Johnston, R. & Slaymaker, T. 2020 Drinking water quality and the SDGs. *npj Clean Water* **3** (1), 37. https://doi.org/10.1038/s41545-020-00085-z.

Bauer, M., Dekker, A., DiGiacomo, P., Greb, S., Gitelson, A., Herlevi, A. & Kuster, T. 2007 *Inland and Nearshore Coastal Water Quality Remote Sensing*. GEOSS Group on Earth Observation, Geneva, Switzerland.

Bhuyar, P., Sundararaju, S., Rahim, M. H. A., Ramaraj, R., Maniam, G. P. & Govindan, N. 2019a Microalgae cultivation using palm oil mill effluent as growth medium for lipid production with the effect of CO_2 supply and light intensity. *Biomass Conversion and Biorefinery*. https://doi.org/10.1007/s13399-019-00548-5.

Bhuyar, P., Hong, D. D., Mandia, E., Rahim, M. H. A., Maniam, G. P. & Govindan, N. 2019b Desalination of polymer and chemical industrial wastewater by using green photosynthetic microalgae, *Chlorella* sp. *Maejo International*

Journal of Energy and Environmental Communication **1** (3), 9–19.

Bhuyar, P., Hong, D. D., Mandia, E., Rahim, M. H. A., Maniam, G. P. & Govindan, N. 2020 Salinity reduction from poly-chem-industrial wastewater by using microalgae (*Chlorella* sp.) collected from coastal region of peninsular Malaysia. *J. Biol. Med. Open Access* **1**, 105.

Bonansea, M., Rodriguez, M. C., Pinotti, L. & Ferrero, S. 2015 Using multi-temporal Landsat imagery and linear mixed models for assessing water quality parameters in Río Tercero reservoir (Argentina). *Remote Sensing of Environment* **158**, 28–41.

Brando, V. E. & Dekker, A. G. 2003 Satellite hyperspectral remote sensing for estimating estuarine and coastal water quality. *Geoscience and Remote Sensing* **41** (6), 1378–1387.

Chander, G., Markham, B. L. & Helder, D. L. 2009 Summary of current radiometric calibration coefficients for Landsat MSS, TM, ETM+, and EO-1 ALI sensors. *Remote Sensing of Environment* **113**, 893–903.

Chapra, M. U. 2015 Innovation and authenticity in Islamic finance. In: *Keynote Forum Lecture Delivered at the Inaugural Session of the Eighth Harvard University Forum on Islamic Finance*, 19–20 April 2008, Harvard Law School, Cambridge, MA, USA.

Chawira, M., Dube, T. & Gumindoga, W. 2013 Remote sensing based water quality monitoring in Chivero and Manyame lakes of Zimbabwe. *Physics and Chemistry of the Earth* **66**, 38–44.

Chen, J. & Quan, W. T. 2012 Using Landsat/TM imagery to estimate nitrogen and phosphorus concentration in Taihu Lake, China. *IEEE J-STARS* **5** (1), 273–280.

Chen, Q., Zhang, Y. & Hallikainen, M. 2007 Water quality monitoring using remote sensing in support of the EU water framework directive (WFD): a case study in the Gulf of Finland. *Environmental Monitoring and Assessment* **124** (1), 157–166.

Christian, H. J., Blakeslee, R. J., Goodman, S. J. & Mach, D. M. 2000 *Algorithm Theoretical Basis Document for the Lightning Imaging Sensor*. Marshall Space Flight Center, NASA, USA.

Concha, J. A. & Schott, J. R. 2014 A model-based ELM for atmospheric correction over Case 2 water with Landsat 8. *Proc. SPIE* **9111** (Ocean Sensing and Monitoring 6), 911112.

CuveWaters 2014 *The Interactive Water Information and Planning Tool (IWIP) of the Cuvelai-Etosha Basin*, 2014 edn. CuveWaters Project, Germany/Namibia.

Dekker, A. G., Vos, R. J. & Peters, S. W. M. 2002 Analytical algorithms for lake water TSM estimation for retrospective analyses of TM and SPOT sensor data. *International Journal of Remote Sensing* **23**, 15–35.

du Plessis, A., Harmse, T. & Ahmed, F. 2014 Quantifying and predicting the water quality associated with land cover change: a case study of the Blesbok Spruit Catchment, South Africa. *Water* **6**, 2946–2968.

El-Saadi, A. M., Yousry, M. M. & Jahin, H. S. 2014 Statistical estimation of Rosetta branch water quality using multi-spectral data. *Water Science* **28**, 18–30.

Goodman, J. A., Lay, M., Ramirez, L., Ustin, S. L. & Haverkamp, P. J. 2020 Confidence levels, sensitivity, and the role of bathymetry in coral reef remote sensing. *Remote Sensing* **12**, 3, 496. https://doi.org/10.3390/rs12030496.

Hafeez, S., Wong, M. S., Ho, H. C., Nazeer, M., Nichol, J., Abbas, S., Tang, D., Lee, K. H. & Pun, L. 2019 Comparison of machine learning algorithms for retrieval of water quality indicators in case-II waters: a case study of Hong Kong. *Remote Sensing* **11** (6), 617.

He, W., Chen, S., Liu, X. & Chen, J. 2008 Water quality monitoring in a slightly-polluted inland water body through remote sensing – case study of the Guanting Reservoir in Beijing, China. *Frontiers of Environmental Science & Engineering in China* **2** (2), 163–171.

Hellweger, F. L., Schlosser, P., Lall, U. & Weissel, J. K. 2004 Use of satellite imagery for water quality studies in New York Harbor. *Estuarine, Coastal and Shelf Science* **61**, 437–448.

Irenosen, O. G., Festus, A. A. & Coolborn, A. F. 2012 Water quality assessment of the Owena Multi-Purpose Dam, Ondo State, Southwestern Nigeria. *Journal of Environmental Protection* **3**, 14–25.

IWRMPJVN 2010 *Integrated Water Resources Management Plan for Namibia*. IWRM Plan Joint Venture Namibia, Ministry of Agriculture, Water and Forestry, Windhoek, Namibia.

Kallio, K. 2000 Remote sensing as a tool for monitoring lake water quality. In: *Hydrological and Limnological Aspects of Lake Monitoring* (P. Heinonen, G. Ziglio & A. Van der Beken, eds), John Wiley & Sons, New York, NY, USA, pp. 237–245.

Kelbert, T. 2016 *Encounters at the Water Point – An Ethnography of the Travelling Model of Community-Based Water Management and Its Application to Rural Water Supply in Namibia*. PhD thesis, University of Cologne, Cologne, Germany.

Kibena, J., Nhapi, I. & Gumindoga, W. 2014 Assessing the relationship between water quality parameters and changes in landuse patterns in the Upper Manyame River, Zimbabwe. *Physics and Chemistry of the Earth* **67–69**, 153–163.

Kim, Y. H., Son, S., Kim, H.-C., Kim, B., Park, Y.-G., Nam, J. & Ryu, J. 2020 Application of satellite remote sensing in monitoring dissolved oxygen variabilities: a case study for coastal waters in Korea. *Environment International* **134**, 105301. https://doi.org/https://doi.org/10.1016/j.envint.2019.105301.

Kolberg, H. 2002 *Preliminary Inventory of Namibia's Wetlands*. Directorate of Scientific Services, Ministry of Environment and Tourism, Windhoek, Namibia. Available from: http://www.drfn. info/arc/pdf/reprints/drfn2979.pdf (accessed 29 August 2014).

Lehmann, F. L. 2010 *Water Quality of the Swakoppoort Dam, Namibia*. Master's thesis, IHF Hydrologie Freiburg, University of Freiburg, Freiburg im Breisgau, Germany.

Li, D. 2009 *A Novel Remote Sensing-Based Water Quality Monitoring System for the Chesapeake Bay Using Landsat 5 Thematic Mapper*. Thomas Jefferson High School for Science and Technology, Alexandria, VA, USA.

Lim, H. S., MatJafri, M. Z., Abdullah, K., Alias, A. N. & Saleh, N. M. 2008 Water quality mapping using remote sensing technique in Penang Straits, Malaysia. *Proc. SPIE* **7105**, 71050A. https://doi.org/10.1117/12.800573.

Maillard, P. & Pinheiro Santos, N. A. 2008 A spatial-statistical approach for modeling the effect of non-point source pollution on different water quality parameters in the Velhas river watershed – Brazil. *Journal of Environmental Management* **86** (1), 158–170.

Masocha, M., Dube, T., Nhiwatiwa, T. & Choruma, D. 2018 Testing utility of Landsat 8 for remote assessment of water quality in two subtropical African reservoirs with contrasting trophic states. *Geocarto International* **33** (7), 667–680.

McMillan, H., Seibert, J., Petersen-Overleir, A., Lang, M., White, P., Snelder, T., Rutherford, K., Krueger, T., Mason, R. & Kiang, J. 2017 How uncertainty analysis of streamflow data can reduce costs and promote robust decisions in water management applications. *Water Resources Research* **53** (7), 5220–5228.

Mendelsohn, J., El Obeid, S. & Roberts, C. 2000 *A Profile of North-Central Namibia*. Environmental Profiles Project, Gamsberg Macmillan Publishers, Windhoek, Namibia.

Mendelsohn, J., Jarvis, A. & Robertson, T. 2013 *A Profile and Atlas of the Cuvelai-Etosha Basin*, 1st edn. RAISON and Gondwana Collection, Windhoek, Namibia.

Mendelsohn, J., Jarvis, A. & Robertson, T. 2016 *A Profile and Atlas of the Cuvelai-Etosha Basin*. John Meinert Printing, Windhoek, Namibia.

NAMWATER 1998 *Guidelines for Evaluating Drinking-Water for Human Consumption with Respect to the Chemical, Physical and Bacteriological Qualities*. Namibia Water Corporation, Windhoek, Namibia.

NamWater 2012 *Management of the Upper Swakop Basin: Key Water Security to the Central Area of Namibia* (G. T. Pazvakawambwa, K. F. Tjipangandjara & E. Chulu, eds), NamWater, Windhoek, Namibia.

NamWater 2013 *Infrastructures – Olushandja Dam*, Namibia Water Corporation Ltd, Windhoek, Namibia.

Nhapi, I. 2009 The water situation in Harare, Zimbabwe: a policy and management problem. *Water Policy* **11** (2), 221–235.

Oberholster, P. J. & Ashton, P. J. 2008 *State of the Nation Report: An Overview of the Current Status of Water Quality and Eutrophication in South African Rivers and Reservoirs* Parliamentary Grant Deliverable, Council for Scientific and Industrial Research (CSIR), Pretoria, South Africa.

Olet, I. 2010 *Water Quality Monitoring of Roxo Reservoir Using Landsat Images and In-Situ Measurements*. MS thesis, International Institute for Geo-Information Science and Earth Observation, Enschede, The Netherlands.

Parida, S., Patsha, A., Bera, S. & Dhara, S. 2017 Spectroscopic investigation of native defect induced electron–phonon coupling in GaN nanowires. *Journal of Physics D: Applied Physics* 50, 275103.

Pasika, S. & Gandla, S. T. 2020 Smart water quality monitoring system with cost-effective using IoT. *Heliyon* **6** (7), e04096. https://doi.org/10.1016/j.heliyon.2020.e04096.

Peterson, K. T., Sagan, V. & Sloan, J. J. 2020 Deep learning-based water quality estimation and anomaly detection using Landsat-8/Sentinel-2 virtual constellation and cloud computing. *GIScience & Remote Sensing* **57** (4), 510–525.

Pu, F., Ding, C., Chao, Z., Yu, Y. & Xu, X. 2019 Water-quality classification of inland lakes using Landsat8 images by convolutional neural networks . *Remote Sensing* 11 (14), 1674

Reza, M. M. 2008 Assessment of suspended sediments concentration in surface waters, using Modis Images. *American Journal of Applied Sciences* **5** (7), 798–804.

Ritchie, J. C., Zimba, P. V. & Everitt, J. H. 2003 Remote sensing techniques to assess water quality. *Photogrammetric Engineering & Remote Sensing* **69** (6), 695–704.

Salama, M. S., Dekker, A., Su, Z., Mannaerts, C. M. & Verhoef, W. 2009 Deriving inherent optical properties and associated inversion-uncertainties in the Dutch Lakes. *Hydrology and Earth System Sciences* **13** (7), 1113–1121.

Schaeffer, B. A., Schaeffer, K. G., Keith, D., Lunetta, R. S., Conmy, R. & Gould, R. W. 2013 Barriers to adopting satellite remote sensing for water quality management. *International Journal of Remote Sensing* **34** (21), 7534–7544.

Schalles, J. F., Gitelson, A. A., Yacobi, Y. Z. & Kroenke, A. E. 1998 Estimation of chlorophyll *a* from time series measurements of high spectral resolution reflectance in an eutrophic lake. *Journal of Phycology* **34**, 383–390.

SDP10 2001 *Water use and Environmental Conditions along the Etaka Canal in Northern Central Namibia*. Summer Desertification Programme, Desert Research Foundation of Namibia, Windhoek, Namibia.

Shuuya, M. K. & Hoko, Z. 2014 Trends and impacts of pollution in the Calueque–Oshakati Canal in north-Central Namibia on water treatment. In: *Combating Water Scarcity in Southern Africa* (J. P. Msangi, ed.), Springer, Dordrecht, The Netherlands, pp. 43–60.

Somvanshi, S., Kunwar, P., Singh, N. B., Shukla, S. P. & Pathak, V. 2012 Integrated remote sensing and GIS approach for water quality analysis of Gomti river, Uttar Pradesh. *International Journal of Environmental Sciences* **3** (1), 62–74.

Tomppo, E., Czaplewski, R. L. & Mäkisara, K. 2002 *The Role of Remote Sensing in Global Forest Assessment: A Remote Sensing Background Paper for Kotka IV Expert Consultation*. Food and Agriculture Organization, Rome, Italy.

Topp, S. N., Pavelsky, T. M., Jensen, D., Simard, M. & Ross, M. R. V. 2020 Research trends in the use of remote sensing for inland water quality science: moving towards multidisciplinary applications. *Water* **12** (1), 169. https://doi.org/10.3390/w12010169.

UN 2011 *Namibia Floods: Situation Report #4*. United Nations, Office of the Resident Coordinator, Windhoek, Namibia.

UNECE 1994 *ECE Standard Statistical Classification of Surface Freshwater Quality for the Maintenance of Aquatic Life.* United Nations Economic Commission for Europe (UNECE), Geneva, Switzerland.

UN-Water 2011 *Policy Brief on Water Quality.* UN-Water, New York, USA. Available from: http://www.unwater.org/downloads/waterquality_policybrief (accessed 12 May 2015).

USGS 2013 *Using the USGS Landsat 8 Product.* US Geological Survey, USA.

Vignolo, A., Pochettino, A. & Cicerone, D. 2006 Water quality assessment using remote sensing techniques: Medrano Creek, Argentina. *Journal of Environmental Management* **81** (4), 429–433.

Wait, A. D., Tuit, C. B. & Maney, J. P. 2020 Forensic sampling practices for oil spills in the marine environment. *Environmental Forensics* **21**, 310–318.

Wang, X. J. & Ma, T. 2001 Application of remote sensing techniques in monitoring and assessing the water quality of Taihu Lake. *Bulletin of Environmental Contamination and Toxicology* **67** (6), 863–870.

Wang, Y., Xia, H., Fu, J. & Sheng, G. 2004 Water quality change in reservoirs of Shenzhen, China: detection using LANDSAT/TM data. *Science of the Total Environment* **328** (1–3), 195–206.

Waxter, M. T. 2014 *Analysis of Landsat Satellite Data to Monitor Water Quality Parameters in Tenmile Lake, Oregon,* Civil and Environmental Engineering. Master's Project Reports 2. Portland State University, OR, USA.

Yang, X. & Jin, W. 2010 GIS-based spatial regression and prediction of water quality in river networks: a case study in Iowa. *Journal of Environmental Management.* **91**, 1943–1951.

Zuccari Fernandes Braga, C., Setzer, A. W. & Drude de Lacerda, L. 1993 Water quality assessment with simultaneous Landsat-5 TM data at Guanabara Bay, Rio de Janeiro, Brazil. *Remote Sensing of Environment* **45** (1), 95–106.

First received 5 April 2020; accepted in revised form 14 October 2020. Available online 29 October 2020

Modelling a rotating biological contactor treating heavy metal contaminated wastewater using artificial neural network

M. Gopi Kiran, Raja Das, Shishir Kumar Behera, Kannan Pakshirajan and Gopal Das

ABSTRACT

The performance of a continuously operated laboratory-scale rotating biological contactor (RBC) was assessed for the removal of heavy metals *viz.* Cu(II), Cd(II) and Pb(II) from synthetic wastewater using artificial neural networks (ANNs). The RBC was inoculated with Sulfate Reducing Bacteria consortium (predominantly *Desulfovibrio* species), and the performance was evaluated at different hydraulic retention times (HRTs) and inlet heavy metal concentrations. A feed-forward back-propagation neural network model was developed using 90 data sets obtained over a period of three months, to predict the removal of heavy metal (HM_{RE}) and COD (COD_{RE}). The predictive capability of the model was evaluated in terms of the coefficient of determination (R) and mean absolute percentage error between the model fitted and actual experimental data, whereas sensitivity analysis was performed on the input parameters by determining the absolute average sensitivity (AAS) values. The higher AAS value of the HRT compared with that of inlet heavy metal concentration suggested that the change of HRT has a significant influence on HM_{RE} and COD_{RE}. Overall, the results obtained from this study demonstrated that ANNs can efficiently predict RBC behaviour with regard to heavy metal and COD removal characteristics under the prevailing operational conditions.

Key words | COD removal, heavy metal removal, modelling, neural network, rotating biological contactor

M. Gopi Kiran
Kannan Pakshirajan
Gopal Das
Centre for the Environment,
Indian Institute of Technology Guwahati,
Guwahati 781 039, Assam,
India

Raja Das
Department of Mathematics, School of Advanced Sciences,
Vellore Institute of Technology,
Vellore, 632 014, Tamilnadu,
India

Shishir Kumar Behera (corresponding author)
Industrial Ecology Research Group, School of Chemical Engineering,
Vellore Institute of Technology,
Vellore 632 014, Tamilnadu,
India
E-mail: *shishir.kb@gmail.com*

Kannan Pakshirajan
Department of Biosciences and Bioengineering,
Indian Institute of Technology Guwahati,
Guwahati 781 039, Assam,
India

Gopal Das
Department of Chemistry,
Indian Institute of Technology Guwahati,
Guwahati 781 039, Assam,
India

HIGHLIGHTS

- Development of a feed-forward back propagation neural network model for an RBC.
- Prediction of heavy metal and COD removal.
- Evaluation of predictive capability in terms of coefficient of determination and mean absolute percentage error.
- Sensitivity analysis by determining the absolute average sensitivity values.

doi: 10.2166/ws.2020.304

GRAPHICAL ABSTRACT

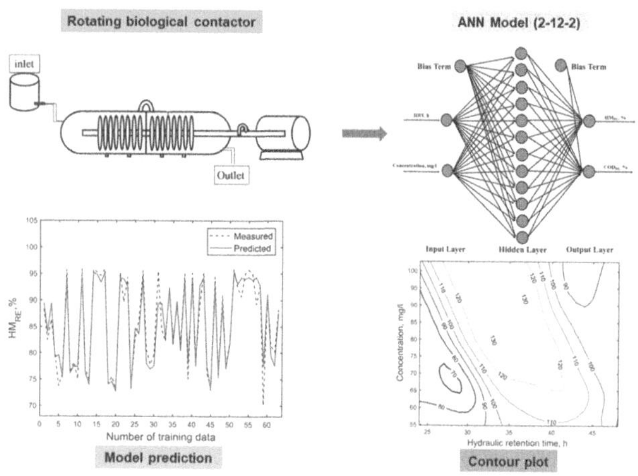

INTRODUCTION

Heavy metals are released into the environment from different industries, such as metallurgy, tanneries, mining, electroplating industries, etc. (Kikot *et al.* 2010). Heavy metals such as Cu, Cd, and Pb discharged from industrial wastewater are toxic at high concentration and, thus, pose serious risk to both human health and the environment. Hence, removal of heavy metals from wastewater before their discharge into the environment is mandatory (Kiran *et al.* 2015).

Compared with the physico-chemical methods, biological methods have been proven to be cost-effective and environmentally friendly technology for heavy metal removal from wastewater (Bai *et al.* 2013; Yang *et al.* 2016a, 2016b; Yajun *et al.* 2019). Different kinds of bioreactor systems have been employed to treat heavy metals present in wastewater. For example, wastewater containing heavy metals (e.g. Cd, Cu, Cr, Zn, Pb and Ni) were treated in suspended growth bioreactors, *viz.* Continuously Stirred Tank Reactors (Gola *et al.* 2020), and Membrane Bioreactor employed for the treatment of textile industry wastewater containing chromium (VI). On the other hand, attached growth bioreactors, *viz.* Packed Bed Bioreactor was employed for the treatment of acid mine drainage containing a variety of heavy metals (Dev *et al.* 2017), and Rotating Biological Contactors (RBCs) for heavy metal removal from synthetic wastewater (Kiran *et al.* 2017a).

In the recent past, RBCs have received great attention due to their capability to treat various types of refractory wastewaters, e.g., coloured industry wastewater (Pakshirajan & Kheria 2012), agrochemical industry wastewater (Vasiliadou *et al.* 2016) and gold mine wastewater (Guadalima & Monteros 2018). The advantages of RBCs over other attached growth systems include but are not limited to: (1) handling specific contaminants *viz.* hydrocarbons, heavy metals, xenobiotics and pharmaceuticals/personal care products, (2) handling high organic loadings and resistance to toxic shocks, (3) requiring low land area, maintenance, energy and start-up costs, and (4) energy generation along with wastewater treatment (Hassard *et al.* 2015).

Recently, the application of artificial neural networks (ANNs) to model and predict the operational efficiency of various biological systems, including wastewater and waste gas treatment systems, has gained remarkable attention (Nair *et al.* 2016; López *et al.* 2017; Antwi *et al.* 2018; Baskaran *et al.* 2019; Lin *et al.* 2020). A neural network can be used to simulate the non-linear input/output dynamics of any process based on the time-series datasets (Menhaj 1998; Chen *et al.* 2019). The primary advantage of ANN over phenomenological/conceptual models is that it does not require information about the complex nature of the underlying process to be explicitly described in mathematical form (Sahoo *et al.* 2005). The ANN model, in this study, was developed using the most widely employed feed-forward back-propagation (BP) algorithm, which uses a gradient descent procedure to minimize the objective function (Rumelhart *et al.* 1986). In other words, in the BP algorithm, the weights between input, output and hidden layers are modified and the process is repeated until the error between the output of the neural network and the desired output is minimized.

López *et al.* (2017) used a two-stage biological waste gas treatment system consisting of a first-stage biotrickling filter (BTF) and second-stage biofilter (BF) for the removal of a gas-phase methanol (M), hydrogen sulphide (HS) and α-pinene (P) mixture. The performance of the gas treatment system was modelled using two multi-layer perceptrons (MLPs) employing the back-propagation algorithm, in order to predict the removal efficiencies of methanol (RE_M), hydrogen sulphide (RE_{HS}) and α-pinene (RE_P). An MLP with the topology 3-4-2 was able to predict RE_M and RE_{HS} in the BTF, while a topology of 3-3-1 was able to approximate the RE_P in the BF. In another study, a three-layered feed-forward back-propagation ANN model was developed to evaluate the performance of an upflow anaerobic sludge blanket reactor in terms of COD removal while treating industrial starch processing wastewater. The proposed Levenberg–Marquardt algorithm based ANN model demonstrated very satisfactory performance during COD removal predictions and simulation (Antwi *et al.* 2018). Baskaran *et al.* (2019) developed an ANN model to predict the performance of a compost biofilter in terms of trichloroethylene (TCE) removal under continuous operation mode. The ANN model used a three-layer feed-forward-based Levenberg–Marquardt algorithm with a topology 3-25-1 to predict the TCE removal with R^2 greater than 0.99 during the model training, validation, and testing. The performance of a laboratory-scale anaerobic bioreactor was modelled by using the feed-forward back-propagation algorithm to optimize the CH_4 content of 60%–70% in biogas obtained from digestion of the organic fraction of municipal solid waste (Nair *et al.* 2016). As evident from the aforementioned reports, ANNs have shown their capability to accurately predict the behaviour of various biological systems by establishing the correspondence between input and output domains. Thus, it is clear that ANN models can be developed for bioreactors with a definite objective and adequate training of time-series data collected from such reactors. However, modelling the performance of RBC or any other reactor for heavy metal removal from wastewater has not been reported so far. This is mainly important considering the non-biodegradable and toxic nature of heavy metals in wastewater such as acid mine drainage (AMD), characterized by its low pH and high sulphate content along with heavy metals.

Hence, this study focused on the feasibility of an ANN model to estimate, predict and simulate experimental results obtained from an RBC treating heavy-metal-contaminated-synthetic wastewater. The objectives of this study were formulated as follows: (i) to create an MLP, by varying the internal network parameters, that would predict the HM_{RE} in the RBC used for heavy metal removal from synthetic wastewater (Kiran *et al.* 2017a), (ii) testing the developed model with data that was not presented to the ANN during training, (iii) to perform sensitivity analysis, analyze and determine the most influencing input parameters (HRT and inlet heavy metal concentrations) for each output (HM_{RE} and COD_{RE}), and (iv) to emphasize the application of ANNs for control of bioreactor input parameters and to address the potential advantages of ANNs for predicting bioreactor performance. To the best of the authors' knowledge, this is the first report that has used ANN to model an RBC treating heavy-metal-contaminated wastewater.

MATERIALS AND METHODS

Experimental

Microorganism and chemicals

Mixed Sulphate Reducing Bacteria (SRB) consortium (predominantly consisting of *Desulfovibrio* species) used in the RBC was acquired from a laboratory-scale upflow anaerobic packed bed reactor treating sulphate-rich wastewater (Kiran *et al.* 2015). During the experiments, sulphate and COD concentration in the influent were adjusted to maintain a COD/SO_4^{2-} ratio of 0.67 ± 0.08. The pH of the solution was adjusted to 7 using 1 N NaOH. The stock solutions of Cd(II), Cu(II), and Pb(II) prepared using $Cd(NO_3)_2 \cdot 4H_2O$, $CuCl_2 \cdot 2H_2O$, and $PbNO_3$, respectively were procured from

Alpha Chemika Co. Ltd, India, Euclid Co. Ltd, India, and Karni Chemicals Co. Ltd, India. All the chemicals used for experiments were of reagent grade and were used without further purification (Kiran *et al.* 2017a).

Reactor set-up and operation

The operational protocol of the laboratory-scale anaerobic RBC (total working volume = 3 L) made up of polymethyl methacrylate material is described elsewhere (Pakshirajan & Kheria 2012). The RBC consists of two identical stages connected in series. Each stage had a working volume of 1.5 L and was equipped with seven discs enclosed with polystyrene mesh and polyurethane foam in which the SRB consortium was immobilized. The disc diameter was 0.16 m with a thickness of 0.0056 m and they were spaced at 0.02 m distance. The submerged surface was 40% (Kiran *et al.* 2017a).

The RBC was operated under continuous mode and was maintained at a temperature of 25 ± 2 °C. The suspended biomass was measured in terms of mixed liquor volatile suspended solids. Samples collected at regular intervals were centrifuged at 8000 × g for 5 min, and the supernatant obtained was analysed for metal, sulphate, sulphide and COD concentrations (Kiran *et al.* 2017a). The metal stock solutions of Cd(II), Cu(II), and Pb(II) of 10,000 mg/l concentration each were prepared using $Cd(NO_3)_2{\cdot}4H_2O$, $CuCl_2{\cdot}2H_2O$, and $PbNO_3$, respectively. The phase-wise (three phases) inlet concentrations of the metals Cd(II) and Pb(II) were chosen as 50, 75 and 90 mg/l, whereas for Cu(II), the concentrations were 100, 150 and 175 mg/l. All these inlet metal concentrations were chosen based on the studies conducted earlier using the same anaerobic biomass containing SRB (Kiran *et al.* 2015, 2017b). Reactor performance in terms of RE_{HM} was evaluated at two different HRTs (24 and 48 h). Each experiment was carried out for a period until three steady-state values of effluent heavy metal concentration at the respective HRT were obtained. All the results presented are averages of duplicate sample analyses. The combined effect of inlet metal and COD concentration and HRT on removal was examined by calculating the RE as given in Equation (1):

$$\%\mathit{Removal\ efficiency} = \frac{C_i - C_o}{C_i} \times 100 \qquad (1)$$

where C_i and C_o are the inlet and outlet concentrations (mg/l), respectively.

Analytical methods

The heavy metal concentration in the samples was determined using an atomic absorption spectrometer (Varian, AA240, The Netherlands). The COD and sulphate concentration was determined by following the closed reflux method and standard turbidimetric method, respectively (APHA 2005). The dissolved sulphide concentration in the liquid samples was determined as per the method described elsewhere (Cord-Ruwisch 1985).

ANN model development

The back-propagation neural network (BPNN) is composed of layers of neurons. The input layer of neurons is connected to the output layer and the training process is undertaken by changing the weights in order to establish a desired input–output relationship. A schematic diagram showing the BPNN with two input nodes, two output nodes, and a single hidden layer of 12 nodes is given in Figure 1. The experimental values did not show any advantage of a double hidden layer over a single layer network. Therefore, a single hidden layer network is used in this work and all the connections have multiplying weights associated with them.

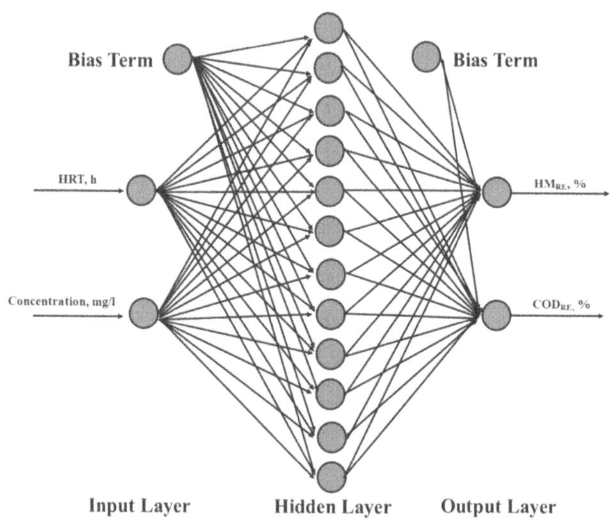

Figure 1 | Schematic of a multilayer perceptron used for predicting RBC behaviour.

The input nodes have a transfer function of unity and the activation functions of the hidden and output nodes are sigmoidal and linear, respectively. The data required for developing a representative and reliable ANN model consisted of the parameters, *viz.*, inlet concentration (mg/l), HRT (h), HM_{RE} (%) and COD_{RE} (%) obtained from steady-state operation of the RBC. The data was normalized and scaled to the range of 0–1 using Equation (2) so as to suit the transfer function in the hidden (sigmoid) and output layer (linear):

$$\hat{X} = \frac{X - X_{min}}{X_{max} - X_{min}} \quad (2)$$

where $\hat{X}$ is the normalized value, and X_{min} and X_{max} are the minimum and maximum values of X respectively.

From the steady-state RBC operation data, the set of 90 data points were divided into training and testing sets; 70% (N_{Tr}: 63) of the data points were used for training the network while the remaining 30% (N_{Te}: 27) were used for testing the developed model. An MLP was formulated to predict the output parameters, HM_{RE} (%) and COD_{RE} (%) using inlet concentration and HRT as the input parameters. Table 1 describes the basic statistical information about the training and test data sets for the RBC. The data used as input are transmitted through the network and a set of output data are obtained. The obtained outputs are compared with the desired output values, and the difference between the desired output and calculated output (error) is used to adjust the weights of the network in order to reduce the magnitude of the error. After training, the ANN accepts new 'unknown' data as input and predicts the corresponding output based on its training. The best values of network parameters, i.e., training count (T_c), number of neurons in the hidden layer (N_H), learning rate (η), momentum term (α) and epoch size (ε) were selected by trial and error. These values are shown in Table 2. The predictive capability of the model was evaluated in terms of the coefficient of determination (R) and the mean absolute percentage error (MAPE). Further details pertaining to the effect of these network parameters and neural network modelling can be found elsewhere. Sensitivity analysis in the form of absolute average sensitivity (AAS) value was computed to determine the most influential input parameter affecting the output of the ANN (Maier & Dandy 1998; Rene *et al.* 2009, 2011).

The predictive modelling work using ANN was carried out using the shareware version of the NN and multivariable statistical modelling software, MATLAB® (Version 9.2.0.538062 [R2017a], Neural Network Toolbox). The specification of the computer used to perform the simulation had the following configurations: Operating system – Windows 8.1 Single Language; processor – Intel(R) Core(TM) i5-4200 CPU @ 1.60 GHz 2.30 GHz; 8.00 GB system memory.

RESULTS AND DISCUSSION

Effect of network parameters and best network topology

The input and output parameters in ANN modelling are usually selected on the basis of the objective of the modelling and the data availability. In order to develop the ANN model for the laboratory-scale RBC in this study, the selection of input parameters *viz.*, HRT and inlet concentration and that of output parameters *viz.*, HM_{RE} (%) and COD_{RE} (%) were decided on the basis of system knowledge and real-life requirements.

The first step of this modelling task was to identify a suitable network topology for the MLP through proper optimization of the network parameters (Maier & Dandy 1998; Sahinkaya 2009). Towards this, T_c and N_H were varied from 0 to 100 by keeping other network parameters such as η (0.75) and α (0.75) at their constant values. The best values of η and α for the developed model were determined by keeping some training parameters constant at their default values (suggested by MATLAB®), and by gradually changing values of the parameters from 0.1 to 0.9, with a step-size increment of 0.1 for each parameter. It was observed that increasing the number of neurons (N_H) from 6 to 12 increased the R value. The optimized values of network parameters for the model are shown in Table 2. The best network topology for the RBC was found to be 2-12-2 (i.e. two input parameters, 12 neurons in one hidden layer and two output parameters). After obtaining the best network topology for the RBC, the connection weights and

Table 1 | Basic statistics of the data used for ANN model development: (a) Cu, (b) Cd, and (c) Pb

(a) Cu

Basic statistics of the training data

Variable	*N*	Mean	Std Dev	Minimum	Maximum	Sum Sq
HRT (h)	63	36.57	12.08	24.00	48.00	13,606.29
Concentration (mg/l)	63	173.28	11.89	150.00	191.31	241,569.85
HM_{RE} (%)	63	95.61	2.31	91.31	99.28	72,675.62
COD_{RE} (%)	63	53.53	15.96	13.73	94.38	28,158.88

Basic statistics of the test data

Variable	*N*	Mean	Std Dev	Minimum	Maximum	Sum Sq
HRT (h)	27	35.56	12.22	24.00	48.00	8,582.21
Concentration (mg/l)	27	166.19	10.10	151.00	186.00	145,067.91
HM_{RE} (%)	27	95.54	1.87	92.10	98.78	47,486.59
COD_{RE} (%)	27	56.26	12.19	28.66	85.43	18,582.62

(b) Cd

Basic statistics of the training data

Variable	*N*	Mean	Std Dev	Minimum	Maximum	Sum Sq
HRT (h)	63	35.43	12.08	24.00	48.00	13,046.11
Concentration (mg/l)	63	78.10	14.67	54.78	101.60	53,185.95
HM_{RE} (%)	63	85.44	8.30	70.19	95.86	59,522.24
COD_{RE} (%)	63	47.95	14.82	17.50	79.25	23,078.86

Basic statistics of the test data

Variable	*N*	Mean	Std Dev	Minimum	Maximum	Sum Sq
HRT (h)	27	40.00	11.53	24.00	48.00	9,926.60
Concentration (mg/l)	27	76.52	13.81	58.10	102.90	33,259.98
HM_{RE} (%)	27	89.41	8.35	66.23	97.38	42,497.02
COD_{RE} (%)	27	54.40	18.34	18.18	79.38	19,480.24

(c) Pb

Basic statistics of the training data

Variable	*N*	Mean	Std Dev	Minimum	Maximum	Sum Sq
HRT (h)	63	36.95	12.06	24.00	48.00	13,787.95
Concentration (mg/l)	63	76.12	14.93	53.35	106.00	51,246.11
HM_{RE} (%)	63	86.39	3.88	78.76	92.74	59,578.72
COD_{RE} (%)	63	56.95	22.73	9.85	95.87	35,290.53

Basic statistics of the test data

Variable	*N*	Mean	Std Dev	Minimum	Maximum	Sum Sq
HRT (h)	27	31.11	11.17	24.00	48.00	6,983.63
Concentration (mg/l)	27	77.94	15.79	53.35	105.00	35,161.19
HM_{RE} (%)	27	85.12	4.79	78.74	92.79	37,990.46
COD_{RE} (%)	27	46.28	21.02	11.47	83.20	16,624.63

Table 2 | Best values of network parameters used for training the network: (a) Cu, (b) Cd, and (c) Pb

Parameters	Cu	Cd	Pb
Training count	10	29	9
Number of neurons in input layer	2	2	2
Number of neurons in hidden layer	12	12	12
Number of neurons in output layer	2	2	2
Learning rate	0.1	0.1	0.1
Momentum term	0.1	0.1	0.1
Error tolerance	0.001	0.001	0.001
Training algorithm	LM	LM	LM
Number of training data set	63	63	63
Number of test data set	27	27	27

Note: LM – Levenberg–Marquardt.

bias term were obtained for the interconnections between different neurons in different layers of the MLP (Table 3). These connection weights determine which input neuron dominates the contribution to a specific hidden neuron, while the sign (+, −) suggests the nature of correlation between an input to a neuron and the output from the neuron. Detailed information on the interpretation of connection weights for neural network models have been discussed in Garson (1991). The biases, which are essentially constant, are an additional input into the next layer and are not influenced by the previous layer but they do have outgoing connections with their own weights. The bias unit ensures that even when all the inputs are zeros, there will still be an activation in the neuron.

Predictive capability of the developed model

The performance of this trained ANN model was assessed by comparing the prediction of the ANN model with that of the experimental values of output parameters. The results were evaluated in terms of the most commonly used error measurement index, MAPE values for HM_{RE} (%) and COD_{RE} (%) during training and testing (Table 4). The measured and model-fitted profiles of HM_{RE} (%) and COD_{RE} (%), during model training and testing, are shown in Figures 2 and 3, respectively. As observed from Figures 4 and 5, the ANN model developed for the RBC showed high R values during both training and testing. The model gave R values between 0.95–0.96 and 0.93–0.96 for heavy metals, and 0.91–0.98 and 0.92–0.98 for COD, respectively, during training and testing. Thus, overall, only <10% of the total deviations could not be mapped by these models during the training and testing process, indicating a strong correlation between predicted and measured values. This phenomenal performance could be attributed to the ability of the BPNN model to capture complex behaviour or trend that existed among the variables obtained from the RBC. The slightly fluctuating heavy metal and COD removal profiles that are usually governed by the microbial activity in the RBC cannot be interpreted by the neural network unless it is integrated with a process-based model. Therefore, any unexpected variations in the removal profiles can affect the ANN's learning capability, eventually leading to a decline in performance of the model (Zamir *et al.* 2011). Similar results on successful application of ANN models for predicting removal of pollutants in different bioreactors are reported in the literature (Antwi *et al.* 2018; Baskaran *et al.* 2020a). Overall, the ANN model developed in this study is found to be well applicable for predicting the HM_{RE} and COD_{RE} of the RBC.

For the RBC under investigation in this study, ANN-based models can provide adequate information on different safe operational regimes in terms of inlet concentrations of the different heavy metals and HRT to reach high HM_{RE} (%) and COD_{RE} (%). These regimes can be represented by two-dimensional contour plots as shown in Figures 6 and 7. These contour plots can be interpreted as follows:

(a) To achieve high HM_{RE} (>90%), the values of independent parameters should be within a specific range, mentioned herein: (i) the allowable inlet heavy metal concentration can be up to 160 mg/l and HRT should be 48 h for Cu (Figure 6(a)), (ii) the allowable inlet heavy metal concentration can be up to 90 mg/l and HRT should be 48 h for Cd (Figure 6(b)), and (iii) the allowable inlet heavy metal concentration can be up to 75 mg/l and HRT should be 48 h for Pb (Figure 6(c)).

(b) To achieve high COD_{RE} (>60%), the various conditions are: (i) the allowable inlet heavy metal concentration can be up to 160 mg/l and HRT should be 48 h for Cu (Figure 7(a)), (ii) the allowable inlet heavy metal concentration can be up to 80 mg/l and HRT should be 48 h for

Table 3 | Connection weights and bias terms of the developed neural network model (2-12-2): (a) Cu, (b) Cd, and (c) Pb

Input to hidden layer

Variable	HID1	HID2	HID3	HID4	HID5	HID6	HID7	HID8	HID9	HID10	HID11	HID12
(a)												
V1	7.82	8.21	9.09	2.11	– 0.79	5.37	3.52	– 3.58	– 5.74	– 10.43	5.97	6.02
V2	– 9.11	3.43	4.46	– 9.03	– 10.47	10.05	– 8.48	8.57	5.36	– 5.28	– 4.53	– 9.20
Bias	– 8.93	– 8.24	– 5.78	5.20	4.04	2.38	4.59	– 2.97	– 5.83	– 2.04	10.94	8.19

Hidden to output layer

	V3	*V4*
HID1	– 0.57	0.92
HID2	– 0.85	2.49
HID3	1.07	– 0.10
HID4	0.30	– 1.30
HID5	0.12	0.68
HID6	0.19	– 1.38
HID7	0.05	1.15
HID8	0.40	– 0.85
HID9	– 1.24	1.09
HID10	– 0.09	– 0.03
HID11	0.59	1.63
HID12	– 0.54	– 1.38
Bias	– 0.33	– 0.60

HID 1 to HID 12 – Hidden layer neurons
Bias – Bias term
Input to the model
V1 HRT, h
V2 Concentration, mg/l

Output of the model
V3 HM_{RE}, %
V4 COD_{RE}, %

Input to hidden layer

Variable	HID1	HID2	HID3	HID4	HID5	HID6	HID7	HID8	HID9	HID10	HID11	HID12
(b)												
V1	– 4.85	– 4.84	1.25	4.39	4.78	– 4.67	1.04	– 1.04	– 5.59	– 1.20	– 4.81	3.59
V2	– 1.41	– 0.60	– 4.94	1.71	– 3.31	1.62	6.43	3.49	0.88	– 4.25	– 1.92	3.05
Bias	4.43	3.97	– 2.72	– 2.36	– 0.94	0.22	1.10	– 1.12	– 1.44	– 2.41	– 2.95	4.08

Hidden to output layer

	V3	*V4*
HID1	– 0.11	0.27
HID2	– 0.17	1.46
HID3	0.22	0.05
HID4	– 1.28	0.26
HID5	0.11	2.09
HID6	0.42	1.19
HID7	0.20	0.12
HID8	– 0.07	0.01
HID9	– 0.34	0.89

HID 1 to HID 12 – Hidden layer neurons
Bias – Bias term
Input to the model
V1 HRT, h
V2 Concentration, mg/l

Output of the model

(*continued*)

Table 3 | continued

Input to hidden layer

Variable	HID1	HID2	HID3	HID4	HID5	HID6	HID7	HID8	HID9	HID10	HID11	HID12
HID10	− 0.62	− 0.60										
HID11	− 1.78	− 1.93										
HID12	− 1.13	− 1.20										
Bias	0.53	0.13										

V3 HM_{RE}, %
V4 COD_{RE}, %

Input to hidden layer

Variable	HID1	HID2	HID3	HID4	HID5	HID6	HID7	HID8	HID9	HID10	HID11	HID12
(c)												
V1	0.38	− 3.91	7.26	− 0.85	3.65	− 0.99	− 0.24	− 0.13	− 10.13	− 6.83	− 9.76	5.73
V2	− 9.74	8.40	6.29	9.43	8.24	− 9.58	− 9.99	9.75	4.66	6.75	− 0.29	− 7.59
Bias	9.01	8.33	− 6.29	2.22	− 4.77	1.53	− 1.20	− 2.42	− 2.82	− 6.39	− 7.87	10.02

Hidden to output layer

	V3	*V4*
HID1	0.57	0.62
HID2	− 0.05	1.18
HID3	0.02	− 2.03
HID4	− 0.09	− 0.37
HID5	− 0.54	0.79
HID6	0.10	0.52
HID7	− 0.85	− 0.59
HID8	0.57	0.38
HID9	− 1.12	− 3.90
HID10	− 0.20	1.06
HID11	0.29	1.94
HID12	− 0.53	0.47
Bias	0.78	− 0.70

HID 1 to HID 12 – Hidden layer neurons
Bias – Bias term
Input to the model
V1 HRT, h
V2 Concentration, mg/l

Output of the model
V3 HM_{RE}, %
V4 COD_{RE}, %

Cd (Figure 7(b)), and (iii) the allowable inlet heavy metal concentration can be up to 80 mg/l and HRT should be 48 h for Pb (Figure 7(c)).

Metal removal and COD removal were observed to be the maximum at HRT of 48 h rather than at 24 h. This is because a long HRT allows the SRB to reduce sulphate to a greater extent by efficient utilization of the carbon source, thereby precipitating the metals as their respective sulphide salts. Compared with cadmium and lead, the higher RE of copper at a higher inlet concentration range can be attributed to its low solubility product value with sulphide, i.e., copper sulphide is least soluble compared with cadmium sulphide and lead sulphide (Kiran *et al.* 2017a). At a low initial metal concentration, precipitation of the insoluble metal due to sulphide produced by the SRB avoids any toxic effect of the metal on the SRB. However, high initial metal concentration level has resulted in a reduced activity of the SRB, leading to low sulphate reduction efficiency and metal removal (Kiran *et al.* 2017b). At a low concentration combination of these metals, the removal of all the metals was the maximum,

Table 4 | Observed mean absolute percentage error (MAPE) during training and testing: (a) Cu, (b) Cd, and (c) Pb

Outputs	HM_{RE} (%)	COD_{RE} (%)
(a)		
Training	0.52	10.71
Testing	0.52	7.32
(b)		
Training	1.93	10.45
Testing	1.89	11.98
(c)		
Training	0.84	14.24
Testing	1.37	16.31

and at a high concentration combination of these metals, the metal RE was slightly lower. Owing to the increase in the metal sulphide (M/S^{2-}) ratio, reduction in the metal RE can be observed at a high concentration combination of metals. In contrast, at a lower value of the M/S^{2-} ratio (<1), sulphide formed due to sulphate reduction is high, which ensures a high metal sulphide precipitation for the metal removal (Villa-Gomez *et al.* 2015).

Sensitivity analysis

To identify the most influential input parameter of the developed model, sensitivity analysis is usually performed by estimating the absolute average sensitivity (AAS) (Maier & Dandy 1998; Rene *et al.* 2009, 2011). The AAS values for Cu_{RE},% were 0.40 and 1.71, respectively, for inlet concentration and HRT and the AAS values for COD_{RE},% were 0.77 and 1.71, respectively. Similarly, the AAS values for Cd_{RE},% were 0.60 and 1.62, respectively, for inlet concentration and HRT and the AAS values for COD_{RE},% were 1.37 and 3.74, respectively. And, the AAS values for Pb_{RE}, % were 0.17 and 0.60, respectively, for inlet concentration and HRT and the AAS values for COD_{RE},% were 1.63 and 1.97, respectively. In all these cases, when compared with inlet concentration, the higher value of HRT has a significant influence on HM_{RE} and COD_{RE}. Sensitivity analysis has been applied to select the key variables affecting the performance of a bioreactor (Rene *et al.* 2009, 2011). Rene *et al.* (2009) examined the performance of a perlite biofilter inoculated with a mixed microbial culture for styrene removal from contaminated air based on two input parameters *viz.*, concentration and flow rate. After sensitivity analysis they found the performance of the biofilter to be largely influenced by the flow rate. In another study, the performance of a monolith bioreactor for styrene removal from contaminated air was examined on the basis of various input parameters *viz.* concentration, G/L (gas/liquid) ratio, and pressure drop. However, after sensitivity analysis they found the performance of the monolith bioreactor to be significantly affected by the concentration followed by the G/L ratio and pressure drop (Rene *et al.* 2011).

Practical implications of ANNs for performance evaluation of different bioreactor systems

Table 5 provides a detailed summary of the different literature reports where neural networks have been used to evaluate the performance of different bioreactor systems, though no serious effort has so far been made to implement this technique for real-time systems. In industrial wastewater treatment plants, ANNs have successfully been implemented for fault detection and diagnosis, plant and instrument monitoring, dynamic forecasting and robust process control (Boger 1992). The models developed based on the laboratory-scale experimental data, where input parameters to the RBC such as inlet heavy metal concentration and HRT were systematically controlled and other physico-chemical and environmental parameters were adequately maintained, might not accurately predict the removal efficiency of a full-scale RBC. Therefore, for full-time applications, the wastewater treatment system should be fitted with online measurement devices to monitor the parameters at a regular interval. This information can be stored in a large database, and the system can be equipped with an automatic control system to maintain the desired values of these process parameters. The trained neural network model can then be integrated with a programmable logic controller (PLC) to ascertain and control the process variables, thereby evaluating the reactor performance on a regular basis. The data obtained from real-time bioreactor operation can be merged with the already existing database, and the ANN model can be trained in offline/online mode, and the connection weights can be updated before integrating it with the PLC. The ANN models, when combined with more prominent applications like genetic algorithms, multi-objective

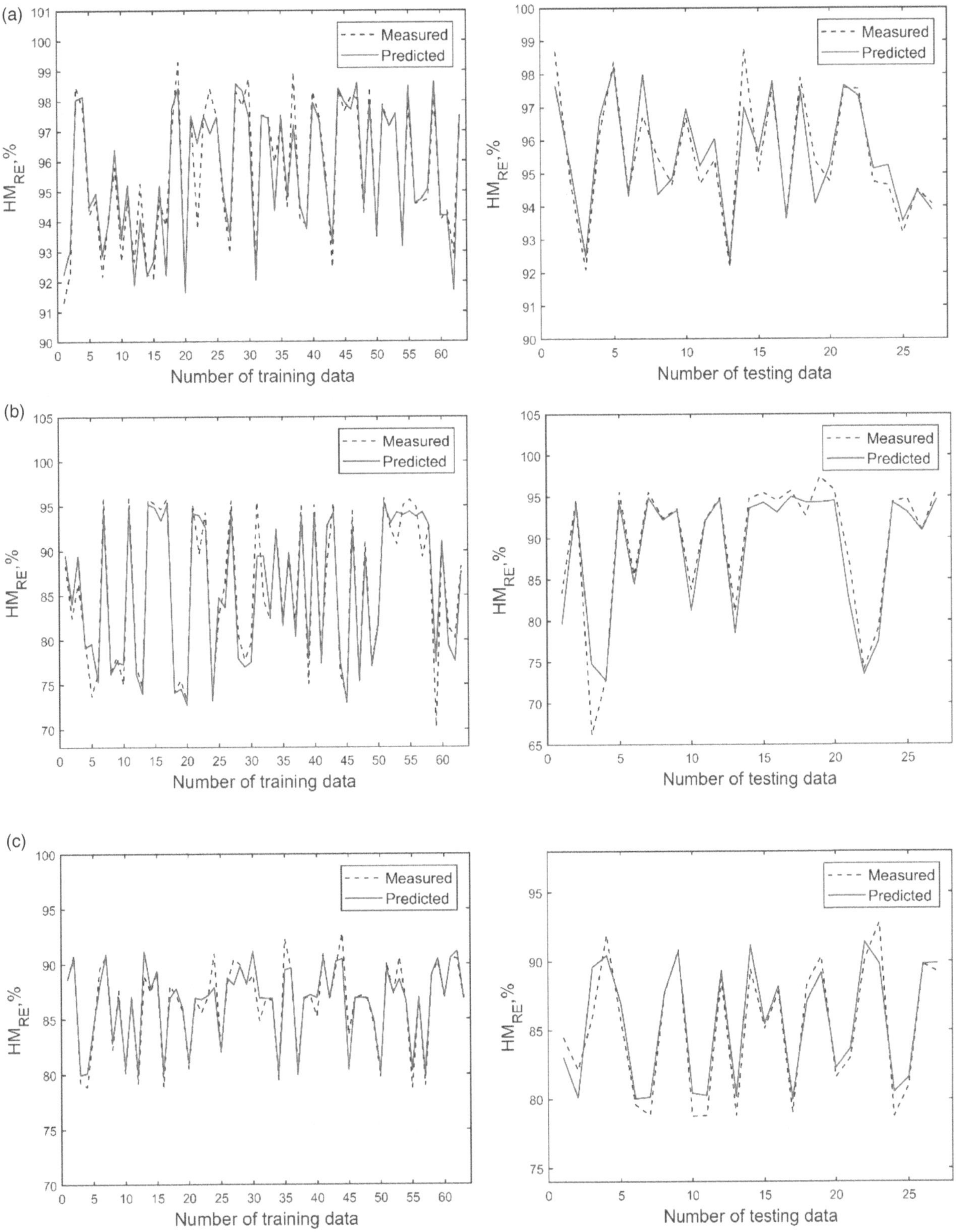

Figure 2 | ANN model predictions of heavy metal removal during training and testing: (a) Cu, (b) Cd, and (c) Pb.

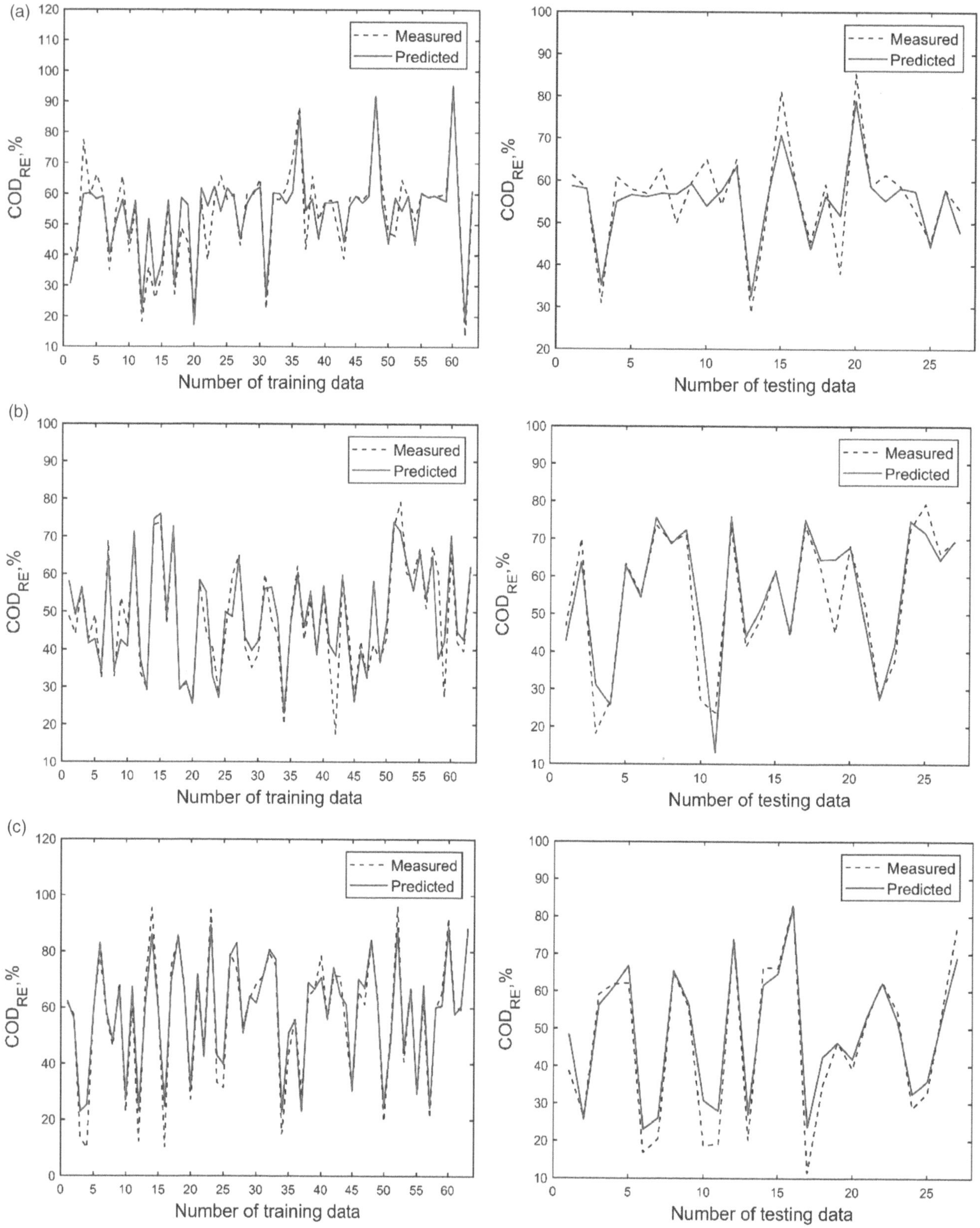

Figure 3 | ANN model predictions of COD removal during training and testing: (a) Cu, (b) Cd, and (c) Pb.

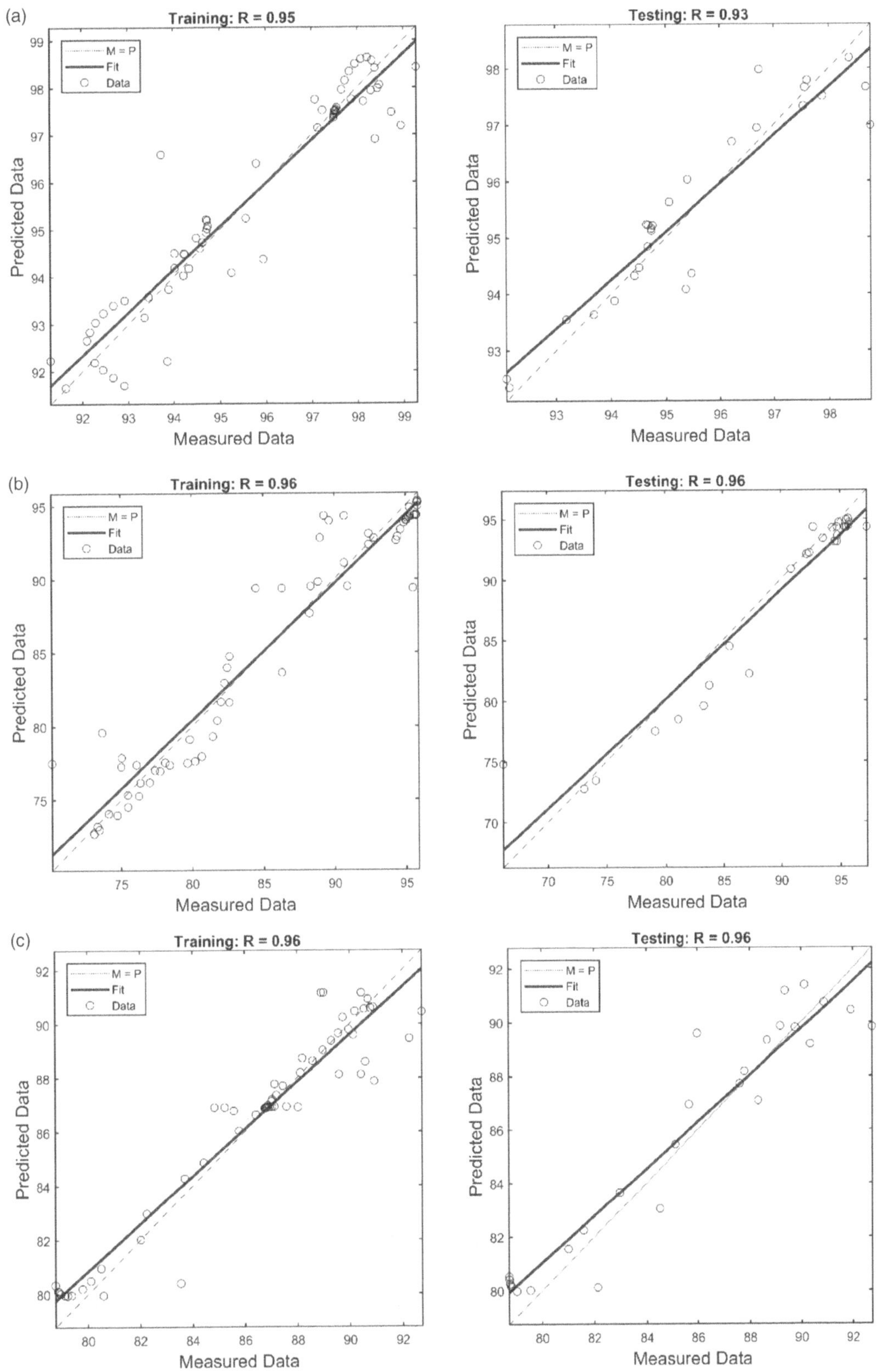

Figure 4 | Comparison of predictions by ANN models with the experimental measurements for heavy metal removal during training and testing: (a) Cu, (b) Cd, and (c) Pb.

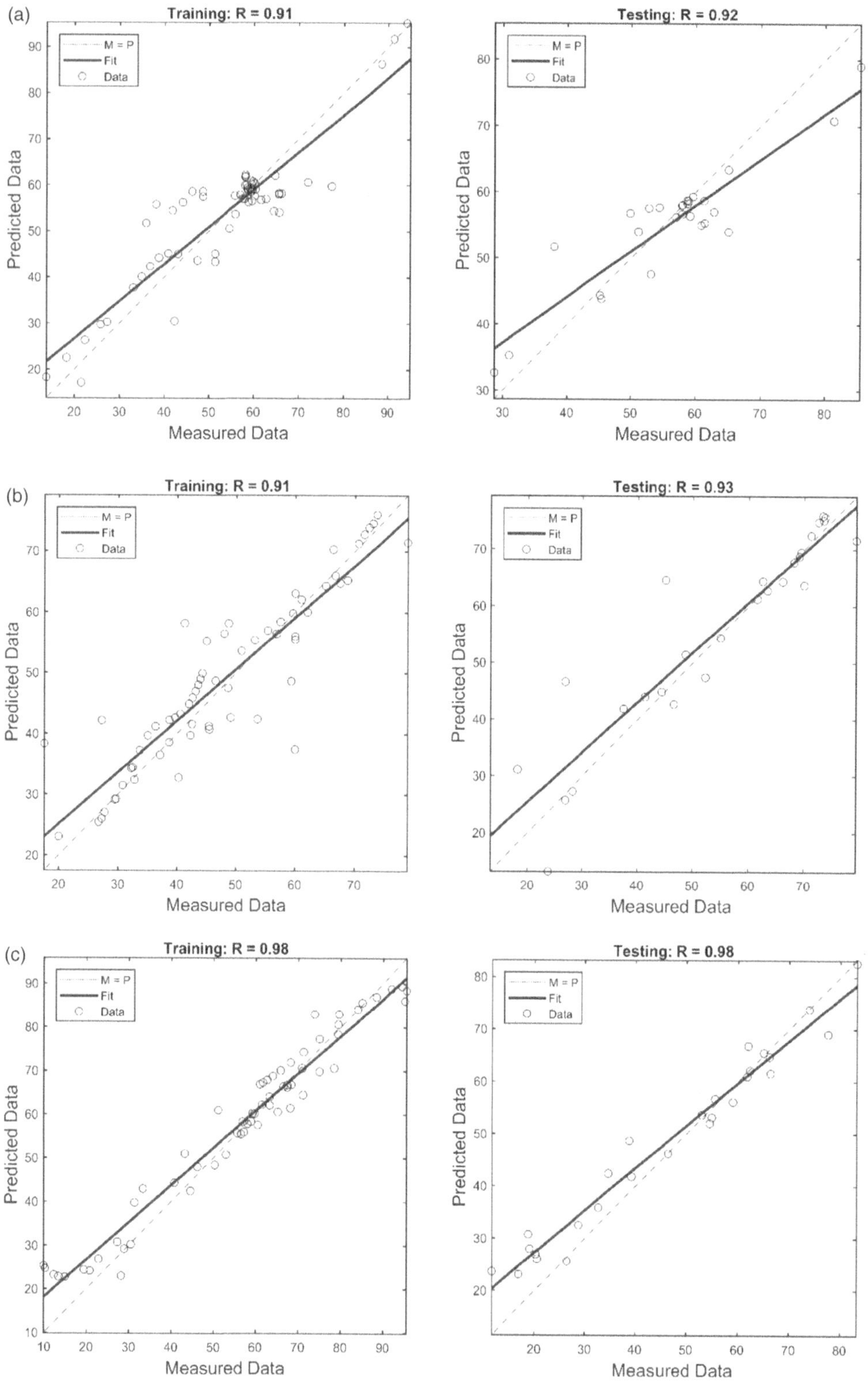

Figure 5 | Comparison of predictions by ANN models with the experimental measurements for COD removal during training and testing: (a) Cu, (b) Cd, and (c) Pb.

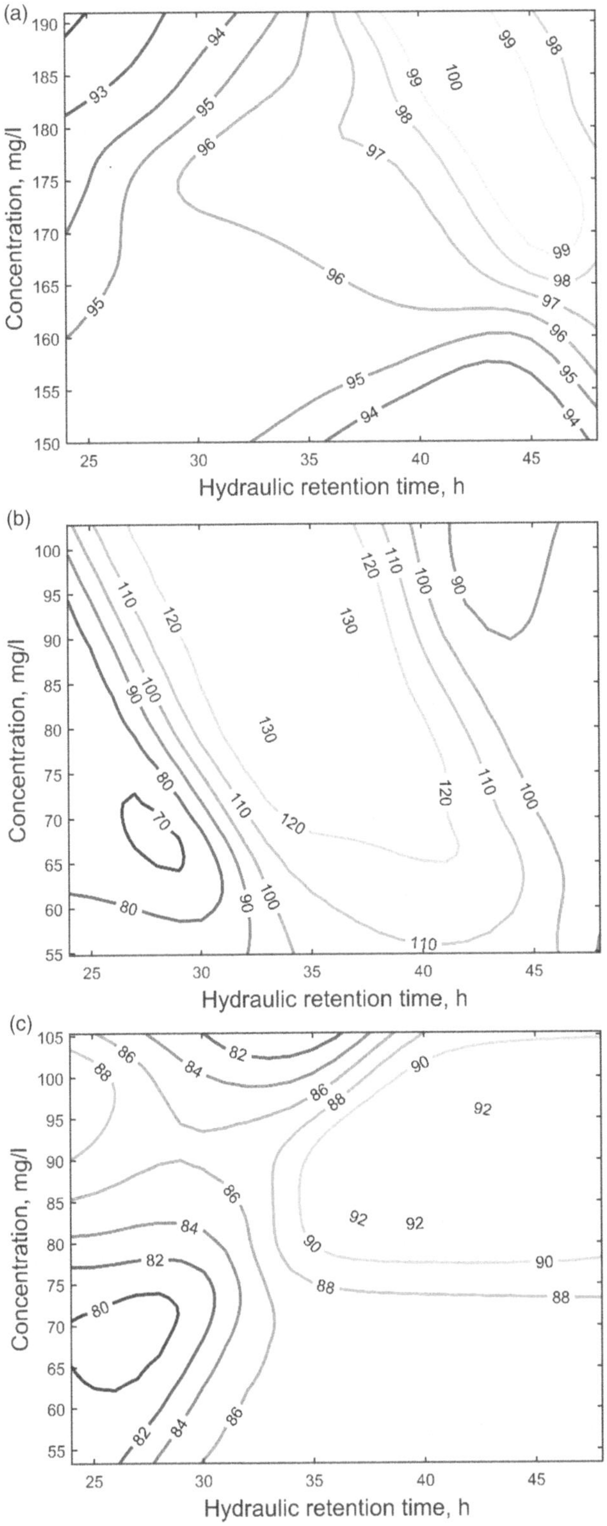

Figure 6 | Contour plots showing the effect of input parameters on the heavy metal removal after ANN training: (a) Cu, (b) Cd, and (c) Pb.

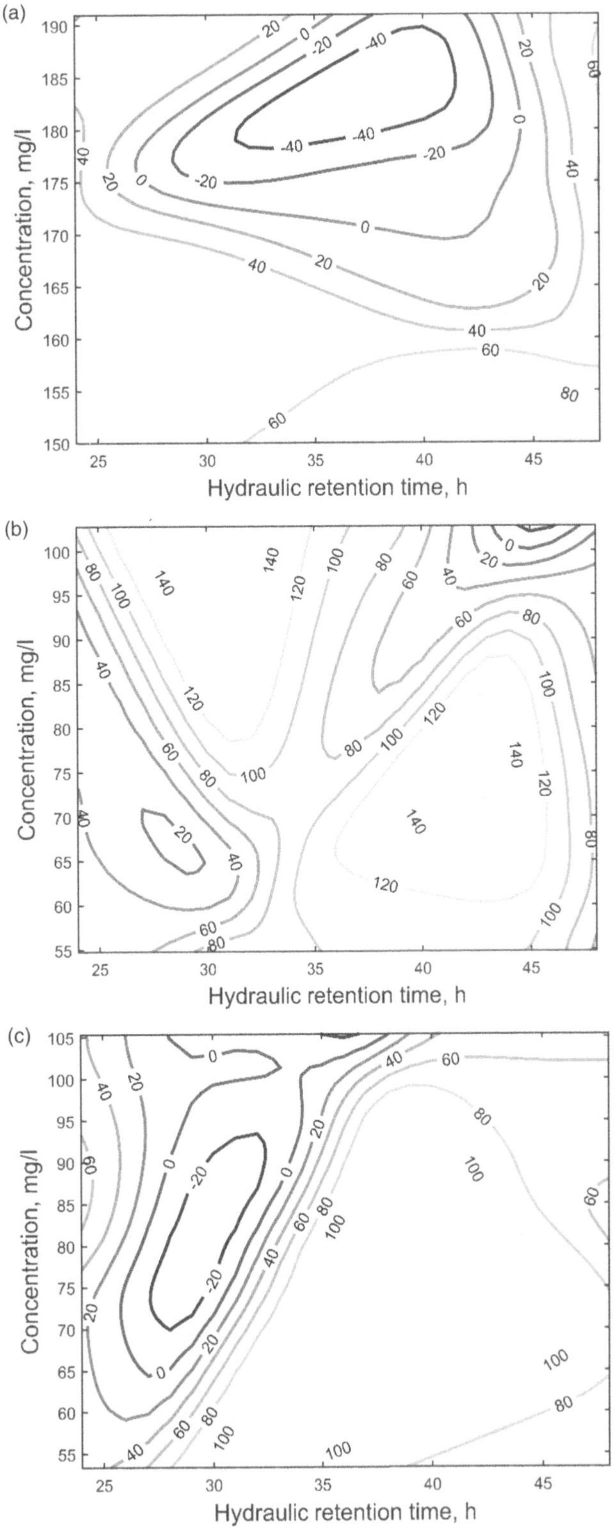

Figure 7 | Contour plots showing the effect of input parameters on the COD removal after ANN training: (a) Cu, (b) Cd, and (c) Pb.

Table 5 | Summary of artificial neural network models developed for different bioreactor systems

Bioreactor type	Type of neural network	Input parameters	Output parameters	Data collection period	Data points used	Network topology	References
Membrane bioreactor	Cascade-forward BP network	COD, SRT, HRT, NH_3, NO_3, TP, flux, pressure	COD, NH_3, NO_3, TP	100 d	30	$8{:}N_H{:}4$	Cinar *et al.* (2006)
FBR[1]	Radial basis function	COD_{in}, HRT	COD_{RE}	NA[2]	NA[2]	$2{:}N_H{:}1$	Rajasimman *et al.* (2009)
Anaerobic bioreactor	LM BP algorithm	pH, MC, TVS, VFA, and biogas volume	CH_4 content in biogas	45 d	NA[2]	$5{:}N_H{:}1$	Nair *et al.* (2016)
Two stage bioreactor (BTF[3] & BF[4])	BP with momentum algorithm	C_M, C_P, and C_{HS}	RE_M and RE_{HS} (BTF[3]) RE_p (BF[4])	Long-term (66 h) and short-term (12 h)	31	3 – 4 – 2 (BTF) 3-3-1 (BF)	López *et al.* (2017)
UASB[5]	LM BP algorithm	pH, COD, NH_4^+, VFA, OLR and biogas yield	COD_{RE}	218 d	218	$6{:}N_h{:}1$	Antwi *et al.* (2018)
BF[4]	LM BP algorithm	C_i, Q and time	C_o	165 d	165	3:25:1	Baskaran *et al.* (2019)
CSTB[6]	LM BP algorithm	C_i, Q and time	C_o	77 d	77	3:15:1	Baskaran *et al.* (2020a)
Airlift bioreactor	LM BP algorithm	C_i, Q and time	C_o	55 d	55	3:12:1	Baskaran *et al.* (2020b)
RBC[7]	LM BP algorithm	C_{HM}, HRT	RE_{HM}	90 d	90	2-12-1	This study

Note: Neural network terminologies: N_H – neurons in the hidden layer; BP – back propagation algorithm.

Input parameters: COD – Chemical Oxygen Demand (mg/l), SRT – Solid Retention Time (d), HRT – Hydraulic Retention Time (d), TP – Total Phosphorus (mg/l), MC – Moisture Content (%), TVS – Total Volatile Solids (mg/l), VFA – Volatile Fatty Acids (mg/l), OLR – Organic Loading Rate (kg COD/m^3.d), C_i – Inlet concentration (g/m^3), Q – Gas flow rate (m^3/h), C_M – Methanol concentration (g/m^3), C_P – α-pinene concentration (g/m^3), C_{HS} – Hydrogen sulphide concentration (g/m^3), C_{HM} – Heavy metal concentration (mg/l).

Output parameters: C_o – Outlet concentration (g/m^3), TP – Total Phosphorus (mg/L), COD_{RE} – Chemical oxygen demand removal efficiency (%), RE_M – Methanol removal efficiency (%), RE_{HS} – Hydrogen sulphide removal efficiency (%), RE_p – α-pinene removal efficiency (%), RE_{HM} – Heavy metal removal efficiency (%).

Superscripts: 1- Fluidized Bed Reactor; 2 – Not available; 3 – Biotrickling Filter; 4 – Biofilter; 5 – Upflow Anaerobic Sludge Blanket reactor; 6 – Continuously Stirred Tank Bioreactor; 7 – Rotating Biological Contactor.

optimization using evolutionary algorithm, multi-objective optimization based on ANN modelling, fuzzy-logic-based control systems, and fuzzy neural optimization softwares would allow better prediction and control of bioreactor performance (Venu Vinod *et al.* 2009; de Medeiros *et al.* 2019).

CONCLUSIONS

An ANN-based model was successfully developed and tested to predict the performance of RBC. The HM_{RE} and COD_{RE} were predicted using inlet concentration and HRT as the inputs to the model. After proper optimization of network parameters, and through vigorous training and testing, the network architecture 2-12-2 was obtained for the RBC. The results from sensitivity analysis showed that changes in HRT significantly affected the HM_{RE} and COD_{RE} of the model. Besides, the conditions to obtain high HM_{RE} and COD_{RE}, as a function of inlet heavy metal concentration and HRT, were also obtained from two-dimensional contour diagrams. The ANN model developed in this work for the RBC would yield more promising results when tested with online monitoring and control systems, thus allowing better access to control process variables and improving the performance of the bioreactor for heavy metal contaminated wastewater treatment.

ACKNOWLEDGEMENTS

Authors thank Centre for Environment, IIT Guwahati, India, for providing the facilities to carry out the bioreactor studies. Raja Das and Shishir Kumar Behera thank VIT, Vellore (India) for providing infrastructural and staff time support to collaborate with researchers from IIT Guwahati.

CONFLICT OF INTERESTS

The authors declare that there is no conflict of interest.

DATA AVAILABILITY STATEMENT

All relevant data are included in the paper or its Supplementary Information.

REFERENCES

Antwi, P., Li, J., Meng, J., Deng, K., Quashie, F. K., Li, J. & Boadi, P. O. 2018 Feed forward neural network model estimating pollutant removal process within mesophilic upflow anaerobic sludge blanket bioreactor treating industrial starch processing wastewater. *Bioresource Technology* **257**, 102–112.

APHA/AWWA/WEF 2005 *Standard Methods for the Examination of Water and Wastewater*, 21st edn. American Public Health Association/American Water Works Association/Water Environment Federation, Washington, DC, USA.

Bai, H., Kang, Y., Quan, H., Han, Y., Sun, J. & Feng, Y. 2013 Bioremediation of copper containing wastewater by sulfate reducing bacteria coupled with iron. *Journal of Environmental Management* **129**, 350–356.

Baskaran, D., Rajamanickam, R. & Pakshirajan, K. 2019 Experimental studies and neural network modeling of the removal of trichloroethylene vapor in a biofilter. *Journal of Environmental Management* **250**, 109385.

Baskaran, D., Sinharoy, A., Paul, T., Pakshirajan, K. & Rajamanickam, R. 2020a Performance evaluation and neural network modeling of trichloroethylene removal using a continuously operated two-phase partitioning bioreactor. *Environmental Technology and Innovation* **17**, 100568.

Baskaran, D., Sinharoy, A., Pakshirajan, K. & Rajamanickam, R. 2020b Gas-phase trichloroethylene removal by *Rhodococcus opacus* using an airlift bioreactor and its modeling by artificial neural network. *Chemosphere* **247**, 125806.

Boger, Z. 1992 Application of neural networks to water and wastewater treatment plant operation. *ISA Transactions* **31** (1), 25–33.

Chen, Y., Yu, G., Long, Y., Teng, J., You, X., Liao, B.-Q. & Lin, H. 2019 Application of radial basis function artificial neural network to quantify interfacial energies related to membrane fouling in a membrane bioreactor. *Bioresource Technology* **293**, 122103.

Çinar, Ö., Hasar, H. & Kinaci, C. 2006 Modeling of submerged membrane bioreactor treating cheese whey wastewater by artificial neural network. *Journal of Biotechnology* **123**, 204–209.

Cord-Ruwisch, R. 1985 A quick method for the determination of dissolved and precipitated sulfides in cultures of sulfate-reducing bacteria. *Journal of Microbiological Methods* **4**, 33–36.

de Medeiros, E. M., Posada, J. A., Noorman, H. & Filhob, R. M. 2019 Modeling and multi-objective optimization of syngas fermentation in a bubble column reactor. *Computer Aided Chemical Engineering* **46**, 1531–1536.

Dev, S., Roy, S. & Bhattacharya, J. 2017 Optimization of the operation of packed bed bioreactor to improve the sulfate and metal removal from acid mine drainage. *Journal of Environmental Management* **200**, 135–144.

Garson, G. D. 1991 Interpreting neural-network connection weights. *AI Expert* **6** (4), 47–51.

Gola, D., Chawla, P., Malik, A. & Ahammad, S. Z. 2020 Development and performance evaluation of native microbial consortium for multi metal removal in lab scale aerobic and anaerobic bioreactor. *Environmental Technology & Innovation* **18**, 100714.

Guadalima, M. P. G. & Monteros, D. A. N. 2018 Evaluation of the rotational speed and carbon source on the biological removal of free cyanide present on gold mine wastewater, using a rotating biological contactor. *Journal of Water Process Engineering* **23**, 84–90.

Hassard, F., Biddle, J., Cartmell, E., Jefferson, B., Tyrrel, S. & Stephenson, T. 2015 Rotating biological contactors for wastewater treatment – a review. *Process Safety and Environmental Protection* **94**, 285–306.

Kikot, P., Viera, M., Mignone, C. & Denoti, E. 2010 Study of the effect of pH and dissolved heavy metals on the growth of sulfate-reducing bacteria by a fractional factorial design. *Hydrometallurgy* **104**, 494–500.

Kiran, M. G., Pakshirajan, K. & Das, G. 2015 Heavy metal removal using sulfate reducing biomass obtained from a lab-scale upflow anaerobic packed bed reactor. *Journal of Environmental Engineering* **142**, C4015010.

Kiran, M. G., Pakshirajan, K. & Das, G. 2017a A new application of anaerobic rotating biological contactor reactor for heavy metal removal under sulfate reducing condition. *Chemical Engineering Journal* **321**, 67–75.

Kiran, M. G., Pakshirajan, K. & Das, G. 2017b Heavy metal removal from multicomponent system by sulfate reducing bacteria: mechanism and cell surface characterization. *Journal of Hazardous Materials* **324**, 62–70.

Lin, H., Dai, Q., Zheng, L., Hong, H., Dend, W. & Wu, F. 2020 Radial basis function artificial neural network able to accurately predict disinfection by-product levels in tap water: taking haloacetic acids as a case study. *Chemosphere* **248**, 125999.

López, M. E., Rene, E. R., Boger, Z., Veiga, M. C. & Kennes, C. 2017 Modeling the removal of volatile pollutants under transient conditions in a two-stage bioreactor using artificial neural networks. *Journal of Hazardous Materials* **324**, 100–109.

Maier, H. R. & Dandy, G. C. 1998 The effects of internal parameters and geometry on the performance of back-propagation neural networks: an empirical study. *Environmental Modelling and Software* **13**, 193–209.

Menhaj, M. B. 1998 *Fundamentals of Neural Networks*. Amirkabir University of Technology Press, Tehran, Iran.

Nair, V. V., Dhar, H., Kumar, S., Thalla, A. K., Mukherjee, S. & Wong, J. W. C. 2016 Artificial neural network based modeling to evaluate methane yield from biogas in a laboratory-scale anaerobic bioreactor. *Bioresource Technology* **217**, 90–99.

Pakshirajan, K. & Kheria, S. 2012 Continuous treatment of coloured industry wastewater using immobilized *Phanerochaete chrysosporium* in a rotating biological contactor reactor. *Journal of Environmental Management* **101**, 118–123.

Rajasimman, M., Govindarajan, L. & Karthikeyan, C. 2009 Artificial neural network modeling of an inverse fluidized bed bioreactor. *International Journal of Environmental Research* **3**, 575–580.

Rene, E. R., Veiga, M. C. & Kennes, C. 2009 Experimental and neural model analysis of styrene removal from polluted air in a biofilter. *Journal of Chemical Technology and Biotechnology* **84**, 941–948.

Rene, E. R., López, M. E., Veiga, M. C. & Kennes, C. 2011 Neural network models for biological waste-gas treatment systems. *New Biotechnology* **29**, 56–73.

Rumelhart, D. E., Hinton, G. E. & Williams, R. J. 1986 Learning representations by back-propagating errors. *Nature* **323**, 533–536.

Sahinkaya, E. 2009 Biotreatment of zinc-containing wastewater in a sulfidogenic CSTR: performance and artificial neural network (ANN) modelling studies. *Journal of Hazardous Materials* **164**, 105–113.

Sahoo, G. B., Ray, C., Wang, J. Z., Hubbs, S. A., Song, R., Jasperse, J. & Seymour, D. 2005 Use of artificial neural networks to evaluate the effectiveness of riverbank filtration. *Water Research* **39**, 2505–2516.

Vasiliadou, I. A., Pariente, M. I., Martinez, F., Melero, J. A. & Molina, R. 2016 Modeling the integrated heterogeneous catalytic fixed-bed reactor and rotating biological contactor system for the treatment of poorly biodegradable industrial agrochemical wastewater. *Journal of Environmental Chemical Engineering* **4**, 2313–2321.

Venu Vinod, A., Arun Kumar, K. & Reddy, G. V. 2009 Simulation of biodegradation process in a fluidized bed bioreactor using genetic algorithm trained feedforward neural network. *Biochemical Engineering Journal* **46**, 12–20.

Villa-Gomez, D. K., Pakshirajan, K., Maestro, R., Mushi, S. & Lens, P. N. L. 2015 Effect of process variables on the sulfate reduction process in bioreactors treating metal containing wastewaters: factorial design and response surface analyses. *Biodegradation* **26**, 299–311.

Yajun, C., Ya, Z., Naiwei, L., Xiaojing, L., Fengfeng, D. & Dongrui, Y. 2019 Biomass production and nutrient removal efficiency of *Suaeda salsa* in eutrophic saline water using a floating mat treatment system. *Water Supply* **19** (1), 254–263.

Yang, J., Liu, J., Wu, C., Kerr, P. G., Wong, P.-K. & Wu, Y. 2016a Bioremediation of agricultural solid waste leachates with diverse species of Cu (II) and Cd (II) by periphyton. *Bioresource Technology* **221**, 214–221.

Yang, J. L., Tang, C. L., Wang, F. W. & Wu, Y. H. 2016b Co-contamination of Cu and Cd in paddy fields: using periphyton to entrap heavy metals. *Journal of Hazardous Materials* **304**, 150–158.

Zamir, M., Halladj, R., Saber, M., Ferdowsi, M. & Nasernejad, B. 2011 Biofiltration of hexane vapor: experimental and neural model analysis. *Clean Soil Air Water* **39**, 813–819.

First received 28 July 2020; accepted in revised form 26 October 2020. Available online 6 November 2020

Waste water reuse in France – social perception of an unfamiliar practice

Patrice Garin, Marielle Montginoul and Benjamin Noury

ABSTRACT

This article presents the results of a local survey conducted in a French wine-growing region where the public authorities were considering irrigating fields and green spaces with treated wastewater. Interviews were conducted face-to-face with 845 potential consumers close to their purchasing place. The majority were in favour of such a project and willing to use recreation areas and purchase local farm products irrigated with treated wastewater. However, we demonstrated the paucity of lay knowledge about treated wastewater. Simple information flyers have a significant effect on attitudes and intended behaviours. Nevertheless 20% of informed respondents said they would no longer buy such irrigated products. Given the lack of confidence in the public authorities' guarantee of food quality, they will be unlikely to change their minds. These unacceptable results to farmers may interest public authorities promoting these projects, still rare in France, in response to the increase in droughts.

Key words | consumer attitude, French local survey, information effect, intended behaviour, social representation, wastewater reuse

Patrice Garin
Marielle Montginoul (corresponding author)
INRAE UMR G-Eau, Montpellier University,
Montpellier,
France
E-mail: *marielle.montginoul@inrae.fr*

Benjamin Noury
INRAE UMR G-Eau, Montpellier University,
Montpellier,
France
and
Société du Canal de Provence,
Aix en Provence,
France
and
Aix Marseille Univ, Université de Toulon,
IMSIC,
Marseille,
France

HIGHLIGHTS

- WWR an unfamiliar practice in France with no anchored social representation.
- Information influences attitudes towards WWR of undecided people.
- Attitudes and intended behaviours driven by disgust, environmental sensitivity, risk perception, and type of use.
- Working women more reluctant than men to WWR; farmers rejecting WWR.
- By communicating, farmers strengthen the social acceptability of their project but with a loss of customers.

doi: 10.2166/ws.2020.242

GRAPHICAL ABSTRACT

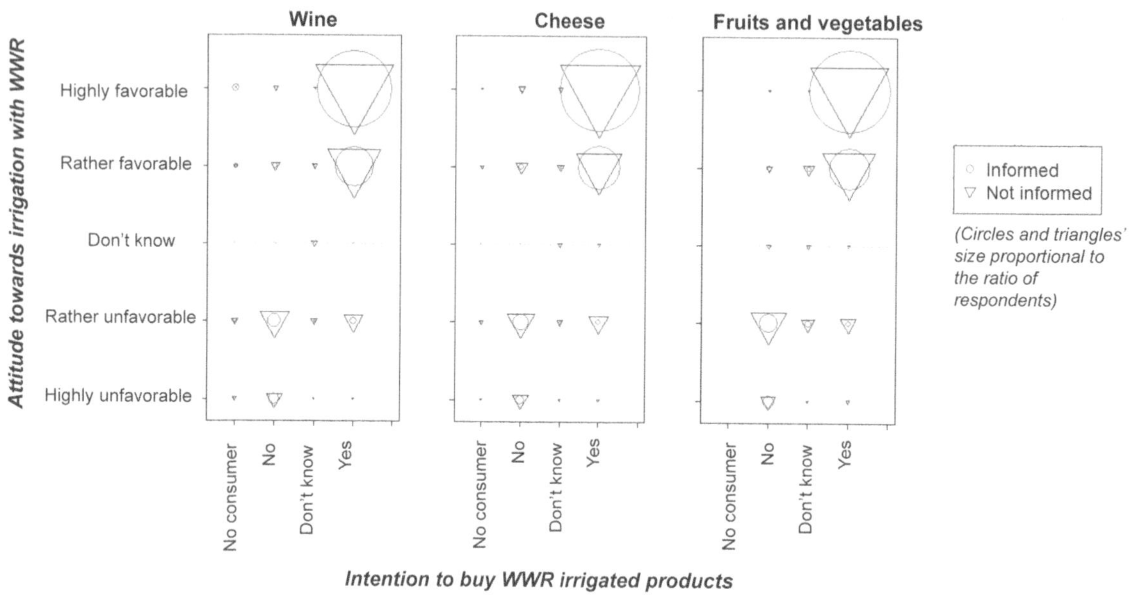

INTRODUCTION

Every second year, half the French departments set up restrictions on use due to water scarcity (Benoit *et al.* 2017). Irrigation is then limited or even prohibited for weeks. Waste Water Reuse (WWR) is thus seen as an interesting solution in these territories, especially as it may be considered as participating in the circular economy and contributing to a so-called virtuous circle (Maurer 2018). However, only 0.2% of treated urban wastewater effluents are reused annually in France. This figure is lower than the European average (2.4%), and a long way from the practices of Italy and Spain, which reuse between 8% and 12% of their treated effluents (Aquarec 2006).

This situation can be explained firstly by the very restrictive regulatory framework. In France, thanks to the 1992 French Water Law, treated wastewater can be utilized but only to irrigate lands, 'provided that its characteristics and manner of use are compatible with the requirements for the protection of public health and the environment' (decree of June 3, 1994). And the Order of June 25, 1994, defines four classes of water quality from A to D. Each class is associated with constraints on use, distance and terrain to be respected according to the crop, the area to be irrigated, and strict technical regulations for sprinkling irrigation – the dominant practice in France (see the following link for an in-depth presentation of the regulation: https://www.arpe-arb.org/files/20171130_2017Rutiliserleseauxtraites.pdf). A survey done by the public institution CEREMA (Centre for Studies and Expertise on Risks, the Environment, Mobility and Development) reports that the main treatment is 'activated sludge' which accounts for 73% of WWR cases (against 44% of all wastewater treatment plants in France), ahead of ponds 23% (against 28%), and membrane processes 3% (against 0.3%)' (Franck-Néel 2020). CEREMA identified only 63 schemes in operation in 2017 in France, including 35 for agricultural purposes, concerning much less than 1‰ of the 1.5 million hectares irrigated annually in the country. Most projects, less than 100 ha, were initiated before the 2000s in rare areas with a really acute lack of water (i.e. islands, coastal zones, etc.), to satisfy some agricultural uses close to wastewater treatment plants.

But this poor practice of WWR is also due to the French situation, where water has not been scarce (thanks also to other solutions to fight against water scarcity, like dams, canals, etc.). This second key reason (both reasons underlined by Lazarova & Brissaud (2007) and Cerceau (2015)) may explain the gradual change that appeared after 2000, with 40% of the new projects considered as addressing quantitative or qualitative water issues at the catchment level. There is now an annual growth rate of 7% in the number of projects (Franck-Néel 2020). WWR is emerging in public and/or national French debates as a way to cope with repeated droughts, especially to supply water to territories not yet covered by irrigation networks, and thus to crops previously not irrigated (wine, lavender, olive trees). For instance, the 'Assises de l'eau', a national forum organized by the National Water Committee, proposed in their conclusions (16 July 2020) to triple the volume of treated WWR.

In this context, French Water Agencies are increasing pilot experiments, to explore the potentialities of this 'new' method of satisfying increasing water needs. The French research and development project SOPOLO ('WWR a possible local solution?') takes up part of these, looking to estimate wastewater demand, especially for irrigation purposes, in a Mediterranean fine-wine-growing territory, Pic Saint Loup, located in the South of France (Figure 1). This vineyard is not irrigated yet. There is no WWR project for agriculture nearby. However, after several severe droughts during the last decade, more and more farmers are complaining about lack of water for irrigation. Recycling water from several sewage treatment plants is one of the options for the local public authority (Communauté de Communes du Grand Pic Saint Loup – CCGPSL), which is a partner in our research activity.

As irrigation can be viewed, from the point of view of economics, as an intermediate resource desired not in itself but for growing products later sold in the market, it seems important to assess the level of acceptance by customers of agricultural products (i.e. households). The aim of this paper is to present the attitudes of consumers, buying their fruit, vegetables and wines in this area, toward a WWR project and their intended behaviour concerning local irrigated foods and recreational areas. These elements will give indicators of the risk of consumers turning away from local production, which would jeopardize the viability of the WWR project.

In the first part, we provide a brief overview of the state of the art of WWR's social acceptability issues; we then present the research method before discussing the main results.

STATE OF THE ART OF WWR SOCIAL ACCEPTABILITY

'Social acceptability' is an 'embarrassing' notion (Barbier & Nadaï 2015), but it was soon acknowledged as being a key element in the development of WWR in the same way as technical issues are (Baumann & Kasperson 1974). During

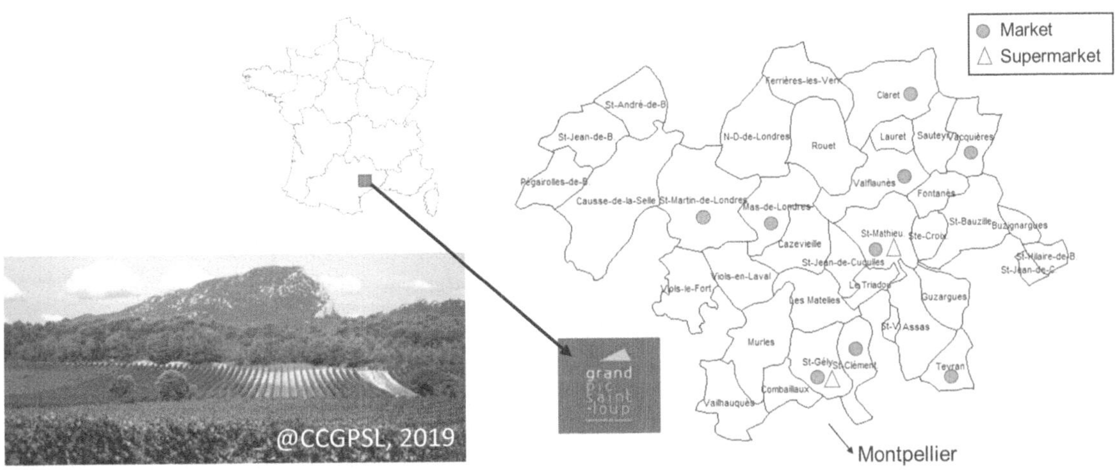

Figure 1 | Pic Saint Loup Territory and location of the survey's sites.

the last 40 years, there has been a lot of research investigating factors associated with public acceptance of recycled water in many countries (Fielding *et al.* 2018), but not in France. In addition, in this country, Berry *et al.* (2016) revealed difficulties related to the rise of controversies concerning all major infrastructure projects (in their case study on WWR, the interviewees refer in particular to the highly controversial project of a new airport 'Notre Dame des Landes'). Water is not exempt from these controversies, whether for the justification of irrigation, the quality of drinking water (Hervé-Bazin 2014), or the public interest in new dams for irrigation (Sibertin-Blanc 2019). France has been marked by health scandals affecting food: mad cow disease, children's milk contaminated with salmonella, avian influenza, swine fever, and meat and cheese affected by *Escherichia coli*. Consumer associations regularly denounce the high levels of pesticide residues on fruit and vegetables, and question the government about emerging pollutants. This context creates a climate of mistrust in food chains that use industrial processes. Our hypothesis is that the treatment of wastewater for agriculture could be seen by consumers as one of these complex industrial processes, susceptible to malfunctions in the elimination of bacteriological, viral or chemical elements. However, WWR has so far received very little media attention in France: the media analysis conducted by Noury *et al.* (2019) for the period 2000–2017 for 50 national, regional and specialized press titles revealed that few press articles (156) mentioned WWR, with only 30% focusing on it. Most of these, published after 2010, mentioned WWR as a way to deal with repeated droughts without the need for new dams, a particularly sensitive subject, and one in conformity with the political trends of circular economy.

The main French reference to public views of water issues is the annual national barometer of opinions from a representative sample of 2,508 people, about water and wastewater services. In the questionnaire, with over 100 questions, there are only three concerning people's acceptance of using recycled water and one on their knowledge of the water cycle (Centre d'Information sur l'Eau & TNS Sofres 2018). However, this national barometer reports general public opinions on the principle of recycling water, measuring 'the extent to which a person believes that supporting the recycled water scheme will deliver positive outcomes' (Nancarrow *et al.* 2008); this is not an opinion on the desirability of a WWR project close to their homes. Nancarrow *et al.* (2008) had already documented a difference between supporting the concept of recycling water, because this attitude is in line with the values people stand for, and accepting personal contact or ingestion of food irrigated with such water, because this involves other intimate factors. More precisely, we wanted the answers to five questions, based on literature reviews (Dolnicar *et al.* 2010; Fielding *et al.* 2018; Etale *et al.* 2020), and the French national survey (Centre d'Information sur l'Eau & TNS Sofres 2018):

- What were the main components of the social representation of WWR in this area, if any?
- Were the inhabitants of this area in favour of irrigating agricultural or recreational areas with recycled water?
- Did their opinions vary depending on the agricultural products or recreational spaces considered?
- Did their opinions change if they were provided with basic information on the principles of water treatment and control?
- Was there consistency between their judgment on the appropriateness of a WWR project and their envisaged purchasing of agricultural products and use of recreational areas irrigated with recycled water?

The French national survey indicated that 52% of the population believed that wastewater was already directly used for drinking water, even though this is strictly forbidden by regulation, and so no treated wastewater is directly used for this purpose. This observation, combined with only the very few number of French WWR projects (no one near to Pic Saint Loup Region), and the low media visibility, led us to assume that the local population, too, had very little knowledge of the subject, and that it was an unfamiliar practice. According to social representation theory (Rateau *et al.* 2011), in this context of unfamiliarity, opinions and beliefs were probably not yet imbedded in knowledge collectively produced by a social group (i.e. a social representation). Identifying the 'central core' of cognitive elements constituting the social representation of wastewater treatment, if any, was therefore of fundamental importance. The central core, anchored in people's minds, is particularly difficult to change (Vergès 1992), even with an information campaign. On the contrary, the other ('peripheral') elements of a social representation can be changed by ordinary exchanges through social networks and by the media. These elements are also sensitive to 'information and awareness-raising' (Noury *et al.* 2019). We intended to test whether positive information on WWR increases the acceptance rate for people who do not yet have deep-rooted beliefs about this topic (Noury *et al.* 2019).

At the national level (2018 survey) (Figure 2), 73% of the population considered water as a 'limited' resource, and 59% feared water shortages in the future; 83% trusted health authorities to control water quality; 86% would accept using tap water from recycled wastewater for their domestic purposes (hygiene, sanitation, cleaning, etc.), and 75% would be willing to consume vegetables irrigated with treated wastewater. This proportion seems to be increasing slowly in France: in 2013, there were 32% who expressed their opposition to the idea of eating fruit and vegetables watered with treated wastewater (Ben Maïd *et al.* 2014), without mention of their motivations. We had no reason to assume that the consumers in the Pic Saint Loup region had a different attitude towards WWR in general. However, we wanted to know if they would have the same judgment if a project were set up in their neighbourhood, bringing them into contact with products irrigated with reclaimed water. We aimed to go a step further in the evaluation of the acceptance rate, by determining the purchasing intended behaviour: would they agree to maintain their purchasing habits for local products (wine, fruit and vegetables, cheese) or utilize recreation areas if they knew they had been irrigated with reclaimed water (all economic aspects neutral, such as prices)? In this research project, we did not have the means to make a taste test with a large panel of consumers. We assumed that stated purchasing intentions would provide a fair approximation of actual behaviour.

For this purpose, we used the theoretical framework proposed by the psychologist Ajzen (1991), who distinguished attitudes, the intended behaviour, and the actual behaviour. This framework has already been successfully used by Nancarrow *et al.* (2008), who established a structural equation model to predict communities' behavioural decisions in relation to the recycling of wastewater for uses involving close personal contact. This framework helped us to select a limited (but observed as relevant in Nancarrow *et al.* 2008) number of factors influencing attitudes, and expected behaviours to test their relevance to our case study and assess their sensitivity to an information action:

- *Disgust*. The extent to which a person experiences the 'yuck' factor at fruit and vegetables irrigated with treated wastewater.

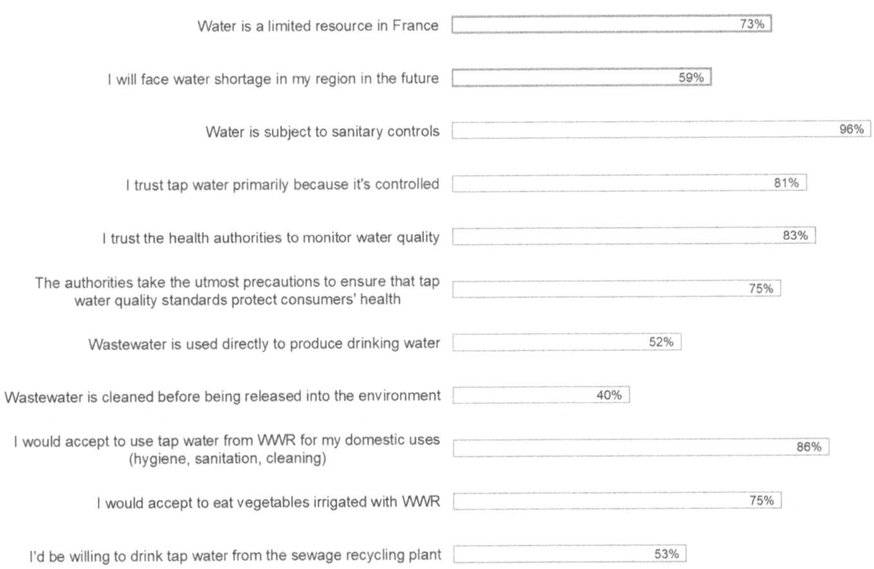

Figure 2 | Key assertions in relation to WWR given by people interviewed in the 2018 French national barometer (Centre d'Information sur l'Eau & TNS Sofres 2018).

- *Perceived sanitary risk.* The sanitary quality of food products is controversial in France, where there have been high-profile health scandals. Moreover media alerts concerning emerging pollutants in water (drug residues, pesticides, plastics, endocrine disruptors…) may be considered by consumers as another source of sanitary risk for WWR.
- *Declared sensitivity to environmental issues*, as part of a perceived social pressure, widely relayed by the media (promotion of recycling, sobriety, local sourcing).
- *Trust.* In France, there is both goodwill towards local agriculture, which supplies local markets, and a strong mistrust of intensive agriculture, which is widely reported in the media. Similarly, there is great confidence in the quality of domestic water but mistrust of private water companies (Hervé-Bazin 2014) and the majority of the population believes that water quality in the environment is poor and deteriorating. For instance, 75% of French people believe in 2018 that 'the authorities take the utmost precautions to ensure that tap water quality standards protect consumers' health'; 85% of French people are confident of tap water quality and drink it, at least occasionally; while 75% consider that the water in the river and aquifers is polluted (Centre d'Information sur l'Eau & TNS Sofres 2018) (see Figure 2). Therefore, we want to measure consumer confidence in the ability of farmers and professionals of the water sector in the Pic Saint Loup area to control the WWR's health risks.
- *Confidence in technology.* We want to test whether general confidence in progress could influence attitudes towards WWR that is very scantily practised in France.
- *Information habits.* This factor has been added with regard to the study of Nancarrow *et al.* (2008) following the works of Dolnicar *et al.* (2010), who found a statistical positive change in the acceptance of recycled and desalinated water when respondents are regular newspaper readers. We thus aim to test whether information habits impact French attitudes to WWR.

METHODS

We conducted a survey in spring 2018 involving 845 consumers, carried out in ten localities belonging to the CCGPSL (Figure 1), near the main markets and supermarkets of the area, before or after they had bought their foodstuffs. Most of them (84%) were inhabitants of the 'Communauté de communes', and 7% were living nearby. Interviews were carried out face-to-face in the street with the help of Master students from Montpellier University, wearing 'Grand Pic Saint Loup community flags' to inspire confidence. All these people were questioned. When people came together, they were either interviewed independently (using two separate interviewers), or only one person was asked to respond. The refusal rates were very low (less than 10%). The sample did not include persons under the age of 15, as children are not responsible for household fruit and vegetable purchases. Apart from this bias, we afterwards checked that the sociodemographic characteristics of our sample of local inhabitants were close to those of the overall population (around 21,000 inhabitants in the area – see the Supplementary Material for more details of this survey).

The questionnaire, which was designed for completion within ten minutes, comprised only closed questions, and was the same regardless of the level of information previously provided. First of all, consumers were asked to give the three words they spontaneously associate with the expression 'treated wastewater' and to rank them in order of importance, following the associative method (Rateau *et al.* 2011) to qualify the social representations of this resource, if any. (Etale *et al.* (2020) asked the same type of question, 'recycled water' and 'desalinated water', then classified the words or expressions into 18 categories, which we tried in our case to follow – adding some categories due to the fact that the question was not exactly the same – and we used it especially in our factorial analysis.) When people were unable to think spontaneously of three-word associations, we did not insist. The rate of people who did not comply with this simple instruction is an indicator of the lack of social mobilization on the subject. A few people asked what treated wastewater was. After giving the word association instructions, we simply told them that it is water from sewage treatment plants.

The questionnaire continued in four different ways:

- A first sample with 'no information' ($N = 204$) simply went on with the questionnaire.

- For the 'neutral information' group ($N = 310$), people received a flyer on which were the characteristics of a WWR project, its objectives, who is involved and for which tasks. There was a very recognizable picture of the Pic Saint Loup at the top of the flyer to associate the project with the site. The rest of the text was not illustrated. The sentences did not include any value judgment (see the Supplementary Material). A student read the main points of the flyer to make sure they were understood. Then the questionnaire went on.
- For the 'persuasive information' group ($N = 176$), the protocol was the same, but the flyer was organized as an advertising medium, highlighting the benefits of WWR through key words on a circle to emphasize the message of recycling.
- Finally, for the 'commitment' group ($N = 155$), students presented to respondents the Pic Saint Loup's map asking the following question: 'are you in favour of short circuits for food products? If yes, put a green sticker on your district of residence, otherwise a red one'. This type of commitment to a cause was meant to put the respondent in a benign disposition with regard to new information in agreement with the cause to which he had just committed himself (Bernard & Joule 2004). The students then presented the flyer of the persuasive communication on WWR. Then the questionnaire went on.

The questionnaire then focused on consumers' acceptance of nine uses in the village they were living in. We insisted on this point to make sure that they felt directly concerned by the project: three 'recreational' uses (golf, roundabout and green spaces) and six 'agricultural' uses (shared gardens, market gardening, orchards, olive trees, vine and meadows for livestock). The uses were chosen among the options studied by the local public water authority, and to reflect the range of proximity of skin or mouth contact with the wastewater. We used a four-item system rated on a Likert scale, from strongly agree to strongly disagree, with no 'neutral' response to force an agreed or disagreed position, but with an option of 'I don't know' for people who had no opinion.

Intended behaviour (purchasing, public garden visits) was assessed through one question per item (wine, fruit and vegetables, goat cheese, parks). Four answers were possible: yes/no/maybe/I don't know (five for wine and cheese with 'not usually buying'). Individuals were invited to rank their environmental sensitivity on a three-item level Likert scale (with the option to say 'don't know'). They were then asked to agree/disagree/partially agree or give no opinion on four aspects (on disgust, on health risks, trust in water actors, trust in technology). They were asked about their media information habits, which was further interpreted in four categories on the following basis:

- Well informed: at least two media read/listened to daily
- Informed: at least one media read/listened to daily
- Mildly informed: at least two media read/listened to once or twice a week
- Poorly or not informed: the other answers
- No reply: when no answer to any media sources.

The questionnaire ended with socio-demographic data.

RESULTS AND DISCUSSION

R1. There is no social representation of 'treated wastewater'

In order to identify the social representation of 'treated wastewater', we followed the hierarchical evocation method proposed by Vergès (1992), who differentiates four areas based on two criteria (the level of interest, and the frequency of occurrence): the 'core' composed of consensual and fundamental elements (characterized by coherence and stability, which are thus resistant to change), the 'contrasting elements' (proposed by few people but with a high rank), and peripheral elements with more personal components (see Table 1 and Montginoul & Vestier (2018) for more details).

The three-word associations with the expression 'treated wastewater' and their rank clearly indicated that citizens did not yet have social representation on it:

- The vocabulary associated with 'treated wastewater' was poor: the rareness indices (words cited only once by one person) and diversity indices (number of different responses) were 0.14 and 0.22 respectively before lemmatization (which consists of grouping similar expressions,

Table 1 | Structural analysis of three words/expressions associated with 'treated wastewater' (green + italic/black/red + bold for expressions positively/neutrally/negatively connoted) (expressed words – no reply excluded) ($N = 845$). Please refer to the online version of this paper to see this table in colour: http://dx.doi.org/10.2166/ws.2020.242

		Level of interest	
		High (rank<2)	*Low (rank ≥ 2)*
Frequency of occurrence	*High (≥ 10%)*	The "core": Wastewater treatment plant (16%)	The "first perimeter": None
	Low (<10%)	The "contrasting elements": *Recycling (9%), environment (7%),* treated wastewater (6%), sanitation (4%), **pollution (4%), dirt (3%),** *clean (3%), water (3%), water saving (3%), positive opinion (2%),* process (2%), **cost (2%),** *potable (2%),* septic tank (1%)	The "second perimeter": Toilet (2%), **disgust (1%),** network (1%), **organic pollutant (1%),** chemical process (1%), **waste (1%),** etc.

like the terms 'connection', 'pipeline', 'water tower' and 'pipe' under the same term 'network', in the second step of the analysis, to build Table 1). For comparison, a similar study conducted on the word 'water' obtained indices of 0.3 and 0.46 (Montginoul & Vestier 2018). Among the sample, 6% were unable to give a single word, 25% could give only one, 47% only two. These figures show that treated wastewater had not been much discussed among the population, which is at the origin of social representation (Rateau *et al.* 2011). This weakness of local knowledge and beliefs on this topic may be explained by the low media coverage of this subject in France.

- 'Treated wastewater' was strongly associated with 'wastewater treatment plant' (Table 1), indicating that citizens related this type of water to its producing entity. This sole strong association with any other, even in the first peripheral perimeter (lower rank but highly cited), indicates interviewers had difficulty in suggesting links. They were able to give only a neutral association, close to a synonym. We can however observe that 9% of the words (and then 21% of respondents) related 'treated wastewater' to WWR, which seems high given the WWR French context.

R2. Citizens see WWR as an opportunity

Contrasting elements (low frequency but high rank in Table 1) highlight different and less common ways of thinking. They refer here mainly to people who saw treated wastewater as an opportunity (recycling, water saving, etc.). There are very few elements negatively connoted (i.e. disgust, pollutions, dirt, cost…) appearing mainly in the 'second peripheral perimeter' (low frequency and low rank), that is for only a few respondents and at a low rank.

R3. Citizens are well intentioned toward WWR, and even more so as the usage is far distant

The survey questioned the attitude of citizens towards nine uses: three 'recreational' uses (golf, roundabout and green spaces) and six 'agricultural' uses (shared gardens, market gardening, orchards, olive trees, wine and meadows for livestock) (Figure 3).

These results support the patterns identified in the literature (Smith *et al.* 2018) on the relationship between attitude and uses: respondents were less likely to use treated wastewater for market gardening (70%) than for watering green spaces (90%). This distinction is also confirmed when consumer behavioural intentions were questioned: 83% would continue to frequent green spaces irrigated with treated wastewater while only 69% would continue to buy from grocers who sell fruit and vegetables irrigated with treated wastewater.

This corroborates the annual French national water survey in which a very large majority of individuals (75% in 2018) were ready to eat fruit and vegetables irrigated with WWR (Centre d'Information sur l'Eau & TNS Sofres 2018).

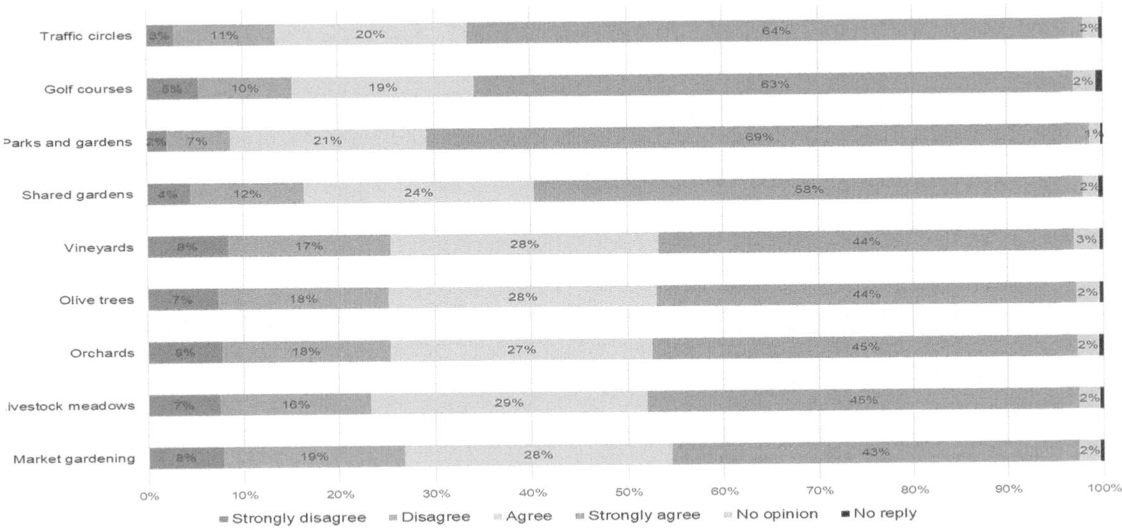

Figure 3 | Attitudes towards different irrigation uses of treated wastewater.

Table 2 | Chi-Square test of independence (p-value < 5% in bold to highlight dependencies) between respondents' characteristics and their intention to consume products from WWR (in case of dependence, the positive or negative link is specified)

	Parks and gardens	Wine	Market gardening
Disgust	−**(0.000)**	−**(0.000)**	−**(0.000)**
Environmental sensitivity	+**(0.000)**	+**(0.001)**	+**(0.000)**
Risk perception	−**(0.003)**	−**(0.000)**	−**(0.000)**
Information	(0.811)	+**(0.007)**	+**(0.001)**
Trust in technologies	+**(0.021)**	(0.106)	(0.142)
Gender	(0.744)	(0.383)	**Women** −**(0.019)**
Age	(0.409)	(0.259)	(0.092)
Information habits	(0.557)	(0.418)	(0.438)

A statistical analysis was conducted to identify the socio-technical variables that could be correlated with these attitudes, among those cited as relevant by various works in other countries (Nancarrow *et al.* 2008; Fielding *et al.* 2018; Smith *et al.* 2018). Table 2 presents results for the irrigation of the three typical uses for which behavioural intentions have been questioned during our survey.

Disgust, environmental sensitivity and risk perception influenced attitudes for all three uses. These results are in line with the synthesis of Fielding *et al.* (2018) in several dozen studies on the most discriminating factors of attitudes for non-potable uses. In this literature review, disgust, and the perception of health risks were cited as the main factors explaining unfavourable attitudes, while the claim of environmental sensitivity characterized favourable opinions. In our study, people who said they were disgusted with the thought of eating a vegetable irrigated with treated wastewater were unfavourable to all WWR projects; those believing there was no risk in using treated wastewater to irrigate vineyards were more favourable than others to all the WWR projects. The more environmentally sensitive the respondents, the more they were in favour of reusing wastewater to irrigate parks and gardens.

In our survey, confidence in technology also acted on WWR attitudes, but the probability of independence is less significant between these two variables. A gender-related effect is also observed but only for market gardening, for which men were more favourable to WWR than women

Table 3 | Attitude, disgust at, and perception of remaining risks of WWR, according to the type of information provided to the interviewees

		No information $N = 204$	**Neutral** $N = 310$	**Persuasive** $N = 176$	**Commitment** $N = 155$
Recreational uses (*p*-value 0.001)	Strongly disagree	2%	3%	3%	5%
	Disagree	11%	7%	6%	13%
	Agree	21%	***16%***	27%	19%
	Strongly agree	64%	***72%***	62%	61%
	No opinion	*1%*	*2%*	*2%*	*1%*
Agricultural uses (*p*-value 0.001)	Strongly disagree	8%	8%	**4%**	8%
	Disagree	22%	12%	17%	18%
	Agree	28%	***23%***	32%	29%
	Strongly agree	37%	***55%***	44%	44%
	No opinion	*3%*	*2%*	*3%*	*2%*
Disgust (*p*-value 0.02)	Yes	16%	14%	14%	14%
	Mitigate	19%	13%	7%	15%
	No	63%	70%	80%	65%
	No opinion	*2%*	*3%*	*0%*	*6%*
Sanitary risks (*p*-value 0.23)	Yes	25%	27%	32%	26%
	Mitigate	26%	18%	18%	19%
	No	41%	47%	43%	41%
	No opinion	*7%*	*7%*	*7%*	*15%*

were. In the synthesis of Fielding *et al.* (2018), the other variables (age, gender, confidence in technology) had varying effects across studies. When they had a significant effect, it was the same as the one observed in our study.

R4. Citizens trust neutral information more than positive information

Information seemed to influence attitudes (Table 3), but not uniformly: it varied according to the type of use and the type of communication.

Information described as neutral significantly strengthened the favourable attitudes for all uses. This result is in line with Fielding *et al.* (2018) who wrote in their synthesis, 'providing factual information about recycled water increases knowledge about, and acceptance of, recycled water'. This impact may be explained by the mention of the health authority's controls exclusively on this communication medium. In the national survey, 83% of the population trusted the authority to control water quality. The so-called persuasive information changed the opinion of the very opposed participants, but only as far as agricultural uses were concerned. This brochure explicitly mentioned agriculture in its slogan 'WWR – a resource for our agriculture', and as a local solution to food insecurity. Finally, in the commitment scheme, participants gave their opinion on local food systems, as a preparatory act, before being subjected to persuasive information. The results do not validate the scientific hypotheses (Bernard & Joule 2004) assuming that the effects of the information could be reinforced by an act of commitment. There was even a slight decrease in positive opinions compared with the control group ('no information'). The relevance of the preparatory act must be questioned.

Information seemed to have some influence on the people who answered 'mitigate' when we asked them if they were disgusted by the idea of eating a fruit or vegetable irrigated with treated wastewater. With information (whatever its nature), the rate of these hesitating persons decreased significantly ($p < 0.02$). We can assume that many among them were completely unaware of the process of recycling water. The flyer was sufficient to convince them that the treatment had 'purified' the water. However, when a person claimed to be disgusted, he or she was in the realm of emotion, not reasoning, and information did not matter. These findings are in agreement with the work of Rozin *et al.* (2015), who studied the disgust at drinking treated wastewater.

Information had less of a reassuring effect on the perception of risk. When people first read the flyers, the rate of persons hesitant about their perception of sanitary risk decreased, but not significantly. If information seemed to reassure some of the undecided, others feared contagion or did not express an opinion. We assume that the multiplication of recent health and food scandals (mad cow disease, meat and milk quality, pesticide residues in food, etc.) aroused a great deal of suspicion of the reassuring information provided by public bodies.

The remaining sanitary risk perception was significantly and positively correlated with (i) disgust at the thought of eating food irrigated with treated wastewater and (ii) the lack of trust in water actors (at the 0.001% threshold). Disgust and risk perception seemed independent of environmental sensitivity, and of general confidence in technology.

Media consumption habits (radio, TV, web, newspapers), too, seemed to influence the effect of information among those who had received useful information through the media; their number increased the probability of responding 'strongly agree' rather than 'agree' (at the 1% threshold) by 25 percentage points. This supports an Australian study (Dolnicar *et al.* 2010) on WWR, which pointed out that the cause of heterogeneity in information responses was rather the experience of information processing than of socio-demographic data.

R5. A strong cohesion between attitudes and behavioural projections, but some apparent contradictions

The study also found a significant correspondence between attitudes and behavioural intentions (p-value = 0.000). People in favour of WWR would not change purchasing habits if they learned that vineyards are irrigated with treated wastewater. A reassuring effect of the information was also noted. Consumers were less hesitant when they had information, but it did not greatly change unfavourable opinions.

The results presented in Table 4 for wine purchases were similar for vegetables or cheeses from breeding that uses fodder watered with wastewater (cf. the graphical abstract). Without information, 22% to 30% of consumers would switch shops if they knew that the shop buys its products from a farmer who irrigates with treated wastewater. With information, this rate was around 20%.

Again, disgust and the perception of persisting sanitary risk played a major role in the intention of purchasing food (p-value 0.01). Two-thirds of those who would change shops (for vegetables, goat cheese, wine) said they were disgusted by the idea of eating such fruit and vegetables; two-thirds also thought there was a health risk with WWR. As our flyers had an effect only on those who were most ignorant of WWR and hesitant about expressing disgust and about risks, purchasing intentions changed little with the information, in the case of this profile of consumers mainly.

Table 4 also reveals complex judgments; 5.8% of the uninformed and 4.6% of the informed were from in favour to very much in favour of water recycling for irrigating food products, as citizens, but not as consumers. Conversely, 8.9% of uninformed and 5.3% of informed persons, opposed to it as citizens, would tolerate it as consumers. In addition, depending on the products, the motivation of between 20% and 25% of those who would change shops was unknown, since they said they were not disgusted, and that they did not perceive risk of contagion. The very limited questionnaire did not allow for a more in-depth analysis of their motivations.

Table 4 | Effect of information on attitudes and purchasing intentions for irrigated wine with treated wastewater (sample without people declaring they never buy wine)

	No information (N = 191)				**Information (N = 556)**			
Attitudes	**No purchase**	**Don't know**	**Purchase**	**Total**	**No purchase**	**Don't know**	**Purchase**	**Total**
Highly unfavourable	6.8%	0.5%	0.5%	7.9%	7.0%	0.5%	0.7%	8.2%
Rather unfavourable	12.6%	2.6%	8.4%	23.6%	8.1%	1.8%	4.6%	14.4%
Rather favourable	3.7%	2.1%	23.0%	28.8%	3.5%	1.1%	23.5%	28.0%
Highly favourable	2.1%	0.5%	33.5%	36.1%	1.1%	0.9%	45.4%	47.3%
Don't know	0.5%	2.6%	0.5%	3.7%	–	0.5%	1.2%	2.1%
Total	25.7%	8.4%	66.0%	100.0%	20.0%	4.7%	75.3%	100.0%

The cohesion between attitudes in favour of WWR for public parks was less consistent with the envisaged frequenting of them if they were irrigated with reclaimed water (see the Supplementary Material). Information did not significantly influence this level of inconsistency. A majority of those in favour of this use of WWR would use the parks if the project were carried out, but the actions of a larger part of the population would not be consistent with their opinions. Among those who would no longer use the parks, three-quarters of the people were in favour of this use. Conversely, three-quarters of the few initial opponents would eventually continue to frequent these places. Feelings of disgust and health risks could not explain these contradictions between attitudes and intended behaviours.

These apparent inconsistencies confirm the need to go beyond the stated intentions of action by interviewees, the level at which most of the work cited in the literature stops (e.g. Nancarrow *et al.* 2008). Between stated intention and actual action there are often great differences, as has been demonstrated regarding other pro-environmental behaviours (de Leeuw *et al.* 2015). It will be necessary to go as far as the observation of actual actions, i.e. here, the actual decision to eat or buy food in a taste test, for example.

Using a factorial analysis (of the 802 respondents who replied to all questions participating in the factorial analysis), we have thus clearly distinguished two opposing subpopulations: half of the total respondents (54%) were in favour to very much in favour of the development of WWR, for all uses. They would not change their consumption habits (more than 80% of them), they were not disgusted (95%), they did not perceive any health risks (67%), they were rather more confident of technological progress (62%), and sensitive to environmental issues (96%). This group is composed of people who had previously received information on WWR (81%), and for 41% of them a neutral kind.

On the other hand, a group representing 17% of the population is made up of people opposed to WWR (both in opinion and intention). Including 22 (out of 23) of the farmers who replied to all questions; this group comprises a majority of retired people (53%), especially more than 75 years old (24%). It includes the more heterogeneous population (cf. the hierarchical clustering – more details in the Supplementary Material), but is highly characterized by persons who have difficulties in associating words with 'treated wastewater' (60% did not give three words, 20% of them did not come up with any word).

The last group comprises 28% of the population (37% belongs to the 'no information' group, of which only 31% was given 'neutral' information); they were resistant to WWR, preferring to stop consuming products that result from the WWR (for more than 50% of them). This group comprised 'mitigated' people (on disgust, health risks, technology, and confidence in decision-makers), and a significant proportion of them (12%) declared themselves not very sensitive to the environment, but were concerned about health problems (the word 'contamination' was cited by 28%). This includes a majority of women (62%), people between 30 and 44 years of age (28%), working people (61%), and those less open to the media (27%). The great majority of them (80%) had a considerable idea of 'treated wastewater' (only 2% did not have anything to say).

R6. A change in purchasing habits with a great economic impact on local producers

The threat of a 20% (26% when no information) drop in potential local buyers would not be acceptable in an agricultural and wine-growing area such as Pic Saint Loup. We did a reconstitution of this work, at the invitation of a mayor. The few winegrowers present confirmed that selling to local customers was essential to the profitability of their business. They stated that they could not accept a 20% reduction in their clientele.

The reduction of current discharges from treatment plants was of interest to drinking water managers (i.e. reducing the risks of aquifer pollution in this karstic area). But the result of this study reinforced other weaknesses in the cost recovery and economic profitability of the WWR project pointed out by other studies conducted in the SOPOLO project. This explains why the Pic Saint Loup community of communes decided not to carry out the project in the near future.

The next research actions on this topic will include (i) a wider range of factors explaining attitudes towards WWR, (ii) an extension of the sample to people with a sound knowledge of wastewater treatment (professionals and students), (iii) an extension of the study of behaviours with tasting tests of products resulting from irrigation with treated wastewater, and (iv) feedback on the behaviour of consumers of agricultural production from some projects currently implemented in France.

CONCLUSIONS

WWR projects in France are still rare, and are only rarely the subject of social acceptability studies. In our survey, disgust and the perception of persisting sanitary risks clearly explained the reluctance of 5% to 30% of our sample of consumers. We found three groups of consumers: the promoters of WWR (50%) significantly more sensitive to the environment, and more confident in the technology and in water managers; the opponents (20%) fearing sanitary risks; and the reluctants partly reassured by prompt information on this little-known subject. WWR project leaders will see this as encouragement to launch intensive targeted information campaigns. But the reluctant group, significantly less open to the media and made suspicious by health scandals, will be difficult to convince. This study needs to be deepened, including taste tests, as some inconsistencies between attitudes, food-buying intentions, disgust, and perceived risks have emerged among a significant minority.

ACKNOWLEDGEMENTS

We would like to thank the Rhone-Mediterranean Water Agency for its financial support of the SOPOLO project, the Pic Saint Loup Community for its encouragement, the students of 2017–2018 class in the first year of the 'Water' Master's program at University of Montpellier who helped to carry out the surveys, and the consumers who agreed to answer.

DATA AVAILABILITY STATEMENT

Data cannot be made publicly available; readers should contact the corresponding author for details.

REFERENCES

Ajzen, I. 1991 The theory of planned behavior. *Organizational Behavior and Human Decision Processes* **50** (2), 179–211.

Aquarec 2006 *Integrated Concepts for Reuse of Upgraded Wastewater*. AQUAREC.

Barbier, R. & Nadaï, A. 2015 Acceptabilité sociale: partager l'embarras (Social acceptability: sharing the embarrassment). *VertigO - la revue électronique en sciences de l'environnement* **15** (3), https://id.erudit.org/iderudit/1035870ar.

Baumann, D. D. & Kasperson, R. E. 1974 Public acceptance of renovated waste water: myth and reality. *Water Resources Research* **10** (4), 667–674.

Ben Maïd, A., Devaux, J. & Thao Khamsing, W. 2014 *Ressources en eau: perception et consommation des Français (Water Resources: French Perception and Consumption)*, Études et documents du CGDD 106. Commissariat général au développement durable, Paris, France.

Benoit, G., Berlizot, T., Bortolotti, S., Brugière, Y., Domallain, D., Jourdier, G., Madignier, M.-L., Mailleau, C., Piaton, H., Pinçonnet, D., Renoult, R. & Rocq, S. 2017 *Eau, agriculture, et changement climatique : Statu quo ou anticipation?* Report no. 16072. Ministère de l'agriculture et de l'alimentation and CGAAER, France. https://agriculture.gouv.fr/sites/minagri/files/cgaaer_16072_2017_rapport.pdf.

Bernard, F. & Joule, R.-V. 2004 Lien, sens et action : vers une communication engageante (Relationship, meaning and action: towards communication with commitment). *Communication et organisation* **24**, 333–345.

Berry, P., Strosser, P. & Berge, M. 2016 *Investigating the Public Acceptance of Wastewater Reuse at the Vendée Greenfield Demonstration Site*. ACTeon.

Centre d'Information sur l'Eau & TNS Sofres 2018 *Baromètre C. I. Eau/TNS Sofres : Les Français et l'eau, enquête nationale 2018 (Barometer C. I. Eau/TNS Sofres: The French and Water, National Survey 2018)*, 22nd edition. Le Centre d'information sur l'eau, Paris, France.

Cerceau, J. 2015 *Acceptabilité sociale des projets de réutilisation des eaux usées (Social Acceptability of Wastewater Reuse Projects)*, NOW MMA Deliverable 4.2. ARMINES/Ecole des Mines d'Alès, France.

de Leeuw, A., Valois, P., Ajzen, I. & Schmidt, P. 2015 Using the theory of planned behavior to identify key beliefs underlying pro-environmental behavior in high-school students: implications for educational interventions. *Journal of Environmental Psychology* **42**, 128–138.

Dolnicar, S., Hurlimann, A. & Nghiem, L. D. 2010 The effect of information on public acceptance – the case of water from alternative sources. *Journal of Environmental Management* **91** (6), 1288–1293.

Etale, A., Fielding, K., Schäfer, A. I. & Siegrist, M. 2020 Recycled and desalinated water: consumers' associations, and the influence of affect and disgust on willingness to use. *Journal of Environmental Management* **261**, 110217.

Fielding, K. S., Dolnicar, S. & Schultz, T. 2018 Public acceptance of recycled water. *International Journal of Water Resources Development* **35** (4), 551–586.

Franck-Néel, C. 2020 *Réutilisation des Eaux Usées Traitées. Le panorama français (Reuse of Treated Wastewater. The French Panorama).* Cerema, France.

Hervé-Bazin, C. 2014 Boire en eaux troubles (Drinking in turbid waters). *Communication et Organisation* **45**, 127–138.

Lazarova, V. & Brissaud, F. 2007 Intérêt, bénéfices et contraintes de la réutilisation des eaux usées en France (Interest, benefits and constraints of wastewater reuse in France). *Revue de l'Eau, l'Industrie et les Nuisances* **299**, 43–53.

Maurer, H. 2018 *L'économie circulaire dans le petit cycle de l'eau : la réutilisation des eaux usées traitées (Circular Economy in the Small Water Cycle: Reuse of Treated Wastewater).* Institut National de l'Economie Circulaire, Paris, France.

Montginoul, M. & Vestier, A. 2018 Smart metering: a water-saving solution? Consider communication strategies and user perceptions first. Evidence from a French case study. *Environmental Modelling & Software* **104**, 188–198.

Nancarrow, B. E., Leviston, Z., Po, M., Porter, N. B. & Tucker, D. I. 2008 What drives communities' decisions and behaviours in the reuse of wastewater. *Water Science & Technology* **57** (4), 485–491.

Noury, B., Pascual Espuny, C. & Ferrand, N. 2019 How can participation and communication support decisions towards wastewater reuse projects? A methodological proposal. In: *Conference on Communication and Environment (COCE)*, 17–21 June, University of British Columbia, Vancouver, Canada.

Rateau, P., Moliner, P., Guimelli, C. & Abric, J.-C. 2011 Social representation theory. In: *Handbook of Theories of Social Psychology*, Volume 2 (P. A. M. Van Lange, A. W. Kruglanski & E. T. Higgins, eds), Sage Publications, London, UK, pp. 477–497.

Rozin, P., Haddad, B., Nemeroff, C. & Slovic, P. 2015 Psychological aspects of the rejection of recycled water: contamination, purification and disgust. *Judgment and Decision Making* **10** (1), 50–63.

Sibertin-Blanc, C. 2019 Formal analysis of the conflictive play of actors regarding the building of a dam. In: *Law, Public Policies and Complex Systems: Networks in Action* (R. Boulet, C. Lajaunie & P. Mazzega, eds), Springer International Publishing, Cham, Switzerland, pp. 177–202.

Smith, H. M., Brouwer, S., Jeffrey, P. & Frijns, J. 2018 Public responses to water reuse – understanding the evidence. *Journal of Environmental Management* **207**, 43–50.

Vergès, P. 1992 L'évocation de l'argent : une méthode pour la définition du noyau central d'une représentation. *Bulletin de Psychologie* **45**, 203–209.

First received 30 June 2020; accepted in revised form 18 September 2020. Available online 30 September 2020

An investigation on the adaptability of residential rainwater harvesting system in Tamil Nadu – techno-economic considerations and the way forward

Mangottiri Vasudevan and Narayanan Natarajan

ABSTRACT

Rainwater harvesting (RWH) is one of the most important means of water conservation today. Although there is tremendous emphasis on the implementation of this versatile technology at various levels, many of us are unaware of its significance and its consequences. This study examines the perception and awareness of people in the state of Tamil Nadu, India, of the RWH system through a questionnaire survey. The survey was conducted to infer basic information on rainwater harvesting features, water quality and conservation, economics of implementation and financial support. The results show that there is an increasing need to promote the usage of harvested rainwater for domestic purposes rather than direct groundwater recharge. The potential risks in the reliable supply of available water sources demand large-scale implementation of RWH especially in apartments. Fewer people are aware of the technical details of operation and maintenance of the RWH system (22.2%) while more people are aware of the general layout of the system (50.5%). Longer experience in using RWH systems (for more than two years) builds confidence in using them for consumptive uses compared with the response from recent adopters. This also demonstrates the scope for a radical change in the perception of water quality by continuous awareness and financial incentives.

Key words | groundwater recharge, rainfall dependency, rainwater harvesting, water quality, water storage tank

Mangottiri Vasudevan (corresponding author)
Department of Civil Engineering,
Bannari Amman Institute of Technology,
Sathyamangalam 638401, Tamil Nadu,
India
E-mail: *devamv@gmail.com*

Narayanan Natarajan
Department of Civil Engineering,
Dr Mahalingam College of Engineering and Technology,
Pollachi 642003, Tamil Nadu,
India

HIGHLIGHTS

- Structured questionnaire survey to realise the awareness of people about improving rainwater harvesting (RWH) infrastructure.
- Suspicion of reliability in available water resources promotes early adoption of RWH systems.
- Experience with RWH makes changes in people's perception towards ensuring water quality and proper usage.
- Major hurdles in RWH: promotion through demonstration, access to financial incentives and implementation of groundwater recharge.

INTRODUCTION

India is a nation with flexible climatic conditions with abnormalities in rainfall throughout the year in different parts of the country. The disparity is primarily observed in the equatorial region, i.e. the southern part of India. The summer season is too dry and precipitation is unpredictable. Tamil Nadu is one of the most urbanised states with 48.45 per cent of its population residing in urban areas. Tamil Nadu gets precipitation from the north-east in addition to the south-west monsoons. Variation in the

doi: 10.2166/ws.2020.249

monsoon has a severe effect on economic life and also the livelihood of its people specifically residing in the rural areas. Thus the state of Tamil Nadu is deficient from the perspective of water resources and hinges extremely on the monsoons. India's approximated overall utilizable water is put at 1.12 billion cubic kilometres (Mohile 2008), but a majority of the scientific evaluations agree that total water requirement will certainly overtake the supply in the future (Chitra & Kantha 2017).

While there are numerous methods to fulfil the water demand in Tamil Nadu, in 2003, a landmark statute on rainwater harvesting (RWH) was set by the then state head in Tamil Nadu. After the high drought years of 2001–2003, the state government's order commanded that all buildings in Tamil Nadu fix RWH structures to store water (*News Minute* 2019). Today, Tamil Nadu is the trend-setter for establishment of rainwater harvesting structures. Out of the 23.92 lakh (2.392 million) buildings in town Panchayats (government, residential, commercial and industrial), 22.94 lakh (2.294 million) have rainwater harvesting facilities (*Hindu Business Line* 2018). RWH is a justifiable means of producing water for oneself, recharging the groundwater and also minimising the overflow throughout the heavy rainfall season. Basic understanding, skills, and a couple of readily available materials can be utilised for the construction of this system. During a stormy period, a person can gather water from his/her rooftop and handle its storage for their benefit. Saved rainwater from rooftops can be made use of for their own residential usage or this can be later led into the wells to surcharge groundwater directly. The quality of water needs to be evaluated prior to usage for drinking purposes. If it satisfies the permissible limits, then the water can be utilised for the water supply of the entire location. If required, sanitation such as chlorination can be used to cleanse the water. The fostering of RWH will certainly increase quickly if the system plan is integrated with the building plan rather than fitting it externally later. RWH has several benefits such as reducing the burden on the municipal water supply, cost effectiveness, minimisation of payment expenses, enhancement of groundwater recharge, fire-fighting under emergency situations, etc.

Numerous studies have actually been carried out in the recent past on RWH in the Indian subcontinent. Bitterman *et al.* (2016) provided a theoretical structure for the measurement of water security in the context of RWH tanks. Mahmood & Hossain (2017) evaluated the expediency of managed residential RWH in south Asian backwoods using remote sensing. Raj (2017) examined the RWH potential of Pallavpuram area in Meerut of Uttar Pradesh using GIS. Ram *et al.* (2018) performed a critical review on the RWH system and its function in a sustainable future for the Indian subcontinent. Hari *et al.* (2018) assessed the RWH potential zones in the Ranga Reddy district of Telangana making use of GIS. Farswan *et al.* (2019) examined the RWH sites in a part of north-west Delhi, India, with the help of geomatic tools. Jain & Jain (2019) evaluated the usefulness of RWH in the Thar Desert of India. Rajasekhar *et al.* (2020) determined the groundwater recharge-based potential RWH sites for sustainable growth of a semiarid region of southern India using geospatial analytic hierarchy process and SCS-CN strategy.

Questionnaire survey is a powerful technique for understanding the perception and awareness of people towards a specific problem or situation pertaining to water supply in society. Several such surveys have been conducted in the recent past. Karim (2010) performed a study to assess the RWH system located in the south-west coastal and arsenic-affected areas of Bangladesh using structured questionnaire surveys. Islam *et al.* (2018) assessed the willingness of people to pay for improved drinking water in a rural area of south-west coastal Bangladesh using survey data collected from 215 households. Al-Mefleh *et al.* (2019) investigated the water management problems and potential solutions in the Al-Mafraq governorate of Jordan. Kosaka *et al.* (2019) investigated the utilisation of water quality monitors at water purification plants throughout Japan via a questionnaire conducted during 2015. Similar other studies were conducted by Proulx *et al.* (2010), Beal *et al.* (2011), Kishida *et al.* (2013), Alsaluli *et al.* (2015), Dismas *et al.* (2018) and Nunes *et al.* (2018).

Although several initiatives on policy-making and management are earmarked for RWH systems in Tamil Nadu, the awareness of the people, especially the student community, towards RWH systems seems to be still limited (Directorate of Town Panchayats 2018). Very few studies have been conducted, to the best of our knowledge, to assess the perception and awareness of the people towards water conservation from agricultural and residential areas.

Moreover, the actual bottlenecks in evaluating the success of promotional activities about the RWH system lie in educating the people on the basic performance features such as cleaning protocols and water quality considerations. The present study aims to address the perception and awareness of the student community towards a futuristic vision of RWH in residential areas. The study also attempts to critically evaluate factors in defining the success of RWH towards the days of growing water demand. As a practical approach, the perception of groundwater recharge in connection with RWH is also analysed to suggest alternative methods of improving effective usage of the harvested water.

METHODOLOGY

To understand the awareness and perceptions of people towards RWH, a structured questionnaire survey was conducted through the online mode due to the inability to meet people personally during the lockdown period in the pandemic situation due to the outbreak of COVID-19. In order to accomplish the stated objectives, we framed 45 questions for the survey under various categories such as demographic details (11%), details about source of water (11%), access to financial support systems (20%), awareness on water quality and treatment (22%), and technical knowhow about operation, troubleshooting and cleaning protocols (36%). All the questions were framed as multiple-choice types with only one option to choose as the best suitable answer (Figure 1).

After framing the questionnaire, the flow chart was divided into eight sections according to the possible answers from the given choices, and suitable branching was provided so that the participants did not need to attend to any irrelevant questions based on the choice of their previous answer. About 47% of the questions were Yes/No types, to make it convenient to draw a conclusion about their perception based on the related questions in every section. This is an important concern in online activity as the average time of close attention is much less. A total of 442 students participated in the survey.

The survey results were analysed after verifying the correctness of the records and elimination of duplicate entries, if any. Since the distribution of the questions was not intended to be framed with a uniform weightage as with the expected output for making use of simple classifications, we considered a dynamic combinatorial analytical approach

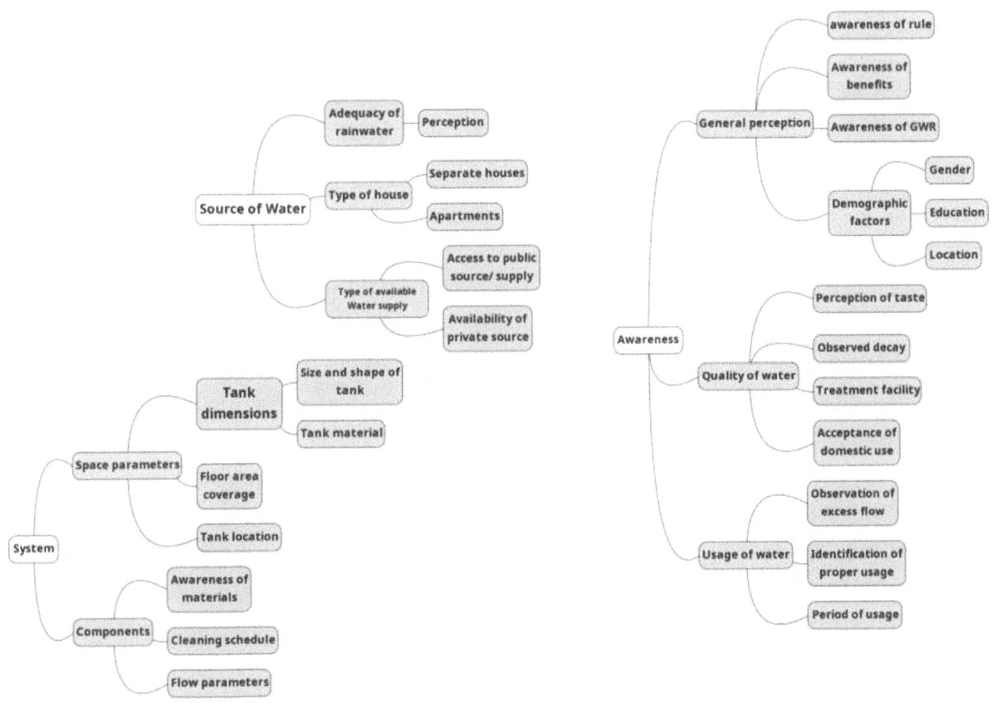

Figure 1 | The outline framework of the questionnaire survey on RWH systems in Tamil Nadu.

with a number of responses as the prime variable. This approach is designed to accommodate the variability in the response of the participants to various sections of the questionnaire and to eliminate the influence of blank responses. The key factors considered in this approach for comparing the frequency are as follows:

i. Source of water against the awareness
ii. Size of the system against financial support received
iii. Demographic variability against awareness of technology for operation and maintenance

Further, each of these six dimensions was subclassified with specific details of the questions while accommodating the intended distribution of the main themes as mentioned above. This was achieved by multi-layer filtering of the responses for each combination of the questions. Finally, the significance of correlation between the selected themes was compared based on the standard error of the mean taken for each subclass of the themes.

RESULTS AND DISCUSSION

The response to the online survey was quite promising during the initial days, especially during the first 4–5 hours after sharing the web link through public access domains and social media platforms. The survey was specifically intended to get a response from the young educated people who are available on online media for various academic/non-academic activities during the lockdown period. We selected the responses pertaining to the state of Tamil Nadu, India, based on the significant contribution of participants involved in this study. Out of the total number of responses, major contributions were from the following seven districts of Tamil Nadu, namely, Coimbatore, Dindigul, Erode, Karur, Namakkal, Tirupur and Salem. Out of the 443 responses, 244 responses were received from rural residents, while 197 responses were received from urban residents. A wider look at the overall response indicates that there is significant variation in the types of responses without any monotonous response and hence, there is a need for a comprehensive analysis of the responses to make any conclusive remarks about their general thought behaviour. As the structured questionnaire focuses on getting all possible alternatives, we identify the major thrust areas to be focused on in the discussion with the view that a deeper evaluation of these responses could help us to draw some practical strategies regarding promotion of RWH in the suburban and peri-urban areas. The discussion follows the sequence of understanding the underlying interactions between (i) source and awareness, (ii) capacity and support, and (iii) technology and education on awareness.

SOURCE, AVAILABILITY AND AWARENESS

Invariably there is a common call for uninterrupted and reliable drinking water supply in the state of Tamil Nadu, especially during the summer season. From the survey, it is observed that out of the total of 442 responses, 54.3% of families depend on a public water supply system either in the form of municipal/corporation supply (52.9%) or public wells (1.4%), while the remaining 45.7% depend on either private bore wells (29.6%), private water supply (9.5%) or private open wells (6.5%) for their daily needs of water (Figure 2). We also observed that about 60.9% of people are staying in separate houses with a single storey and 30.1% are living in separate houses that are multi-storey. Of course, this may be a temporary effect during the situation of lockdown as people are staying at their residential homes rather than at work places (hostel staff quarters, rental houses etc.).

The availability of an RWH system at their place of stay was the most critical parameter in this study as we intend to realise the extent of awareness in terms of installation and practice of usage. For the people who depend on a public source of water for their daily domestic usage, about 51.6% have an RWH system installed in houses, while only 31.2% of the people who rely on private sources have actually implemented an RWH system (Figure 3). This is an interesting observation as one may assume there is either sufficient water availability (reliability on the system) or some socio-economic limitations in going for a RWH. As the study mainly includes six districts of Tamil Nadu, the average annual rainfall during the last five years (2015–2020) showed an

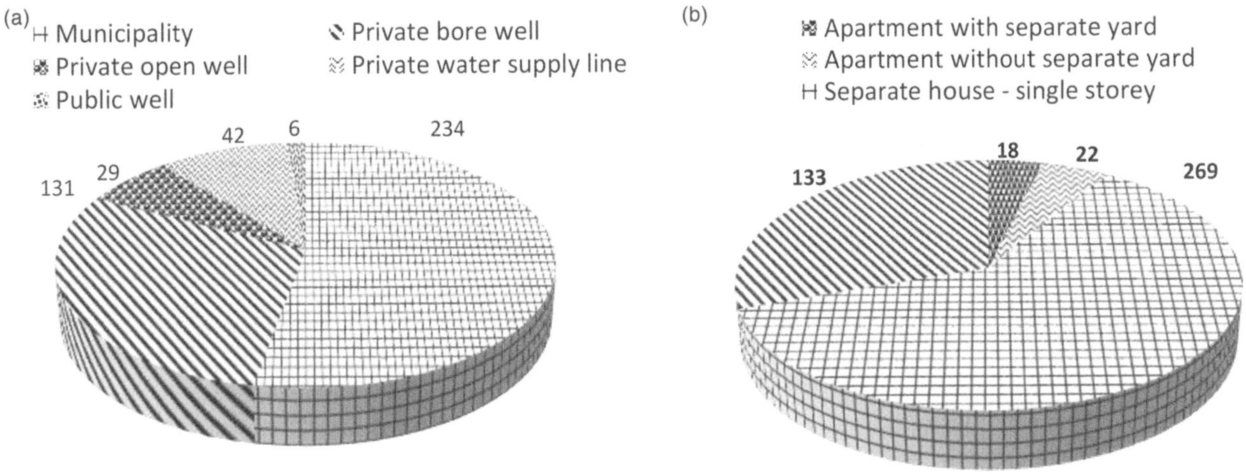

Figure 2 | The distribution of responses regarding the basic infrastructure for water accessibility in terms of (a) types of water supply and (b) type of houses.

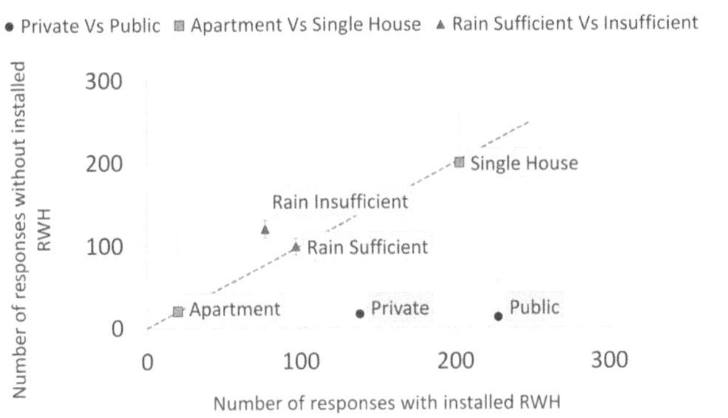

Figure 3 | Comparison of the variation in responses related to the availability of an RWH system in combination with the basic infrastructure for water accessibility (type of water supply, type of house, perception on the adequacy of rainwater).

increasing trend (57% ± 23%) compared with the previous years pertaining to recent anomalies in the local climate and rainfall features. This has resulted in achieving reliable water availability through various public and private water distribution systems. Since the majority of the respondents are young educated people, an increasing scope of awareness of water conservation practices can be inferred. To understand more about it, we have followed the systematic breakdown of the questionnaire. The response to the question on the reliability of rainwater for meeting their annual water demands was just equal (49.9% agree and 50.1% disagree). However, the percentages of people who have actually implemented RWH are only 21.9% and 17.4% for the above-mentioned conditions (agree/disagree). This strongly suggests that one key factor in promoting RWH collectively in the expanding townships could be a clinical awareness mission. On the other side, people who live in separate houses (both single story and multi-storey) have implemented RWH (45.9%) much more than those living in apartments (4.5%).

The details of the influence of availability of reliable source and pertinent infrastructure on the perception of people and installation of an RWH system can be inferred from Figure 3. Based on the discrete statistical measures, the availability of installed RWH systems in the households is more dependent on the type of houses and concern for rainwater adequacy rather than the available source of drinking water in their premises. The standard error on the mean is quite high for housing type (110.5 ± 90.5) when compared with type of supply (98.5 ± 23.5) and people's concern on rainwater adequacy (98.25 ± 10.25).

AWARENESS, QUALITY CONTROL AND USAGE

The importance of RWH is, perhaps, generally understood to be in devising a reliable system capable of storing rainwater, but it is equally important to make it available for recharging the groundwater at all possible locations. In this context, the water quality is of prime consideration to identify the best usage of collected water as well as to make it safe for future usage. The survey devoted a large number of questions to this aspect. As the target group was students, understanding their general response on water quality is very crucial in evaluating their awareness of resource conservation and reuse. The majority of the participants (93.7%) are aware of the Government ordinance stating that installation of the RWH system is compulsory in all new buildings in Tamil Nadu. However, considering the availability of installed systems, the highest usage of collected water is gardening including groundwater recharge (18.8%), followed by domestic usage such as washing and flushing (17.5%), and cooking and drinking (5.8%) (Figure 4). It is also observed that people who have experience in using an RWH system (for more than two years) consume a greater quantity of the harvested water for cooking (4.0%), washing (9.9%) and gardening (9.9%), compared with those recent users (less than two years) (2.2%, 9.0% and 9.0% respectively). Based on the responses from rural households, 51.63% of them possess an RWH system in their houses. Furthermore, 45.08% of the residents from rural households are aware that RWH is mandatory for a house but they do not have one. This shows that awareness of RWH is quite significant among the rural public in Tamil Nadu, and the mandate of the state rule insists more people adopt RWH during construction of new buildings.

Another important aspect of consumer awareness is about the technical details. While the majority of people are aware of the general dimensions and space requirements of the RWH storage tanks, most of the participants did not prefer to respond to the questions pertaining to the operation and maintenance of the storage tank, filter unit, pump and pipeline system. The survey results on awareness of the cleaning methods and frequency of cleaning the components are addressed only by 22.2% of the participants, while about 50.5% of participants answered for the general layout of the system. It is to be understood that, though being a simple system for installation and operation, people are not much aware about the technicality of the system unless it becomes an emergency. This is one of the reasons for getting widely distinctive responses for the questions from the 'experience' category and 'awareness of rule' category.

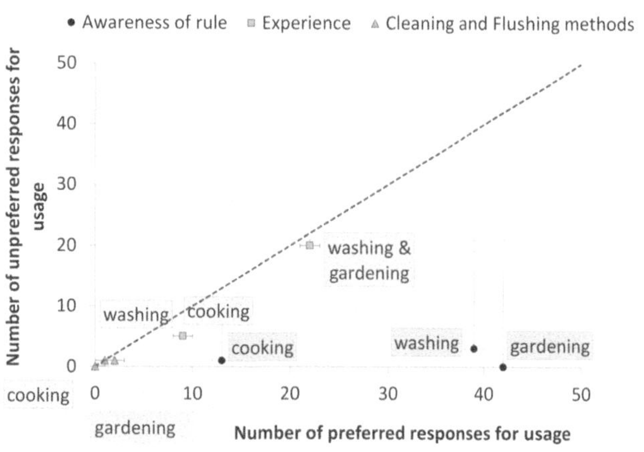

Figure 4 | Comparison of the responses related to the usage of water in combination with the awareness parameters (awareness of rule, cleaning and flushing methods and experience).

The comparative response of awareness, experience and details of technicalities on the effective usage of harvested rainwater is shown in Figure 4. Based on the standard error on the mean, the highest variability (16.3 ± 21) is observed in the outdoor usage considering the awareness of the rule as the dependent variable. This also suggests analysis of the specific design features to improve the quality of the stored water so that it can be promoted for regular domestic usage.

CAPACITY, QUALITY, EXPENSE AND SUPPORT

The specific details of the installed RWH systems collected from the survey are very useful in understanding the investment potential of the local residents as well as their perspective on the reliability and financial feasibility. The questions regarding basic usage and location were answered by all the participants, while the questions pertaining to the dimensions, material and components were answered only by 44.0% of the participants.

It is observed that only 16.2% of houses have the overhead tanks for the RWH system while 27.3% of houses have storage tanks at ground level or underground level (Figure 5). This information is closely related to the type of the house as discussed above. However, it may not be possible to assume at this stage that only those who have an overhead tank may use a pumping system for domestic usage, while others may prefer direct access through taps or use for gardening purposes. This is clear based on the

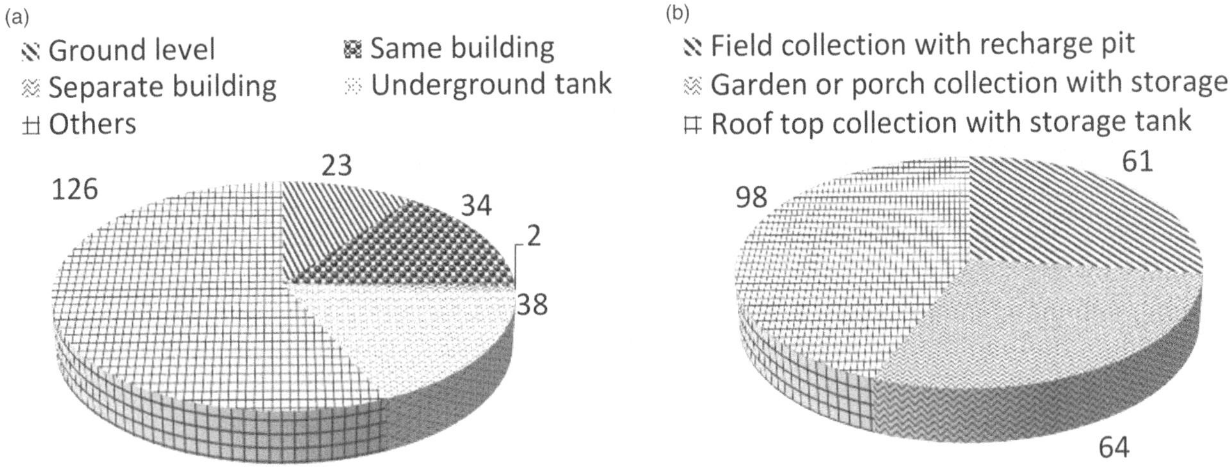

Figure 5 | The distribution of responses regarding the space availability for water storage in terms of (a) location of the storage tank and (b) point of collection of rainwater.

observation of the actual usage of the collected water. The majority of people used the harvested water for domestic purposes either collected from the rooftops (44.0%) or from the porch/garden (28.7%). About 27.4% of people are directly using water for groundwater recharge. It is to be understood that the values mentioned here are normalised for the positive response cases based on the availability of the RWH system, which is 50.5% of the study population.

Some basic features of the general perception of people about the construction of the storage tank were revealed in this survey. The most preferred shape of the storage tank is rectangular (57.2%) and the most preferred material for the construction of the tank is concrete masonry (54.1%) followed by stone masonry (23.5%) (Figure 6). This is based on the size of the tank and the flexibility of operations including cleaning and flushing requirements. However, 35.7% people reported cylindrical geometry, which is clear evidence of the increasing trend towards plastic overhead tanks (17.4%). This particular combination has put us forward towards analysing the data in relation to the quality

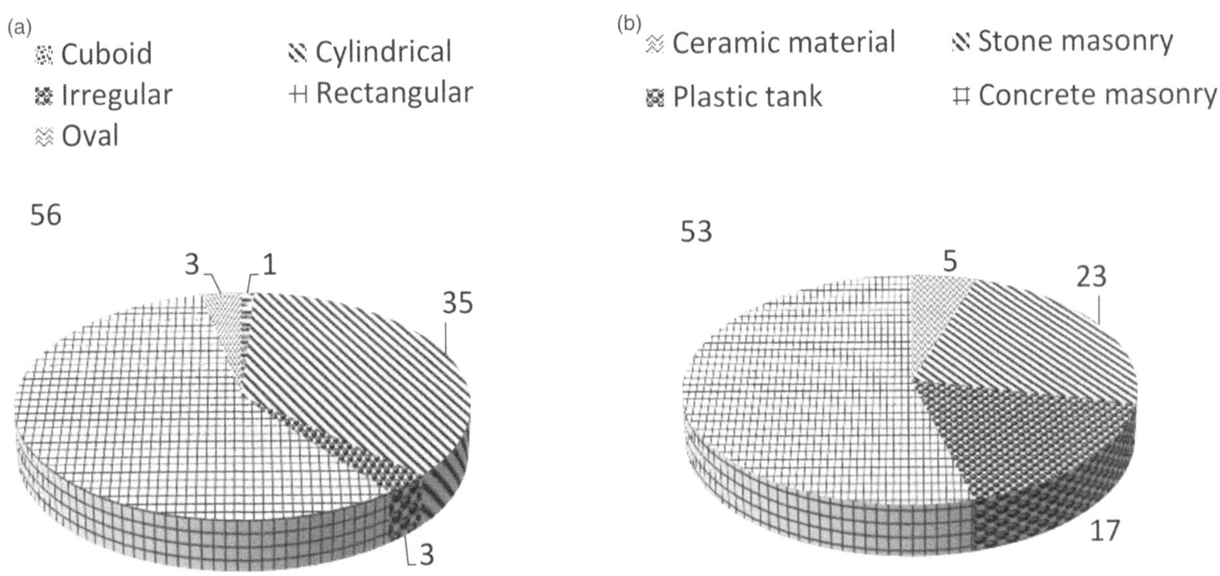

Figure 6 | The distribution of responses regarding types of storage facility available in terms of (a) shape and (b) material.

of the stored water as well as the financial adequacy to select the required type of material and dimensions for the RWH system.

The financial conditions are analysed under the heading 'support', which denotes the awareness and accessibility of various Government schemes to provide incentives for the beneficiaries for promoting RWH systems widely. We compared the installed capacity of the RWH systems and the total investment (including capital cost and operational and maintenance costs) with the awareness of incentives offered by the state and central governments from time to time. The number of houses having a large RWH system (4.0% of the selected group) (defined based on a storage tank capacity which is greater than 2,000 L) have received incentives, while a comparatively larger fraction of people (9.4%) are not aware of such schemes or did not attempt (1.3%) to receive such grants (Figure 7). However, for the small RWH systems (having storage tank capacity less than 2,000 L), about 10.8% of the houses received financial offers, while a large number of people are either not aware of (10.3%) or have not received them (7.6%).

In the case of a total expenditure greater than Rs. 10,000 (INR) for having an RWH system at home, more people (8.1%) enjoyed a financial offer compared with people having higher capacity systems. Since the individual responses are mutually exclusive, we have observed that this contradiction may be due to the difference in the construction method or investments in the materials and additional components such as filter and plumbing units. In contrast, most of these 'expensive' people are reported to be unaware of any financial schemes (10.3%) compared with the same group with larger tanks. Similarly, only 7.2% of the people who have total expenses less than Rs. 10,000 (INR) have accessed financial incentives, compared with the fairly large number of people who were unaware (23.8%) or have not received the financial assistance (14.3%). This particular result points to the specific case of marginal farmers and businessmen who were not effectively informed about the attractive offers by accepting a suitable RWH system at their home. It is also important to note that the standard error on the mean is lowest in this comparison, showing nearly equal distribution between the given options. This definitely proves the diverse nature of the population and hence, the conclusions drawn are to be carefully applied to the whole population.

The other important aspect of the survey was to infer the specifications of the installed RWH system in order to draw some kind of a model design for the selected samples (Table 1). As mentioned before, out of the shortlisted participants, only 43.4% provided the technical specifications of their installed systems. These observations imply a conservative approach to space utility rather than conservation of an invaluable resource at hand. Matching with the previous results, it is clear that most of the houses are single storey,

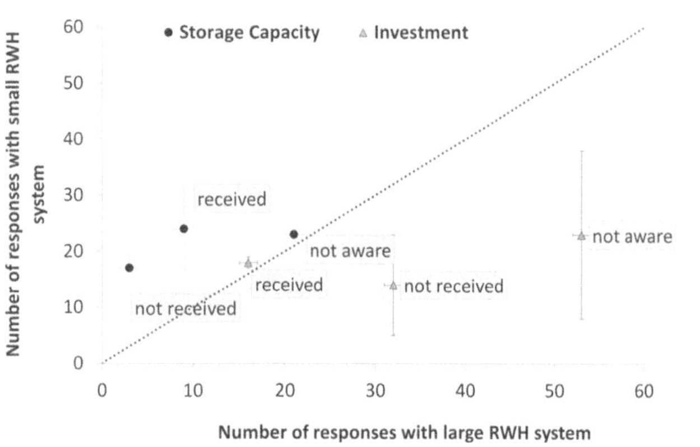

Figure 7 | Comparison of the responses related to the capability of expansion of RWH in combination with the limiting parameters (storage capacity and financial status).

Table 1 | Details of the technical features of storage tank for the harvested water

Storage capacity of the tank	No. of responses	Material of the storage tank	No. of responses
Capacity < 1,000 L	34	Ceramic material	5
Capacity = 1,000 L to 2,000 L	31	Stone masonry	23
Capacity = 2,000 L to 5,000 L	18	Plastic tank	17
Capacity > 5,000 L	15	Concrete masonry	53
Space occupancy on the floor		**Depth of the storage tank**	
Space occupied <25%	33	Depth < 50 cm	12
Space occupied = 25%–50%	21	Depth = 50–100 cm	39
Space occupied > 50%	44	Depth > 100 cm	47

occupied by low-to-medium income groups of people (based on the expenses parameters), and may not have sufficient space to accommodate storage tanks. When it comes to the quality of the stored water, only 21.7% people presented their viewpoints. Most of them agree that stored water is to be treated before direct consumption (69.4% of the responses) either by boiling (32.7%), carbon filter (6.1%) or by RO filter (30.6%).

The cleaning aspects of the RWH are very important to maintaining the required quality of stored water with longer shelf life. It is observed that people are quite aware of the need to flush the first receiving rain without taking it to the storage tank. About 84.7% of people responded that the rainwater collecting area (e.g. rooftop) has been cleaned at least once before taking into storage. About 88.8% of people reported that the storage tank is flushed at least once before the initial storage. Similarly 65.3% of people have cleaned and replaced their filter units at least once. These are some good results indicating the evolution of awareness coming from experience, and not from the regular promotional activities by various agencies through the popular media.

GENDER, TECHNOLOGY AND AWARENESS

Being an educational survey, the responses were further scrutinised for realising the overall attitude and perception of the young generation towards conservation philosophy. In addition, the influence of external features on the thinking behaviour were analysed with respect to gender. The results showed that about 19.4% of the male participants are sensitive towards water quality and are aware of the cleaning schedule and mechanisms, while only 8.2% of the female participants showed interest in water quality and cleaning components of the RWH system. On a similar note, about 19.4% of the male participants showed their interest and knowledge in the technical features of the RWH system, while the positive response from the female participants was only 7.1%. However, with respect to awareness of groundwater recharging or gardening systems, high positive responses were observed from both male (38.8%) and female (22.4%) participants (Figure 8). This could be an indication of the general perception of people of the

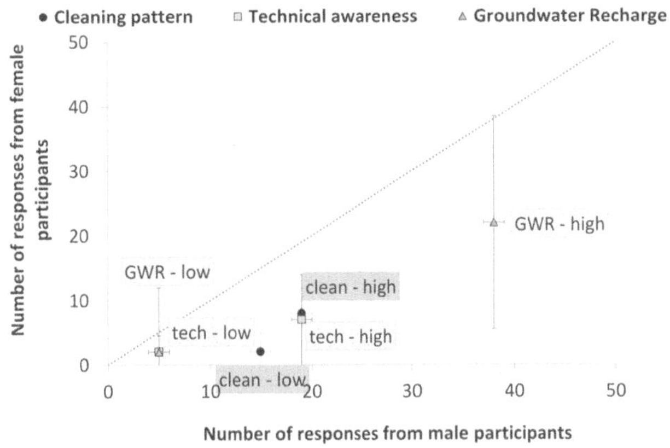

Figure 8 | Comparison of the responses related to the gender variations in combination with the technical awareness parameters (knowledge of technical features, awareness of the cleaning protocol and methods of groundwater recharge).

possible utility of harvested rainwater. Generally, people from agricultural backgrounds are more aware about agricultural RWH, which may have influenced them to choose that response. It also indicates that more comprehensive awareness and promotion are required to extend the activities to make residential GWR successful.

A general observation from the figure indicates comparatively low values of standard error on the mean within the three selected classes, which is of course not truly representative of the actual population. This is because, on average, there is a huge amount of missing data in this part of the questionnaire from both male (44.8%) and female (27.1%) participants.

CHALLENGES IN PROMOTION

Apart from the encouraging indications derived from the responses, there are some critical issues which may be indicative of the existing gap in the success of promotional activities, or can serve as directional information towards future activities. We can generally address them in two categories: quality, and utility and economics. Based on the general psychology of the participants in this survey, we recommend addressing these issues in a more comprehensive way to ensure the success of any promotional activities towards environmental development and resource conservation.

The quality of storage is a major concern for most of the people. Even though many vendors suggest no treatment, people are in general more sceptical about the physical, chemical and biological characteristics of the drinking water. It is to be understood that the entire aspect of water quality in this survey has been derived only based on the physical appearance and taste of the water, and no test results were referred to ensure the quality of the stored water. As mentioned earlier, about 69.4% of the responding people believe in some kind of treatment, either boiling or filtration. This shows that the problem is more psychological and technical. However, there are many reports that water quality can be preserved in completely closed storage tanks, especially for RWH systems. In order to promote RWH in rural and semi-urban areas, the survey suggests that proper test results are to be demonstrated to the public while educating on the general advantages of RWH systems.

The second prominent issue observed in the study was about the awareness of the holistic utility of harvested water. The actual reasoning behind opting for groundwater recharge is not completely understood in this survey, although it is appreciated as a great initiative by the majority of the participants (Figure 9). However, a few cases reported a negative or null effect of recharging due to RWH systems, which may need specific attention to employ other conservation means to improve the efficiency. But in any case, the first and foremost point in advocating about RWH is to educate people that it is the purest form of naturally available water and should be effectively diverted for its prime use, i.e. consumptive use. When groundwater recharge persists in being problematic in some areas by virtue of local hydro-geological conditions, it would be better to assign proper storage provisions to conserve the water for domestic uses.

The next important issue in considering the lack of awareness is, of course, in its financial commitment. It is quite surprising to see that a large number of people still fear that owning an individual RWH system may cause additional financial liability (Figure 10). In contrast to the perception, there are many active schemes that share financial liability by exemption from taxes, providing incentives etc. Although several schemes are promoted by the Government of India from time to time, people are to be educated about the availability of incentives and technical support. Lately, Prime Minister Krishi Sinchayee Yojana (PMKSY) has been promoting throughout India an amalgamation of various previous schemes of similar nature such as the Accelerated Irrigation Benefit Programme (AIBP) of the Ministry of Water Resources, River Development & Ganga Rejuvenation, Integrated Watershed Management Programme (IWMP) of the Department of Land Resources (DoLR) and On Farm Water Management (OFWM) of the Department of Agriculture and Cooperation (DAC). In essence, we feel that an integrated public–private partnership for RWH has to emerge as for waste management or groundwater management. Although they are primarily focusing on agricultural RWH, water is considered as a global element of ecology and deserves all attention for its timely management and preservation.

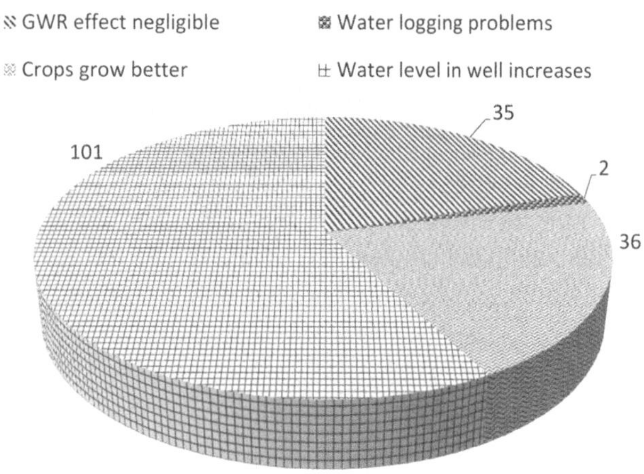

Figure 9 | The distribution of responses regarding the perception on beneficial use of harvested rainwater.

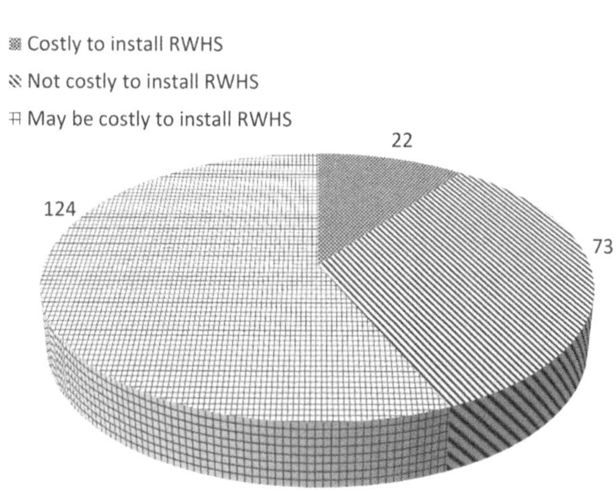

Figure 10 | The distribution of responses regarding the perception of economic liability for installing an individual RWH system in a residential area.

The survey results indicate the growing need to create awareness about RWH without getting a financial burden by accessing various government schemes for technical and financial support. Although 31.9% of the responding people accept that it is not costly, a lion's share (54.2%) of the people think that it may be costly to install individual RWH at the place of residence (Figure 10). In any case, it is important to infer that creating awareness in the young age group of the population has a high degree of acceptance, flexibility, adaptability and replicability, which makes them future water leaders.

CONCLUSION

The present study investigates the socio-economic feasibility and technical awareness of residential rainwater harvesting systems among a group of undergraduate students. A structured questionnaire survey was conducted through the online medium and the responses were analysed for the combinatorial effects of the awareness of the source of water, size of the system, financial support received and the demographic variability and the awareness of technology for operation and maintenance against the perception for improving RWH usage. Some of the salient observations from the study are given briefly here.

- The RWH system is popular where more people are depending on public sources of water for their daily domestic usage (51.6%) than the people who rely on private sources (31.2%).
- The people who live in separate houses have implemented RWH (45.9%) much more than those living in apartments (4.5%).
- Longer experience in using RWH systems (for more than two years) builds confidence in using them for consumptive uses compared to recent adopters. However, most of the people use the harvested rainwater for gardening and groundwater recharge rather than direct consumption.
- Fewer people are aware of the technical details of operation and maintenance of the RWH system (22.2%) while more people are aware of the general layout of the system (50.5%). The awareness of operation and maintenance of the RWH system showed a positive trend with increase in experience, showing that experience speaks better than external promotion.
- A majority of the people use harvested water for domestic purposes either collected from rooftops (44.0%) or from the porch/ garden (28.7%) while about 27.4% people prefer direct groundwater recharge.
- General awareness about financial assistance through incentives vests with people having large RWH systems (defined based on the storage tank capacity as well as the initial investment). However, a larger fraction of the people having small RWH systems received financial assistance which is, however, small compared with the larger systems.
- The awareness about the quality of water is purely radical based on the general belief that water needs to be treated before use (21.7%). However, for the majority of the people, there was no observable decay in the water quality for domestic use (81.6%).
- The difference in gender showed wide differences in the awareness of technicalities of the system where male participants showed higher familiarity (19.4%) than the female participants (7.1%). However, both genders showed high positive responses (38.8% and 22.4% respectively) towards groundwater recharging using harvested rainwater.
- The concern for improved water quality from RWH storage tanks is highly objective as more people (69.4%) prefer there to be treatment before direct consumption. It is suggested that proper test results are to be demonstrated to the public while educating on the general advantages of the RWH systems.
- Advocacy on groundwater recharge has to be approached with caution to improve the priority-wise usage of the harvested rainwater. In essence, we feel that an integrated public–private partnership for RWH has to emerge with due consideration for eco-socio-economic variability.

ACKNOWLEDGEMENT

The authors thank the students, staff and management of their organisations for providing necessary support for the conduct of the study. The authors also thank the participants of the survey who devoted their time for a noble cause.

DATA AVAILABILITY STATEMENT

Data cannot be made publicly available; readers should contact the corresponding author for details.

REFERENCES

Al-Mefleh, N. K., AlAyyash, S. M. & Khaled, F. A. B. 2019 Water management problems and solutions in a residential community of Al-Mafraq city, Jordan. *Water Supply* **19** (5), 1371–1380.

Alsaluli, A., Ahmed, A. & Davies, J. 2015 Public engagement in integrated urban water management in Saudi Arabia: teachers' perceptions in relation to water awareness. *Water Supply* **15** (4), 871–880.

Beal, C., Stewart, A. R., Spinks, A. & Fielding, K. 2011 Using smart meters to identify social and technological impacts on residential water consumption. *Water Supply* **11** (5), 527–533.

Bitterman, P., Tate, E., Van Meter, K. J. & Basu, N. B. 2016 Water security and rainwater harvesting: a conceptual framework and candidate indicators. *Applied Geography* **76**, 75–84.

Chitra, M. & Kantha, B. L. 2017 An analysis on supply of water in Tamil Nadu. *Shanlax International Journal of Economics* **5** (3), 48–54.

Directorate of Town Panchayats 2018 Rain Water Harvesting. The Directorate of Town Panchayats (14 November). Available from https://www.tn.gov.in/dtp/rainwater.htm (accessed 23.06.2020).

Dismas, J., Mulungu, D. M. M. & Mtalo, F. W. 2018 Advancing rainwater harvesting as a strategy to improve water access in Kinondoni municipality, Tanzania. *Water Supply* **18** (3), 745–753.

Farswan, S., Vishwakarma, C. A., Mina, U., Kumar, V. & Mukherjee, S. 2019 Assessment of rainwater harvesting sites in a part of north-west Delhi, India using geomatic tools. *Environmental Earth Sciences* **78**, 329.

Hari, D., Reddy, K. R., Vikas, K., Srinivas, N. & Vikas, G. 2018 Assessment of rainwater harvesting potential using GIS. *IOP Conference Series: Materials Science and Engineering* **330**, 012119.

Hindu Business Line 2018 Tamil Nadu leads in rainwater harvesting. *The Hindu Business Line* (16 July). Available from: https://www.thehindubusinessline.com/specials/india-file/tamil-nadu-leads-in-rainwater-harvesting/article24436406.ece# (accessed 23.6.2020).

Islam, M., Akber, M. A. & Islam, A. 2018 Willingness to pay for improved drinking water in southwest coastal Bangladesh. *Water Supply* **19** (1), 1–10.

Jain, N. & Jain, T. K. 2019 Rainwater harvesting – the case study related to India's Thar Desert. *SSRN*, https://ssrn.com/abstract = 3327743.

Karim, M. R. 2010 Assessment of rainwater harvesting for drinking water supply in Bangladesh. *Water Supply* **10** (2), 243–249.

Kishida, N., Konno, Y., Nemoto, K., Amitani, T., Maki, A., Fujimoto, N. & Akiba, M. 2013 Recent trends in microorganism-related off-flavor problems in drinking water treatment systems in Japan. *Water Supply* **13** (5), 1228–1235.

Kosaka, K., Koike, Y., Miyabayashi, Y., Saito, K., Asami, M., Sasaki, M., Sato, S. & Akiba, M. 2019 National survey of utilisation of continuous water quality monitors in water supply systems in Japan. *Water Supply* **19** (5), 1347–1353.

Mahmood, A. & Hossain, F. 2017 Feasibility of managed domestic rainwater harvesting in South Asian rural areas using remote sensing. *Resources, Conservation and Recycling* **125**, 157–168.

Mohile, A. D. 2008 *Comprehensive Mission Document of National Water Mission, Volume 2*. Ministry of Water Resources (MOWR), Government of India, New Delhi, India. http://mowr.gov.in/sites/default/files/Mission_Doc_Vol22880755143_0.pdf.

News Minute 2019 16 yrs after landmark rainwater harvesting rule in TN, Chennai thirsts for water. *The News Minute* (20 June). Available from: https://www.thenewsminute.com/article/16-yrs-after-landmark-rainwater-harvesting-rule-tn-chennai-thirsts-water-103979 (accessed 23.6.2020).

Nunes, L. G. C. F., Soares, A. E. P., da Silva, J. K. & da Silva, S. R. 2018 Rational water use indicators for public schools in Recife, Brazil. *Water Supply* **19** (1), 207–215.

Proulx, F., Rodriguez, M. J., Sérodes, J. B. & Miranda, L. F. 2010 Factors influencing public perception and use of municipal drinking water. *Water Supply* **10** (3), 472–485.

Raj, S. 2017 Assessing the rainwater harvesting potential of Pallavpuram area in Meerut, Uttar Pradesh. In: *Sustainable Smart Cities in India* (P. Sharma & S. Rajput, eds), Springer, Cham, Switzerland, pp. 199–211.

Rajasekhar, M., Gadhiraju, S. R., Kadam, A. & Bhagat, V. 2020 Identification of groundwater recharge-based potential rainwater harvesting sites for sustainable development of a semiarid region of southern India using geospatial AHP, and SCS-CN approach. *Arabian Journal of Geosciences* **13**, 24.

Ram, H., Rakshith, N. & Birawat, K. K. 2018 A critical review on rainwater harvesting and case study – way ahead to sustainability. *International Journal for Scientific Research and Development* **6** (3), 2114–2119.

First received 30 June 2020; accepted in revised form 24 September 2020. Available online 13 October 2020

Agricultural fertiliser from brewery effluent – the recovery of nutrients from the biomass of activated sludge and high rate algal pond treatment systems

Richard P. Taylor, Clifford L. W. Jones and Richard K. Laubscher

ABSTRACT

The disposal of waste biomass generated from biological wastewater treatment plants is a costly process and poses environmental threats to the receiving environment. This study aimed to determine the suitability of algae and waste activated sludge (WAS) produced from a brewery effluent treatment system as a fertiliser in agriculture. The change in soil characteristics and the growth of a crop fertilised with algae or WAS was compared with a conventional inorganic fertiliser. Swiss chard plants (*Beta vulgaris*) fertilised with anaerobically digested (AD) algae or WAS had a significantly higher mean biweekly yield (5.08 ± 0.73 kg/m^2) when compared with the inorganic fertiliser control (3.45 ± 0.89 kg/m^2; $p < 0.0001$). No difference was observed in the soil's physical fertility when algae or WAS were applied to the soil ($p > 0.05$). The nitrogen applied to the soil from algae and WAS biomass appeared to leach out of the soil less than the nitrogen supplied by inorganic fertilisers. The application of WAS or algae on soil increased the soil's sodium concentration and sodium absorption ratio from 774.80 ± 13.66 mg/kg to 952.17 ± 34.89 mg/kg and 2.91 ± 0.04 to 3.53 ± 0.13, respectively. Regulations on the application of algae or WAS on agricultural soils should be altered to consider the limit values for sodium.

Key words | agriculture, anaerobic digestion, biosolids, brewery effluent, high rate algal ponds, waste activated sludge (WAS)

Richard P. Taylor (corresponding author)
Clifford L. W. Jones
Department of Ichthyology and Fisheries Science, Rhodes University, P.O. Box 94, Grahamstown, 6140, South Africa
E-mail: *richardtaylorwater@gmail.com*

Richard K. Laubscher
Institute for Environmental Biotechnology, Rhodes University, P.O. Box 94, Grahamstown, 6140, South Africa

HIGHLIGHTS

- Swiss chard production was similar when fertilised with fertiliser, waste activated sludge or algae.
- Anaerobically digested waste activated sludge and algae can replace inorganic fertiliser.
- Waste activated sludge and algae increased the soil's sodium absorption ratio.
- Soil application rates of sludge should consider sodium limit values.

INTRODUCTION

Organic and inorganic pollutants are converted into biomass during the treatment of wastewater in biological systems such as high rate algal ponds (HRAP) and activated sludge (AS; Tchobanoglous *et al.* 2003). This biomass (biosolids) is separated from the treated effluent and disposed of. The disposal of waste biomass is a costly process and poses environmental threats if it is not carried out correctly (Hospido *et al.* 2010). Tchobanoglous *et al.* (2003) state that the 'management of the solids and concentrated contaminants removed by effluent treatment has been, and continues to be, one of the most difficult and expensive problems in the field of wastewater engineering'. Practices and technologies need to be

doi: 10.2166/ws.2020.256

developed that allow the sustainable disposal of biomass, especially where the recovery or reuse of resources trapped in the biomass are considered (Singh & Agrawal 2008). Wastewater biosolids are nutrient-rich organic solids which have the potential to be used for fertiliser, soil amendment or energy production (Dobhal & Singhal 2017; Elalami *et al.* 2020).

Biosolids can supplement or replace commercial fertilisers as they contain nitrogen, phosphorus and various micro-nutrients such as copper, zinc, molybdenum, boron, calcium, iron, magnesium, and manganese (Singh & Agrawal 2008; Engida *et al.* 2020). The organic matter content can be used to increase the cation exchange capacity (CEC) of the soil and can aid in amending soils that are carbon-depleted (Engida *et al.* 2020), and thus can potentially improve crop production. For example, the yield of maize, amaranthus, cowpea and crossandra fertilised with sewage sludge was the same as, or higher than, that of plants fertilised with a commercial inorganic fertiliser (Chitdeshwari *et al.* 2002; Engida *et al.* 2020). The yield of barley fertilised with waste activated sludge (WAS) applied to soils at 15 t/ha dry weight was significantly higher than barley fertilised with a commercial inorganic fertiliser (Antolín *et al.* 2005). After four years of annual application of WAS from a sewage treatment facility, the soil's CEC, total organic carbon and available nitrogen increased significantly; however, so did the heavy metal concentration in the harvested grain (Antolín *et al.* 2005).

The majority of studies have reported an improvement in physical properties when WAS is applied to soils (Epstein 1975; Tsadilas *et al.* 1995; Lu *et al.* 2012). The application of WAS to soils increases its organic matter content, which aids in stabilising the soil structure by increasing interparticulate cohesion within aggregates and enhancing their hydrophobicity (Diacono & Montemurro 2010). Municipal solid waste applied to the soil, every two years, increased soil aggregate stability by 29%, thus increasing its resistance to erosion (Annabi *et al.* 2007). The application of sewage sludge to the soil has been shown to improve its water holding capacity, porosity and the bulk density of the soil (Epstein 1975; Ojeda *et al.* 2003).

Anaerobic digestion is a recommended stabilisation step prior to the land disposal of waste sludge as it allows the recovery of carbon into an energy source (biogas) and can increase the availability of nutrients to plants (Tchobanoglous *et al.* 2003). This process results in the conversion of protein-bound nitrogen to ammonia which can be utilised by plants and can increase the nutrient availability of biosolids. Anaerobically digested WAS had higher phosphorus availability than heat-dried WAS (Lyberatos *et al.* 2004). To date, there is only one publication which compares the use of AD to increase the nutrient availability of WAS to plants (Warman & Termeer 2005). Sludge pre-treated using AD resulted in a higher yield of *Zea mays* when compared with sludge that was composted (Warman & Termeer 2005). No other current literature has compared the use of AD to possibly increase the fertiliser value of effluent-grown algae and to document the subsequent effect on soil fertility.

High rate algal ponds and AS are effective brewery effluent treatment technologies which both produce a biomass than can be used in agriculture. The suitability of these biomasses as a fertiliser needs to be assessed and can aid in deciding which technology is favourable in a particular situation. Algae have resilient cell walls which can decrease their decomposition rate and thus reduce the availability of nutrients to plants when compared with WAS (Markou *et al.* 2012). Algal biomass cultured in piggery wastewater significantly increased the growth of wheat when compared with unfertilised treatments (Jenkins *et al.* 2017), however this was a pot experiment and only carried out for four weeks. This is the first study which compares the use of algae and WAS as a fertiliser replacement, where both the algae and the WAS are produced from effluent treatment systems that have been used to treat the same effluent.

Aims and objectives

The aim of this study was to determine the suitability of algae and WAS produced from a brewery effluent treatment system as a fertiliser in agriculture. This was done by comparing the change in soil characteristics and the growth of a crop fertilised with algae or WAS with a conventional inorganic fertiliser. This study will also determine what effect an AD pre-treatment step may have on the fertiliser quality and the fertility of the receiving soil.

The objectives of this study were to:

- compare plant growth and soil characteristics of plots fertilised with WAS, algae or inorganic fertiliser;

- compare plant growth and soil characteristics between soils fertilised with anaerobically digested and non-anaerobically digested algae or WAS; and
- determine the suitability of anaerobically digested and non-anaerobically digested WAS or algal biomass as an inorganic fertiliser replacement.

MATERIALS AND METHODS

Untreated brewery effluent was screened through a 500 μm drum filter (Autrex Industrial Screening, serial no. A 140/02, model no. R 015) to remove solids and then anaerobically digested in an up-flow anaerobic sludge blanket reactor. After AD the effluent was polished in a commercial AS system or a pilot HRAP system.

The majority of the post-AD effluent (800 m^3/d) was treated in an AS system consisting of a rectangular aeration basin and clarifier operated by an independent commercial company (Proxa Pty Ltd, South Africa). The aeration basin was operated at a hydraulic retention time (HRT) between 0.4 and 0.5 days, while the dissolved oxygen concentration was maintained around 0.8 mg/l. Effluent was decanted from the aeration basin into a clarifier operated at an HRT between 0.3 and 0.4 days, which separated the suspended solids from the treated effluent. Settled sludge from the clarifier was also disposed of into the municipal sewage works, and this process was called 'sludge wasting'. This was done to maintain the desired concentration of micro-organisms in the aeration basin of 350–400 ml/l volume of settleables. Sludge wasting was performed once or twice a week depending on the volume of settleables in the aeration basin, and was done when the volume of settleables increased above 450 ml/l and was stopped when it decreased to 300–350 ml/l.

Effluent entering the HRAP had undergone AD and stabilisation in a primary facultative pond (PFP) operated at a HRT of four days. Effluent from the PFP was decanted into two identical parallel HRAP systems, consisting of two ponds in series. The first pond of each system was 25 cm deep, with a surface area of 14.8 m^2 and a volume of 3,700 l. Effluent decanted from the first pond to the second pond, which was 11.5 cm deep, with a surface area of 15 m^2 and a volume of 1,700 l. Each pond had a stainless-steel paddle wheel which continuously stirred the effluent and kept the algal cells in suspension. Paddle wheels moved HRAP effluent at an approximate velocity of 4.15 m/s and 6.10 m/s in ponds 1 and 2 respectively. Both algal systems were fed 1,800 l/d of effluent, which equated to an HRT of three days. The HRAPs were fed continuously during daylight hours (08:00–17:00), while feed stopped during the remaining 15 h. Post-HRAP effluent, from both systems, flowed into two algal settling cones which separated the algal biomass from the treated effluent.

The settled algae and WAS slurries were used directly as a fertiliser or anaerobically digested and then used as a fertiliser in the crop production trial.

TREATMENTS

Five treatments were prepared by adding the following soil amendments/fertilisers to a sandy loam top-soil (10% silt, 20% clay, 70% sand; Macvicar *et al.* 1977): (1) un-digested algae (algae); (2) anaerobically digested algae (AD-algae); (3) un-digested WAS (WAS); (4) anaerobically digested WAS (AD-WAS); and (5) a commercial inorganic fertiliser (inorganic-fertiliser) that served as a reference control (Hygrotech Pty Ltd, South Africa; registration number K5709; Act 36 of 1947). These fertilisers were applied to the soil at a nitrogen application rate equivalent to 80 kg/ha (Laboski & Peters 2012), two days before the soil was planted, and they were mixed into the top five centimetres of the soil.

Two semi-continuously-fed anaerobic digesters were used to digest the settled WAS and algal biomass. Both were made of plastic drums with a total volume of 220 l, a head space of 60 l and an operating volume of 160 l. They were stirred with a submersible mixer (Sobo, WP 400M, South Africa) for five minutes every half an hour, and were situated in a temperature-controlled room (37 °C). The digesters were seeded with sludge obtained from a biogas-producing up-flow anaerobic sludge blanket reactor at Ibhayi Brewery. They were fed once a day following the removal of the equivalent volume of digestate. The settled algae and WAS fed to the digester had a total solids content of 25 g/l. Each digester was fed 9.5 l of either algae or WAS per day resulting in a feeding rate of 1.5–2.0 g of TS/$l_{reactor}$/d and an HRT of 16 d. These two digesters were operated for 60 days, before the digestate was collected and applied to the soil as a fertiliser.

EXPERIMENTAL SPECIES, SYSTEM AND IRRIGATION

Swiss chard (*Beta vulgaris* cv. Fordhoek giant) seedlings were purchased from a commercial nursery (Moorland Seedlings Pty Ltd, Humansdorp, South Africa). They were planted at a density of 16 plants per square metre, in raised beds (1.0×1.0 m, with a soil depth of 0.7 m), that were filled with the amended soils. Each soil amendment treatment was randomly assigned to three raised beds, such that the treatments were replicated three times with a replicate consisting of a single raised bed.

They were irrigated once a day, except during rain or directly thereafter (Laboski & Peters 2012), and received a total of 377 mm of water over the 13-week growth trial (275 mm irrigation and 102 mm of rain). Plants were irrigated via a drip irrigation system receiving a maximum of 5 mm of water per irrigation to minimise leaching (Laboski & Peters 2012).

DATA COLLECTION

A sample of the un-digested and anaerobically digested WAS and algae that were applied to the soil were subject to elemental analysis (inductively coupled plasma mass-spectrometry and X-ray fluorescence) at an independent laboratory (Central Analytical Facilities, Stellenbosch University, South Africa).

At the beginning of the trial, ten plants were randomly taken from the population of seedlings that were used for the experiment, and were weighed to determine the mean wet starting mass (0.1 g). After five weeks, the plants were ready for harvesting, when all the large, fully expanded leaves from each plant were removed and the wet weight weighed (0.1 g). This was repeated every two weeks until the experiment was terminated, after 13 weeks. At the end of the trial, all the above-ground biomass was harvested and wet weight weighed (0.1 g).

At the beginning of the trial three plants were randomly chosen and used for leaf chemical analysis. These plants were not used in the experiment due to the destructive nature of the sampling. At the end of the trial leaves from each plot were collected and used for chemical analysis. All samples were analysed for aluminium, calcium, copper, iron, manganese, magnesium, nitrogen, phosphorus, potassium, sodium and zinc concentrations at an analytical laboratory (inductively coupled plasma mass-spectrometry; Cedara, Department of Agriculture, Kwa-Zulu Natal, South Africa).

Daily temperature and rainfall data were recorded using a rainfall gauge and a thermometer (Hanna, HI 991300, United Kingdom) which were situated next to the experimental area.

Infiltration rate was measured in each replicate at the beginning and end of the trial. Infiltration rates were determined using a 30-cm-diameter ring infiltrometer placed 15 cm into the soil with 12 cm protruding above the surface. Five litres of water were poured into the ring infiltrometer and the time taken for the water to drain into the soil was recorded (ISO 2017). Infiltration rate was then calculated (Equation (1), ISO 2017):

$$\text{Infiltration rate (cm/min)} = (\text{volume of water added/surface area})/\text{time} \quad (1)$$

Air-filled porosity (AFP), bulk density and water holding capacity were measured in each replicate at the beginning and end of the trial (ISO 2017). The apparatus used was a 110 mm plastic pipe with an end cap that had four 3.0 mm holes drilled into it. The pipe was bored into the soil to get an undisturbed soil sample. A gauze was placed over the top of the vessel and submerged in water to just above the surface of the soil, for 60 min. The holes in the bottom were then sealed and the vessel was moved into a tray, where the holes were unblocked. The vessel was left to drain for 30 min and the amount of water collected was measured. Air-filled porosity was calculated using Equation (2) (ISO 2017). Directly after the AFP test the vessel was placed in a drying oven at 105 °C and allowed to dry for a minimum of 24 h, until a constant mass was achieved. Water holding capacity was calculated using Equation (3)

(ISO 2017), and bulk density was then calculated using Equation (4) (ISO 2017).

$$\text{Air filled porosity (\%)} = (\text{volume drained/volume of soil}) \times 100 \quad (2)$$

$$\text{Water holding capacity (\%)} = ((\text{wet weight} - \text{dry weight})/\text{volume}) \times 100 \quad (3)$$

$$\text{Bulk density (g/cm}^3) = \text{dry weight/volume} \quad (4)$$

Soil aggregate stability was measured, in each replicate, at the beginning and end of the experiment using five grams of 2–5 mm aggregates (Le Bissonnais 1996). Samples were placed in distilled water and allowed to stand for ten minutes. The distilled water was then removed with a pipette and the aggregates transferred onto a 0.05 mm sieve which was immersed in ethanol and shaken five times with a gentle regular helical rotation movement. The >0.05 mm aggregates on the sieve were collected, dried at 40 °C, and then gently sieved using a column of six sieves: 2.00, 1.00, 0.50, 0.20, 0.10, and 0.05 mm (Le Bissonnais 1996). The aggregate stability was represented by the mean weight diameter (MWD) of aggregates and was calculated using Equation (5) (Le Bissonnais 1996):

$$\text{Mean weight diameter} = \sum (d \times m)/100 \quad (5)$$

where d was the mean diameter between the two sieves (mm) and m was the weight fraction of aggregates remaining on the sieve (%).

The chemical analysis of the soil was determined at the start of the trial ($n = 5$). At the end of the trial, a sample was taken from the bed of each replicate for soil chemical analysis. Soil samples were sent to a commercial analytical laboratory and analysed for pH, cation exchange capacity (CEC), carbon, calcium, copper, potassium, phosphorus, sodium, magnesium, manganese, nitrogen and zinc (Ambic-2-extractable and KCl-extractable; Cedara, Department of Agriculture, Kwa-Zulu Natal, South Africa). The sodium adsorption ratio (SAR) of the soil was calculated using Equation (6), where sodium, calcium, and magnesium are expressed in milliequivalents per litre (meq/l), obtained from a saturated paste soil extract (Qadir *et al.* 2003):

$$\text{Sodium adsorption ratio } Na^{+} \div \sqrt{\frac{Ca^{2+} + \quad + Mg^{2+}}{2}} \quad (6)$$

Samples of anaerobically digested and un-digested WAS or algae applied to the soil were tested for *Escherichia coli*. Similar analyses were repeated on soil and leaf samples taken from each replicate every four weeks at the Ibhayi Brewery laboratories (IS 17994 method; ISO 2014).

STATISTICAL ANALYSIS

The experimental design allowed for: (1) a multifactor analysis of variance (ANOVA) where the treatments included two soil-amendments (algae and WAS), both of which were either subject to AD or left un-digested; and (2) a one-way ANOVA that included a comparison of the four treatments described above and a fifth inorganic-fertiliser treatment that acted as the reference-control. If no significant interactions were observed between soil-amendment factor and/or AD pre-treatment factor (multifactor ANOVA), then the statistical analysis generated from the one-way ANOVA/Kruskal–Wallis ANOVA was used. All analyses were carried out at $p < 0.05$ and, when differences were found, a Tukey's multiple range analysis was used at $p < 0.05$. Data collected over the course of the trial were compared using multifactor repeated measures or one-way ANOVA or a non-parametric Mauchly's sphericity test, if ANOVA assumptions were not met ($p < 0.05$). All data were checked for equality of variance and for the normal distribution of the residuals using Levene's test and a Shapiro–Wilk plot of the residuals, respectively. If the assumptions were not met, then the data were log or square-root transformed and checked for equal variance and normal distribution of residuals. If the assumptions were still not met, a non-parametric Mann–Whitney U test or a Kruskal–Wallis ANOVA was used to compare the data between treatments. All analyses were performed using a statistical software package (Statistica Version 10, StatSoft Inc., Tulsa, USA). Statistical analysis on pH data was done using hydrogen ion concentration.

RESULTS

WAS and algal biomass applied to the soil were within the heavy metal limit for application to agricultural land (Table 1; DWAF 1997). WAS biomass had a chromium concentration between 243.5 and 223.3 mg/kg, which was below the land application limit. Algal biomass had a four-times higher chromium concentration (1,043.7–1,057.1 mg/kg) than WAS biomass (Table 1). Mercury concentrations in algal and WAS biomasses were below 3.5 mg/l. All results came back negative for the presence of *E. coli* in algae, sludge, soil and on the plant leaves.

Harvested Swiss chard biomass was not influenced by an interaction between fertiliser treatment and time (repeated measures ANOVA, $F_{(12,30)} = 0.85$, $p = 0.6045$) or by an interaction between biomass type (algae vs WAS) and AD pre-treatment (repeated measures multifactor ANOVA, $F_{(3,24)} = 1.54$, $p = 0.2288$). However, there was a significant difference in biomass harvested between fertiliser treatments, with a higher average biomass harvested from algae and WAS fertiliser treatments when compared with the inorganic-fertiliser treatment (Figure 1; repeated measures ANOVA, $F_{(4,10)} = 48.62$, $p < 0.0001$). The mean fortnightly yield from the inorganic-fertiliser treatment was 3.45 ± 0.89 kg/m^2 while algae and WAS treatments had a similar yield of 5.08 ± 0.73 kg/m^2 (Figure 1).

The chemical concentration of plant leaves was not influenced by an interaction between factors of biomass type (WAS vs algal) and AD pre-treatment (multifactor ANOVA, $p > 0.05$). Similarly, no significant differences were observed in leaf elemental concentration between biomass type and AD pre-treatment (multifactor ANOVA, $p > 0.05$). There was also no significant difference in the nitrogen concentration of Swiss chard leaves cultivated under the five fertiliser treatments at the end of the trial (Table 2); although the leaf nitrogen concentration in all the algae and WAS treatments increased significantly from the start to the end of the trial, those of the inorganic-fertiliser treatment did not increase significantly compared with those from those at the start (Table 2; Kruskal–Wallis, $p > 0.65$, $F_{(5,20)} = 11.80$, $p = 0.0376$). The potassium and phosphorus concentrations of Swiss chard plants were similar between all fertiliser treatments (Table 2; Kruskal–Wallis, $p > 0.05$). The iron concentration in leaves from the inorganic-fertiliser treatment was significantly lower than that of leaves from the algae and WAS treatments (Table 2; Kruskal–Wallis, $p > 0.65$, $F_{(5,20)} = 15.06$, $p = 0.0101$). There was no difference in the concentration of iron in the leaves of Swiss chard plants subject to the algae or WAS fertiliser treatments (Table 2). Except for iron, there was no significant difference in all the other nutrient (calcium, magnesium, sodium, zinc, copper, manganese and aluminium)

Table 1 | Elemental composition of WAS, algae and inorganic fertiliser applied to soil and limits for disposal of sludge on agricultural land (DWAF 1997)

Parameter	Limit	Inorganic fertiliser	Algae	AD-algae	WAS	AD-WAS
Nitrogen (g/kg)	N/A	68	53.9	51.3	52.5	50.4
Phosphorus (g/kg)	N/A	42	15.5	13.2	20.5	21.4
Potassium (g/kg)	N/A	208	5.4	6.6	4.6	3.5
Cadmium (mg/kg)	15.7	–	0.6	1.8	2.6	0.9
Cobalt (mg/kg)	100.0	–	6.5	4.0	2.0	2.5
Chromium (Cr^{3+}) (mg/kg)	1,750.0	–	1,043.7	1,057.1	243.5	223.3
Copper (mg/kg)	50.5	22	38.1	41.6	36.3	48.8
Mercury (mg/kg)	10.0	–	<3.5	<3.5	<3.5	<3.5
Nickel (mg/kg)	200.0	–	59.2	63.1	45.4	55.3
Lead (mg/kg)	50.5	–	10.9	19.7	10.3	21.4
Arsenic (mg/kg)	15.0	–	2.0	8.6	3.2	7.6
Selenium (mg/kg)	15.0	–	0.3	1.3	0.8	1.4

Anaerobically digested algae (AD-algae); anaerobically digested WAS (AD-WAS).

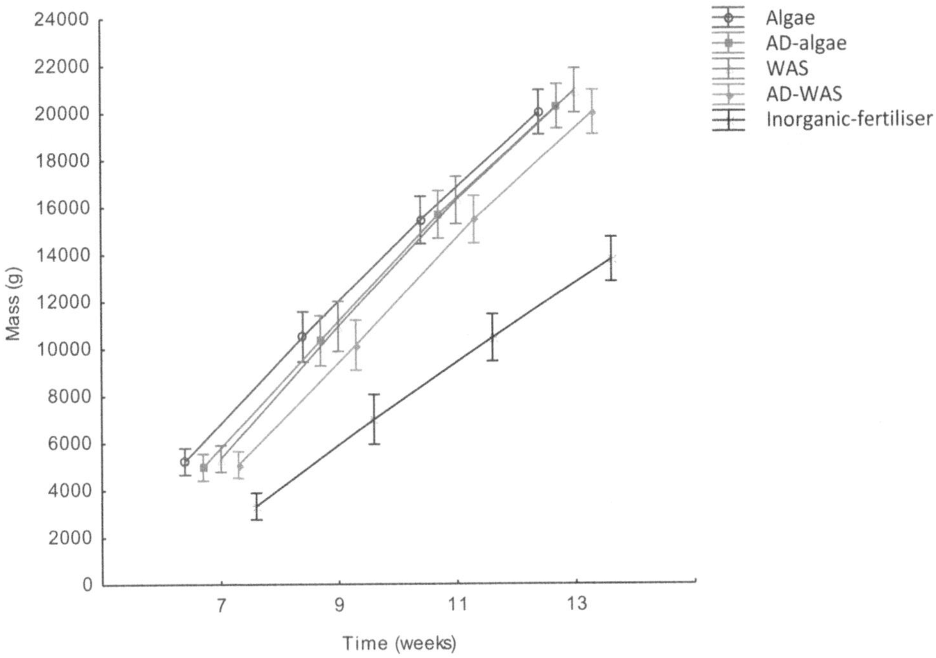

Figure 1 | Cumulative leaf (mean ± 95% confidence interval) biomass harvested fortnightly from Swiss chard plants subject to the experimental fertiliser treatments (repeated measures ANOVA, $F_{(4, 10)} = 48.62$, $p < 0.0001$); anaerobically digested algae (AD-algae), anaerobically digested WAS (AD-WAS).

Table 2 | The mean (±standard error) leaf chemical concentration for Swiss chard plants grown using different fertilisers, after 13 weeks and at the start of the trial

Parameter	Start	Algae	AD-algae	WAS	AD-WAS	Inorganic-fertiliser	*F/H*	*P* value
Nitrogen (g/kg)	39.24 ± 0.55^{a}	50.84 ± 01.04^{b}	51.73 ± 1.45^{b}	50.56 ± 1.65^{b}	52.45 ± 1.67^{b}	41.19 ± 6.82^{ab}	$H = 11.80$	0.0376
Calcium (g/kg)	8.12 ± 0.32^{a}	9.30 ± 00.77^{a}	8.29 ± 0.28^{a}	8.25 ± 0.31^{a}	8.41 ± 0.38^{a}	8.70 ± 0.58^{a}	$F = 0.91$	0.5042
Magnesium (g/kg)	7.50 ± 0.17^{a}	8.78 ± 00.43^{a}	8.64 ± 0.16^{a}	8.78 ± 0.19^{a}	8.77 ± 0.42^{a}	8.44 ± 0.39^{a}	$H = 9.27$	0.0988
Potassium (g/kg)	50.83 ± 0.90^{a}	59.65 ± 02.79^{a}	61.39 ± 4.15^{a}	59.57 ± 0.81^{a}	58.55 ± 2.61^{a}	56.20 ± 2.79^{a}	$H = 8.71$	0.1212
Sodium (g/kg)	28.27 ± 1.76^{a}	30.13 ± 2.00^{a}	30.55 ± 1.55^{a}	30.47 ± 1.44^{a}	30.93 ± 1.47^{a}	31.10 ± 1.74^{a}	$H = 10.87$	0.0541
Zinc (mg/kg)	61.56 ± 1.54^{a}	39.74 ± 05.44^{b}	34.00 ± 0.89^{b}	37.79 ± 2.78^{b}	36.89 ± 2.38^{b}	42.10 ± 4.55^{b}	$H = 12.10$	0.0334
Copper (mg/kg)	52.98 ± 3.21^{a}	7.37 ± 00.91^{b}	6.34 ± 0.64^{b}	6.27 ± 0.52^{b}	6.13 ± 0.60^{b}	6.98 ± 0.97^{b}	$H = 11.89$	0.0363
Manganese (mg/kg)	88.29 ± 2.20^{a}	99.42 ± 12.85^{a}	85.48 ± 3.35^{a}	88.18 ± 5.21^{a}	86.88 ± 6.34^{a}	93.52 ± 5.39^{a}	$H = 1.875$	0.8685
Iron (mg/kg)	0.11 ± 0.10^{a}	0.31 ± 00.03^{b}	0.28 ± 0.03^{b}	0.32 ± 0.03^{b}	0.29 ± 0.03^{b}	0.12 ± 0.02^{a}	$H = 15.06$	0.0101
Phosphorus (g/kg)	8.39 ± 0.18^{a}	2.62 ± 00.17^{b}	2.75 ± 0.21^{b}	2.98 ± 0.09^{b}	2.63 ± 0.27^{b}	2.91 ± 0.23^{b}	$H = 11.82$	0.0374
Aluminium (g/kg)	0.78 ± 0.03^{a}	0.28 ± 00.02^{b}	0.26 ± 0.03^{b}	0.26 ± 0.03^{b}	0.27 ± 0.03^{b}	0.22 ± 0.02^{b}	$H = 12.43$	0.0294

Anaerobically digested algae (AD-algae), anaerobically digested WAS (AD-WAS).
Superscripts (a, b) in the same row represent significantly different treatment means (one-way ANOVA/Kruskal–Wallis, $p < 0.05$).

concentrations in the leaves between all five fertiliser treatments (Table 2; one-way ANOVA/Kruskal–Wallis, $p > 0.60$). The final aluminium, zinc, copper and phosphorus concentrations of the leaves were significantly lower than their starting concentrations, for all fertiliser treatments (Table 2; Kruskal–Wallis, $p < 0.05$).

The physical properties of the soil were not influenced by an interaction between factors of biomass source and AD pre-treatment (multifactor ANOVA, $p > 0.05$). After 13 weeks, there was also no significant difference in the bulk density, porosity, water holding capacity, infiltration rate and mean weight diameter of soils subject to the five fertiliser treatments (Table 3; one-way ANOVA/Kruskal–Wallis, $p > 0.05$). The combined mean porosity, water holding capacity and mean weight diameter of the soils were 33.90% ± 0.71%, 22.57% ± 0.72% and 1.57 ± 0.27 mm, respectively (Table 3).

All measured soil chemical concentrations were not influenced by an interaction between biomass type and AD pre-treatment (multifactor ANOVA, $p > 0.05$). The zinc concentration of the soil was significantly higher in the algae-fertilised soils when compared with WAS-fertilised soils (Table 4; multifactor ANOVA, $F_{(1,8)} = 85.51$, $p = 0.0001$). No other significant difference was observed between soil chemical concentration and biomass source or AD pre-treatment step (multifactor ANOVA, $p > 0.05$).

The soil pH, exchangeable acidity and CEC were similar between all fertiliser treatments (Table 4; one-way ANOVA/Kruskal–Wallis, $p > 0.05$). The soil phosphorus and potassium concentrations were higher in inorganic-fertilised soils compared with algae or WAS fertiliser soils (Table 4; one-way ANOVA, $F_{(5,14)} = 28.35$, $p < 0.0001$). After 13 weeks, the calcium, copper, magnesium and manganese concentrations of the soil were similar between all five fertiliser treatments (Table 4; one-way ANOVA/Kruskal–Wallis, $p > 0.05$). Soil fertilised with inorganic fertiliser had a higher potassium concentration than soils fertilised with algae or WAS (Table 4; one-way ANOVA, $F_{(5,14)} = 4.20$, $p = 0.0156$). The SAR and sodium concentrations were significantly higher in algae and WAS fertiliser soils when compared with soils fertilised with inorganic fertiliser (Table 4; one-way ANOVA, $p < 0.05$). Algae and WAS fertilised soils had a similar sodium concentration and SAR with a combined mean of 952.17 ± 34.89 mg/kg and 3.53 ± 0.13, respectively (Table 4). Inorganic fertiliser and WAS-fertilised soils had a similar zinc concentration (Table 4; one-way ANOVA $p > 0.05$).

At the end of the trial, the carbon (54.46 ± 0.92 g/kg) and nitrogen (3.42 ± 0.05 g/kg) concentrations of soils fertilised with algae or WAS were significantly higher than those of soils fertilised with inorganic fertiliser (C: 50.01 ± 1.43 and N: 3.12 ± 0.06 g/kg; one-way ANOVA, $p < 0.05$). For both elements, the concentrations of soils fertilised with AD-WAS or algae were significantly lower than those of soils fertilised with undigested WAS or algae (multifactor ANOVA, $P < 0.05$; Table 5).

DISCUSSION

Biosolids have the potential to supplement or replace commercial fertilisers as they contain nitrogen, phosphorus and various micro-nutrients needed to support plant growth (Singh & Agrawal 2008). In this study, the yield of Swiss chard plants cultivated in soil fertilised with WAS or algae was significantly higher than the yield from inorganic-fertiliser treatments. Similarly, the yields of various crops fertilised with WAS were similar to crops fertilised with a commercial fertiliser (Chitdeshwari *et al.* 2002; Warman & Termeer 2005; Singh & Agrawal 2008). However, barley fertilised with sewage sludge had a lower yield when compared with inorganic fertiliser, but the addition of inorganic potassium to the sludge increased the yield of barley, indicating that the sludge was deficient in potassium (Miah *et al.* 1999). The present study shows that WAS and algae can be utilised as an inorganic fertiliser replacement when applied to the soil at the same nitrogen loading rate and can even increase crop yield.

Table 3 | The mean (±standard error) starting and final physical characteristics of soils subject to the different fertiliser treatments (one-way ANOVA/Kruskal–Wallis, $p < 0.05$)

Parameter	Start	Algae	AD-algae	WAS	AD-WAS	Inorganic-fertiliser	*F/H* value	*P* value
Bulk density (g/cm^3)	1.42 ± 0.01	1.41 ± 0.06	1.44 ± 0.03	1.35 ± 0.06	1.38 ± 0.03	1.42 ± 0.04	$F = 2.303$	0.0764
Porosity (%)	36.17 ± 0.67	33.48 ± 1.57	34.08 ± 2.19	36.61 ± 1.62	33.75 ± 0.88	31.55 ± 0.72	$F = 1.020$	0.4301
Water holding capacity (%)	22.43 ± 0.68	23.01 ± 0.98	20.49 ± 1.35	22.79 ± 1.65	24.99 ± 2.49	21.58 ± 0.99	$F = 0.947$	0.4691
Infiltration rate (cm/min)	0.80 ± 0.02	0.82 ± 0.11	0.92 ± 0.10	0.95 ± 0.04	0.72 ± 0.03	0.82 ± 0.05	$H = 8.367$	0.2127
Mean weight diameter (mm)	1.56 ± 0.09	1.55 ± 0.04	1.67 ± 0.03	1.49 ± 0.05	1.56 ± 0.06	1.59 ± 0.10	$F = 0.788$	0.5782

Anaerobically digested algae (AD-algae), anaerobically digested WAS (AD-WAS).

Table 4 | The mean (±standard error) chemical characteristics of soils subject to the different fertiliser treatments

Parameter	Start	Algae	AD-algae	WAS	AD-WAS	Inorganic-fertiliser	*F/H* value	*P* value
pH	7.60 ± 00.03^{a}	7.63 ± 00.051	7.53 ± 00.041	7.65 ± 00.031	7.64 ± 00.051	7.57 ± 00.051	$F = 0.67$	0.6502
Exchangeable acidity (cmol(+)/kg)	0.09 ± 00.00^{a}	0.09 ± 00.011	0.09 ± 00.011	0.10 ± 00.011	0.09 ± 00.011	0.10 ± 00.00^{a}	$H = 1.69$	0.8900
CEC (cmol(+)/kg)	26.52 ± 00.34^{a}	27.97 ± 01.251	29.57 ± 03.891	25.90 ± 00.451	28.23 ± 00.791	24.22 ± 00.361	$H = 10.23$	0.0691
Phosphorus (mg/kg)	37.50 ± 01.21^{a}	55.93 ± 03.03^{b}	57.26 ± 03.07^{b}	58.37 ± 03.79^{b}	56.21 ± 02.36^{b}	79.79 ± 03.61^{c}	$F = 28.35$	<0.0001
Potassium (mg/kg)	618.00 ± 17.87^{a}	691.33 ± 52.27^{a}	651.33 ± 19.38^{a}	653.33 ± 3.80^{a}	686.33 ± 32.91^{a}	826.33 ± 50.60^{b}	$F = 4.20$	0.0156
Calcium (g/kg)	4.49 ± 00.04^{a}	4.79 ± 00.08^{a}	4.64 ± 00.071	4.51 ± 00.121	4.75 ± 00.091	4.64 ± 00.151	$F = 1.78$	0.1812
Copper (mg/kg)	1.16 ± 00.08^{a}	1.33 ± 00.12^{a}	1.17 ± 00.071	1.27 ± 00.121	1.43 ± 00.121	1.20 ± 00.061	$F = 1.22$	0.3499
Magnesium (mg/kg)	518.80 ± 17.06^{a}	544.00 ± 16.46^{a}	516.67 ± 22.591	506.67 ± 09.351	515.67 ± 29.541	517.00 ± 18.011	$F = 0.37$	0.8632
Manganese (mg/kg)	30.80 ± 00.97^{a}	51.67 ± 02.91^{b}	47.33 ± 02.19^{b}	48.67 ± 01.45^{b}	46.33 ± 06.36^{b}	53.00 ± 03.51^{b}	$H = 12.28$	0.0311
Sodium (mg/kg)	774.80 ± 13.66^{a}	1,001.67 ± 38.97^{b}	943.67 ± 33.07^{b}	940.67 ± 39.87^{b}	922.67 ± 27.65^{b}	782.00 ± 13.32^{a}	$F = 12.72$	0.0001
Sodium absorption ratio	2.91 ± 00.04^{a}	3.66 ± 00.13^{b}	3.51 ± 00.13^{b}	3.54 ± 00.12^{b}	3.40 ± 00.13^{b}	2.90 ± 00.08^{a}	$F = 11.31$	0.0002
Zinc (mg/kg)	4.16 ± 00.21^{a}	14.83 ± 01.41^{b}	12.53 ± 00.92^{b}	5.10 ± 00.50^{a}	5.53 ± 00.46^{a}	4.10 ± 00.40^{a}	$H = 15.27$	0.0093

Anaerobically digested algae (AD-algae), anaerobically digested WAS (AD-WAS), cation exchange capacity (CEC).
Values in the same row represented by a different superscript symbol (a, b) represent significantly different treatment means (one-way ANOVA/Kruskal–Wallis, $p < 0.05$).

Table 5 | The mean (±standard error) carbon and nitrogen concentrations of soils fertilised with the algae and WAS treatments

Parameter	Algae	AD-algae	WAS	AD-WAS	$F_{(1,8)}$ value	*P* value
Carbon (g/kg)	58.70 ± 1.04[a]	54.82 ± 1.05[b]	57.03 ± 0.81[a]	55.29 ± 0.76[b]	9.221	0.0161
Nitrogen (g/kg)	3.64 ± 0.04[a]	3.47 ± 0.06[b]	3.66 ± 0.04[a]	3.53 ± 0.06[b]	7.569	0.0250

Anaerobically digested algae (AD-algae), anaerobically digested WAS (AD-WAS).
Superscripts (a, b) in the same row represent significantly different treatment means (multifactor ANOVA, $p < 0.05$).

Heavy metals can accumulate in the leaf tissue of plants and render them unsafe for consumption when WAS is used as a fertiliser. Of the heavy metals tested, iron was the only metal present at a higher concentration in Swiss chard leaves from WAS- and algae-fertilised treatments when compared with the inorganic-fertiliser treatment. A single application of WAS to the soil, at 112 t/ha dry weight, increased the copper, zinc, iron and chromium concentrations of snap beans and flax plants (Dowdy *et al.* 1978; Tsakou *et al.* 2002). Land application of sludge has been shown to increase the foliar cadmium, chromium, copper, iron, manganese and zinc concentrations of agricultural crops (Hernández *et al.* 1991; Warman & Termeer 2005; Singh & Agrawal 2008). The acidity of the soil influences the accumulation of heavy metals in crop tissue due to their increase in solubility as pH decreases; therefore it is advisable only to apply waste sludge to soils with a pH above 7.0 (Hernández *et al.* 1991; DWAF 1997). The accumulation of heavy metals in plant tissue varies between plant species, with Swiss chard assimilating more soil iron into its leaves than cabbage, kale, potato, red beet and cauliflower (Itanna 1998; Itanna 2002). The land application of algae and WAS from a brewery effluent treatment process increased the foliar iron concentration, however the concentration was still well within the limits for human consumption. The foliar heavy metal concentration of crops grown in sludge-fertilised soils needs to be continuously monitored if practiced commercially as there is a possibility of heavy metal contamination, especially in acidic soils.

Biosolids have the ability to improve the soil's physical characteristics and are thus commonly used for soil reclamation. There was no difference in the physical properties of the soil receiving algae, WAS or inorganic fertiliser. The literature reports an improvement in the soil's physical properties when sludge is applied (Tsadilas *et al.* 1995; Singh & Agrawal 2008; Diacono & Montemurro 2010). The organic matter added to the soil from WAS improved the bulk density, porosity and water holding capacity of the soil, after one year (Ojeda *et al.* 2003; Diacono & Montemurro 2010). Organic matter concentration of the soil increases when sludge is applied to soil and this aids in stabilising soil structure by increasing inter-particle cohesion within aggregates and increasing their hydrophobicity, thus increasing soil structural stability (Diacono & Montemurro 2010). The improvement in soil structural stability is normally only noticed one to two years after the application of sludge (Sodhi *et al.* 2009; Tejada *et al.* 2009). During this study no differences were noticed in the soil's physical characteristics and structure due to the short period of the trial (13 weeks). However, after prolonged application (>two years) an increase in the soil's physical structure would be expected (Ojeda *et al.* 2003; Tejada *et al.* 2009).

When applying sludge to soils the possibility of contaminating them with heavy metals exists. Of the heavy metals analysed, zinc was the only element that was higher in algae-fertilised soils when compared with soils fertilised with inorganic fertiliser. Various authors have reported an increase in soil iron, copper, zinc, manganese nickel, chromium, cadmium and lead concentrations when WAS is applied to soil (Hernández *et al.* 1991; Singh & Agrawal 2008; Lu *et al.* 2012). Zerzghi *et al.* (2010) applied anaerobically digested WAS to desert soils for 20 years and reported no detrimental effect on the soil's heavy metal content or soil fertility. Soils containing a high clay content decrease the mobility of heavy metal into plant tissue, however once all absorption sites are saturated this benefit is redundant (Wuana & Okieimen 2011). It is important to apply sludge from WWTPs to soil below the permitted application rate of 8.0 t/ha/y and to ensure that its heavy metal content is always within the limits for land application (DWAF 1997; EPA 1999). If this is carried out, in conjunction with regular soil monitoring procedures, the land application of sludge from WWTPs should not have any detrimental effect on the soil's chemical fertility (Zerzghi *et al.* 2010).

Sodium contamination of agricultural soils is the leading cause of rendering soils unsuitable for agriculture (Qadir *et al.* 2003; Muyen *et al.* 2011). In this study, the sodium concentration and SAR of soils fertilised with algae or WAS were significantly higher than those of the soils fertilised with inorganic fertiliser. Karami *et al.* (2012) applied various organic wastes (animal manure, waste sludge and plant derived organic wastes) to soils and, after six months, they recorded a decrease in the soil's SAR. This was due to an increase in soil magnesium and calcium concentration and not a decrease in soil sodium concentration (Karami *et al.* 2012). The land application of animal manures to calcareous soils with a sodium concentration greater than 0.9 g/l resulted in an increase in soil sodium concentrations (Bernal *et al.* 1993). The application of sewage sludge to potting soil that was planted with olive trees resulted in an increase in soil sodium concentrations, SAR and decreased plant growth due to osmotic and sodium stress (Gascó & Lobo 2006). The application of brewery-effluent-grown algae and WAS to land increased the soil's SAR, which could be accompanied by a deterioration in the soil's structural stability and crop yield.

In South Africa there are no sodium guidelines or limits for the disposal of waste sludge on agricultural land. In this study, the soil's sodium concentration increased when WAS or algae was applied to it. Future research should monitor the influence of WAS on the soil's SAR and sodium concentration, as prolonged application could lead to soil salinisation, rendering it unsuitable for agriculture (Gascó & Lobo 2006; Muyen *et al.* 2011). Thus, regulations on the application rates of sludge to agricultural soils should also consider limit values for sodium (not only heavy metals) to avoid the accumulation of sodium in the soil and consequent deterioration in the soil's fertility.

The use of inorganic fertilisers in agriculture results in the leaching of nitrate and ammonia, which can lead to eutrophication of surrounding water bodies (Withers *et al.* 2014). After only 13 weeks, the nitrogen concentration of soils fertilised with algae or WAS was significantly higher than that of soil fertilised with inorganic fertiliser. At the beginning of the trial all fertilisers were applied at the same application rate. Nitrogen is removed from the soil by plant uptake, by leaching into surrounding water bodies or by transformation into a gaseous form (N_2, NH_3, NO, NO_2, N_2O; Withers & Lord 2002; Withers *et al.* 2014). More nitrogen was assimilated into plant biomass in the algae- and WAS-fertilised treatments as plants grown in these treatments had a greater biomass production and similar leaf nitrogen content when compared with inorganic fertilised plants. Therefore, a greater portion of nitrogen must have left the soil via leaching or outgassing to the atmosphere in inorganic-fertiliser treatments when compared with WAS- or algae-fertilised treatments.

The amount of nitrogen leached from soils is dependent on the form of nitrogen supplied, the crop utilisation rate and the irrigation regime, while the mass of nitrogen lost to the atmosphere is influenced by soil temperature, oxygen concentration, moisture and the amount of available organic carbon and nitrogen (Bremner 1997; Signor & Cerri 2013). Nitrogen leaching is predominant in carbon-deficient soils containing more inorganic nitrogen than crops can utilise (Bremner 1997; Signor & Cerri 2013). The nitrogen in the inorganic-fertiliser treatment was present in the form of ammonia and nitrate, whereas in the WAS and algae treatments it was present as ammonia and other organic compounds (Hygrotech Pty Ltd, South Africa). Since ammonia and nitrate are highly soluble in water (Hollister *et al.* 2013), more nitrogen was probably lost to leaching from the inorganic-fertiliser treatment when compared with the algae- and WAS-fertilised treatments where 50–60% of the nitrogen applied to the soil was in an insoluble particulate form. Walsh *et al.* (2012) reported greater nitrate and ammonia leaching from a sandy clay loam fertilised with ammonium nitrate when compared with anaerobically digested sludge. However, Schröder *et al.* (2010) concluded that the extent of nitrogen leaching from grassland soils was not influenced by the nitrogen source (organic vs inorganic), the main factor being balancing nitrogen supply with crop demand. More nitrogen was retained in the WAS- or algae-fertilised soils than in the inorganic-fertiliser soils because all nitrogen supplied by the inorganic fertiliser was in a soluble form (ammonia and nitrate) while 50–99% of the nitrogen supplied by the WAS and algae was in an insoluble form.

Sludge-derived organic fertiliser can be advantageous over inorganic fertilisers as a portion of its nitrogen is in an insoluble form and only becomes available to plants after microbiological decay in the soil, thus resulting in less nitrogen being leached into surrounding water bodies (Walsh *et al.* 2012; Withers *et al.* 2014). Future studies should determine the effect wastewater derived sludge has on the metabolic community structure and activity of a soil and collect and analyse the leachate from the soil to determine the amount of nitrogen lost via leaching and to the atmosphere.

During AD, insoluble particulate bound nitrogen is converted to ammonia, which can be utilised by plants (Tchobanoglous *et al.* 2003). There was no difference in the crop yield and chemical leaf concentration of Swiss chard plants grown in soils fertilised with anaerobically digested or non-anaerobically digested algal or WAS biomass. Heat-dried WAS that was previously subject to AD had a higher concentration of plant available inorganic phosphorus than non-anaerobically digested heat-dried WAS (Lyberatos *et al.* 2004). The yield of *Zea mays* from sludge fertiliser treatments was higher in soils fertilised with AD-WAS when compared with soils fertilised with untreated WAS (Warman & Termeer 2005). The application of sewage sludge increased the soil's microbial activity, respiration and enzyme activities, indicating the soil's microbes are able to degrade the sludge into constituents that are available for plant uptake (Maiti *et al.* 1992). The microbial community present in the soil was able to degrade insoluble particulate bound nutrients into soluble plant-available nutrients, resulting in no difference in crop yield between soils fertilised with anaerobically digested or non-anaerobically digested biomass (Antolín *et al.* 2005).

Anaerobic digestion is able to recover a portion of the carbon in waste biomass into a fuel source (methane) before it is applied to the soil. The carbon concentration of soils fertilised with AD sludge or algae was significantly lower than soils fertilised with undigested sludge or algae. During AD the carbon content of organic matter is converted into methane and carbon dioxide, thus decreasing its carbon concentration (Tchobanoglous *et al.* 2003). Anaerobic digestion is a recommended pre-treatment step prior to the land disposal of waste sludge as the process can decrease its odour and pathogen count, as well as the greenhouse gas emissions from further decomposition in the soil (EPA 1999; Warman & Termeer 2005; Singh & Agrawal 2008). The cost of construction, operation and maintenance must be weighed up against the benefit of biogas recovery for an anaerobic pre-treatment step to be viable. The AD stabilisation step is beneficial as it allows the recovery of energy from waste solids, does not have any negative effects on plant growth or soil fertility and decreases the amount of carbon applied to the soil.

CONCLUSION

These waste products can be utilised as an inorganic fertiliser replacement when applied to the soil at the same nitrogen loading rate and can even increase crop yield. The nitrogen applied to the soil from algal and WAS biomass appeared to leach out of the soil less than the nitrogen (ammonia and nitrate) supplied by inorganic fertilisers. These biomasses offer a good alternative to inorganic fertilisers as the slow nutrient release will aid in reducing the nutrient contamination conventional inorganic fertiliser-based agriculture has on surrounding water bodies. Anaerobic digestion is a viable pre-treatment step for sludge and algal biomasses as it recovers a portion of the carbon as an energy source and does not decrease their quality as fertilisers. The application of algae generated from a brewery effluent treatment system increased the soil sodium concentration, so it is recommended that regulations on the application rate of sludge to agricultural land should consider the limit values for sodium to avoid its accumulation in the soil and deterioration in soil fertility.

ACKNOWLEDGEMENTS

The financial support from the Department of Science and Technology in South Africa is acknowledged (DST/CON 0268/2015; DST/CON 0196/2017). Thanks are also due to Ibhayi Brewery (SAB Ltd) and its staff for their in-kind financial support and assistance throughout the project.

DATA AVAILABILITY STATEMENT

All relevant data are available from an online repository or repositories: https://drive.google.com/drive/u/0/folders/11cQMEncGsvjlfwW-PftJTR6uER17kGCq.

REFERENCES

Annabi, M., Houot, S., Francou, C., Poitenaud, M. & Le Bissonais, Y. 2007 Soil aggregate stability improvement with urban composts of different maturities. *Soil Science Society of American Journal* **71**, 413–423.

Antolín, M. C., Pascaul, I., García, C., Polo, A. & Sánchez-Díaz, M. 2005 Growth, yield and solute content of barley in soils treated with sewage sludge under semiarid Mediterranean conditions. *Field Crops Research* **94**, 224–237.

Bernal, M. P., Roig, A. & García, D. 1993 Nutrient balances in calcareous soils after application of different rates of pig slurry. *Soil Use and Management* **9**, 9–14.

Bremner, J. M. 1997 Sources of nitrous oxide in soils. *Nutrient Cycling in Agroecosystems* **49**, 7–16.

Chitdeshwari, T., Savithri, P. & Mahimairaja, S. 2002 Effect of sewage biosolid composts on the yield of crops. *Indian Journal of Environmental Protection* **21**, 911–912.

Diacono, M. & Montemurro, F. 2010 Long-term effects of organic amendments on soil fertility: a review. *Agronomy for Sustainable Development* **30**, 401–422.

Dobhal, P. & Singhal, P. 2017 Studies on influence of blue green algae as biofertilizers on *Lycopersicon esculentum* (tomato) plants. *Research Journal of Agriculture and Forestry Sciences* **5** (5), 14–16.

Dowdy, R. H., Larson, W. E., Titrud, J. M. & Latterell, J. J. 1978 Growth and metal uptake of snap beans grown on sewage sludge-amended soil: a four-year study. *Journal of Environmental Quality* **7**, 252–257.

DWAF 1997 *Permissible Utilisation and Disposal of Sewage Sludge.* Department of Water Affairs and Forestry, Pretoria, South Africa.

Elalami, D., Monlau, F., Carrere, H., Abdelouahdi, K., Charbonnel, C., Oukarroum, A., Zeroual, Y. & Barakat, A. 2020 Evaluation of agronomic properties of digestate from macroalgal residues anaerobic digestion: impact of pretreatment and co-digestion with waste activated sludge. *Waste Management* **108**, 127–136.

Engida, T., Mekonnen, A., Wu, J. M., Xu, D. & Wu, Z. B. 2020 Review paper on beverage agro-industrial wastewater treatment plant bio-sludge for fertilizer potential in Ethiopia. *Applied Ecology and Environmental Research* **18**, 33–57.

EPA 1999 *Biosolids Generation, Use, and Disposal in the United States*, EPA530-R-99-009 United States Environmental Protection Agency (EPA), Washington, DC, USA.

Epstein, E. 1975 Effect of sewage sludge on some soil physical properties. *Journal of Environmental Quality* **4**, 139–142.

Gascó, G. & Lobo, M. C. 2006 Composition of a Spanish sewage sludge and effects on treated soil and olive trees. *Waste Management* **27**, 1494–1500.

Hernández, T., Moreno, J. I. & Costa, F. 1991 Influence of sewage sludge application on crop yields and heavy metal availability. *Soil Science and Plant Nutrition* **37**, 201–210.

Hollister, C. C., Bisogni, J. J. & Lehmann, J. 2013 Ammonium, nitrate, and phosphate sorption to and solute leaching from biochars prepared from corn stover (*Zea mays* L.) and oak wood (*Quercus* spp.). *Journal of Environmental Quality* **42**, 137–144.

Hospido, A., Carballa, M., Moreira, M., Omil, F., Lema, J. M. & Feijoo, G. 2010 Environmental assessment of anaerobically digested sludge reuse in agriculture: potential impacts of emerging micropollutants. *Water Research* **44**, 3225–3233.

ISO 2014 *Water Quality – Criteria for the Establishment of Equivalency between Microbiological Methods*, ISO 17994. International Organisation for Standardization, Geneva, Switzerland.

ISO 2017 *Physical Properties of Soils*, ISO/TC 190. International Organization for Standardization, Geneva, Switzerland.

Itanna, F. 1998 Metal concentrations of some vegetables grown with industrial liquid waste at Akaki, Ethiopia. *SINET Ethiopian Journal of Science* **21**, 133–144.

Itanna, F. 2002 Metals in leafy vegetables grown in Addis Ababa and toxicological implications. *Ethiopian Journal of Health Development* **16**, 295–302.

Jenkins, S. N., Paredes-Valdivia, C. A., Waite, I. S., Ayre, J., Moheimani, N., Barton, L. & Abbott, L. K. 2017 Assessing the suitability of microalgae biomass produced from piggery waste as a fertiliser. *Animal Production Science* **57** (12), 2480–2480.

Karami, A., Homaee, M., Afzalinia, S., Ruhipour, H. & Basirat, S. 2012 Organic resource management: impacts on soil aggregate stability and other soil physico-chemical properties. *Agriculture, Ecosystems and Environment* **148**, 22–28.

Laboski, C. A. M. & Peters, J. B. 2012 *Nutrient Application Guidelines for Field, Vegetable, and Fruit Crops in Wisconsin.* Cooperative Extension Publishing, Madison, WI, USA.

Le Bissonnais, Y. 1996 Aggregate stability and assessment of soil crustability and erodibility: I. Theory and methodology. *European Journal of Soil Science* **47**, 425–437.

Lu, Q., He, Z. L. & Stoffella, P. J. 2012 Land application of biosolids in the USA: a review. *Applied and Environmental Soil Science* **2012**, 201462.

Lyberatos, G., Klimantos, P., Vassilakis, I. & Kornaros, M. 2004 On the impact of land application of secondary sludge. *Water, Air and Soil Pollution: Focus* **4**, 349–357.

Macvicar, C. N., De Villiers, J. M., Loxton, R. F., Verster, E., Lambrechts, J. J. N., Merryweather, F. R., Le Roux, J., Van Rooyen, T. H. & Harmse, H. J. 1977 *Soil Classification: A Binomial System for South Africa.* Department of Agricultural Technical Services, Pretoria, South Africa.

Maiti, P. S., Sah, K. D., Gupta, S. K. & Banerjee, S. K. 1992 Evaluation of sewage sludge as a source of irrigation and manure. *Journal of the Indian Society of Soil Science* **40**, 168–172.

Markou, G., Angelidaki, I. & Georgakakis, D. 2012 Microalgal carbohydrates: an overview of the factors influencing carbohydrates production, and of main bioconversion technologies for production of biofuels. *Applied Microbiology and Biotechnology* **96**, 631–645.

Miah, M. Y., Chiu, C.-Y., Hayashi, H. & Chino, M. 1999 Barley growth in response to potassium fertilization of soil with long term application of sewage sludge. *Soil Science and Plant Nutrition* **45**, 499–504.

Muyen, Z., Moore, G. A. & Wrigley, R. J. 2011 Soil salinity and sodicity effects of wastewater irrigation in South East Australia. *Agricultural Water Management* **99**, 33–41.

Ojeda, G., Alcañiz, J. M. & Ortiz, O. 2003 Runoff and losses by erosion in soils amended with sewage sludge. *Land Degradation and Development* **14**, 563–573.

Qadir, M., Steffens, D., Yan, F. & Schubert, S. 2003 Sodium removal from a calcareous saline–sodic soil through leaching and plant uptake during phytoremediation. *Land Degradation and Development* **14**, 301–307.

Schröder, J. J., Assinck, F. B. T., Uenk, D. & Velthof, G. L. 2010 Nitrate leaching from cut grassland as affected by the substitution of slurry with nitrogen mineral fertilizer on two soil types. *Grass and Forage Science* **65**, 49–57.

Signor, D. & Cerri, C. E. P. 2013 Nitrous oxide emissions in agricultural soils: a review. *Pesquisa Agropecuária Tropical* **43**, 322–338.

Singh, R. P. & Agrawal, M. 2008 Potential benefits and risks of land application of sewage sludge. *Waste Management* **28**, 347–358.

Sodhi, G. P. S., Beri, V. & Benbi, D. K. 2009 Soil aggregation and distribution of carbon and nitrogen in different fractions under long-term application of compost in rice–wheat system. *Soil Tillage Research* **103**, 412–418.

Tchobanoglous, G., Burton, F. L. & Stensel, D. H. 2003 *Wastewater Engineering: Treatment and Reuse*. McGraw-Hill, New York, USA.

Tejada, M., Hernandez, M. T. & Garcia, C. 2009 Soil restoration using composted plant residues: effects on soil properties. *Soil Tillage Research* **102**, 109–117.

Tsadilas, C. D., Matsi, T., Barbayiannis, N. & Dimoyiannis, D. 1995 Influence of sewage sludge application on soil properties and on the distribution and availability of heavy metal fractions. *Communications in Soil Science and Plant Analysis* **26**, 2603–2619.

Tsakou, A., Roulia, M. & Christodoulakis, N. S. 2002 Growth of flax plants (*Linum usitatissimum*) as affected by water and sludge from a sewage treatment plant. *Bulletin of Environmental Contamination and Toxicology* **68**, 56–63.

Walsh, J. J., Jones, D. L., Edwards-Jones, G. & Williams, A. P. 2012 Replacing inorganic fertilizer with anaerobic digestate may maintain agricultural productivity at less environmental cost. *Journal of Plant Nutrition and Soil Science* **175**, 840–845.

Warman, P. R. & Termeer, W. C. 2005 Evaluation of sewage sludge, septic waste and sludge compost applications to corn and forage: Ca, Mg, S, Fe, Mn, Cu, Zn, and B content of crops and soils. *Bioresource Technology* **96**, 1029–1038.

Withers, P. J. A. & Lord, E. I. 2002 Agricultural nutrient inputs to rivers and groundwaters in the UK: policy, environmental management and research needs. *Science of the Total Environment* **282–283**, 9–24.

Withers, P. J. A., Neal, C., Jarvie, H. P. & Doody, D. G. 2014 Agriculture and eutrophication: where do we go from here? *Sustainability* **6**, 5853–5875.

Wuana, R. A. & Okieimen, F. E. 2011 Heavy metals in contaminated soils: a review of sources, chemistry, risks and best available strategies for remediation. *ISRN Ecology* **2011**, 402647.

Zerzghi, H., Brooks, J. P., Gerba, C. P. & Pepper, I. L. 2010 Influence of long-term land application of Class B biosolids on soil bacterial diversity. *Journal of Applied Microbiology* **109**, 698–706.

First received 27 July 2020; accepted in revised form 24 September 2020. Available online 13 October 2020

Cultivation of *Chlorella vulgaris* in anaerobically digested gelatin industry wastewater

G. C. Blanco, M. J. Stablein and G. Tommaso

ABSTRACT

This work aimed to study the effect of using anaerobically digested gelatin industry wastewater as a culture medium for the cultivation of *Chlorella vulgaris* microalgae in bubble column photobioreactors (PBRs). Batch experiments were carried out to determine the growth kinetics by inoculating microalgae in wastewater prepared with different dilutions and supplemented with Bold's Basal Medium (BBM). From the values of the saturation constants ($K_S = 50.25$ mgN-$NH_4^+ \cdot L^{-1}$) and substrate inhibition ($K_I = 28.12$ mgN-$NH_4^+ \cdot L^{-1}$) obtained in the adjustment to the Andrews kinetic model ($R^2 = 0.817$), the PBRs achieved specific maximum growth rates (μ_{max}) of 0.343 d^{-1}, biomass productivity of 0.141 $g \cdot L^{-1} \cdot d^{-1}$, lipid content of 12.45%, lipid productivity of 17.63 $mg \cdot L^{-1} \cdot d^{-1}$ and instantaneous ammoniacal nitrogen consumption rates of 20.06 and 14.22 $mg \cdot L^{-1} \cdot d^{-1}$. The addition of wastewater to the culture medium provided an increase in biomass productivity of 57.45% in relation to the negative control. The results obtained demonstrate the high efficiency of *C. vulgaris* in the removal of nitrogenous compounds and the potential of using anaerobically digested gelatin industry wastewater in the production of microalgae biomass.

Key words | biomass production, *Chlorella vulgaris*, gelatin industry, kinetic modelling, lipids, wastewater post-treatment

G. C. Blanco
G. Tommaso (corresponding author)
Laboratory of Environmental Biotechnology, Department of Food Engineering, University of São Paulo, 225, Duque de Caxias Norte, Pirassununga, São Paulo 13635-900, Brazil
E-mail: *tommaso@usp.br*

M. J. Stablein
Department of Agricultural and Biological Engineering, University of Illinois at Urbana-Champaign, 1304 W Pennsylvania Avenue, Urbana, IL 61801, USA

HIGHLIGHTS

- The use of wastewater in bubble column photobioreactors (PBRs) provided a 57.45% increase in biomass productivity (PB).
- Andrews model successfully predicted the growth kinetics of *C. vulgaris* in wastewater ($R^2 = 0.817$).
- Ammoniacal nitrogen (N-NH_X) was preferred compared with nitrate (N-NO_3) during the operation of photobioreactors.
- The first-order decay model with residual successfully predicted the kinetics of nitrate (N-NO_3) consumption.
- Two first-order decay models with residuals in series better represented the ammoniacal nitrogen (N-NH_X) depletion.

INTRODUCTION

Contamination of water bodies from consumption of natural resources and unsustainable practices is of increasing importance. Advanced wastewater treatment processes that can simultaneously replenish these resources and minimize environmental contamination are essential to our society. Conventional anaerobic methods aimed at removing carbonaceous material are not effective in removing nutrients, therefore, there is a need in many cases to install post-treatment units (Mai *et al.* 2018). In this context, the cultivation of microalgae has been shown to be an alternative to tertiary treatment, since, in

doi: 10.2166/ws.2020.263

addition to the ability to absorb potentially toxic nutrients and metals present in aquatic environments, they accumulate lipids, produce oxygen, and capture the carbon dioxide present in the atmosphere. The yields of these products vary according to the species and can be induced by system disturbances, resulting in cellular stress, as well as through the adaptation of certain cultivation strategies (Tan & Lee 2016).

Despite the low nitrogen content in the composition of microalgal biomass, ranging between 1% and 10%, this element is essential in the regulation of metabolic pathways and is the most critical nutrient for the growth of microalgae after carbon (Pistorius *et al.* 1978). Some algae assimilate inorganic nitrogen in different forms such as nitrate ($N\text{-}NO_3^-$), nitrite ($N\text{-}NO_2^-$), and ammoniacal nitrogen ($N\text{-}NH_x$). During the assimilation process, the inorganic nitrogen compounds pass through the cytoplasmic membrane where enzymatic reductions occur to form ammonium ($N\text{-}NH_4^+$), which is later converted to organic compounds, such as the necessary peptides, amino acids, proteins, enzymes, and nucleic acids for the growth of biomass (Li *et al.* 2019).

The assimilation of ammoniacal nitrogen requires less energy, and it is the preferred nitrogen source of microalgae (Maestrini *et al.* 1986). According to Li *et al.* (2019), if ammoniacal nitrogen, nitrate, or nitrite are present in the culture medium, the microalgae will consume ammoniacal nitrogen first, then nitrite, and finally nitrate. However, the excess of ammoniacal nitrogen in the culture medium presents negative aspects such as volatilization and the inhibition of microbial growth, and thus, nitrate becomes the most used form for the cultivation of microalgae, as it is chemically stable in aquatic environments (Florencio & Vega 1983).

Several studies have cultivated microalgae using anaerobically treated wastewater from the food industry; however, few studies have addressed the inhibitory effect of ammoniacal nitrogen and the kinetic modelling of microbial growth based on consumption of nitrogen compounds in bubble column photobioreactors (PBRs) when fed with this type of wastewater. The wastewater produced by the gelatin industry has high loads of proteins, sulphides, and organic nitrogen. Wastewater of this type, when treated by anaerobic reactors, presents wastewater with high concentrations of ammoniacal nitrogen ($N\text{-}NH_x$), and this is due to the degradation of proteins present in the organic material (Ramsay & Pullammanappallil 2001). During the degradation of protein-rich materials, proteins are hydrolyzed to peptides and subsequently to amino acids that are rapidly fermented by fermentative acidogenic bacteria and converted into short-chain volatile acids, carbon dioxide, and ammoniacal nitrogen. Each type of wastewater generated in food processing leads to specific considerations for the choice of treatment technology employed (Janosz Rajczyk 1993).

Kinetic models describe the growth of microorganisms and the rate of substrate consumption during operation, providing relevant data for the design of high-efficiency photobioreactors (PBRs). According Eze *et al.* (2018), microalgae growth and the rate of algae biomass accumulation with constant light intensity, temperature, and homogeneous mixing is typically determined by substrate and nutrient availability. Thus, a robust kinetic model is essential to predict substrate consumption, cell growth, and optimize operating conditions for the cultivation of microalgae in wastewater. Several mathematical models have been adjusted to try to describe the growth of microorganisms in PBRs. The kinetic models normally applied in studies of microalgae specific growth rates (μ), using one of the three main substrates (inorganic carbon, nitrogen, or phosphorus) as functions, are the Monod and Andrews models (Chojnacka & Marquez-Rocha 2004). Generally, the growth rates of microorganisms (μ) in the Monod models are limited by the substrate availability, whereas the Andrews model growth rates are limited by the excess of substrate in the culture medium that inhibits cell growth (Eze *et al.* 2018). Therefore, the concentration of nutrients can determine the maximum specific growth rate of microalgae (μ_{max}). In this sense, this work seeks to evaluate the viability of cultivating microalgae species *Chlorella vulgaris* in anaerobically digested gelatin industry wastewater in order to determine the kinetic parameters of microbial growth and substrate consumption, the biomass productivity, the efficiency of the removal of nitrogen compounds ($N\text{-}NH_x$ and $N\text{-}NO_3^-$), and the lipid productivity.

METHODS

Inoculum

The inoculum of the microalgae *Chlorella vulgaris* clone BMAK D1 was provided by the Microalgae Engineering

Laboratory of the University of São Paulo of Lorena School of Engineering (EEL/USP). The culture was maintained and propagated under aseptic conditions using 250 mL conical flasks in the Environmental Biotechnology Laboratory of the University of São Paulo of Animal Sciences (FZEA/USP) culture collection in modified Bold's Basal Medium (BBM). The inoculum was kept under controlled temperature (26 ± 2 °C) and photoperiod (24/0 hours, light/dark) conditions with an estimated light intensity of 4,600 lumens per square metre (lm/m^2) (Zorn *et al.* 2017).

Anaerobically digested gelatin industry wastewater

The wastewater used in this work came from the UASB (Upflow Anaerobic Sludge Blanket) reactor of the Gelita Brasil Ltda industry located in Mococa, SP. The collection took place in July 2019 and was performed in triplicate. The collected wastewater was stored in 5.0 L gallons and frozen. Analysis of crude and filtered COD, total Kjeldahl nitrogen (N-NTK), ammoniacal nitrogen ($N\text{-}NH_x$), nitrite ($N\text{-}NO_2^-$), nitrate ($N\text{-}NO_3^+$), organic nitrogen ($N\text{-}N_{org}$), total nitrogen ($N\text{-}N_T$), total phosphorus (P_T), and sulphates (SO_4^{2-}) were performed according to the methods described in *Standard Methods for the Examination of Water and Wastewater* (APHA 1998), while the dissolved oxygen (DO) analysis was performed using a branded probe (SOP model YSI 55). The physicochemical characterization of wastewater was done in triplicate.

In order to control the displacement of the chemical balance of free ammonia ($N\text{-}NH_3$) to ammonium ion ($N\text{-}NH_4^+$) and hydrogen sulphide (H_2S) to hydrogen sulphide (HS^-), the pH of the wastewater was adjusted to 7.5. The adjustment was made by adding aliquots of dilute sulfuric acid (0.05 M) and quantifying the pH values by means of a properly calibrated benchtop digital pH meter (Tecnal model TEC-03MP). In order to prevent light from being blocked due to the high solids concentration and turbidity of the wastewater, the effluent was filtered through a filter system composed of a Kitasato flask and a Büchner funnel with 0.45 μm filter membrane (Unifil ETQ) before the inoculation.

After filtration of the wastewater, the sterilization was carried out by adding $1.0\ mL{\cdot}L^{-1}$ of sodium hypochlorite (5%). The solution was manually homogenized with a glass rod and allowed to stand for 12 hours for complete elimination of microorganisms (species control). A higher concentration of sodium hypochlorite than conventional sterilization methods was used to promote irreversible oxidation of sulphides. After this period, an indicator test was performed to verify the absence of chlorine in the solution. In cases where residual chlorine was observed, the solution was neutralized with sodium thiosulfate in order to prevent possible interference in the growth of microalgae.

Finally, the wastewater was supplemented with modified Bold's Basal Medium (BBM) to optimize the growth of microorganisms and to supply the nutritional deficiencies of the effluent. Table 1 presents the physicochemical characterization of the anaerobically digested gelatin industry wastewater and the composition of the basal medium used in the supplementation. The wastewater physicochemical parameters were determined experimentally and are presented as averages with Standard Deviation (SD), while the modified BBM is presented with standard composition values.

Calibration curve

The microalgae growth was monitored based on a calibration curve ($R^2 = 0.9996$) designed to convert the

Table 1 | Physicochemical characterization of the wastewater collected in the UASB reactor from the gelatin industry and theoretical composition of the modified Bold's Basal Medium (BBM)

Parameters of modified Bold's Basal Medium	Composition ($mg{\cdot}L^{-1}$)	Parameters of wastewater	Composition ($mg{\cdot}L^{-1}$)
$NaNO_3$	750	DQO_{Crude}	170.65 ± 3.61
K_2HPO_4	75	$DQO_{Filtered}$	122.20 ± 5.66
KH_2PO_4	175	OD	0.18 ± 0.05
$NaHCO_3$	75	N-NTK	359.98 ± 1.81
$CaCl_2.2H_2O$	5	$N\text{-}NH_4^+$	281.90 ± 3.11
$Mg.SO_4.7H_2O$	20	$N\text{-}NO_3^-$	0.22 ± 0.01
EDTA	50	$N\text{-}NO_2^-$	10.44 ± 0.30
$FeCl_3.6H_20$	4	$N\text{-}N_T$	370.64 ± 2.13
H_3BO_3	4	$N\text{-}N_{ORG}$	78.08 ± 4.92
$ZnSO_4.7H_2O$	1.5	P_T	2.05 ± 0.06
$Cl_2Mn.7H_2O$	0.24	SO_4^{2-}	521.91 ± 14.96
$CuSO_4.5H_2O$	0.24	pH	7.75 ± 0.07
$CoCl_2.6H_2O$	0.06		
Na_2MoO_4	0.20		

absorbance (abs) results to biomass concentration ($g{\cdot}L^{-1}$). The wavelength used in quartz cuvette absorbance readings was 680 nm (Guimarães *et al.* 2019), determined using a HACH model DR 2800 spectrophotometer. The curve was generated by operating the batch column photobioreactor with modified Bold's Basal Medium (BBM) (Table 1). When the stationary growth phase was reached, a concentrated aliquot (stock solution) of the reactor was collected and diluted in ten different concentrations, and the absorbance of each curve point was quantified. At the end of the growth period, the remaining volume in the photobioreactor was collected to determine the biomass concentration of the stock solution, relating it to the absorbances obtained in the readings of each dilution. A mathematical relationship was established between the detected signal and the analyte concentration using simple linear regression through which the regression ($a = 0.718$ and $b = -0.0054$) and correlation ($R^2 = 0.9996$) coefficients were obtained. The analysis of errors and accuracy of the standard curve was performed from the Standard Deviation (SD) of each sample data and the Adjusted R-Square (R^2). Plotting of the calibration curve was performed using *Origin 8.0* software.

Inoculum adaptation

Before the kinetic test, the inoculum was adapted to the wastewater. The adaptation process was performed by sequential batches in a 1.0 L bubble column photobioreactor (PBR). The PBR used was fed with increasing concentrations of ammoniacal nitrogen ($C_1 = 7.05$ $mg{\cdot}L^{-1}$, $C_2 = 14.09$ $mg{\cdot}L^{-1}$, $C_3 = 28.19$ $mg{\cdot}L^{-1}$, $C_4 = 56.38$ $mg{\cdot}L^{-1}$) and kept in operation until the exponential growth phase and an approximate biomass concentration of 1.2 $g{\cdot}L^{-1}$ in each condition was reached. Then, the batches were discontinued, and the biomass was left for one hour to settle to the bottom of the reactor. After sedimentation of the biomass, 70% of the useful volume (supernatant) of the photobioreactor was removed and refilled with diluted wastewater supplemented with BBM. Once the maximum substrate concentrations tolerated by the inoculum were reached, i.e. the substrate concentration where no inhibition and cell death occurred, the inoculum was collected and recycled to the next stages of the algal adaptation and growth experiment.

Microbial growth kinetics assays

The objective of this assay was to evaluate microalgae growth by monitoring the cumulative biomass production (X) and to determine the optimal substrate concentration (S_{opt}). For the kinetic assay, six different concentrations of ammoniacal ($N\text{-}NH_x$) nitrogen were studied: $C_1 = 7.05$ $mgN\text{-}NH_4^+{\cdot}L^{-1}$, $C_2 = 14.09$ $mgN\text{-}NH_4^+{\cdot}L^{-1}$, $C_3 = 28.19$ $mgN\text{-}NH_4^+{\cdot}L^{-1}$, $C_4 = 56.38$ $mgN\text{-}NH_4^+{\cdot}L^{-1}$, $C_5 = 112.76$ $mgN\text{-}NH_4^+{\cdot}L^{-1}$ and $C_6 = 169.14$ $mgN\text{-}NH_4^+{\cdot}L^{-1}$. Nutrient solutions composed of Bold's Basal Medium (BBM) were prepared as described in Table 1 and added to the reaction flasks to optimize the activity of the microorganisms. Reactor pH was monitored and controlled at 7.5 by the addition of dilute sulfuric acid (0.05 M) to prevent sulphide formation and ammonia nitrogen volatilization due to ammonium ion conversion ($N\text{-}NH_4^+$) to free ammonia ($N\text{-}NH_3$). Finally, three negative reaction flasks were prepared with the same inoculum in Bold's Basal Medium (BBM) without wastewater.

Subsequently, the flasks were incubated in batch at a temperature of 26 °C at 200 rpm in a shaker incubator. This equipment was kept in an air-conditioned room and isolated from sunlight. The light used as a power source by microalgae was provided exclusively by 24-hour 6 W LED lamps, resulting in an estimated illuminance of 4,600 lumens per square metre (lm/m^2) (Zorn *et al.* 2017). The microalgae were cultivated without aerators, no CO_2 (carbon dioxide cylinders) feed, and instead, pneumatic agitation was performed. Thus, the only carbon source for the microorganisms was the organic carbon contained in the wastewater, sodium bicarbonate of the basal medium, and the CO_2 present in the atmosphere introduced by the headspace, due to the constant shaking of the bottles (complete mixture). The incubation lasted seven days until biomass production ceased.

Bubble column photobioreactors (PBRs)

The microalgae were grown in a system composed of three batch bubble column photobioreactors with a total volume of 1.2 L each, equipped with external lighting and pneumatic agitation. Bubble column PBRs are characterized by their high volumetric gas transfer coefficients. The bubbling of gases at the bottom of the column provides efficient use of carbon dioxide and the effective removal of dissolved oxygen. This system provides more efficient operational control over culture conditions and essential growth parameters than open systems, reducing the chance of contamination and the invasion of competing microorganisms. In this type of photobioreactor, the growth of microalgae is normally limited by other variables, such as light intensity and nutrient availability.

Each photobioreactor was composed of a glass column with a 5 mm wall thickness, an internal diameter of 65 mm, a height of 450 mm, and a useful volume of 1.0 L. The aeration system consisted of a diffuser, located at the base of each column, connected by a silicone hose to a diaphragmatic air compressor with a flow rate of 4 L·min^{-1} for the dispersion of air microbubbles in the system. Figure 1 shows the schematic drawing of the bubble column PBRs used.

Kinetic modelling of microbial growth

The biomass production profiles of each condition were adjusted to the modified Gompertz kinetic model (Equation (1)) described by Mansouri (2017), using *Origin 8.0* software:

$$y = C \times \exp\left\{-\exp\left[\frac{\mu_{max} * \exp(1)}{C} \times (\lambda - t) + 1\right]\right\} \quad (1)$$

In this expression, Y is the concentration of biomass at time t (g·L^{-1}); C is the asymptote of $\ln(X/X_o)$ at time t; μ_{max} is the maximum specific growth rate (d^{-1}); λ is the latency phase duration time (d); t is the batch reaction time (d); and exp is the Euler number.

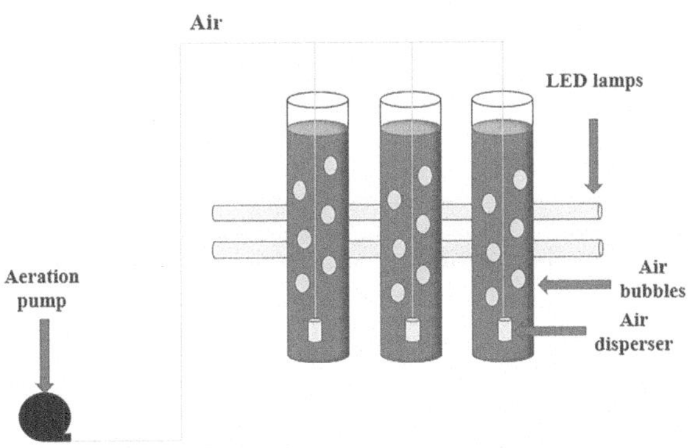

Figure 1 | Schematic drawing of the bubble column photobioreactors (PBRs).

Determining the maximum specific growth rates (μ_{max}), a graph was constructed that relates the initial substrate concentrations (S) to the maximum specific growth rates (μ_{max}) of each condition, and the adjustments to known kinetic models were tested by obtaining kinetic parameters and optimal substrate concentration (S_{opt}). As substrate inhibition was observed, the adjusted kinetic model used was the Andrews model (Equation (2)), and the S_{opt} calculated by Equation (3). The adjustment to the Andrews model was performed in the *Origin 8.0* software.

$$\mu = \mu_{max} * \frac{S}{K_s + S + (S^2/K_i)} \quad (2)$$

$$S_{opt} = \sqrt{K_s * K_i} \quad (3)$$

In these expressions, μ is the specific growth rate (d^{-1}); K_s is the saturation constant defined by the Monod equation (mgN-NH_4^+·L^{-1}); K_i is the substrate inhibition constant (mgN-NH_4^+·L^{-1}); and S_{opt} is the optimal substrate concentration for the operation of bubble column photobioreactors (mgN-NH_4^+·L^{-1}).

To estimate the parameters of the models, the Levenberg–Marquardt (LM) algorithm was used to adjust the expressions of the equations to the experimental results. The accuracy of the predictive models was investigated using the Adjusted R-Square (R^2) and the Standard Error (SE).

Kinetic modelling of substrate consumption

The consumption of ammoniacal nitrogen (N-NH_x) and nitrate (N-NO_3^-) were adjusted to the first-order kinetic expression and to the modified first-order kinetic expression, presented in Equation (4) and (5) respectively. Both adjustments were performed using the *Origin 8.0* software.

$$S = S_o + A\,e^{(-k/t)} \quad (4)$$

$$S = S_R + (S_i - S_R)\,e^{-kt} \quad (5)$$

In this expression, S is the substrate concentration (mg·L^{-1}); S_o is the initial substrate concentration (mg·L^{-1}); S_R is the residual substrate concentration (mg·L^{-1}); t is time (d); and k is the decay rate constant (d^{-1}). From the k results obtained in the kinetic modelling, the instantaneous substrate consumption rates (r_S) were calculated using Equation (6):

$$r_S = -k.S_o \quad (6)$$

Harvest

Biomass was harvested by flocculation, using a 2.0 L decanter. Flocculation was carried out with wastewater at the outlet of the bubble column photobioreactors (PBRs), adding 2.0 mL·L^{-1} of the aluminium sulphate flocculant solution (0.16 M) for decanting the biomass. After about 30 minutes at rest, the phases separated with the concomitant decantation of the biomass in the lower phase of the beaker.

After separating the phases, excess water was removed from the surface, and then, the decanted biomass was filtered using a filtration system composed of a Kitasato flask and a Büchner funnel with qualitative filter paper (Unifil ETQ - 185 mm).

The wet biomass was uniformly distributed over a watch glass and taken to dry in an oven at a temperature of 55–60 °C for a period of 24 hours and subsequently placed in desiccators in order to stabilize the mass at room temperature (Zorn *et al.* 2017). After reaching constant mass, the biomass was crushed using a porcelain and pistil gral until it reached a homogeneous powder.

Lipids extraction

The method used in the extraction of lipids via organic solvents was developed following the methodology by Zorn *et al.* (2017). This methodology promotes a higher lipid yield during the extraction process and is characterized by the use of microalgal biomass with 64% humidity, a chloroform:methanol:water ratio of 5.7:3:1, a total of 33 mL of solvents, and only 70 minutes of ultrasound.

Biomass and lipid productivity

The biomass concentration was quantified by spectrophotometry and the lipid content by extraction with organic solvents. These values were leveraged for calculation of the biomass productivity (P_B) through Equation (7) and the lipid productivity (P_L) through Equation (8). The calculation of the lipid productivity was performed by multiplying the biomass productivity (P_B) by the mass fraction of lipids (F_M). The calculation of the mass fraction of lipids (F_M) was done using Equation (9):

$$P_B = \frac{(X_F - X_O)}{(t_F - t_O)} \quad (7)$$

$$P_L = P_B * F_M \quad (8)$$

$$F_M = \frac{L_E}{X_E} \quad (9)$$

In these expressions, P_B is the biomass productivity (g·L^{-1}·d^{-1}); P_L is the lipids productivity (g·L^{-1}·d^{-1}); X_F is the final concentration of biomass (g·L^{-1}); X_O is the initial concentration of biomass (g·L^{-1}); F_M is the mass fraction of lipids present in the biomass (%); L_E is the mass of lipids extracted (g); X_E is the amount of biomass used in the extraction (g); and t is the time (d).

Statistical analysis

To statistically confirm which conditions had the greatest influence on the biomass production process (response variable), ANOVA (analysis of variance) was used. The ANOVA was performed using *Minitab* software version 17 and a 95% confidence interval (p-value = 0.05). The biomass productivity (P_B) data, used in the statistical analysis, were calculated (Equation (7)) from the biomass concentration results obtained by spectrophotometry during the monitoring of the cultures in the bubble column photobioreactors (PBRs).

RESULTS AND DISCUSSION

Microbial growth kinetics

The biomass production results obtained in each condition of the kinetic assay were adjusted to the modified Gompertz model to obtain the maximum specific growth rates (μ_{max}). Each condition tested (C_1, C_2, C_3, C_4, C_5, and C_6) has different concentrations of ammoniacal nitrogen (N-NH_x), as described in the section above on 'Microbial growth kinetics assays'. Considering that cell death occurred in condition C_6, the experimental data for microbial growth were not obtained, and therefore, no adjustment was made. The non-linear adjustments performed for each condition of the kinetic assay are shown in Figure 2.

Table 2 was then created for better visualization of the maximum specific growth rate (μ_{max}) and latency phase duration time (λ) results, obtained from Equation (1) and presented in the section above on 'Kinetic modelling of microbial growth'. The analysis of the errors of the kinetic parameters was performed using Standard Error (SE) and the Adjusted R-Square (R^2), while the analysis of the errors of the initial concentrations of ammoniacal nitrogen (N-NH_x) was performed using the Standard Deviation (SD).

Despite the absence of ammoniacal nitrogen (N-NH_x) in the control experiment, there was cell growth in this condition because the nitrogen source was present in the form of nitrate (N-NO_3^-), due to supplementation with Bold's Basal Medium (BBM). By interpreting the data, it is possible to observe that the increase of substrate concentration (N-NH_x) in the culture medium provided an increase in the maximum specific growth rates (μ_{max}). However, conditions with ammoniacal nitrogen concentrations higher

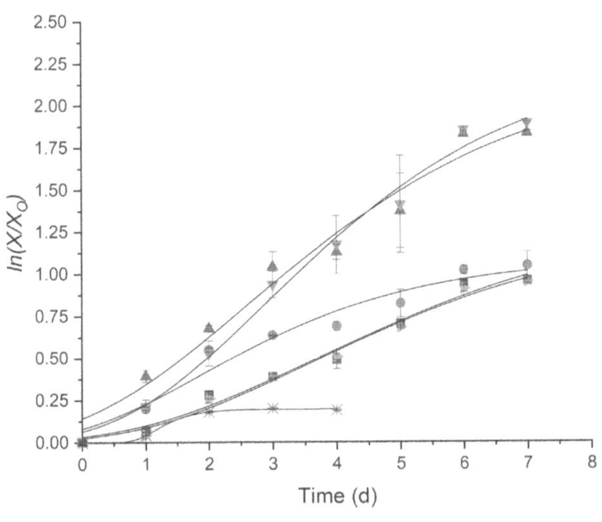

Figure 2 | Growth of *Chlorella vulgaris* microalgae under different concentrations of ammoniacal nitrogen (N-NH_x): Condition 1 (■); Condition 2 (●); Condition 3 (▲); Condition 4 (▼); Condition 5 (*); Control (►); modified Gompertz model (–).

Table 2 | Kinetic parameters obtained through non-linear adjustments (modified Gompertz kinetic model) and initial ammoniacal nitrogen concentration (N-NH_x) of each condition

Condition	N-NH_x (mgN-NH_4^+·L^{-1})	μ_{max} (d^{-1})	λ (d)	R^2
Control	0.00 ± 0.00	0.179 ± 0.026	0.97 ± 0.27	0.9879
C_1	7.05 ± 0.08	0.177 ± 0.018	0.89 ± 0.31	0.9832
C_2	14.09 ± 0.16	0.240 ± 0.055	0.10 ± 0.43	0.9716
C_3	28.19 ± 0.31	0.316 ± 0.047	0.03 ± 0.42	0.9733
C_4	56.38 ± 0.62	0.363 ± 0.035	0.63 ± 0.24	0.9904
C_5	112.76 ± 1.24	0.211 ± 0.029	0.80 ± 0.04	0.9986
C_6	169.14 ± 1.87	0.000 ± 0.000	0.00 ± 0.00	0.0000

than 56.8 mgN-NH_4^+·L^{-1} presented a decrease in the maximum specific growth rate (μ_{max}), indicating inhibition due to excess substrate (Azov & Goldman 1982).

Analyzing the results of the latency phase duration time (λ), it can be observed that the time was relatively low under all conditions, proving that the inoculum adaptation process to the wastewater was efficient. It is noted that the control showed the largest latency phase among the tested conditions and that as the availability of ammoniacal nitrogen in the medium increased, the latency phase (λ) was reduced. This is likely due to the fact that microalgae have higher affinity to ammoniacal nitrogen compared with oxidized nitrogen compounds, which require a higher energy consumption in the assimilation process (Maestrini *et al.* 1986). However, conditions C_4 and C_5 presented considerably higher ammoniacal nitrogen concentrations and demonstrated a gradual increase in the latency phase.

Thus, after evaluating the behavior of maximum specific growth rates (μ_{max}) and latency phase duration time (λ), the Andrews model (Equation (2)) was proposed and adjusted to describe the behavior of microorganisms. From the model, the kinetic parameters μ_{max}, K_s, and K_i were determined to be 1.27 d^{-1}, 50.25 mg·L^{-1}, and 28.12 mg·L^{-1}, respectively. The fit to the Andrews model is illustrated in Figure 3.

Since the Andrews kinetic model adjusted satisfactorily to microbial growth ($R^2 = 0.872$) and there was inhibition by substrate excess, the optimal substrate concentration ($S_{opt} = 37.59$ mgN-NH_4^+·L^{-1}) was determined (Equation (3)). Tam & Wong (1996) also performed experiments to evaluate the influence of ammoniacal nitrogen concentration on the

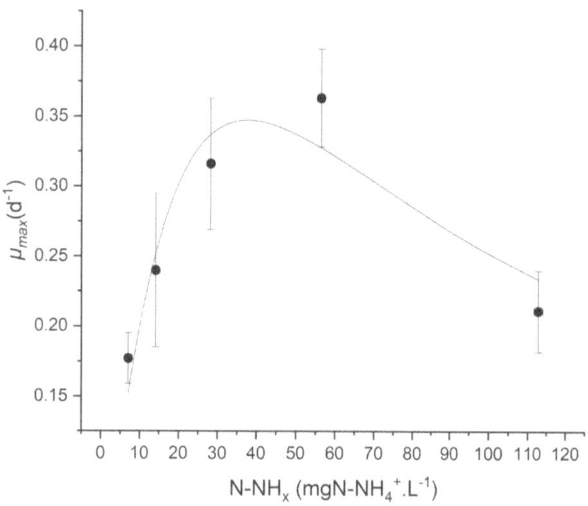

Figure 3 | Maximum specific growth rate (μ_{max}) of microalgae in relation to the concentration of ammoniacal nitrogen (N-NH$_x$) present in wastewater (Andrews kinetic model): maximum specific growth rate (●); Andrews model (–).

growth of *C. vulgaris* microalgae. The assays were performed in synthetic culture media with different concentrations of ammoniacal nitrogen (exclusive source of nitrogen in the medium). By contrast, the current study cultivated microalgae photoautotrophically in conical reactor flasks with a useful volume of 400 mL, operated in batch without CO_2 feed and with pneumatic agitation.

Comparing the results of the optimal ammoniacal nitrogen concentrations (S_{opt}), it is possible to observe that the concentration obtained by Tam & Wong (1996) was approximately six times higher. This variation is predictable because in the test by Tam & Wong (1996), pH was 7.0 and ammonium chloride was used as the nitrogen source. Therefore, due to the displacement of the chemical equilibrium of the reaction, all ammoniacal nitrogen was present in the ionic form of ammonium (N-NH_4^+), which is less toxic to microalgae (Li *et al.* 2019). However, in the kinetic test performed in the present work, the pH was adjusted and monitored at 7.5 and industrial wastewater was used as a nitrogen source. Consequently, a portion of the ammoniacal nitrogen (N-NH_x) was present as free ammonia (N-NH_3), which is toxic to the metabolism of microorganisms (Li *et al.* 2019). According to Azov & Goldman (1982), if the culture pH is close to or above 8.0, algal photosynthesis is inhibited by free ammonia concentrations of 28 mg·L^{-1}. This explains why Tam & Wong (1996) observed that *Chlorella vulgaris* still maintained some growth at 1,000 mg·L^{-1} ammoniacal nitrogen concentrations and reported an operating S_{opt} that was six times higher.

The pH of the cultures was monitored daily during the kinetic assay, increasing between 0.1 to 0.4 per day. The variations were corrected to the initial value of 7.5 by adding aliquots of sulfuric acid (0.05 M). The constant variation in the pH of the culture medium is explained by the photosynthetic fixation of CO_2 resulting in the shift of pH to alkalinity (Chojnacka & Marquez-Rocha 2004). Previous reports have stated that normally in mixotrophic crops, the final pH of the assays will depend on the dominant, autotrophic or heterotrophic metabolism, and it generally remains constant throughout the processes (Kong *et al.* 2011). Therefore, due to the constant increase in pH, it can be suggested that the predominant metabolism during microalgae growth in the kinetic assay was autotrophic metabolism.

Regarding the maximum specific growth rate (μ_{max}) values obtained in both the optimal substrate concentrations of this project (0.363 d^{-1}) and the work of Tam & Wong (1996) (0.236 d^{-1}), the significant difference between the results is justified. There was better hydrodynamic behavior of the vials within the shaker (complete mix) compared with the conical flasks with pneumatic agitation, eliminating dead zones, and favoring system homogenization, as well as mass transfer between substrate and microorganisms (Muharam *et al.* 2018).

The dissolved oxygen (DO) of the reactor flasks was also monitored daily during the kinetic test cultures, presenting no changes in relation to their initial concentration (DO = 5.69 mgO_2·L^{-1}). Thus, it is possible to state that the proposed headspace degassing strategy was efficient.

Cultivation in bubble column photobioreactors (PBRs)

Kinetic modelling of microalgae growth

After obtaining the kinetic parameters for microbial growth (see section 'Microbial growth kinetics') and determining the optimum substrate concentration for operation, the column bubble photobioreactors (PBRs) were started in batch. The microbial growth curves were fitted to the modified Gompertz kinetic model by the *Origin 8.0* software. Adjustments to the growth of microalgae in bubble column

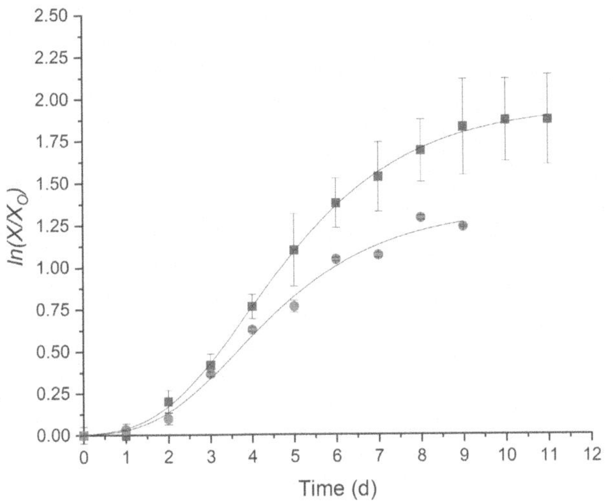

Figure 4 | Growth of *Chlorella vulgaris* microalgae in bubble column photobioreactors (PBRs) fed with wastewater and with Bold's Basal Medium (control): wastewater (■); control (●); modified Gompertz model (–).

photobioreactors (PBRs) fed with wastewater and with Bold's Basal Medium (control) are illustrated in Figure 4.

From the nonlinear adjustments of the kinetic models, the maximum specific growth rates (μ_{max}) and the duration of the latency phase (λ), obtained in the preparation of Table 3, were obtained for better visualization and interpretation of the results. In addition to the kinetic parameters obtained during kinetic modelling, the biomass productivity (P_B) of the photobioreactors was also calculated and is presented in Table 3. The analysis of the errors of the kinetic parameters was performed using Standard Error (SE) and the Adjusted R-Square (R^2), while the analysis of the errors of biomass productivity (P_B) was performed using the Standard Deviation (SD).

It can be observed that the maximum specific growth rates (μ_{max}) of the cultures in the photobioreactors, operated under the optimal conditions determined by the kinetic test of microbial growth, reached values very close (0.34 d^{-1}) to those obtained in condition C_4 of the kinetic test (0.36 d^{-1}). Therefore, it is possible to state that the mathematical modelling of the process (section 'Microbial growth kinetics') was coherent and provided an optimized operation of the PBRs, corroborating the result of Eze *et al.* (2018).

Analyzing the results presented in Tables 3 and 2, it is possible to observe that there was a longer duration of the latency phase (λ) in the cultures grown in the bubble column photobioreactors (PBRs). This increase in the duration of the latency phase (λ) is justified by the fact that the pneumatic agitation system of the bubble column photobioreactors was not as efficient as the shaker in the homogenization and mass transfer processes. Thus, it is noted that the bubble-column-type photobioreactors did not operate as ideal batch mix reactors, resulting in a longer adaptation period for the microorganisms, corroborating the result of López-Rosales *et al.* (2019) and Muharam *et al.* (2018).

Comparing the results of μ_{max} and X_F obtained, it is possible to verify that in the microalgae cultivations in bubble column photobioreactors (PBRs), the addition of ammoniacal nitrogen (N-NH_x) provided an increase in both the maximum specific growth rate (0.34 d^{-1}) and the production of biomass (1.83 g·L^{-1}) with respect to the negative control. This is attributed to the greater affinity of microorganisms to nitrogen in reduced forms (N-NH_x), which lessens the energy expended to reduce and assimilate oxidized nitrogen compounds (N-NO_2^- and N-NO_3^-). Therefore, although it is necessary to dilute the effluent, the increase in the maximum specific growth rates (μ_{max}) and the production of biomass (X_F) demonstrate the potential of cultivating *C. vulgaris* in anaerobically digested wastewater from food industries.

Mansouri (2017) modelled the growth of photoautotrophically grown *C. vulgaris* microalgae in 20 L airlift photobioreactors, and they used an artificial culture medium (BG-11) with only nitrate (1.5 g·L^{-1}) as a nitrogen source. In this work, the kinetic model that provided the best fit was the Baranyi model ($R^2 = 0.989$); however, the

Table 3 | Biomass productivity and kinetic parameters obtained through non-linear adjustments (modified Gompertz model) in the growth of microalgae in bubble column photobioreactors (PBRs)

Condition	N-NH_x (mgN-NH_4^+·L^{-1})	P_B (g·L^{-1}·d^{-1})	μ_{max} (d^{-1})	λ (d)	R^2
Control	0.0 ± 0.00	0.081 ± 0.005	0.256 ± 0.024	1.68 ± 0.23	0.992
PBRs with wastewater	44.30 ± 3.47	0.141 ± 0.021	0.343 ± 0.009	1.77 ± 0.08	0.999

modified Gompertz model also showed a good Adjusted R-Square ($R^2 = 0.954$). In adjusting to the Baranyi model, Mansouri (2017) obtained a maximum specific growth rate (μ_{max}) of $0.741\ d^{-1}$ and biomass production (X_F) of $1.81\ g{\cdot}L^{-1}$. The value of μ_{max} reported by Mansouri (2017) was higher than that obtained in the present study, while the production of biomass (X_F) reached nearly equal values. The difference in the maximum specific growth rates (μ_{max}) is justified because the cultures were performed in two different photobioreactor configurations (bubble column and airlift) and used a high concentration of nitrate ($1.5\ g{\cdot}L^{-1}$). Airlift photobioreactors provide a more uniform and efficient distribution of air bubbles in the system because of the internal tube installed in this reactor type, which reduces the bubble sizes and directs the gas flow. Consequently, this favors the mass transfer between microorganisms and substrate, as well as the assimilation of carbon dioxide, and thus, a higher maximum specific growth rate can be achieved (Wang *et al.* 2012).

Statistical analysis of biomass productivity

The cell growth results, obtained by spectrophotometry during operation of the bubble column photobioreactors (PBRs), allowed for calculation of biomass productivities (P_B) with Equation (7). With these values (Table 3), analysis of variance (ANOVA) was performed using the *Minitab 17* software, considering a confidence interval of 95% (p-value = 0.05).

Interpreting the ANOVA table, it can be seen that the p-value (0.014) was lower than the confidence interval, that is, there was a statistically significant difference between the biomass productivity averages for the bubble column photobioreactors (PBRs). This confirms the hypothesis that the addition of anaerobically digested wastewater from the food industry results in higher biomass productivity, and the increase in biomass productivity using wastewater was 57.45%.

Kinetic modelling of nitrogen consumption

During the operation of the bubble column photobioreactors (PBRs) fed with wastewater, the concentrations of ammoniacal nitrogen ($N\text{-}NH_x$) were quantified. The results of ammoniacal nitrogen consumption, shown in Figure 5, were adjusted to the first-order kinetic expressions (Equation (4)) and as two decay phases were observed during the consumption of ammoniacal nitrogen, two modified first-order kinetic expressions (Equation (5)) in series were also adjusted.

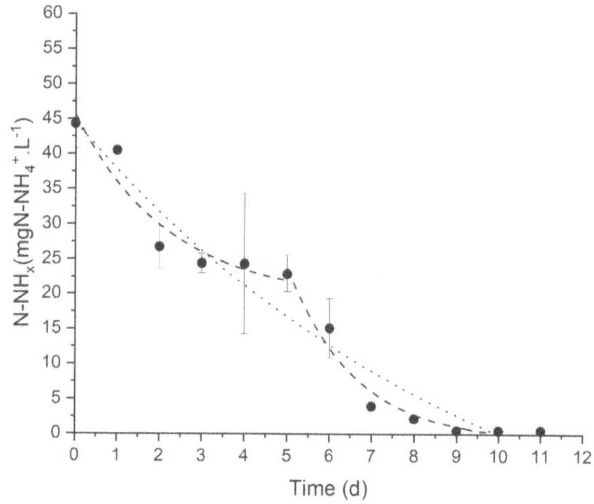

Figure 5 | Adjustment of ammoniacal nitrogen ($N\text{-}NH_x$) consumption in bubble column photobioreactors (PBRs). Subtitle: Experimental data (●); first-order kinetic expression (….); modified first-order kinetic expression (- - -).

The first decay phase was from day 1 to day 5, where microalgae still assimilated nitrate, and the second phase of decay was from day 5 to 10, where preference was given to nitrogen in the reduced form ($N\text{-}NH_x$). The second phase demonstrated the interruption of nitrate assimilation, coincidentally resulting in lack of adjustment of the model during this period. The preference for ammoniacal nitrogen began at day 5 due to the greater availability of nitrate in the medium as compared with ammoniacal nitrogen in the beginning of the batch, which favored the mass transfer between substrate and microorganism. Thus, two modified first-order kinetic expression adjustments were proposed for the phases of $N\text{-}NH_x$ consumption, named as Phase 1 and Phase 2 (before and after the interruption of nitrate assimilation), respectively. Based on the kinetic parameters obtained, Table 4 is provided to better visualize these results. The analysis of the errors of the kinetic parameters was performed using Standard Error (SE) and the Adjusted R-Square (R^2).

Analyzing the Adjusted R-Square (R^2), it is possible to verify that the modified first-order kinetic expression adjusted satisfactorily to the two phases of substrate consumption in the bubble column photobioreactors (PBRs).

Table 4 | Kinetic parameters of ammoniacal nitrogen consumption (N-NH_x)

Kinetic model	Phase	k (d^{-1})	r_s ($mg \cdot L^{-1} \cdot d^{-1}$)	R^2
First-order kinetic expression	Single phase	0.100 ± 0.052	4.430 ± 2.302	0.9375
Modified first-order kinetic expression	Phase 1	0.453 ± 0.283	20.058 ± 12.531	0.8637
Modified first-order kinetic expression	Phase 2	0.616 ± 0.156	14.219 ± 3.601	0.9529

Although the adjustment to single phase achieved a good Adjusted R-Square (R^2), the analysis of substrate consumption by parts proved to be more adequate to explain the consumption of nitrogen forms present in the bulk liquid. Regarding the ammoniacal nitrogen decay rate constants (k), it was observed that from the fifth day on when the nitrate consumption was interrupted (Figure 6), the microalgae channelled all their energy into the assimilation of ammoniacal nitrogen, and consequently, there was an increase in the decay rate constant from 0.453 to 0.616 d^{-1}. However, the initial substrate concentration (C_o) was lower in phase 2, and although the k values increased, the instantaneous substrate consumption rates (r_S) decreased. It is important to note, however, that both r_S results were considerably superior to those obtained by Choi & Lee (2013) in their study.

Interpreting the behavior of the substrate decay and cell growth curves shown in Figure 6, it is possible to observe that during the first five days, the microalgae consumed nitrate and ammoniacal nitrogen. During the period between days 3 and 5, the consumption of both ammoniacal nitrogen and nitrate was impaired. After the fifth day, the consumption of nitrate completely ceased, while ammoniacal nitrogen was assimilated until the stationary phase of growth of the microorganisms. This behavior confirms the results obtained by Scherholz & Curtis (2013), who cultivated photoautotrophic cultures of *Chlorella vulgaris* and *Chlamydomonas reinhardtii* in 1.5 L photobioreactors containing substrate concentrations of 0.0135 gN-$NH_4^+ \cdot L^{-1}$ with chloride and nitrate as ions (NH_4Cl and $NH_4NO_3^-$). In this study, when ammoniacal nitrogen and nitrate were supplied simultaneously, the microalgae *Chlorella vulgaris* and *Chlamydomonas reinhardtii* gave preference to the consumption of ammoniacal nitrogen, inhibiting the consumption of nitrate. However, the observation of Li *et al.* (2019) that microalgae will first consume ammoniacal nitrogen, then nitrite, and finally nitrate could not been verified.

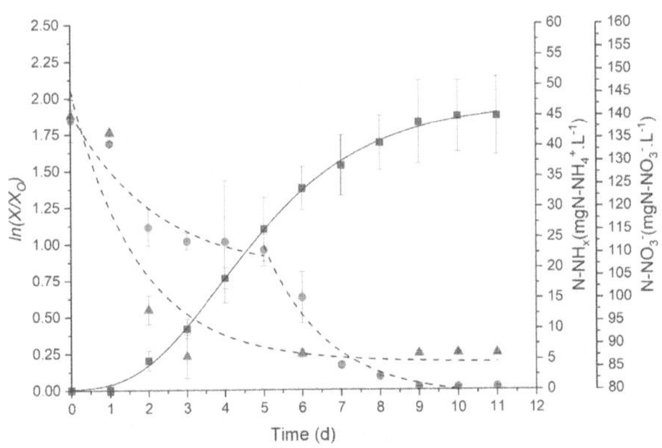

Figure 6 | Biomass growth and substrate consumption in bubble column photobioreactors (PBRs) fed with wastewater: growth experimental data (■); ammoniacal nitrogen experimental data (●); nitrate experimental data (▲); modified Gompertz model (–); modified first-order kinetic expression (- - -).

Florencio & Vega (1983) monitored the concentrations of ammoniacal nitrogen and nitrate during the growth of microalgae, since ammoniacal nitrogen also inhibited the use of nitrate and nitrite. Tischner & Lorenzen (1979) stated that depending on the amount of ammoniacal nitrogen added to the medium, this inhibition can last for a long period until the initial rate of nitrate uptake is reached again. This series of results confirms the suspicion about ammoniacal nitrogen uptake and reduction of nitrate expressed by Pistorius *et al.* (1978).

Total lipid content and lipid productivity

After harvesting the biomass, the total lipid content was determined, and the lipid productivity of the bubble column photobioreactors (PBRs) was calculated. The total lipid content was 0.1245 grams, representing a mass fraction of 12.45%, while the lipid productivity was 17.63 $mg \cdot L^{-1} \cdot d^{-1}$. The mass fraction of accumulated lipids obtained in the present work can be compared with the results presented by Safi *et al.* (2014), who considered that microalgae of the species *C. vulgaris* accumulate between 5.0% and 58.0% of lipids. Thus, it can be determined that the mass fraction obtained (12.45%) in the current study was relatively low. This fact is justified by the theory described by Tan & Lee (2016), who stated that the concentration of accumulated lipids is inversely proportional to the availability of nitrogen in the culture medium. Therefore, as the concentration of total nitrogen present in the medium was extremely high during and also after the growth period, the content of total lipids accumulated by the microalgae was low. Thus, the results suggest that microalgae grown under ideal conditions produce large amounts of biomass and present low neutral lipid content, while microalgae grown with a shortage of nutrients accumulate high levels of neutral lipids and grow slowly.

Matos *et al.* (2015) cultivated microalgae of the species *Chlorella* sp. in 4.0 L inverted conical bioreactors, operated in batch, with constant aeration, and at a temperature of 27 °C. The bioreactors were fed with artificial culture medium (BBM) and a desalination concentrate. The basal medium had a concentration of 250 $mg{\cdot}L^{-1}$ of nitrate, while the desalination concentrate had 1.35 $mg{\cdot}L^{-1}$ of ammonium and 30 $mg{\cdot}L^{-1}$ of total nitrogen. In this study, a maximum total lipid percentage of 11.46% and maximum lipid productivity of 19.6 $mg{\cdot}L^{-1}{\cdot}d^{-1}$ were achieved, and these values are similar to those obtained in the present project.

Guimarães *et al.* (2019) cultivated microalgae *Chlorella vulgaris* in 5.0 L bubble column photobioreactors under constant lighting and aeration. The photobioreactors (PBRs) were fed with domestic effluent collected from the aerobic biological reactor in the effluent treatment station at the Lorena Engineering School at the University of São Paulo. The wastewater had a concentration of 10.88 mg of N-NO_3^-. In this study, the accumulated lipid content was 16.4%, and the lipid productivity was 0.0126 $g{\cdot}L^{-1}{\cdot}d^{-1}$. The lipid content was higher than that obtained in the present project due to the low availability of nitrogen in the domestic effluent; however, the lipid productivity was lower because although the lipid content percentage was high, the biomass productivity was low (0.0552 $g{\cdot}L^{-1}{\cdot}d^{-1}$). This phenomenon is explained by the fact that the availability of substrate is a limiting step in the balance of growth and production of fatty acids, and thus, the biomass and fatty acid synthesis pathways compete for the samc substrates (Tan & Lee 2016).

In the study by Guimarães *et al.* (2019), biomass production ended, and the nutrient availability was reduced while the photobioreactors were kept in operation for a few more days in order to promote the stress of microorganisms. The difference from the present project is notable, as the batch was interrupted at the exact moment when the microorganisms reached the stationary growth phase. Even after the stagnation of biomass production, there was approximately 80 $mg{\cdot}L^{-1}$ of nitrate, which prevented the microorganisms from being exposed to critical conditions of low substrate availability in the culture environment.

CONCLUSION

The cultivation of microalgae *C. vulgaris* in bubble column photobioreactors (PBRs) fed with wastewater nutrients proved to be a viable and efficient alternative for both biomass production and the removal of nitrogenous compounds from water, especially ammoniacal nitrogen (N-NH_x). The Andrews model satisfactorily fitted the biomass growth data by presenting a high Adjusted R-Square (R^2), correctly describing the microbial kinetics and providing the optimized process operation. It can be observed that the use of anaerobically digested gelatin industry wastewater resulted in higher specific growth rates and higher biomass productivity, however, the accumulated lipid content per gram of biomass was low despite lipid productivity having been reasonable. Future studies can further elaborate on the potential benefits and limitations of using food industry and process wastewaters for simultaneous nutrient recovery and algal biomass production at larger scale and different cultivation system configurations in order to evaluate system viability, economics, and sustainability.

DATA AVAILABILITY STATEMENT

All relevant data are included in the paper or its Supplementary Information.

REFERENCES

APHA/AWWA/WEF 1998 *Standard Methods for the Examination of Water and Wastewater*, 20th edn. American Public Health Association/American Water Works Association/Water Environment Federation, Washington, DC, USA.

Azov, Y. & Goldman, J. C. 1982 Free ammonia inhibition of algal photosynthesis in intensive cultures. *Applied and Environmental Microbiology* **43** (4), 735–739.

Choi, H. J. & Lee, S. M. 2013 Performance of *Chlorella vulgaris* for the removal of ammonia-nitrogen from wastewater. *Environmental Engineering Research* **18** (4), 235–239.

Chojnacka, K. & Marquez-rocha, F.-J. 2004 Kinetic and stoichiometric relationships of the energy and carbon metabolism in the culture of microalgae. *Biotechnology* **3** (1), 21–34.

Eze, V. C., Velasquez-Orta, S. B., Hernández-García, A., Monje-Ramírez, I. & Orta-Ledesma, M. T. 2018 Kinetic modelling of microalgae cultivation for wastewater treatment and carbon dioxide sequestration. *Algal Research* **32** (1), 131–141.

Florencio, F. J. & Vega, J. M. 1983 Utilization of nitrate, nitrite and ammonium by *Chlamydomonas reinhardii*: photoproduction of ammonium. *Planta* **158** (4), 288–293.

Guimarães, D. H. P., Marino, V. F., Blanco, G. C., dos Santos, W. R. & de Moraes Gomes Rosa, M. T. 2019 Viability of the uses of effluent for cultivation of Chlorella vulgaris in order to integrate the generation of biofuels with the treatment of water. In: *Proceedings of the 4th Brazilian Technology Symposium (BTSym'18): Emerging Trends and Challenges in Technology* (Y. Iano, R. Arthur, O. Saotome, V. V. Estrela & H. J. Loschi, eds), Springer, Cham, Switzerland, pp. 347–360.

Janosz Rajczyk, M. 1993 Fermentation of food industry wastewater. *Water Research* **27** (7), 1257–1262.

Kong, W.-B., Song, H., Cao, Y., Yang, H., Hua, S. F. & Xia, C. G. 2011 The characteristics of biomass production, lipid accumulation and chlorophyll biosynthesis of *Chlorella vulgaris* under mixotrophic cultivation. *African Journal of Biotechnology* **10** (55), 11620–11630.

Li, X., Li, W., Zhai, J., Wei, H. & Wang, Q. 2019 Effect of ammonium nitrogen on microalgal growth, biochemical composition and photosynthetic performance in mixotrophic cultivation. *Bioresource Technology* **273**, 368–376.

López-Rosales, L., Sánchez-Mirón, A., Contreras-Gómez, A., García-Camacho, F., Battaglia, F., Zhao, L. & Molina-Grima, E. 2019 Characterization of bubble column photobioreactors for shear-sensitive microalgae culture. *Bioresource Technology* **275**, 1–9.

Maestrini, S. Y., Robert, J.-M., Leftley, J. W. & Collos, Y. 1986 Ammonium thresholds for simultaneous uptake of ammonium and nitrate by oyster-pond algae. *Journal of Experimental Marine Biology and Ecology* **102** (1), 75–98.

Mai, D. T., Kunacheva, C. & Stuckey, D. C. 2018 A review of posttreatment technologies for anaerobic effluents for discharge and recycling of wastewater. *Critical Reviews in Environmental Science and Technology* **48** (2), 167–209.

Mansouri, M. 2017 Predictive modeling of biomass production by *Chlorella vulgaris* in a draft-tube airlift photobioreactor. *Advances in Environmental Technology* **2** (3), 119–126.

Matos, Â. P., Morioka, L. R. I., Sant'Anna, E. S. & França, K. B. 2015 Teores de proteínas e lipídeos de *Chlorella* sp. cultivada em concentrado de dessalinização residual (Protein and lipid contents of *Chlorella* sp. grown in residual desalination concentrate). *Ciência Rural* **45** (2), 364–370.

Muharam, Y., Ismail, D. & Wirya, A. 2018 Process optimization of microalgae cultivation in a bubble-column photobioreactor. *IOP Conference Series: Earth and Environmental Science* **105**, 012002.

Pistorius, E. K., Funkhouser, E. A. & Voss, H. 1978 Effect of ammonium and ferricyanide on nitrate utilization by *Chlorella vulgaris*. *Planta* **141** (3), 279–282.

Ramsay, I. R. & Pullammanappallil, P. C. 2001 Protein degradation during anaerobic wastewater treatment: derivation of stoichiometry. *Biodegradation* **12** (4), 247–256.

Safi, C., Zebib, B., Merah, O., Pontalier, P.-Y. & Vaca-Garcia, C. 2014 Morphology, composition, production, processing and applications of *Chlorella vulgaris*: a review. *Renewable and Sustainable Energy Reviews* **35**, 265–278.

Scherholz, M. L. & Curtis, W. R. 2013 Achieving pH control in microalgal cultures through fed-batch addition of stoichiometrically-balanced growth media. *BMC Biotechnology* **13** (1), 39.

Tam, N. F. Y. & Wong, Y. S. 1996 Effect of ammonia concentrations on growth of *Chlorella vulgaris* and nitrogen removal from media. *Bioresource Technology* **57** (1), 45–50.

Tan, K. W. M. & Lee, Y. K. 2016 The dilemma for lipid productivity in green microalgae: importance of substrate provision in improving oil yield without sacrificing growth. *Biotechnology for Biofuels* **9** (1), 255.

Tischner, R. & Lorenzen, H. 1979 Nitrate uptake and nitrate reduction in synchronous *Chlorella*. *Planta* **146** (3), 287–292.

Wang, B., Lan, C. Q. & Horsman, M. 2012 Closed photobioreactors for production of microalgal biomasses. *Biotechnology Advances* **30** (4), 904–912.

Zorn, S. M. F. E., Pedro, G. A., Amaral, M. S., Loures, C. C. A. & Silva, M. B. 2017 Avaliação dos fatores envolvidos na extração de lipídios da biomassa da microalga *Chlorella minutíssima* via solventes (Evaluation of the factors involved in the extraction of lipids from the biomass of the microalgae *Chlorella minutíssima* via solvents). *Holos* **2**, 66–79.

First received 29 June 2020; accepted in revised form 1 October 2020. Available online 14 October 2020

Effect of pH on nutrient removal and crop production of hydroponic systems treating brewery effluent

Richard P. Taylor, Clifford L. W. Jones and Mark Laing

ABSTRACT

The use of a crop to remove nutrients from brewery effluent and the influence of pH on these removal rates was evaluated. Cabbage (*Brassica oleracea*) was grown in recirculating hydroponic systems fed with post-anaerobically digested brewery effluent (BE) either subject to pH adjustment (6.5–7.0) or unaltered pH (8.0–8.5). These were compared with cabbages grown in water only and in a inorganic fertiliser nutrient solution (NS). Hydroponic systems fed with pH adjusted BE removed significantly more nitrogen and phosphorus than systems fed with pH unadjusted BE ($p < 0.05$). The final weight of cabbages from the pH adjusted BE systems were 6.7 times greater than cabbages from the pH unadjusted BE systems, whereas pH adjustment had no influence on cabbage weight in the water-only and NS treatments. Anaerobically digested BE that is not pH adjusted is not a suitable water and nutrient source for the hydroponic production of cabbages. However, pH adjustment of BE renders it more suitable for hydroponic crop production with hydroponic systems decreasing dissolved inorganic nitrogen, ammonium, phosphate and chemical oxygen demand concentrations by 72.8, 31.8, 98.5 and 51.0%, respectively. Hydroponic systems can be used to treat post-anaerobically digested BE to a similar standard obtained by conventional activated sludge treatment system.

Key words | activated sludge, beneficiation, brewery effluent, hydroponics, nutrient recovery

Richard P. Taylor (corresponding author)
Clifford L. W. Jones
Department of Ichthyology and Fisheries Science, Rhodes University,
P.O. Box 94, Grahamstown, 6140,
South Africa
E-mail: *richardtaylorwater@gmail.com*

Mark Laing
Department of Plant Pathology,
University of KwaZulu-Natal,
Pvt Bag X01, Pietermaritzburg, 3209,
South Africa

HIGHLIGHTS

- Effluent pH adjustment increased cabbage mass.
- Effluent pH adjustment increased nutrient removal by cabbage plants.
- Effluent pH adjustment increased the nitrogen concentration of cabbage leaves.
- Hydroponic systems can be used to remove nutrients from pH adjusted effluent.

INTRODUCTION

Effluent, from the beer making process, at Ibhayi brewery in Port Elizabeth is treated onsite for reuse in non-production activities or discharge to the municipal sewer. Effluent is treated via anaerobic digestion (AD), activated sludge (AS) micro- and ultra-filtration. The AS treatment process removes nitrogen and decreases the chemical oxygen demand (COD) from the effluent. Activated sludge is an energy intensive treatment process, as agitators are required to provide oxygen to the aerobic micro-organisms responsible for effluent treatment (Simate *et al.* 2011). Waste sludge biomass is also produced during the AS treatment process which needs to be disposed of at cost to the company and the environment (Simate *et al.* 2011). There is a need to identify alternative energy efficient, low waste production systems for effluent treatment that can be used to replace AS.

doi: 10.2166/ws.2020.330

Constructed wetlands are capable of providing the same treatment capabilities as AS systems (Kadlec & Wallace 2009). Constructed wetlands are normally planted with wetland plants that have little to no value and the produced plant biomass needs to be harvested and disposed of. The possibility exists to use hydroponic crop production systems to treat post-anaerobically digested (post-AD) brewery effluent to the same quality as post-AS effluent. The plants should remove nitrogen and phosphorus from the effluent and convert it into a plant biomass, which can be sold to cover the running costs of the treatment plant. This would also minimise the waste produced by an effluent treatment plant.

Brewery effluent (BE) can successfully be used as an irrigation source in irrigated crop production; however, there are still uncertainties regarding how suitable it is for this purpose (Taylor *et al.* 2018; Worku *et al.* 2018). Brewery effluent that has undergone AD followed by stabilisation in a primary facultative pond has been found suitable to support the growth of lettuce tomatoes and cabbage (Power & Jones 2016; Taylor *et al.* 2018). Taylor *et al.* (2018) found that post-anaerobically digested BE was suitable to irrigate cabbages grown in the soil, however yield was 13% lower when compared with cabbages irrigated with an inorganic fertiliser solution. The main limitations of using brewery effluent in crop production systems are the high conductivity (EC) and alkalinity of the effluent (Power & Jones 2016; Taylor *et al.* 2018).

The high sodium content of BE poses a major problem when it is used as a source of water or nutrients for crop production (Sweeney & Graetz 1991; Sou *et al.* 2013; Riera-Vila *et al.* 2019). Brewery effluent has a salinity of 2,700 $\mu S/cm^2$, and an EC at this level may have negative effects on crop growth and yield. Jones *et al.* (2014) concluded that in order to successfully utilise BE as a source of nutrients for plant growth the salinity issue of BE needs to be addressed. The general effect of salinity is a reduced growth rate and yield of most crops (Shannon & Grieve 1999). At low to medium concentrations this is primarily due to osmotic effects because of the reduced osmotic potential between the root plasma and soil water (Munns & Termaat 1986; Jacoby 1994; Mahjoor *et al.* 2016). Severity of salinity response is species specific and is also mediated by environmental factors such as humidity, temperature, wind, light and air pollution (Shannon *et al.* 1994). High temperatures and low humidity increase the effects of salinity (Shannon & Grieve 1999; Munns & Tester 2008). Salinity may cause ion toxicities and nutrition deficiencies, depending on the composition of the saline solution (Epstein & Bloom 2005). Chow *et al.* (1990) found that the K^+ requirements for shoot growth are higher under high salinities than low salinities. High concentrations of Na and Cl may accumulate in the leaves and cause 'scorching and firing' of the leaves (Shannon & Grieve 1999; Zorb *et al.* 2019). Calcium deficiencies are common when there is a high Na content in the soil water (Zorb *et al.* 2019). Not all salinity effects are bad, and spinach yields have been shown to increase in low to moderate salinity (Osawa 1963; Zorb *et al.* 2019). Cabbage heads are more compact at low salinities but are less compact as salinity increases (Osawa 1961).

The pH of irrigation waters affects the availability of nutrients to plants and thus the growth and yield of plants (Lucas & Davis 1961; Tyson *et al.* 2007). The precipitation of Fe^{2+}, Mn^{2+}, phosphate, Ca^{2+} and Mn^{2+} to insoluble and unavailable salts can happen when the pH of a nutrient solution (NS) rises to above 7.5 (Tyson *et al.* 2007). Power & Jones (2016) grew tomatoes on BE with or without pH adjustment. They found that adjusting the pH to 6.5 increased the height and yield of tomato plants by 100%. This is important because the pH of anaerobically digested BE is normally above 7.5 (Power & Jones 2016; Taylor *et al.* 2018). Therefore, the pH of BE will reduce the availability of nutrients to plants and should be adjusted in order to make all the nutrients available to plants. Lucas & Davis (1961) and Epstein & Bloom (2005) both suggested that the optimal pH range for most plants is between five and seven. The pH of the irrigation water will therefore need to be manipulated to pH 6 to ensure that all the nutrients in BE are made available to plants.

Brewery effluent is not an ideal NS for hydroponic crop production but it has been shown to support the growth of tomatoes, lettuce and cabbage (Jones *et al.* 2014). The potential exists to use hydroponic crop production systems in replacement of conventional AS systems. This would allow the recovery of nutrients into a valuable byproduct and also decrease the energy and carbon usage of the effluent treatment system as hydroponic systems do not require large agitators and the plants are able to sequester carbon from the atmosphere during the effluent treatment process.

Aims and objectives

The aim of this study was to determine what nutrients cabbage is able to remove from BE, the influence of pH on this removal rate, and the effect of pH on crop growth and plant health in a hydroponic production system. Cabbage was grown on three different NSs, in a recirculating hydroponic system, where the pH of each NS was or was not adjusted. The objectives were to determine:

(1) the effect of NS type and pH on the health, growth and leaf chemical composition of cabbage plants grown in a hydroponic production system;
(2) the effect of pH on the removal of nutrients and elements from NSs by cabbages plants grown in a hydroponic production system.

MATERIALS AND METHODS

Experimental species

Cabbage (*Brassica oleracea* cv. Star 3301) was used as the test crop because it has similar pH requirements and nutrient requirements as a wide range of vegetables (Liu *et al.* 2015). Two hundred cabbage seedlings (Starke Ayres Pty Ltd, South Africa) were purchased from a commercial seedling supplier (Moorlands Seedlings Pty Ltd, Humansdorp). Of these, 90 similar size seedlings were used for this experiment.

Treatments

Three irrigation water sources were used that included post-PFP BE, tap water (water-only) and a conventional irrigation source consisting of an inorganic fertiliser and tap water (NS). The pH of each irrigation water source was either adjusted to 6.5 with 98% sulphuric acid (Protea Chemicals Pty Ltd, South Africa) or left unadjusted. This resulted in six irrigation treatments (Table 1). The

Table 1 | Six treatments (T1–T6) that were used to irrigate cabbage plants

Water source	pH not adjusted	pH adjusted to 6.5
Primary facultative pond (PFP)	T1	T2
Municipal water (water only)	T3	T4
Municipal water + inorganic fertiliser (NS)	T5	T6

conventional irrigation source was comprised of commercially available inorganic fertiliser (Hygrotech Pty Ltd, South Africa; Registration number K5709; Act 36 of 1947), and calcium nitrate with a composition of 11.7% nitrogen and 16.6% calcium by weight, mixed at a mass ratio of 1:0.8 and dissolved in municipal water to achieve an EC of 1,800 μm. Cabbage was grown in each treatment for 12 weeks.

Experimental system

The experiment was carried out in 18 identical recirculating hydroponic growing systems each containing five pots. Each treatment was replicated three times with a replicate consisting of an entire hydroponic system.

The system (Figures 1 and 2) was a variation of the Dutch bucket hydroponic system (Roberto 2005). Each channel was made from a 1,500 mm long 160 mm diameter polyvinylchloride (PVC) pipe with five 110 mm holes drilled in the top to accommodate the 120 mm pots (Power & Jones 2016, Figures 1 and 2). Each pot was filled with 13 mm washed gravel, which served as a physical support but provided no nutritional benefit to the plant (Power & Jones 2016).

Below each channel the NS was stored in a 25 L plastic bucket and an 18 watt submersible pump (Resun, Model: SP-2500, China) fed the NS through a 15 mm irrigation line along the length of the channel (Figure 2). A 5 mm tube, fitted with a microvalve, connected the 15 mm main line to each of the gravel-filled pots (Power & Jones 2016). The microvalue was used to ensure even distribution of NS to each pot (Power & Jones 2016).

The closed, recirculating system was similar to the one used by Power & Jones (2016). Briefly. a 20 mm drain hole

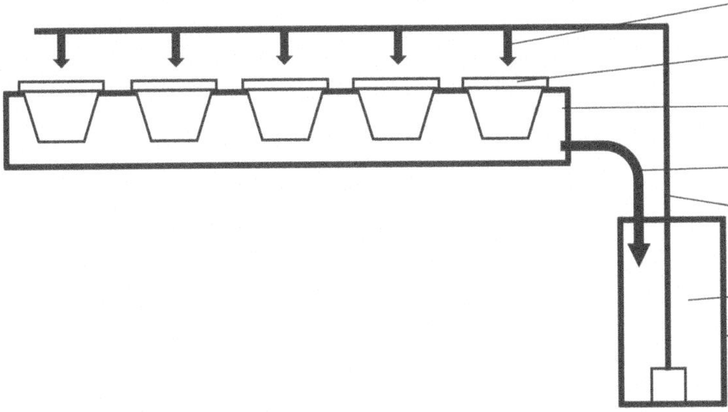

Figure 1 | Cross-section of hydroponic systems use to grow cabbages (not drawn to scale).

Figure 2 | Experimental hydroponic systems at the start of the trial.

was cut into the end of each channel. A 20 mm PVC pipe was fitted into this hole and extended both into and out of the 160 mm channel. A 20 mm plastic elbow joint was fitted on either end of the 20 mm PVC pipe. The elbow inside the channel faced upwards to raise the water level in the channels to 50 mm and to create a submerged root zone for the plants. The outer elbow faced downwards, towards the NS reservoir. The outer elbow was connected to the reservoir by a 20 mm plastic hose (Figure 2).

At the start of the trial, one plant was planted in each pot filled with gravel. The NS in each hydroponic system was replaced every seven days or when the water level in the reservoir was less than 25% full, whichever came first.

Data collection

Water quality parameters of different treatments were recorded before being placed in the reservoirs and prior to replacement with a new water source. Temperature, pH and electrical conductivity (EC) of the different treatments were recorded using an electronic probe (Hanna, HI 991300, United Kingdom). Dissolved oxygen was measured using a handheld probe (Oxyguard, Handy Polaris, Denmark). COD, ammonium, nitrite, nitrate and phosphate of the irrigation solution were recorded using a spectrophotometer (Merck Spectroquant Pharo 100 spectrophotometer, product number 100706, Darmstadt, Germany) and commercially available test kits, using standard methods (Merck Pty Ltd, products: 1.14559.0001, 1.14752.0001, 1.14776.0001, 1.09713.0001, 1.14842.0001, 1.14895.0001). Each sample was filtered through an 8 μm pore filter paper prior to analysis.

Total inorganic nitrogen (TIN) was calculated by using Equation (1):

$$\text{Total inorganic nitrogen} = [N - NH_4^-] + [N - NO_2^-] + [N - NO_3^-] \quad (1)$$

At the beginning of the trial, the mass of each plant planted in each pot was recorded. At the end trial the mass of each plant was also recorded. The chlorophyll concentration of cabbage leaves were estimated using a handheld meter (CCM-200 Plus Chlorophyll Content

Meter, Opti-Sciences Inc., USA). Chlorophyll concentration index (CCI) readings were recorded at the start of the trial and every four weeks until the end of the experiment, on the uppermost fully expanded leaf of each plant. At the end of the trial plant leaves from each replicate were sent to a commercial analytical laboratory (BemLab Pty Ltd, Strand, South Africa). These samples were analysed for N, P, Na, Cl, K, Al, Ca, Cu, Fe, Mn, Mg and Zn from each treatment, with a sample consisting of plant leaves from one hydroponic system. Daily temperature data were recorded using a thermometer (Hanna, HI 991300, United Kingdom).

Statistical analysis

Treatment means were compared using a one-way or multifactor analysis of variance (ANOVA) and a Tukey multiple range analysis at $p < 0.05$. Data collected over the course of the trial were compared using a one-way or multifactor repeated measures ANOVA ($p < 0.05$). All data were checked for equality of variance and for the normal distribution of the residuals using Levene's test and a Shapiro–Wilk plot of the residuals, respectively. If the assumptions were not met then the data were log or square-root transformed and checked for equal variance and normal distribution of residuals. If the assumptions were still not met, a non-parametric Mann–Whitney *U*-test or a Kruskal–Wallis ANOVA was used to compare the data between treatments. All analyses were performed using the Statistica (version 10) software package (StatSoft Inc., Tulsa, USA).

RESULTS

Water quality

The average water temperature of the hydroponic systems was 21.07 °C and ranged between 18.8 and 24.1 °C during the experiment. At the start, post-PFP BE had the highest pH (8.36 ± 0.03) followed by tap water (7.69 ± 0.04) and then NS (7.34 ± 0.06; Table 2). The pH in all systems increased over time (Tables 2 and 3). The mean pH of the effluent systems (8.71 ± 0.06), prior to replacement, was higher than the NS (8.07 ± 0.04) or water-only treatments (8.07 ± 0.05; Table 3). Just before replacement, the pH of effluent unadjusted treatments (9.06 ± 0.06) was higher than effluent adjusted treatments (8.36 ± 0.05; Table 3).

Effluent treatments had the highest starting conductivity (3,069 ± 43 μS/cm^2) followed by the NS (1,839 ± 41 μS/cm^2) and the water-only treatments (579 ± 9.0 μS/cm^2; Table 2). The conductivity of BE treatments increased from 2,914 ± 47 to 3,223 ± 39 μS/cm^2 when the pH was adjusted (Table 2). The conductivity of BE and water-only treatments did not change while the conductivity of NS treatments

Table 2 | Mean (± standard error) water quality parameters of fresh irrigation solutions

Parameter	PFP	PFP*	NS	NS*	Water-only	Water-only*	F/H	*p*-value
pH	8.36 ± 0.03	6.54 ± 0.10	7.34 ± 0.06	6.72 ± 0.07	7.69 ± 0.04	6.57 ± 0.07		
Conductivity (μS/cm^2)	2,864 ± 39^a	3,208 ± 39^a	1,888 ± 15^b	1,924 ± 16^b	567 ± 8.8^c	594 ± 9.2^c	H = 194.5	0.0001
DO (mg/l)	6.27 ± 0.09^a	6.21 ± 0.08^a	7.20 ± 0.05^b	7.03 ± 0.07^b	7.11 ± 0.07^b	7.05 ± 0.06^b	F = 1.372	0.2559
COD (mg/l)	213 ± 6.7^a	216 ± 6.1^a	17 ± 0.41^b	18 ± 0.55^b	17 ± 0.39^c	15 ± 0.19^c	H = 161.6	0.0001
TIN (mg/l)	57.6 ± 2.5^a	54.8 ± 2.7^a	40.7 ± 1.7^b	40.6 ± 1.7^b	10.2 ± 0.41^c	10.0 ± 0.36^c	H = 178.6	0.0001
NH_4-N (mg/l)	36.5 ± 2.3^a	36.9 ± 2.3^a	17.2 ± 0.85^b	17.5 ± 0.84^b	1.6 ± 0.11^c	1.7 ± 0.10^c	H = 190.4	0.0001
NO_2-N (mg/l)	0.85 ± 0.04^a	0.80 ± 0.05^a	0.04 ± 0.01^b	0.04 ± 0.01^b	0.03 ± 0.01^b	0.03 ± 0.01^b	H = 148.0	0.0001
NO_3-N (mg/l)	20.2 ± 0.50^a	21.1 ± 0.69^a	23.4 ± 0.92^b	23.1 ± 0.92^b	8.3 ± 0.32^c	8.5 ± 0.35^c	H = 182.9	0.0001
PO_4-P (mg/l)	27.9 ± 0.51^a	27.3 ± 0.70^a	28.1 ± 0.63^a	28.8 ± 0.55^a	12.4 ± 0.56^b	12.1 ± 0.47^b	H = 163.0	0.5683
Cl (mg/l)	188 ± 5.0^a	194 ± 5.1^a	105 ± 2.0^b	108 ± 1.2^b	103 ± 1.6^b	101 ± 1.3^b	H = 156.2	0.0001
Na (mg/l)	614 ± 9.3^a	627 ± 9.1^a	152 ± 2.5^b	154 ± 3.2^b	145 ± 2.8^b	144 ± 2.6^b	H = 147.6	0.0001

Values in the same row represented by a different superscript letter$^{(a,b,c,d)}$ represent means that are significantly different (multifactor ANOVA/Kruskal–Wallis, $p < 0.05$). Treatments marked with *were subject to pH adjustment using sulphuric acid. Primary facultative pond (PFP), NS (NS), chemical oxygen demand (COD) dissolved oxygen (DO).

Table 3 | Mean (± standard error) water quality parameters of old irrigation solutions

Parameter	Discharge limits#	AS effluent	PFP	PFP*	NS	NS*	Water-only	Water-only*	F-value	*p*-value
pH	5.5–9.5	8.40 ± 0.20	9.06 ± 0.06	8.36 ± 0.05	8.09 ± 0.04	8.04 ± 0.04	8.14 ± 0.04	7.99 ± 0.05		
Conductivity (μS/cm^2)	1,500	3,385 ± 41	3,184 ± 56^a	3,203 ± 41^a	916 ± 30^b	979 ± 19^b	523 ± 8.7^c	541 ± 7.2^c	H = 189.2	0.0001
DO (mg/l)		1.07 ± 0.03	6.80 ± 0.16	6.78 ± 0.12	7.05 ± 0.11	6.76 ± 0.11	7.04 ± 0.12	6.83 ± 0.12	F = 0.6037	0.5487
COD (mg/l)	75	118 ± 4.3	106 ± 3.6^a	106 ± 2.7^a	22.6 ± 0.69^b	24.0 ± 0.83^b	21.3 ± 0.70^b	20.8 ± 0.69^b	H = 148.8	0.0001
TIN (mg/l)		20.3 ± 0.9	23.1 ± 0.49^a	14.9 ± 0.58^b	11.2 ± 0.95^b	11.5 ± 0.93^b	6.25 ± 0.32^c	6.66 ± 0.51^c	H = 138.9	0.0001
NH_4-N (mg/l)	3	0.34 ± 0.48	0.63 ± 0.22	0.56 ± 0.21	0.66 ± 0.24	0.61 ± 0.24	0.32 ± 0.07	0.25 ± 0.04	H = 9.544	0.0893
NO_2-N (mg/l)		0.07 ± 0.01	0.10 ± 0.01	0.10 ± 0.01	0.07 ± 0.01	0.09 ± 0.01	0.08 ± 0.01	0.08 ± 0.02	H = 8.941	0.1114
NO_3-N (mg/l)	15	19.8 ± 1.9	22.4 ± 0.63^a	14.2 ± 0.58^b	10.8 ± 0.75^c	0.46 ± 0.77^c	5.99 ± 0.29^d	5.84 ± 0.29^d	H = 141.9	0.0001
PO_4-P (mg/l)	10	10.3 ± 1.5	26.5 ± 0.72^a	18.6 ± 0.60^b	16.6 ± 0.72^b	17.3 ± 0.61^b	9.76 ± 0.48^d	9.09 ± 0.45^d	H = 158.4	0.0001
Cl (mg/l)			227 ± 5.6^a	224 ± 7.3^a	133 ± 2.8^b	132 ± 2.4^b	123 ± 1.7^b	125 ± 1.9^b	H = 140.3	0.0001
Na (mg/l)			723 ± 4.9^a	727 ± 5.4^a	179 ± 3.5^b	179 ± 3.8^b	170 ± 3.4^b	172 ± 2.9^b	H = 133.6	0.0001

Values in the same row represented by a different superscript letters[a,b,c,d] represent means that are significantly different (multifactor ANOVA/Kruskal–Wallis, $p < 0.05$).
Treatments marked with *were subject to pH adjustment using sulphuric acid. Activated sludge (AS), primary facultative pond (PFP), NS (NS), chemical oxygen demand (COD), dissolved oxygen (DO). Values in the AS effluent column where not included in statistical analysis. # Discharge limits into a natural water resource (DWAF 1998).

decreased (Tables 2 and 3). The pH adjustment of irrigation solutions had no effect on the conductivity of old irrigation solutions (Table 3).

Dissolved oxygen was not influenced by an interaction between pH regime and water source for fresh or old NSs (Multifactor ANOVA, $p > 0.05$; Tables 2 and 3). The initial DO levels where significantly higher in NS and water-only treatments than that in any of the BE treatments (Multifactor ANOVA, $F_{(2,192)} = 99.72$, $p < 0.0001$). The DO of old irrigation solutions was similar in all hydroponic systems (Multifactor ANOVA, $F_{(2,192)} = 0.72$, $p = 0.49$). COD decreased in BE treatments from a mean of 215 ± 6.4 to 106. ± 3.2 mg/l (Tables 2 and 3). In NS and water-only treatments the COD increased from a mean of 15.3 ± 0.38 to 22.2 ± 0.72 mg/l (Tables 2 and 3). Hydroponic systems were able to decrease COD concentrations to similar concentrations observed in post-AS effluent (Table 3). The mean COD of AS-treated effluent was 16.4 ± 3.9 mg/l higher than effluent treated in hydroponic systems (Table 3). However, none of the systems was able to decrease effluent COD to discharge standards (Table 3).

The BE systems had the highest starting TIN (56.2 ± 2.6 mg/l) followed by NS (40.7 ± 1.7 mg/l) and water-only (10.1 ± 0.39 mg/l; Table 2). Total inorganic nitrogen decreased in all experimental systems, with the BE systems having the highest end TIN, followed by NS and then water-only (Kruskal–Wallis, $H_{(5,198)} = 130.06$, $p < 0.0001$; Table 3). The pH adjusted BE systems had a lower mean end TIN (14.89 ± 0.55 mg/l) compared with the unadjusted BE systems (23.1 ± 0.49 mg/l); this was not observed in NS of water-only systems (Table 3). The same trend was observed for nitrate, because nitrate made the bulk of TIN (Table 3). The mean TIN and nitrate concentration of effluent treated in pH adjusted hydroponic systems were 5.36 ± 0.76 and 5.61 ± 1.22 mg/l lower than effluent treated in the AS system (Table 3). However, neither of the systems could consistently decrease the concentration of nitrate in brewery effluent to discharge standards (Table 3).

Effluent systems had the highest starting ammonium concentration (36.7 ± 2.3 mg/l) followed by NS (17.4 ± 0.85 mg/l) and then water-only (1.67 ± 0.11 mg/l; Table 2). The ammonium concentration of old irrigation solutions was similar between all experimental systems (Kruskal–Wallis, H = 9.54, $p = 0.09$; Table 3). Both the AS and hydroponic systems were able to decrease effluent ammonium concentration to within the limits for discharge into a natural water resource (Table 3). The nitrite concentration in new irrigation solutions was higher in BE systems (0.83 ± 0.05 mg/l) while NS and water-only systems (0.04 ± 0.01 mg/l) had similar starting nitrite concentrations (Table 2). All experimental systems had similar old irrigation solution nitrite concentrations (Kruskal–Wallis, $H = 8.94$, $p = 0.11$; Table 3).

Effluent and NS systems had the highest starting phosphate concentration (28.0 ± 0.60 mg/l), while water-only systems had the lowest (12.2 ± 0.52 mg/l; Table 2). At replacement the BE systems had the highest phosphate concentration followed by NS and then water-only systems (Table 3). The pH adjusted BE systems had lower end phosphate concentrations (18.6 ± 0.60 mg/l) than the pH unadjusted BE systems (26.5 ± 0.72 mg/l; Table 3). The mean phosphate concentration of AS-treated effluent was 8.29 ± 1.03 mg/l lower than effluent treated in the pH adjusted hydroponic systems (Table 3). However, neither system could consistently decrease the phosphate concentration of BE to discharge standards (Table 3).

Effluent hydroponic systems had a higher starting sodium concentration than water-only or NS systems (Table 2). The sodium concentration increased in all systems with BE systems having the highest end sodium concentration followed by NS and water systems (Kruskal–Wallis, $H = 133.60$, $p < 0.0001$; Table 3). The pH adjustment of hydroponic systems had no influence on the sodium concentration of the old NSs (Table 3). The mean starting chloride concentration of BE systems (191 ± 5.1 mg/l) was higher than NS and water only systems (104 ± 1.5 mg/l). The chloride concentration increased in all systems with BE systems having the highest end chloride concentration of 226 ± 6.5 mg/l (Tables 2 and 3). Nutrient solution and water only systems had a combined mean end chloride concentration of 128 ± 2.2 mg/l (Table 3).

Plant productivity

The final weights of cabbages were influenced by an interaction between the pH regimes and the water sources (Multifactor ANOVA, $F_{(2,12)} = 85.50$, $p < 0.0001$; Figure 3).

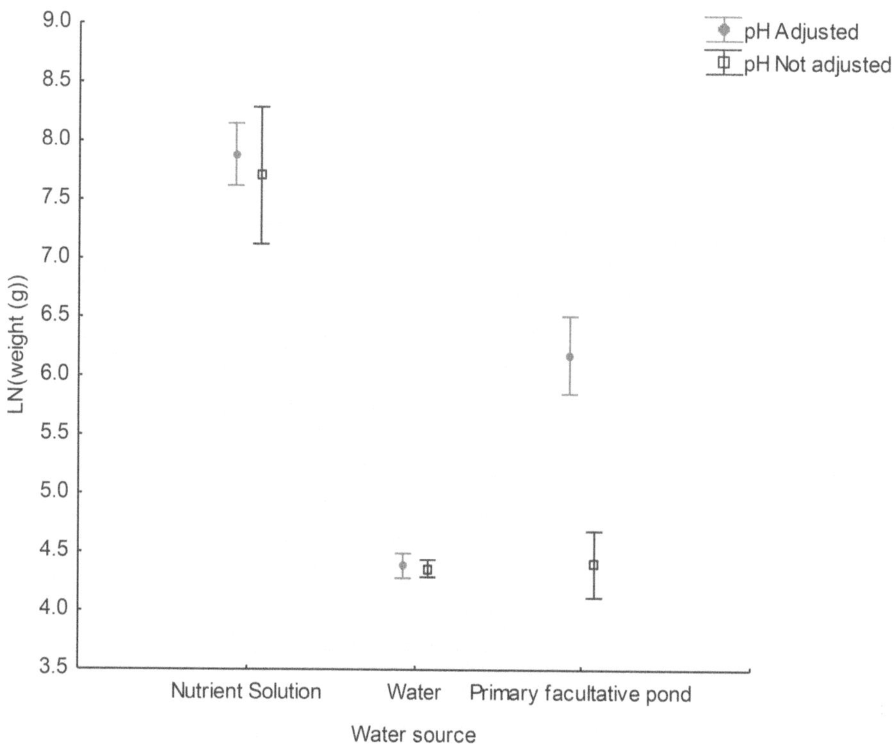

Figure 3 | Mean (± 95% confidence interval) log transformed individual weight of cabbage plants subject to the various irrigation solutions, over the 12 week experiment (multifactor ANOVA, $F_{(2,12)} = 85.50$, $p < 0.0001$).

Cabbages from the pH adjusted BE systems were significantly bigger than cabbages from the pH unadjusted BE systems, whereas pH adjustment had no influence on cabbage size in the water only and NS treatments (Figure 3). After 12 weeks, cabbages from the NS systems were bigger than cabbages from all other treatments (Figure 3). Similarly, the CCI of cabbages was influenced by an interaction between the pH regime and the water source (Multifactor repeated measures ANOVA, $p < 0.05$, Figure 4). The CCI of cabbages grown in water only and pH unadjusted BE hydroponic systems decreased over time (Figure 4). The CCI of cabbages grown in NS and pH adjusted BE systems increased over time, with NS grown cabbages having the highest CCI throughout the trial (Figure 4).

Plant chemical composition

The Na leaf concentration was highest in BE grown cabbages (21.9 ± 0.41 g/kg) followed by NS (9.86 ± 0.42 g/kg) and water-only grown cabbages (3.06 ± 0.62 g/kg; Kruskal–Wallis, $H_{(5,18)} = 15.74$, $p = 0.008$; Figure 5). Cabbages grown in pH adjusted BE systems had a lower Na concentration (14.8 ± 0.54 g/kg) than cabbages grown in BE unadjusted systems (29.1 ± 0.27 g/kg; Figure 5). The leaf concentration of all the measured macro- and micronutrients were significantly higher in cabbages grown in pH adjusted systems compared with plants grown in pH unadjusted BE systems, with the exception of Al, Cl and Zn (Table 4). The N concentration of cabbage leaves was similar between BE and NS grown cabbages (29.2 ± 0.5 g/kg) with water-only grown cabbages having the lowest leaf N concentration (7.9 ± 0.2 g/kg; Table 4). Nutrient solution grown cabbages had the highest P and K leaf concentration followed by BE and then water-only grown cabbages (Table 4). The leaves of BE grown cabbages had lower Mg and Mn concentrations than NS and water-only grown cabbages (Table 4). Cabbages grown in pH unadjusted BE systems had lower leaf Fe and Cu concentrations than cabbages from all other experimental systems (Table 4). Effluent and NS grown cabbages had higher leaf Cl and Al concentrations than water-only grown cabbages (Table 4).

DISCUSSION

Water quality

The post-PFP BE had a high alkalinity as it took about three millilitres of 98% sulphuric acid to decrease the pH of 25 l BE from 8.41 ± 0.27 to 6.5. The pH in all pH adjusted hydroponic systems increased to a pH between 8.0 and 9.0 after seven days. The pH adjustment of BE was only done once, at the beginning of each replacement because constant pH adjustment would increase the already high conductivity of the BE systems thus putting more osmotic stress on the plants. The high alkalinity of BE is generated from the addition of sodium hydroxide to BE before it goes through

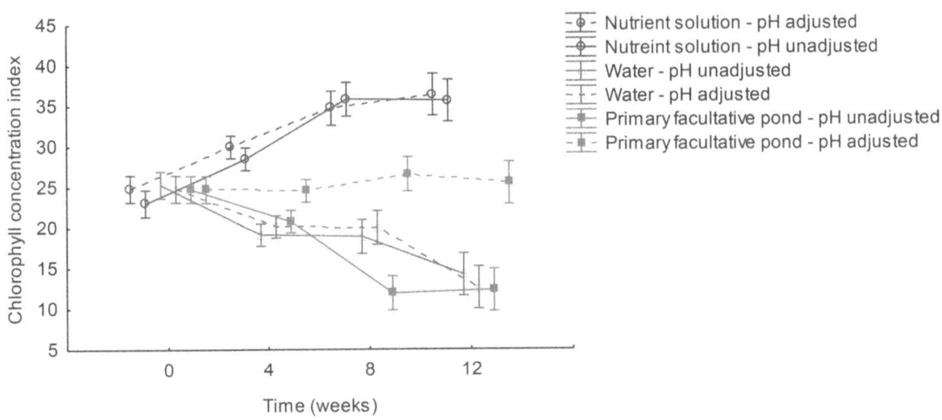

Figure 4 | Mean (± 95% confidence interval) chlorophyll concentration index of cabbage plants subject to the various irrigation solutions, over the 12 week experiment (multifactor repeated measures ANOVA $F_{(15,36)} = 34.76$, $p < 0.0001$).

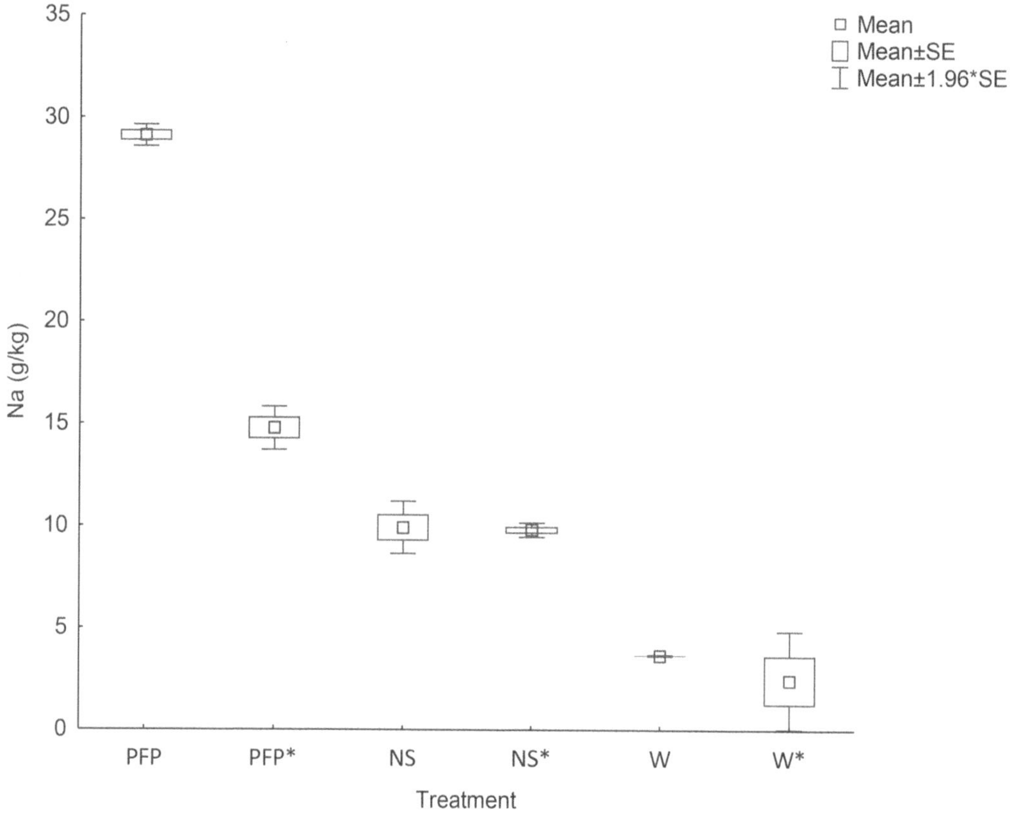

Figure 5 | Sodium concentration of cabbage leaves from plants subject to the various irrigation solutions (Kruskal–Wallis, $H_{(5,18)} = 15.74$, $p = 0.008$). Nutrient solution (NS), water-only (W), primary facultative pond (PFP). The irrigation water of treatments marked with *were subject pH adjustment using sulphuric acid.

Table 4 | Mean (± standard error) chemical concentration of cabbage leaves from plants subject to the experimental irrigation solutions

Element	PFP	PFP*	NS	NS*	Water-only	Water-only*	F/H	*p*-value
Aluminium (mg/kg)	50.7 ± 0.6^{a}	41.1 ± 5.4^{a}	34.4 ± 1.6^{ab}	33.7 ± 1.9^{ab}	15.9 ± 2.1^{b}	17.0 ± 1.2^{b}	H = 13.30	0.0207
Chloride (g/kg)	7.7 ± 0.1^{a}	7.3 ± 0.3^{a}	7.4 ± 0.4^{a}	7.5 ± 0.4^{a}	5.4 ± 0.3^{b}	5.5 ± 0.4^{b}	F = 0.42	0.6660
Copper (mg/kg)	0.2 ± 0.1^{a}	2.9 ± 0.6^{b}	3.0 ± 0.1^{b}	3.0 ± 0.1^{b}	2.7 ± 0.2^{b}	2.6 ± 0.3^{b}	H = 10.37	0.0655
Iron (mg/kg)	25.6 ± 2.1^{a}	66.0 ± 9.9^{b}	70.3 ± 5.5^{b}	67.0 ± 4.0^{b}	40.5 ± 0.1^{ab}	40.7 ± 1.2^{ab}	H = 14.44	0.0130
Nitrogen (g/kg)	27.7 ± 0.1^{a}	32.8 ± 0.2^{b}	28.2 ± 1.1^{a}	28.0 ± 0.7^{a}	8.2 ± 0.1^{c}	7.5 ± 0.3^{c}	H = 14.82	0.0112
Magnesium (g/kg)	1.7 ± 0.1^{a}	2.8 ± 0.1^{b}	4.0 ± 0.2^{c}	4.7 ± 0.5^{c}	8.7 ± 0.1^{d}	8.6 ± 0.2^{d}	H = 15.83	0.0073
Manganese (mg/kg)	8.9 ± 0.7^{a}	16.9 ± 1.3^{b}	27.9 ± 2.2^{c}	28.7 ± 2.0^{c}	30.1 ± 0.8^{c}	31.3 ± 1.2^{c}	F = 4.00	0.0466
Phosphorous (g/kg)	1.7 ± 0.1^{a}	4.3 ± 0.1^{b}	4.6 ± 0.1^{c}	4.7 ± 0.1^{c}	2.1 ± 0.1^{d}	2.2 ± 0.1^{d}	F = 255.64	0.0001
Potassium (g/kg)	10.4 ± 0.1^{a}	25.7 ± 0.4^{b}	32.4 ± 1.0^{c}	33.3 ± 0.5^{c}	15.4 ± 0.1^{d}	15.1 ± 0.2^{d}	F = 442.67	0.0001
Zinc (mg/kg)	36.9 ± 3.7^{ac}	42.8 ± 2.0^{a}	37.6 ± 0.8^{a}	36.3 ± 0.9^{ab}	35.4 ± 0.4^{bc}	34.8 ± 0.4^{bc}	F = 2.02	0.1756

Values in the same row represented by a different superscript letters$^{(a,b,c,d)}$ represent means that are significantly different (multifactor ANOVA/Kruskal–Wallis, $p < 0.05$). Treatments marked with *were subject to pH adjustment using sulphuric acid. Primary facultative pond (PFP), NS (NS).

AD and the generation of carbonate, bicarbonate and ammonium alkalinity during AD (Van Rensburg *et al.* 2003; Power & Jones 2016). The pH of anaerobic supernatants has been shown to increase with aeration of the liquor (Musvoto *et al.* 2000; Power & Jones 2016). The high alkalinity of BE is a concern when using BE in hydroponic crop production systems because it is difficult to maintain a stable pH between 6.5 and 7.0 in hydroponic systems (Power & Jones 2016).

One of the main functions of the AS process is to decrease the COD effluents. The hydroponic systems were successful in decreasing the COD of BE. However, the AS and hydroponic systems were not able to decrease the COD of BE to discharge standards. The main mechanisms for COD removal in constructed wetlands include sedimentation, filtration, adsorption, plant uptake and microbial metabolism (El-Khateeb *et al.* 2009; Ong *et al.* 2009; Fan *et al.* 2013). The plants and associated micro-organisms in the hydroponic systems were able to reduce the COD of the BE, indicating that hydroponic systems could potentially be used to replace AS systems for the treatment of post-AD brewery effluent.

The AS treatment process is also responsible for the removal of ammonium and inorganic nitrogen contained in wastewater. The hydroponic systems were able to decrease the ammonium concentration of brewery effluent to discharge standards and decrease the TIN concentration of the effluent, with the same trend observed in the AS system. The standard operating procedures of the AS system at Ibhayi brewery automatically maintains DO levels between 0.5 and 1.2 mg/l (Mclean 2015). These DO concentrations allow an oxygen concentration gradient within the flocs. Nitrifying bacteria occur in the outer layer, where aerobic conditions are present while denitrifiers reside in the middle anoxic zone of the flocs (Law *et al.* 2012). This allows nitrification and denitrification to occur simultaneously, resulting in the removal of ammonium via oxidation and nitrate via denitrification to nitrogen gas (Mekinia *et al.* 2009). The oxic conditions present in the hydroponic systems allowed for the nitrification of ammonium to nitrate, after which the plants were responsible for the removal of nitrate via assimilation (Kadlec & Wallace 2009; Sun *et al.* 2019). Hydroponics systems were successful in decreasing the TIN and ammonium concentration of brewery effluent further indicating their potential in a brewery effluent treatment process.

Plant productivity

Brewery effluent is not a complete NS and may contain certain properties that inhibit crop growth. Cabbages grown in the NS hydroponic systems grew faster and had a higher end weight than BE grown cabbages, since commercial hydroponic solutions are designed to promote plant growth. Jones *et al.* (2014) and Power & Jones (2016) found that irrigation with post-PFP BE reduced tomato and lettuce growth when compared with plants irrigated with a hydroponic NS. They concluded that the pH and high conductivity were the major factors contributing to this. The post-PFP BE contained higher concentrations of ammonium nitrogen and similar P concentrations when compared to the NS used in this experiment. Ammonium toxicity increases because the proportion of its cationic form increases as the pH increases (Britto & Kronzucker 2002; Bar-Yosef *et al.* 2009; Borgognone *et al.* 2012). Ammonium toxicity is linked with specific nutrient uptake restrictions in crops, specifically other cations such as Ca, P and Mg (Britto & Kronzucker 2002; Bar-Yosef *et al.* 2009; Borgognone *et al.* 2012). Brewery effluent grown cabbages had lower Ca, P and Mg leaf concentrations than NS or water-only grown cabbages. This suggests that ammonium toxicity may have played a role in putting stress on the BE irrigated plants thus reducing their growth. However, the ammonium in BE is unstable and the oxic conditions in the hydroponic systems allowed for the biological nitrification of ammonium to nitrate (Tyson *et al.* 2007; Sun *et al.* 2019). This results in a decrease in ammonium toxicity and in ammonium generated alkalinity (Gallert *et al.* 1998; Britto & Kronzucker 2002). Therefore, it is probably not ammonium generated alkalinity or toxicity that caused the reduction in growth of BE grown cabbages. However, the pH of hydroponic systems still rose thus indicating that other factors such as the carbon acid/base system were contributing to the high alkalinity observed in the hydroponic systems.

It is probably not the lack of nutrients in BE that resulted in the decreased plant growth but other properties such as high conductivity and pH (Jones *et al.* 2014; Power & Jones 2016). The conductivity and salinity of most effluents limit their use in irrigated crop production (Pescod 1992; Muyen *et al.* 2011). For the first five days after planting all cabbage seedlings planted in BE hydroponic systems showed signs of wilting while cabbages from the NS and Water-only systems did not. Wilting is a sign of osmotic stress in plants and this reflects the conductivity and/or salinity of the BE used to irrigate the plants (Epstein & Bloom 2005; Castillo 2011). The cabbages stopped wilting and regained their regular shape five days after planting. Plants have the ability to adjust to moderate levels of salinity via various physiological adaptions, such as ion compartmentalization, osmotic adjustment, succulence, selectivity and uptake of ions, enzyme responses, and the balance of nutrient uptake (Munns 2002; Castillo 2011; Pandolfi *et al.* 2012). Initially, the cabbage seedlings were not accustomed to the conductivity levels of BE and took a few days to adjust to the high conductivity of BE systems.

The salinity of irrigation waters between 2,000 and 3,000 $\mu S/cm^2$ causes a decrease yield in most crops (Shannon & Grieve 1999). The threshold salinity of cabbages is 1,800 $\mu S/cm^2$ and the conductivity of BE (3,068 ± 43 $\mu S/cm^2$) should result in a 10–20% decrease in yield (Shannon & Grieve 1999). This is primarily due to the decrease in the osmotic potential between the root plasma and extracellular water (Munns & Termaat 1986; Jacoby 1994). This means that plants have to spend more energy to take up water, which increases respiration and has negative effects on growth (Munns & Termaat 1986; Jacoby 1994; Mahjoor *et al.* 2016). The impact of salinity is species specific and is also mediated by environmental factors such as humidity, temperature, wind, light and air pollution (Shannon *et al.* 1994). The conductivity of post-PFP BE puts stress on crops, which could be expected to be accompanied by a decrease in growth and yield.

The bioavailability of most plant nutrients are affected by pH with their availability decreasing drastically at pH values above 7.5 (Lucas & Davis 1961; Tyson *et al.* 2007). The pH adjustment of BE significantly increased the growth, yield and CCI of cabbages. Chlorophyll concentration index is related to the chlorophyll content of a leaf and can be used to directly measure the photosynthetic integrity and potential of a leaf (Moran *et al.* 2000; Hailemichael *et al.* 2016). The yield and CCI of cabbages grown in pH unadjusted BE systems and water-only systems (nutrient limited) were similar. Brewery effluent with no pH adjustment is not a suitable irrigation source for hydroponic production of cabbages. The high pH of BE in pH unadjusted systems decreased the availability of nutrients to cabbages, resulting in cabbages growing at the same rate as nutrient limited cabbages.

The effect of pH on the availability of micronutrients to cabbage plants became more pronounced, when looking at the leaf chemical composition. The leaf concentration of all the measured macro- and micronutrients was higher in cabbages grown in pH adjusted systems compared with pH unadjusted BE systems, with the exception of aluminium (Table 4). Most micronutrients become less available to plants when the pH rises above 6.5 (Lucas & Davis 1961; Epstein & Bloom 2005; Tyson *et al.* 2007). Plants often display nutritional disorders when grown under saline environments and these disorders can become even more pronounced when the pH of the environment in unfavourable (Grattan & Grieve 1999).

There was a marked difference in the Na concentration of cabbage leaves subject to the various irrigation treatments. The leaves of effluent grown cabbages had a substantially higher sodium concentration compared with that of cabbages grown in NS and water-only. The sodium concentration of plant tissue has been found to increase as the sodium concentration of irrigation waters increases (Glenn *et al.* 1999; Silveira *et al.* 2009; Diaz *et al.* 2013). Sodium puts osmotic stress on plants and can reduce the availability of certain nutrient to plants such as Ca, K and Mg (Grattan & Grieve 1999; Zorb *et al.* 2019). High concentrations of sodium in saline waters increases the ratio of Na to nutrient cations such as Ca, K and Mg, resulting in increased Na uptake and decreased nutrient cation uptake (Maas & Grieve 1987; Alam *et al.* 1989; Ehret *et al.* 1990). Plants have to spend more energy on obtaining water and nutrients, thus reducing the amount of energy spent on growth therefore, reduced yield (Ehret *et al.* 1990; Grattan & Grieve 1999; Mahjoor *et al.* 2016). The interaction between salinity and nutrient deficiencies is extremely complex and is influenced by crop type, soil type and the environment in which the crops are grown (Grattan & Grieve 1999).

CONCLUSION

The pH adjustment of post-PFP BE had a major influence on the growth, health and chemical composition of cabbage plants grown in hydroponic systems. The final weight of cabbages from the pH adjusted BE systems were 6.7 times greater than cabbages from the pH unadjusted BE systems, whereas pH adjustment had no influence on cabbage weight in the water-only and NS treatments. The macro-

and micronutrient concentrations of cabbage leaves increased when the pH of post-PFP BE was adjusted to 6.5 at the start of each irrigation cycle. Post-PFP BE that is not pH adjusted is not a suitable water and nutrient source for the hydroponic production of cabbages; however, pH adjustment of BE renders it much more suitable. The pH adjustment of BE resulted in increased nutrient removal by hydroponic systems when compared to systems fed pH unadjusted BE. Hydroponic systems fed with post-PFP brewery adjusted to pH 6.5 were able to decrease dissolved inorganic nitrogen, ammonium, phosphate and COD concentrations from 54.8, 36.9, 27.3 and 216 mg/l to 14.9, 0.56, 18.6 and 105 mg/l; respectively

The high alkalinity of BE is a concern, firstly for deceasing the availability of nutrients in BE, and, secondly for making it hard to maintain a pH range between 6.5 and 7.0 to optimise the availability of nutrients to the plants. Continual pH adjustment would increase the conductivity of the BE, putting more osmotic stress onto the irrigated plants. The generation of alkalinity needs to be fully understood and technologies or practices need to be investigated that can reduce the alkalinity of BE for it to be successfully used in hydroponic crop production. Hydroponic crop production systems can be used to treat post-anaerobically digested BE to a similar standard obtained by AS treatment and future research should be aimed at identifying the most suitable crop species for this purpose.

ACKNOWLEDGEMENTS

The financial support from the Levenstein bursary at Rhodes University and the Water Research Commission (K5/2284) in South Africa is acknowledged. Thankyou to SAB Ibhayi Brewery and its staff for their in-kind financial support and assistance throughout the project.

DATA AVAILABILITY STATEMENT

All relevant data are available from an online repository or repositories. https://drive.google.com/drive/u/0/folders/1hCSOrlf3AZVmR8ISEgMIRzWcj-b4GfEC.

REFERENCES

Alam, S. M., Naqvi, S. S. M. & Amzi, A. R. 1989 Effect of salt stress on growth of tomato. *Pakistan Journal of Scientific and Industrial Research* **32**, 110–113.

Bar-Yosef, B., Mattson, N. S. & Lieth, H. J. 2009 Effects of NH_4:NO_3:urea ratio on cut roses yield, leaf nutrients content and proton efflux by roots in closed hydroponic system. *Scientia Horticulturae* **122**, 610–619.

Borgognone, D., Colla, G., Rouphael, Y., Cardarelli, M., Rea, E. & Schwarz, D. 2012 Effect of nitrogen form and NS pH on growth and mineral composition of self-grafted and grafted tomatoes. *Scientia Horticulturae* **149**, 61–69.

Britto, D. T. & Kronzucker, H. J. 2002 NH_4^+ toxicity in higher plants: a critical review. *Journal of Plant Physiology* **159**, 567–584.

Castillo, N. 2011 *A Hydroponic Approach to Evaluate Responses to Salinity Stress in Cotton*. Doctor of Philosophy Thesis, Texas Technology University, Lubbock, USA.

Chow, W. S., Ball, M. C. & Anderson, J. M. 1990 Growth and photosynthetic responses of spinach to salinity: implications of K^+ nutrition for salt tolerance. *Australian Journal of Plant Physiology* **17**, 563–578.

Diaz, F. J., Benes, S. E. & Grattan, S. R. 2013 Field performance of halophytic species under irrigation with saline drainage water in the San Joaquin Valley of California. *Agricultural Water Management* **118**, 59–69.

DWAF 1998 *General Limits for Discharge Into A Water Resource. Revision of the General Authorizations in Terms of Section 39 of the National Water Act, National Gazette No. 26187*. Department of Water Affairs and Forestry, Pretoria, South Africa.

Ehret, D. L., Redmann, R. E., Harvey, B. L. & Cipywnyk, A. 1990 Salinity-induced calcium deficiencies in wheat and barley. *Plant Soil* **128**, 143–151.

El-Khateeb, M. A., Al-Herrawy, A. Z., Kamel, M. M. & El-Gohary, F. A. 2009 Use of wetlands as post-treatment of anaerobically treated effluent. *Desalination* **245**, 50–59.

Epstein, E. & Bloom, A. 2005 *Mineral Nutrition of Plants: Principles and Perspectives*. Sinauer Associates, Sunderland, USA.

Fan, J., Zhang, B., Zhang, J., Ngo, H. H., Guo, W., Liu, F., Guo, Y. & Wu, H. 2013 Intermittent aeration strategy to enhance organics and nitrogen removal in subsurface flow constructed wetlands. *Bioresource Technology* **141**, 117–122.

Gallert, C., Bauer, S. & Winter, J. 1998 Effect of ammonia of the anaerobic digestion of protein by a mesophilic and thermophilic biowaste population. *Applied Microbiology and Biotechnology* **50**, 495–501.

Glenn, E. P., Brown, J. J. & Blumwald, E. 1999 Salt tolerance and crop potential of halophytes. *Plant Sciences* **18**, 227–255.

Grattan, S. R. & Grieve, C. M. 1999 Salinity-mineral nutrient relations in horticultural crops. *Scientia Horticulurae* **78**, 127–157.

Hailemichael, G., Catalina, A., Gonzalez, M. R. & Martin, P. 2016 Relationships between water status, leaf chlorophyll content and photosynthetic performance in tempranillo vineyards. *South African Journal of Enology and Viticulture* **37** (2), 149–156.

Jacoby, B. 1994 Mechanisms involved in salt tolerance by plants. In: *Handbook of Plant and Crop Stress* (M. Pessarakli ed.). Marcel Dekker, New York, pp. 97–123.

Jones, C. L. W., Britz, P. J., Cilliers, A., Laubscher, R., Power, S. & Scheepers, R. 2014 *Environmentally Sustainable Beneficiation of Brewery Effluent: Algal Ponding, Constructed Wetland, Hydroponic Vegetables and Aquaculture (Final Draft)*. Report to the Water Research Commission (K5/2008), South Africa.

Kadlec, R. H. & Wallace, S. 2009 *Treatment Wetlands*, 2nd edn. CRC Press, Taylor & Francis Group, Florida, United States of America.

Law, Y., Liu, Y., Pan, Y. & Yuan, Z. 2012 Nitrous oxide emission from wastewater treatment processes. *Philosophical Transactions of the Royal Society* **367**, 1265–1277.

Liu, G. D., Simonne, E. H., Morgan, K. T. & Hochmuth, G. J. 2015 Soil and fertilizer management for vegetable production in Florida. In: *Nutrient Management of Vegetable and row Crops Handbook* (K. T. Morgan ed.). Institute of food and agricultural sciences, University of Florida, Florida, USA.

Lucas, R. E. & Davis, J. F. 1961 Relationships between pH values of organic soils and availabilities of 12 plant nutrients. *Soil Science* **92**, 177–182.

Maas, E. V. & Grieve, C. M. 1987 Sodium-induced calcium deficiency in salt-stressed corn. *Plant Cell & Environment* **10**, 559–564.

Mahjoor, F., Ghaemi, A. A. & Golabi, M. H. 2016 Interaction effects of water salinity and hydroponic growth medium on eggplant yield, water-use efficiency, and evapotranspiration. *International Soil and Water Conservation Research* **4**, 99–107.

Mclean, O. 2015 Effluent plant operator, Ibhayi Brewery, Port Elizabeth, South Africa, 11 May 2015.

Mekinia, J., Stensel, H. D., Czerwionka, K., Drewnowski, J. & Zapero, D. 2009 Nitrogen transformations and mass balances in anaerobic/anoxic/aerobic batch experiments with full-scale biomasses from BNR activated sludge systems. *Water Science and Technology* **60**, 2463–2470.

Moran, J. A., Mitchell, A. K., Goodmanson, G. & Stockburger, K. A. 2000 Differentiation among effects of nitrogen fertilization treatments on conifer seedlings by foliar reflectance: a comparison of methods. *Tree Physiology* **20**, 1113–1120.

Munns, R. 2002 Comparative physiology of salt and water stress. *Plant Cell & Environment* **25**, 239–250.

Munns, R. & Termaat, A. 1986 Whole-plant responses to salinity. *Plant Physiology* **13**, 143–160.

Munns, R. & Tester, M. 2008 Mechanisms of salinity tolerance. *Annual Review of Plant Biology* **59**, 651–681.

Musvoto, E. V., Wentzel, M. C. & Ekama, G. A. 2000 Integrated chemical-physical process modelling–II. Simulating aeration treatment of anaerobic digester supernatants. *Water Research* **34**, 1868–1880.

Muyen, Z., Moorea, G. A. & Wrigley, R. J. 2011 Soil salinity and sodicity effects of wastewater irrigation in South East Australia. *Agricultural Water Management* **99**, 33–41.

Ong, S. A., Uchiyama, K., Inadama, D. & Yamagiwa, K. 2009 Simultaneous removal of colour, organic compounds and nutrients in azo dye-containing wastewater using up-flow constructed wetland. *Journal of Hazardous Materials* **165**, 696–703.

Osawa, T. 1961 Studies of salt tolerance of vegetable crops in sand cultures, II leafy vegetables. *Japanese Society for Horticulture* **30**, 48–56.

Osawa, T. 1963 Studies of salt tolerance of vegetable crops with special reference to osmotic effects and specific ion effects. *Japanese Society for Horticulture* **32**, 211–223.

Pandolfi, C., Mancuso, S. & Shabala, S. 2012 Physiology of acclimation to salinity stress in pea (*Pisum sativum*). *Environmental and Experimental Botany* **84**, 44–51.

Pescod, M. B. 1992 *Wastewater Treatment and Use in Agriculture – FAO Irrigation and Drainage Paper 47*. FAO, Rome, Italy.

Power, S. D. & Jones, C. L. W. 2016 Anaerobically digested brewery effluent as a medium for hydroponic crop production – the influence of algal ponds and pH. *Journal of Cleaner Production* **139**, 167–174.

Riera-Vila, I., Anderson, N. O., Hodge, C. F. & Rogers, M. 2019 Anaerobically-digested brewery wastewater as a NS for substrate-based food production. *Horticulture* **5** (2), 1–12.

Roberto, K. 2005 *How to Hydroponics*, 4th edn. Future garden press, New York, USA.

Shannon, M. C. & Grieve, C. M. 1999 Tolerance of vegetable crops to salinity. *Scientia Horticulturae* **78**, 5–38.

Shannon, M. C., Grieve, C. M. & Francois, L. E. 1994 Whole plant response to salinity. In: *Plant Environment Interactions* (R. E. Wilkinson ed.). Marcel Dekker Inc., New York, USA.

Silveira, J. A. G., Araujo, S. A. M., Lima, J. P. M. S. & Viegas, R. A. 2009 Root and leaves display D contrasting osmotic adjustment mechanisms in response to NaCl-salinity in Atriplex nummularia. *Environmental and Experimental Botany* **66**, 1–8.

Simate, G. S., Cluett, J., Lyuke, S. E., Musapatika, E. T., Ndlovu, S. & Walubita, L. F. 2011 The treatment of brewery wastewater for reuse: state of the art. *Desalination* **273**, 235–247.

Sou, M. Y., Mermoud, A., Yacouba, H. & Boivin, P. 2013 Impacts of irrigation with industrial treated wastewater on soil properties. *Geoderma* **200–201**, 31–39.

Sun, S., Gao, L., He, S., Huang, J. & Zhou, W. 2019 Nitrogen removal in response to plants harvesting in two kinds of enhanced hydroponic root mats treating secondary effluent. *Science of The Total Environment* **670**, 200–209.

Sweeney, D. W. & Graetz, D. A. 1991 Application of distillery waste anaerobic digester effluent to St. Augustine grass. *Agriculture, Ecosystems and Environment* **33**, 341–351.

Taylor, R. P., Jones, C. L. W., Laing, M. & Dames, J. 2018 The potential use of brewery effluent as a water and nutrient source in irrigated crop production. *Water Resources and Industry* **19**, 47–60.

Tyson, R. V., Simmone, E. H., Davis, M., Lamb, E. M., White, J. M. & Treadwell, D. D. 2007 Effect of NS, nitrate-nitrogen concentration, and pH on nitrification rate in perlite medium. *Journal of Plant Nutrition* **30**, 901–913.

Van Rensburg, P., Musvoto, E. V., Wentzel, M. C. & Ekama, A. 2003 Modelling multiple mineral precipitation in anaerobic digester liquor. *Water Research* **37**, 3087–3097.

Worku, A., Tefera, N., Kloos, H. & Benor, S. 2018 Bioremediation of brewery wastewater using hydroponics planted with vetiver grass in Addis Ababa, Ethiopia. *Bioresources and Bioprocessing* **5**, 39.

Zorb, C., Geilfus, C. M. & Dietz, K. J. 2019 Salinity and crop yield. *Plant Biology* **21** (1), 31–38.

First received 27 July 2020; accepted in revised form 3 November 2020. Available online 19 November 2020

Water reuse: dairy effluent treated by a hybrid anaerobic biofilm baffled reactor and its application in lettuce irrigation

K. A. Santos, T. M. Gomes, F. Rossi, M. M. Kushida, V. L. Del Bianchi, R. Ribeiro, M. S. M. Alves and G. Tommaso

ABSTRACT

There is a synergy between the large quantities of organics-rich effluents generated by the dairy industry and the continually increasing water needs for crop irrigation. In this sense, this study aimed at evaluating the effect of decreasing the hydraulic retention time (HRT) on the stability and efficiency of a hybrid anaerobic biofilm baffled reactor (HABBR) treating simulated fat- and salt-rich dairy wastewater, followed by its agricultural reuse. The reactor was monitored over 328 days, during which 72, 24, and 12 h were the hydraulic detention times. After achieving steady-state, the reactor presented organic matter removal greater than 90% and produced biogas with 41 ± 23%, 53 ± 3%, and 64 ± 12% of methane for HRTs of 72, 24, and 12 h, respectively. The best process performance was observed for an HRT of 24 h, and thus, a lettuce culture was irrigated with the treated effluent. The irrigation was performed in five different treatments, for which the amount of treated effluent added to tap water varied from 0 to 100%. Both the effluent and the harvested vegetables were evaluated for microbial contamination. Apart from the 75% effluent supply condition, there were no losses in leaf mass or area observed; instead, there was an increase of these parameters for the 25% and 50% effluent supply treatment. The use of dairy effluent treated by the HABBR allowed for microbiologically safe food production. Therefore, the process offered both potential cost reduction for fertilizers, preservation of water resources, and a renewable energy source.

Key words | agro-industrial effluent, anaerobic digestion, hybrid anaerobic biofilm baffled reactor, *Lactuca sativa*

K. A. Santos
T. M. Gomes
F. Rossi
M. M. Kushida
R. Ribeiro
G. Tommaso (corresponding author)
Laboratory of Environmental Biotechnology, Department of Food Engineering, Faculty of Animal Science and Food Engineering,
University of São Paulo,
Duque de Caxias Norte, 225, Pirassuunga SP,
Brazil
E-mail: *tommaso@usp.br*

V. L. Del Bianchi
Bioprocess Laboratory, Engineering and Food Technology Department,
São Paulo State University,
São José do Rio Preto,
Brazil

M. S. M. Alves
Centre of Biological Engineering,
University of Minho,
Braga,
Portugal

HIGHLIGHTS

- The anaerobic baffled reactor was stable during treatment of fat-rich salt dairy wastewater.
- The methane content was enhanced with hydraulic retention time reduction.
- The use of dairy effluent reduced the needs related to N fertilizer by up to 50%.
- The use of drinking water for lettuce irrigation has been reduced by 50%.
- The effluent did not present a risk of contamination in lettuce production.

doi: 10.2166/ws.2020.276

GRAPHICAL ABSTRACT

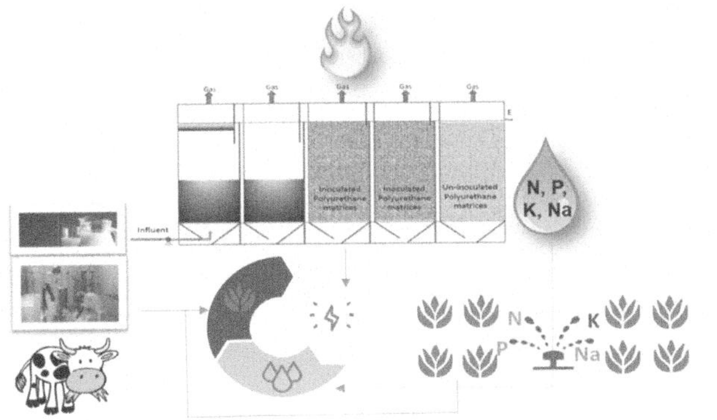

INTRODUCTION

The milk processing industry is one of the most important sectors in the food industry. However, as a consequence, there is a large volume of fat-rich and salty effluents generated. Anaerobic digestion has many advantages, in addition to treating wastewater, because it also produces a methane-rich biogas that can be recovered and used as a fuel, providing 20% of energy requirements from the dairy factory and reducing the total carbon footprint emissions by 13%, according to Stanchev *et al.* (2020). Although several studies have highlighted advantages of using anaerobic digestion for the treatment of dairy effluents (Demirel *et al.* 2005; Karadag *et al.* 2015), the effective treatment of such effluents remains a challenge due to its high lipid content and the inhibitory effect that long-chain fatty acids may present in certain concentrations (Alves *et al.* 2009).

The anaerobic baffled reactor (ABR) is a simple and efficient configuration for many wastewater treatments. The main advantage is the natural capability of phase separation, allowing acidogenesis and methanogenesis optimization (Cohen *et al.* 1982; Weiland & Rossi 1991). Consequently, different microbial groups can develop under favorable conditions. This configuration also has longer biomass retention time as an advantage, in addition to the resistance to organic and hydraulic shocks (Barker & Stuckey 1999).

Proposed by Bachmann *et al.* (1985) and extensively studied since then (Zhu *et al.* 2015; Soh *et al.* 2020), ABRs have been widely used for treating several types of effluent, either in their original configuration or presenting some optimizations. ABR are used in the treatment of sanitary effluent (Nasr *et al.* 2009), fish meal wastewater (Putra *et al.* 2020), vinasse from corn ethanol production (Sayedin *et al.* 2018), slaughterhouse wastewater (Al Smadi *et al.* 2019), vegetable waste (Gulhane *et al.* 2017), as well as effluents from oil refineries (Arvin *et al.* 2019) or heavy oil that produce water with high salt concentrations and poor nutrients (Ji *et al.* 2009).

Although very versatile, this reactor presents some limitations. According to Sayedin *et al.* (2018), the main conventional ABRs limitations are related to low biomass growth rate and biomass washout. Problems related to sludge washout are commonly reported in literature when granular reactors are used to treat dairy effluents (Passegi *et al.* 2012). One option to overcome such problems is the granular or flocculent biomass combination with fixed films. These reactors are called hybrid biofilm reactors and they were defined by Büyükkamaci & Filibeli (2002) as a combination of the upflow anaerobic sludge blanket (UASB) in the lower part, with a filter in the upper part. According to the authors, hybrid reactors may provide the advantages of the UASB and anaerobic film reactors, reducing their limitations. According to Karadag *et al.* (2015), hybrid biofilm reactors have been widely applied for dairy wastewaters treatment. Gomes *et al.* (2011) have observed organic matter removal efficiencies greater than 90% using an hybrid UASB reactor equipped with a superior polyurethane bed for dairy effluents treatment, even with high volumetric organic loads, such as 16 kg·m^{-3}·d^{-1}. The concept of hybrid biofilms reactors can easily be applied to ABRs, due to its natural compartmentalization. In this way, Fujihira *et al.* (2018) introduced a polyurethane bed inside an ABR with four chambers treating a solid/lipid-rich wastewater, while Nguyen *et al.* (2020) included a fixed-bed stage outside an ABR used for synthetic dyeing effluents treatment.

Even with an elevated performance, anaerobic reactors have not been able to produce effluents that meet such strict standards required for treated effluent discharge in water bodies, both in relation to residual organic matter and nutrient concentrations. The complete treatment is expensive because it aims to remove nutrients and salts, thus including tertiary and advanced level (thus including tertiary and advanced level), and small- to medium-sized dairy plants have therefore frequently encountered difficulties in controlling their emissions. Based on this scenario, the anaerobically treated effluents reuse would be advantageous for resources recycling, eliminating the need for a further post-treatment. In fact, the reuse of anaerobically treated wastewaters for crop irrigation can be an alternative for small wastewater treatment plants, which saves resources related to moisture and nutrient provision (Bame *et al.* 2014; Gomes *et al.* 2015).

Although dairy wastewater is a moisture and nutrient source, its salinity may restrict the agricultural reuse (Matsumoto *et al.* 2012). On the other hand, it is important to consider that different cultures may show different behavior related to salinity tolerance and nutrient uptake rates (Oliveira *et al.* 2011; Bourazanis *et al.* 2016). Lettuce (*Lactuca sativa L*) is considered 'moderately sensitive' to the presence of salts (Ayers & Westcot 1999) and is a vegetable known for its high vitamin A content. It adapts well to greenhouse cultivation and has a low cost of production with a short production cycle (Filgueira 2008). However, the needs for lettuce irrigation are high. These characteristics make this cultivar a candidate for irrigation with agro-industrial effluents.

Besides crop yields, another paramount aspect of agricultural reuse is the microbiological security of produced vegetables. The major concern is focused on the farmer health care, as well as agricultural products consumers, especially in the case of vegetables grown on the ground and consumed raw. Most studies that consider agro-industrial reuse have been carried out to evaluate the quality of effluents and post-harvested vegetables in relation to physic-chemical parameters, and further studies are required on their microbiological quality (WHO 2006).

Considering a circular economy scenario implementation, through water reuse in agriculture, this research has aimed to present data on a modified ABR, a hybrid anaerobic baffled biofilm reactor (HABBR), treating simulated fat-rich and salt dairy wastewater, posteriorly reused for irrigation of lettuce crops. Regarding the reactor operation, the objective was to verify the hydraulic retention time (HRT) influence, and hydraulic and organic shocks application on the overall organic matter removal efficiency. Our hypothesis was that, if properly operated, the HABBR could combine advantages from suspended biomass use to the treatment of fatty rich wastewaters, as preconized by Alves *et al.* (2009), while also preventing biomass losses, as observed by Gomes *et al.* (2011), providing a solid-free effluent as the necessary, considering irrigation purposes. The treated effluent from the best operational conditions was used to irrigate a lettuce crop to verify the influence on the amount of effluents used on the crop yield as well as on the biological safety of those produced vegetables. The results presented here have proved that the anaerobic digestion may be a key technology to an integrated approach to the Food–Energy–Water Nexus, thus providing security for the following generations.

METHODS

HABBR: configuration, operation, and monitoring

Reactor operation

A bench-scale HABBR was monitored in three operational conditions, namely OC1, OC2, and OC3, which were differentiated according to the applied HRT, which were 72, 24, and 12 h, respectively. The HABBR reactor (Figure 1) was constructed using acrylic glass, with five chambers with a total volume of 25 L. The

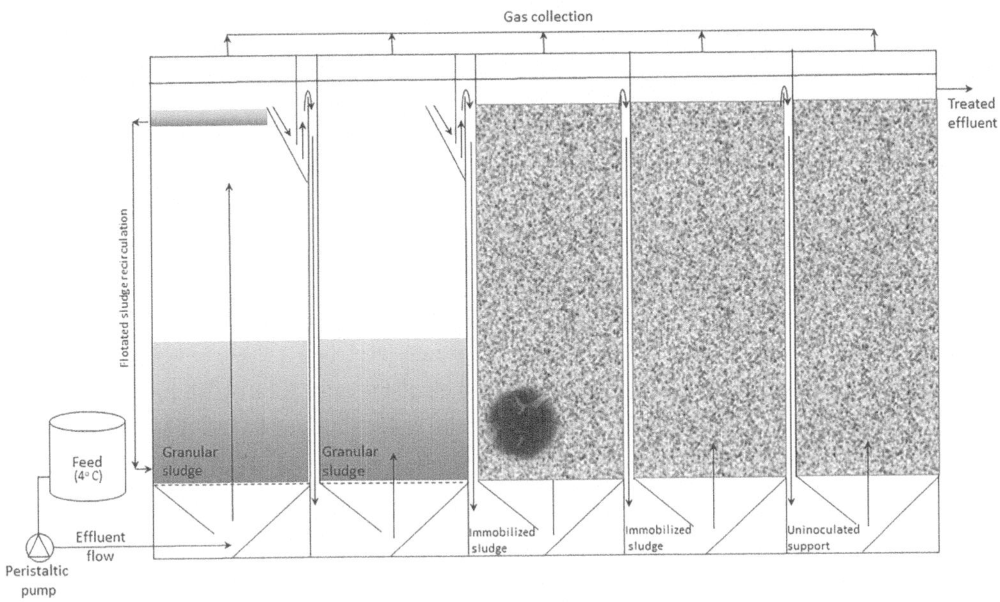

Figure 1 | Scheme of the HRT – HABBR.

first and second chambers contained granular biomass, and cubic polyurethane foam matrices (1 cm side length) were used as a support for biomass immobilization in chambers 3 and 4. Chamber 5 was provided only with polyurethane matrices. The granular inoculum was obtained from an efficient UASB reactor treating effluent from starch production. The reactor was kept inside a temperature-controlled chamber (37 ± 2 °C) to achieve temperature stability. During OC2, a hydraulic and organic load shock experiment was carried out to verify the possibility of decreasing the HRT from 24 to 12 h. The procedure was performed three times and consisted of reducing the HRT to 12 h for 36 h, subsequently subjecting the system to 24 h of HRT for 72 h. In all conditions, the HRT was maintained by a peristaltic pump (Gilson Mini plus) with four channels connected to ports distributed equidistantly across the reactor's width.

Substrate

Formulated dairy effluent with organic matter concentration expressed as chemical oxygen demand (COD) of 3 g·L^{-1}, oil and grease of 310 mg·L^{-1}, and a conductivity of 4.5 mS·cm^{-1} was used to feed the HABBR. The effluent was supplemented with 1 g.L^{-1} sodium bicarbonate ($NaHCO_3$), and its pH was corrected with a 10% hydrochloric acid (HCl) solution to a range within pH 7.0–7.5. The formulated wastewater composition was based on the characterization previously performed by Cichello *et al.* (2013), which is presented in Table 1. The composition of macronutrient and micronutrient solutions were prepared according to Zhender *et al.* (1980). The substrate reservoir was maintained at a temperature below 4 °C to minimize biochemical reactions. Before entering the reactor, the substrate was heated to 37 ± 2 °C.

Reactor monitoring

The reactor was monitored twice a week. Organic matter concentration expressed as COD, Kjeldahl nitrogen (total and

Table 1 | Formulated dairy wastewater composition

Component	Concentration
Skim powdered milk	2 g COD·L^{-1}
Milk cream (35% fat)	1 g COD·L^{-1}
Peracetic acid	0.0181 ml·**L**$^{-1}$
Hydrogen peroxide (33%)	0.028 ml·**L**$^{-1}$
Nitric acid	0.39 ml·**L**$^{-1}$
Macronutrients solution	1.8 ml·**L**$^{-1}$
Micronutrients solution	1 ml·**L**$^{-1}$

ammoniacal) concentration solids (gravimetric method), phosphorus (spectrophotometric method), sodium (using flame photometer), and chloride (using inductively coupled plasma–atomic emission spectrometry – ICP–AES), were performed according to *Standard Methods for the Examination of Water and Wastewater* (APHA 2012). The pH value was measured with a calibrated potentiometer. Bicarbonate alkalinity and total volatile acid content were determined by titration, according to the methods proposed by Ripley *et al.* (1986) and Dillalo & Albertson (1961), respectively. The biogas composition was determined by gas chromatography, using a thermal conductivity detector (Shimadzu) equipped with a Carboxen 1010 PLOT column, 30 m × 0.53 mm, according to recommendations contained at Rosero-Henao *et al.* (2019).

Agricultural reuse

Agricultural reuse was performed using simulated dairy effluent treated in the HABBR after achieving a steady state during the second operational condition. The experiment was conducted in a greenhouse with 100 m^2, in polycarbonate, and a gable roof. It consisted of a randomized block using factorial design 5 × 4 with four replicates. The treatments (T) consisted of different proportions of tap water (TW) and the anaerobically treated dairy wastewater (ANE): T1, 0% ANE +100% TW; T2, 25% ANE +75% TW; T3, 50% ANE +50% TW; T4, 75% ANE +25% TW; and T5, 100% ANE +0% TW. The lettuce seedlings were placed in 2.5-L pots on 24 April 2015. The predominant soil in the region, classified as Rhodic Hapludox, was used to fill the pots (EMBRAPA 1999). A composite sample was taken and sent to the Laboratory of Agricultural Sciences/ZAZ/FZEA (Table 2).

The characteristics presented in Table 2 were defined for lettuce crop fertilization, as suggested by Raij *et al.* (1996). The pots were filled with 3 kg of soil, which were previously adjusted adding 2.00 $t \cdot ha^{-1}$ of limestone (100%), 100 $kg \cdot ha^{-1}$ of N, 400 $kg \cdot ha^{-1}$ of phosphorus pentoxide (P_2O_5), and 100 $kg \cdot ha^{-1}$ of potassium oxide (K_2O), supplied by ammonium nitrate (NH_4NO_3) (34% of N), simple superphosphate (18% of P_2O_5 and 16% of Ca and 8% S), and potassium chloride (KCl) (60% of K_2O). The N fertilization was 100, 75, 50, 25, and 0% for T1, T2, T3, T4, and T5, respectively, and it was complementary to the addition of the treated effluents.

Irrigation management was based on the replacement of estimated crop evapotranspiration (ETc). This was calculated using a reduced Class A evaporation pan. The evaporation measure was multiplied by the culture coefficient (Kc) proposed by Marouelli *et al.* (2008) for different crop development stages. The correction coefficient (Kp) used for the reduced pan inside the greenhouse was 1, as recommended by Farias *et al.* (1994). The frequency of irrigation adopted was 2 days, in which the water volume was applied manually with the help of a graduated bottle.

The seedlings were harvested 41 days after transplanting. The analyzed parameters were the wet and dry weight of roots and leaves and the leaf area measured by LI-Cor model LI 3100. The data were subjected to a wide analysis range. For situations in which there were significant differences, the means were subjected to regression analysis.

Table 2 | Result of soil analysis used to fill the experimental portions

pH ($CaCl_2$)	P (res) ($mg \cdot dm^{-3}$)	S ($mg \cdot dm^{-3}$)	K (res) ($mmolc \cdot dm^{-3}$)	Ca ($mmolc \cdot dm^{-3}$)	Mg ($mmolc \cdot dm^{-3}$)	H + Al ($mmolc \cdot dm^{-3}$)	OM ($g \cdot kg^{-1}$)
5.5	11	20	2.1	21	9	44	11
SB	CTC	V	B	Cu	Fe	Mn	Zn
($mmolc \cdot dm^{-3}$)	($mmolc \cdot dm^{-3}$)	(%)	($mg \cdot dm^{-3}$)	($mg \cdot dm^{-3}$)	($mg \cdot dm^{-3}$)	($mg \cdot dm^{-3}$)	($mg \cdot dm^{-3}$)
32	76	42	0.58	0.5	10	1.5	0.5
Total sand	Clay	Loam					
($g \cdot dm^{-3}$)	($g \cdot dm^{-3}$)	($g \cdot dm^{-3}$)					
640	341	19					

OM, organic matter; SB, sum of bases; CTC, cation exchange capacity; V, base saturation.

Sample of leaves were washed with HCl solution (0.1%) and dried in a forced circulation oven (65 °C) until reaching constant weight. Subsequently, they were processed in a mill and sent of nutritional diagnosis analyses for the macronutrients N, K, Ca, S, P, Mg, and Na in accordance with Malavolta *et al.* (1997).

Microbial analyses

Effluents

Samples of 20 mL of the treated effluent were collected aseptically in a sterile container and analyzed weekly for 4 weeks. Decimal dilutions in saline peptone water were prepared according to the estimated contamination and analyzed for the presence of total coliforms and *Escherichia coli presence*, which were determined by inoculation, in duplicate, in Compact Dry EC plates (Nissui Pharmaceutical Co., Ltd, Tokyo, Japan). Heterotrophic bacteria were analyzed by inoculation on to the surface of Plate Count Agar (PCA), and incubated at 35 ±2 °C for 48 h (Vanderzant & Splittstoesser 2001; Silva *et al.* 2010).

Plants

Cellulose sponges soaked in buffered peptone water were used to aseptically recover microorganisms from the surface of lettuce leaves before harvesting them to sample the many conditions of soil contact, composition, and hydrophobicity. Decimal dilutions in saline peptone water were prepared according to the estimated sample contamination, which were determined by inoculation performed in duplicate. The total quantification of total mesophilic counts (PCA plates at 35 ±2 °C for 48 h), psychrotrophic (PCA plates at 7 ±2 °C for 10 days), molds, and yeasts (Dichloran Rose Bengal Chloramphenicol Plate (DRBC) at 25 ±2 °C for 10 days) were performed according to Vanderzant & Splittstoesser (2001) and Silva *et al.* (2010). Total coliforms and *E. coli* were analyzed by inoculation on Compact Dry EC plates (Nissui Pharmaceutical Co. Ltd, Tokyo, Japan) at 35 ±2 for C/24 hours. The quantification of *Staphylococcus aureus* was performed by inoculation on Compact Dry X-SA plates (Nissui Pharmaceutical Co. Ltd, Tokyo, Japan) at 35 ±2 °C for 24 h. The presence of *Salmonella* spp. was verified according to the Association of Official Analytical Chemists International (AOAC) Official Method 2003.09 (AOAC International 2007), using the BAX System PCR Assay for *Salmonella* (Du Pont Qualicon Co.) as certified by the AOAC Research Institute No 100.201.

Statistical analysis

Data regarding agricultural reuse were subjected to multiple analysis. For situations in which there were significant differences indicated by the Tukey test (5% significance level), the means were subjected to regression analysis. The software used was SISVAR 5.3 (Ferreira 2011). The results from the microbial analyses were analyzed by the Kruskal–Wallis statistical test, often used to test null hypothesis where all treatments have equal distribution functions against alternative hypothesis that at least two treatments present different distribution functions. The null hypothesis have argued that the effluent use and concentration in the irrigation medium did not statistically interfere in the microbiological contamination. The alternative hypothesis have argued that treatments generate different results.

RESULTS AND DISCUSSION

Reactor startup and monitoring

Figure 2(a) shows the efficiency of removing organic matter in each HRT over the entire operating time. The average concentration of organic matter expressed in COD was 3,078 mg·L^{-1} at the reactor inlet and 273 mg·L^{-1} at the system outlet, which represented an efficiency of 91 ± 2.4%. An average filtered COD of 2.4 g·L^{-1} was observed at the entrance and 0.23 g·L^{-1} at the output, which represented COD removal efficiency of 91 ± 3% in relation to the filtered sample. The average methane fraction in the analyzed samples throughout the operation was 46 ± 23%. The observed results are in the same range of values observed by Langenhoff *et al.* (2000) who evaluated the performance of an ABR with eight compartments and total volume of 10 L. The reactor was fed with diluted semi-skimmed milk with COD of

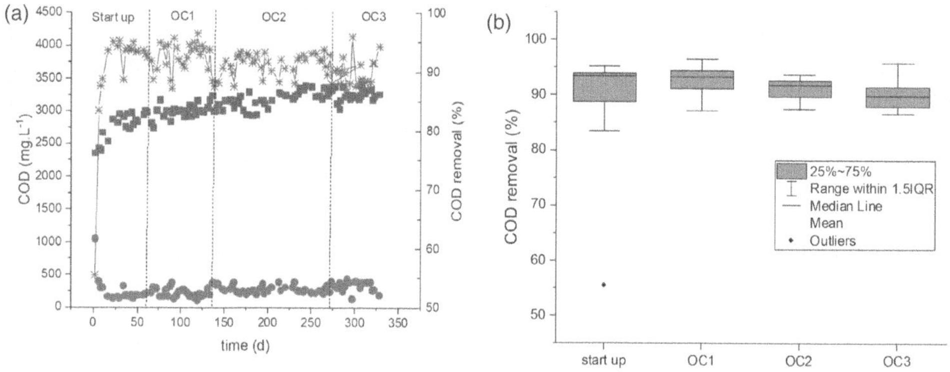

Figure 2 | Values of organic matter concentration expressed in COD and removal efficiency along the startup, operational condition 1 (OC1), operational condition 2 (OC2) and operational condition 3 (OC3): affluent COD (■), effluent COD (●) and COD removal efficiency (Ú) (a); boxplot graphic for COD removal efficiency obtained for the startup, OC1, OC2 and OC3 (b).

$0.5\ g{\cdot}L^{-1}$ and it was operated with HRT ranging between 80 and 10 h. For all applied HRT, COD removals with values above 80% were achieved.

The reactor showed stable operation throughout all conditions studied, which could be observed through consistent bicarbonate alkalinity, as well as rising methane content in the biogas. The reactor robustness can also be stated, taking into account the low variability in the values obtained for COD removal, as shown in Figure 2(b).

Start up and first operational condition

The reactor was operated over 149 days with a hydraulic detention time (TDH) of 72 h and an organic loading rate (OLR) of $1\ g.L^{-1}{\cdot}d^{-1}$, comprising the reactor startup and OC1. During the first 28 days, the organic matter removal efficiency went from 51% to 90%. The production of bicarbonate alkalinity was verified since the beginning of the operation; however, the startup period was considered concluded within 59 days of operation due to the consistency of such production and because the organic matter removal had a variation below 2% (Seborg *et al.* 1989). The results are similar to those observed by Jürgensen *et al.* (2018), who studied an ABR (HRT of 1.6 d) preceded by an continuous stirred tank reactor, which served as a buffer tank. According to the authors, the system startup treating dairy wastewater (organic loading rate (OLR) ranging from 1.25 and $4.50\ g.L^{-1}{\cdot}d^{-1}$, total HRT of 9.2 days) lasted 90 days, after which 82% of organic matter removal was achieved. After 60 days, Gomes *et al.* (2011) achieved, on average, 90% of organic matter removal in a hybrid UASB reactor treating dairy wastewater fed with an OLR of $1.1\ g.L^{-1}{\cdot}d^{-1}$.

At the end of OC1, the HABBR showed removal averages of organic matter of 92 ± 3% and 92 ± 2% (referring to the results of the last 90 days of operation) for unfiltered and filtered samples, respectively. The methane fraction in the biogas was, on average, 41 ± 23%. The results were comparable to those verified by Wang *et al.* (2004) who applied a TDH of 96 h in an ABR treating glucose-based wastewater with COD of $2.5\ g{\cdot}L^{-1}$. The authors observed that COD removal efficiency was 94% at the end of the experiment. The HABBR showed stability throughout OC1, with an average production of alkalinity to bicarbonate of 266 ± 126 mg $CaCO_3{\cdot}L^{-1}$. At the end of this condition, a ratio between intermediate alkalinty and partial alkalinity (IA/PA) of 0.1 was verified at the reactor output, which reinforces the reactor stability. Riplay *et al.* (1986) proposed to determine the relationship between intermediate alkalinity (related to the volatile acids) and partial alkalinity (related to bicarbonate) to address the anaerobic processes stability. Considering this, values higher than 0.3 indicated the occurrence of disturbances in the process. Biomass flotation occurred due to the applied oil and grease load, which was $0.105\ g\ COD{\cdot}L^{-1}{\cdot}d^{-1}$. However, each chamber was equipped with deflectors to biomass containment, which prevented sludge displacement to other compartments.

Second operational condition

The HABBR was monitored during 105 days with a TDH of 24 h and OLR of 2.96 $g.L^{-1}.d^{-1}$ during OC2. The average COD removal efficiency was 91 ± 1.8% for unfiltered and filtered samples of 91 ± 2.0%. Biogas presented a methane fraction of 53 ± 27%. The reactor was operated with stability, with an average alkalinity production of 380 ± 32 mg $CaCO_3 \cdot L^{-1}$. After 12 days at this operational condition, a biomass flotation in the first chamber due to an oil and grease load of 0.31 g $COD \cdot L^{-1} \cdot d^{-1}$ was observed. The floated biomass was grayish in color. This phenomenon was previously verified by Gomes *et al.* (2011) while operating a hybrid UASB treating dairy effluents (OLR ranging from 0.5 to 15.8 $g \cdot L^{-1} \cdot d^{-1}$). The authors verified a biomass flotation when the OLR of 4.6 g $COD \cdot L^{-1} \cdot d^{-1}$ was applied, which is higher than in OC2 when considering the entire reactor volume. Nonetheless, it is important to consider that ABR resemble a sequence of complete mix reactors in series (Barker & Stuckey, 1999). In the first chamber of the HABBR, the OLR was 9.5 $g \cdot L^{-1} \cdot d^{-1}$. Thus, clogging problems have occurred due to suspended biomass displacement to the third compartment, which contained adhered biomass. Bottom discharges and floated biomass removal were sufficient to solve such problems that did not occur during the entire operation. According to Alves *et al.* (2000), anaerobic biomass is susceptible to long-chain acids acclimation, which reverses the inhibitory effects of such molecules. Thus, when adapted to a higher concentration of oil and grease, the HABBR biomass was able to be degraded in such fraction, avoiding this substrate accumulation or its degradation intermediates (i.e. long-chain fatty acids), consequently impairing biomass flotation.

After 90 days of operation, a hydraulic and organic load shock experiment was performed. The experiment consisted of reducing detention time to 12 h for 36 h, and subsequently submitting the reactor to detention time of 24 h for 72 h. The procedure was repeated three times, and according to the monitored performance during this process, the alkalinity values were comparable to those verified in the OC1, with average production values of 346 ± 26 mg $CaCO_3 \cdot L^{-1}$. The values of the IA/PA ratio were lower than 0.3 for the three sequential shocks applied, which indicated that the HABBR quickly absorbed shocks without suffering disturbances due to accumulations of volatile acids. The average removals of organic matter were 85 ± 4.5%, 85 ± 4%, and 83 ± 4.8% for the three sequential shocks applied. The biogas composition showed an increase in the methane fraction, which was, on average, 64 ± 1.3%. During application periods that lasted 24 h, between the shock applications, the removal of organic matter was, on average, 90.5 ± 0.5%. Manariotis & Grigoropoulus (2002) studied the performance of an ABR containing three compartments and treating synthetic effluent with an OLR ranging from 0.303 and 0.662 kg $COD \cdot L^{-1} \cdot d^{-1}$. For HRT of 24 h and 12 h, the authors obtained an efficiency COD removal of 83% and 80%, respectively.

Third operational condition

During the third experimental condition, the system was operated for approximately 71 days, with a hydraulic detention time of 12 h and an OLR of 5.92 $g.L^{-1}.d^{-1}$. During this period, the average methane fraction in biogas was 62 ± 12%. The reactor showed consistent production of alkalinity, which was, on average, 355 ± 55 mg $CaCO_3 \cdot L^{-1}$, with IA/PA of 0.2.

Unfortunately, during this condition the reactor constantly presented problems, such as liquid leakage and a partial biogas outlet clogging due to the obstruction caused by the biomass. There was a biomass migration to the fourth compartment, and the treatment using minimum operational maintenance was not viable. For the HABBR operation, with part of its immobilized biomass, the system's floated sludge management must be constant. In this way, Passeggi *et al.* (2012) presented an interesting configuration of UASB reactor equipped with a scum extraction device and a lamella settler. According to the authors, the problems associated with the lipid hydrolysis low-rate lipid and foam accumulation have been resolved by extracting the scum, and 90% removal of organic matter has been verified in their system, which was on an industrial scale. This solution could easily be applied to the; however, if the intention is to keep the operation as simple as possible, conditions similar to those applied at OC2 should be maintained, if the effluent has comparable oil and grease concentrations.

Agricultural reuse

The research results on water reuse in agricultural crops are different in relation to production maintenance and quality of crops irrigated with treated effluents (Ayoub *et al.* 2016; Shahrivar *et al.* 2019), mainly due to cultivated species variability, the characteristics of these waters, and the specificities of each cultivation site, such as climate and soil. Thus, the application of the HABBR effluent to lettuce irrigation is presented and discussed in the following paragraphs.

The fresh and dry mass of lettuce roots, as well as the concentration of K, Ca, and P in leaf tissue, showed no significant changes among the tested treatments. On the other hand, the dry and fresh leaves mass values, leaf areas, and the values of N, Na, Mg, and S in the leaf tissue were influenced by effluent dosage, with a correlation coefficient greater than 0.92 (Figures 3 and 4).

Thus, an increase in the values of such biometric parameters was observed when T2 and T3 were applied. T2 showed a 37% and 6% increase for fresh and dry leaf mass, respectively, and a 37% increase in leaf areas compared to T1. The observed results are different from those observed by Oliveira *et al.* (2011), who investigated different lettuce cultivars exposed to five levels of water salinity. They observed a linear reduction in the leaf number, leaf area, and mass with an increasing salt concentration. Cáceres *et al.* (2015) applied the nitrified liquid phase from an anaerobic digester fed with cow manure to a lettuce crop without any nutrient supplementation. Comparing the present crop yields to a crop that received standard fertilization revealed no differences.

When the effluent supply was 75%, leaf weights and areas decreased compared to to all other treatments. On the other hand, when the effluent supply was 100% (T5), an increase in weight and leaf area was observed (Figure 4), suggesting that the input of organic matter provided by T5 offset the salt stress caused by it, which was not observed when the effluent supply was only 75%. The organic matter soil supply, especially in tropical soils, can bring physical, chemical, and biological benefits to crop development and, in the case of vegetables, such increase is even more beneficial (Zandonadi *et al.* 2014).

Better yields are attributed to plants submitted to irrigation with organic leachates, not only because of the high concentration of nutrients but also because of the presence of humic substances (Tejada *et al.* 2008; Singh *et al.* 2010). The interaction between humic substances application with different concentrations of sodium chloride in wheat cultivation were observed by Asik *et al.* (2009). Under saline stress, the application of humus (1 g soil·kg^{-1}) have resulted in dry mass increase due to the nutrients input increment.

The uptake of N, Na, and S in lettuce leaf tissue with increasing effluent doses between T1 and T5 were 42%, 238%, and 242%, respectively (Figure 4). The increase of N caused by the augment of the effluent dose was reflected in the absorption of these nutrients by the plant. Nevertheless, it did not cause yield enhancement, as expected. This result must be related to the high salinity levels provided by the effluent and also to the N form. The N concentration in the effluent was divided between the organic form (49.46%) and $NH4^+$ (50.54%). In a vast bibliographic review Andrews *et al.* (2013) concluded that the N form available to plants can affect seed germination time and rate, leaf expansion and function, dry matter partition between shoot and root, and root architecture. The magnitude of these effects depends not only on N supply, but also on environmental factors.

Britto & Kronzucker (2002) stated that a high NH_4^+ concentration can be phytotoxic to fertirrigated crops, and competes with the absorption of other cations due to the need to maintain the electroneutrality in the plant. Results of this competition can be verified by the reduction of Mg in the lettuce leaf when increasing effluent doses are applied (Figure 4(c)). In this way, the Mg concentration (Figure 4(c)) presents the same trend depicted in Figure 3 for the leaf mass and area. In plant metabolism, magnesium is a component of the chlorophyll molecule (vital for photosynthesis) and is involved in enzymatic activation, protein synthesis, and translocation of synthesized compounds (Guo *et al.* 2016).

As for the Na and S concentrations, in addition to the N concentration, the increase in dairy effluent dose increases availability of these elements to lettuce plants (Figure 4). Pereira *et al.* (2011) evaluated the impact on the soil–plant system that used treated sewage effluents in tropical conditions for cultivatating citrus plants and they verified the benefits in soil productivity and acidity. However, in

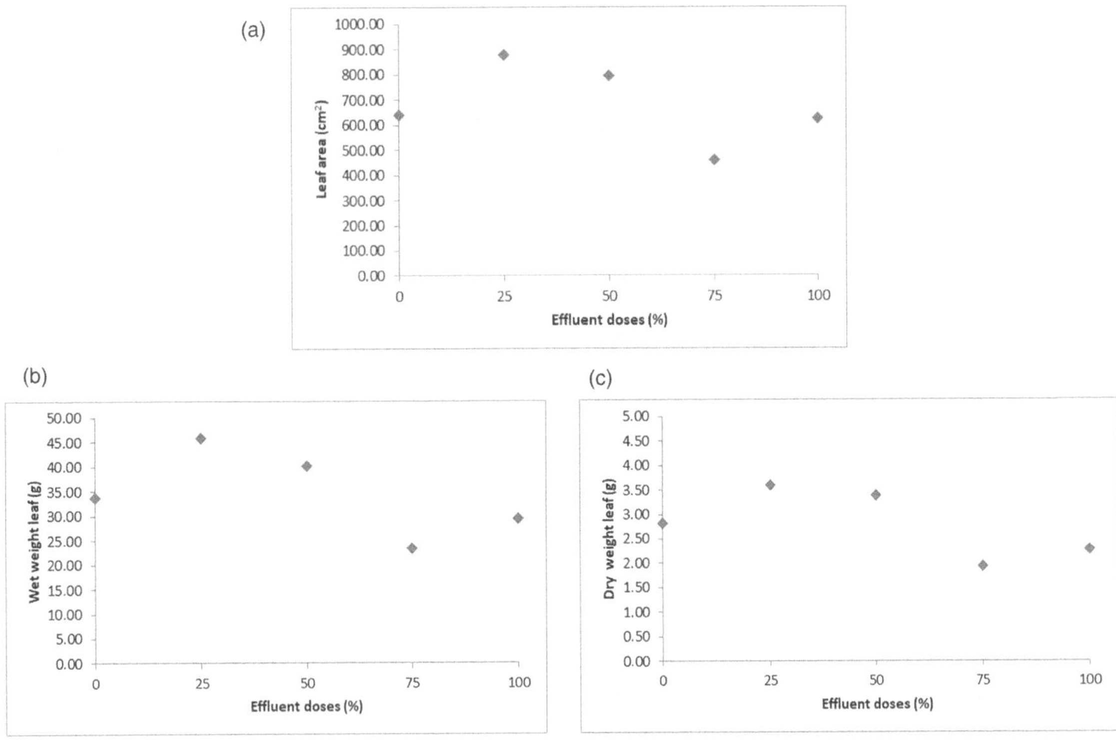

Figure 3 | Effect of different proportions of dairy effluent, 0%, 25%, 50%, 75% and 100% (T1, T2, T3, T4, and T5), on the leaf area (a) and wet (b) and dry (c) weight leaves of the lettuce. Significant at $P < 0.05$.

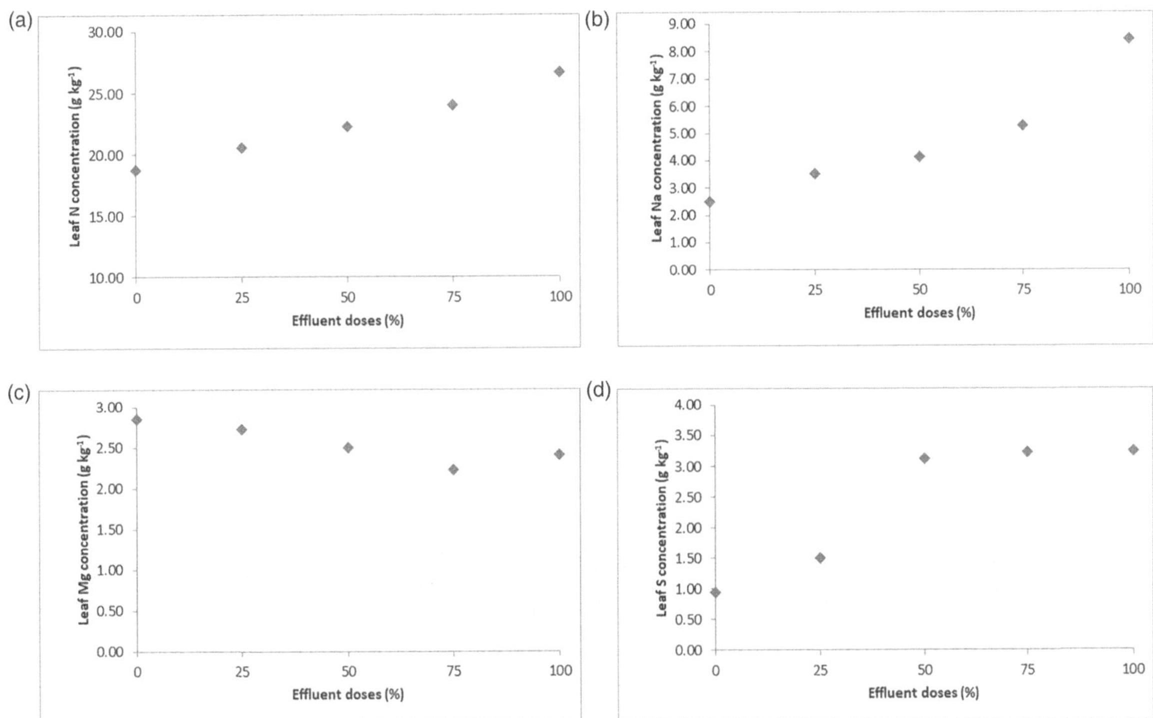

Figure 4 | Effect of different proportions of dairy effluent, 0%, 25%, 50%, 75%, and 100% (T1, T2, T3, T4, and T5), on the concentration of N (a), Na (b), Mg (c), and S (d) in leaf tissue. Significant at $P < 0.05$.

treatments with irrigation sheeting superior to plant demand, there were nutritional imbalances due to the high Na and S concentration.

The results of applying the effluent from HABBR to lettuce cultivation were promising, allowing savings of up to 50% in N fertilizer. The agricultural reuse technique can also complement the wastewater treatment with high organic and nutrient concentrations, avoiding the necessity of reaching the high quality parameters required when discharges in water sources surface are considered. The research should continue and focus the assessment on nutrient dynamics in plant and soil in relation to the N source provided by these treatments, as well as the high salt content.

Microbial analyses

Microbial analyses were performed to verify the contamination potential of the reactors for food security of the produced vegetables, because the inoculum from the anaerobic reactors can often come from domestic wastewater treatment plants, which also present potential contamination by pathogenic microorganisms. Although the HABBR was fed with simulated dairy effluent, thus theoretically free of microorganisms, it may not be the case for industrial full-scale situations, where the level of asepsis in the dairy products can be very high and where such production units do not mix their domestic effluents with their industrial effluents.

Effluents

The results for counting heterotrophs and total coliforms in the effluent over a 4-week period are shown in Table 3. *E. coli* contamination was not detected in any of the effluent samples, which were, therefore, within the limits recommended by the World Health Organization (WHO). The presence of fecal coliforms is indicative of poor effluent quality and, according to the WHO, the legislation restricts the presence of *E. coli* in irrigation water to a maximum of 3.0 log colony forming units (CFU)/100 mL$\cdot^{-1}$ (WHO 2006).

During the 4 weeks of lettuce cultivation, the effluent produced by the HABBR showed similar concentrations, without statistical differences, for total coliforms and for heterotrophic bacteria, as shown in Table 3. Heterotrophic bacteria, as well as total coliforms, are microorganisms naturally present in water samples and, by analogy, in effluents. Their presence does not create immediate health risks; however, values above 6.0 log CFU·100 mL^{-1} (WHO 2006) may represent contamination risks to the culture that has been irrigated with such source. For the present analysis, the results of counting these microorganisms remained within acceptable limits.

Table 3 | Microbiological analysis results for the anaerobically treated effluent

Microbiological analyses (log CFU·g^{-1})	Week 1	2	3	4
Heterotrophic bacteria	4.9^{A}	5.0^{A}	5.9^{A}	5.3^{A}
Total coliforms	4.7^{A}	4.9^{A}	5.1^{A}	3.8^{A}

CFU – colony forming unity.
Matching upper case letters in the lines do not differ statistically by the Kruskal–Wallis test ($P < 0.05$).

Plants

Table 4 shows the mean values among the four replicates of lettuce microbiological characteristics produced under different doses of treated dairy effluent.

The presence of *E. coli* and *Samonella sp.* has not been detected in any of the samples, meaning that they adhere to the requirements of Brazilian legislation that regulates microbiological standards for fresh vegetables. The law requires the absence of *Salmonella sp.* in 25 g of food, and the limit for thermotolerant coliforms is 2 log CFU at

Table 4 | Results of microbiological analyses on lettuces irrigated with different effluent concentrations

Microbiological analyses (log CFU/g)	Effluent dosage (%) 0	25	50	75	100
Mesophilic bacteria	4.9^{A}	5.3^{A}	4.3^{A}	5.9^{A}	5.4^{A}
Total coliforms	$<1^{A}$	$<1^{A}$	1.6^{A}	3.5^{A}	3.0^{A}
Molds and yeasts	3.6^{A}	3.6^{A}	3.9^{A}	3.3^{A}	3.6^{A}
Psychrotrophic bacteria	2.9^{A}	3.3^{A}	3.7^{A}	3.9^{A}	4.1^{A}
Staphylococcus aureus	2.1^{A}	2.6^{A}	3.0^{A}	3.0^{A}	3.0^{A}

Matching upper case letters in the lines do not differ statistically by the Kruskal-Wallis test ($P < 0.05$).

$45\ ^{\circ}C \cdot g^{-1}$ (BRASIL 2001). From the statistical point of view, the null hypothesis that the five treatments did not have significant difference in relation to the total counts of mesophilic, psychrotrophic, *S. aureus* aerobes, total coliforms, molds, and yeasts was accepted at a 5% level of significance, inferring that irrigating lettuce crops with 25%, 50%, 75%, or 100% effluent yields leads to a microbiological contamination statistically equal to irrigation with only water. It is important to note that, in addition, the current results remained within the accepted limit considered safe for human consumption in Brazil (BRASIL 2001). Thus, it is possible to state that even irrigation with a 100% effluent did not represent significant increase in microbiological load compared to irrigation with drinking water alone. This is in accordance with the data obtained in the effluent analyzes previously presented (Table 3), demonstrating that it is safe to use the effluent from the dairy treated in ABBR.

CONCLUSIONS

The HABBR has been proven to be a reliable option for fatty-rich salty wastewater treatment, as demonstrated by the consistent alkalinity production, high organic matter removal (above 90%), and rich methane biogas production. HRT of 24 h provided the optimal operational conditions among those tested, although hydraulic shocks using 12 h of HRT were well tolerated.

When carefully conducted, the agricultural reuse of the anaerobic-treated effluent could guarantee yields of the lettuce crops. The 50% dose of dairy anaerobically treated effluent resulted in the best conditions for residue utilization in agriculture, presenting an equal lettuce production without nutritional disorders and a reduction in 50% of necessary nitrogenous mineral fertilizer.

Throughout the irrigation time, the treated effluent did not show potential for microbiological contamination, proven by the results of the analysis carried out on the plants, where the count of ended mesophilic aerobes, psychrotrophic, *S. aureus*, molds and yeasts, and total coliforms were below levels allowed by legislation, thus guaranteeing food safety.

ACKNOWLEDGEMENTS

The authors are grateful to Coordenação de Aperfeiçoamento de Pessoal de Nível Superior – CAPES for the grants to L. K. A. Santos (Finace Code 001) and to M. M. S. Alves (PVE 123/2012, process no. 23038.009805/2012-39), as well for the financial support.

DATA AVAILABILITY STATEMENT

All relevant data are included in the paper or its Supplementary Information.

REFERENCES

Al Smadi, B. M., Al-Hayek, W. & Abu Hajar, H. A. 2019 Treatment of Amman slaughterhouse wastewater by anaerobic baffled reactor. *Int. J. Civ. Eng.* **17**, 1445–1454.

Alves, M. M., Mota Vieira, J. A., Alvares Pereira, R. M., Pereira, M. A. & Mota, M. 2000 Effects on lipids and oleic acid on biomass development in anaerobic fixed-bed reactors. Part II: Oleic acid toxicity and biodegradability. *Water Res.* **35**, 264–270.

Alves, M. M., Pereira, M. A., Sousa, D. Z., Cavaleiro, A. J., Picavet, M., Smidt, H. & Stams, A. J. M. 2009 Waste lipids to energy: how to optimize methane production from long-chain fatty acids (LCFA). *Microb. Biotechnol.* **2**, 538–550.

Andrews, M., Raven, J. A. & Lea, P. J. 2013 Do plants need nitrate? The mechanisms by which nitrogen form affects plants. *Annals Appl. Biol.* **163**, 174–199.

APHA/AWWA/WEF 2012 *Standard Methods for the Examination for Water and Wastewater*, 22nd edn. American Public Health Association/American Water Works Association/ Water Environment Federation, Washington DC, p. 1496.

Arvin, A., Hosseini, M., Amin, M. M., Darzi, G. N. & Ghasemi, Y. 2019 A comparative study of the anaerobic baffled reactor and an integrated anaerobic baffled reactor and microbial electrolysis cell for treatment of petrochemical wastewater. *Biochem. Eng. J.* **144**, 157–165.

Asik, B. B., Turan, M. A., Çelik, H. & Katkat, V. 2009 Effects of humic substances on plant growth and mineral nutrients uptake of wheat (Triticum durum cv. Salihli) under conditions of salinity. *Asian J. Crop Sci.* **1**, 87–95.

Association of Official Analytical Chemists International 2007 Official methods of analysis. 18th Edition, Associations of Analytical Chemists, International, Washington DC.

Ayers, R. S. & Westcot, D. W. 1999 *Water Quality for Agriculture* (H. R. Tradução de Gheyi, J. F. Medeiros & F. A. V.

Damasceno, eds). v. 29 revisado 1 (FAO). UFPB, Campina Grande, p. 218.

Ayoub, S., Al-Shdiefat, S., Rawashdeh, H. & Bashabsheh, I. 2016 Utilization of reclaimed wastewater for olive irrigation: effect on soil properties, tree growth, yield and oil content. *Agric. Water Manage.* **176**, 163–169.

Bachmann, A., Beard, V. L. & McCarty, P. L. 1985 Performance-characteristics of the anaerobic baffled reactor. *Water Res.* **19**, 99–106.

Bame, I. B., Hughes, J. C., Titshall, L. W. & Buckley, C. A. 2014 The effect of irrigation with anaerobic baffled reactor effluent on nutrient availability, soil properties and maize growth. *Agric. Water Manage.* **134**, 50–59.

Barker, D. J. & Stuckey, D. C. 1999 A review of soluble microbial products in wastewater treatment systems. *Water Res.* **33** (14), 3063–3082.

Bourazanis, G., Roussos, P. A., Argyrokastritis, I., Kosmas, C. & Kerkides, P. 2016 Evaluation of the use of treated municipal waste water on the yield, oil quality, free fatty acids' profile and nutrient levels in olive trees cv Koroneiki, in Greece. *Agric. Water Manage.* **163**, 1–8.

BRASIL 2001 Ministério da Saúde. Agência Nacional de Vigilência Sanitária – Anvisa. 2001 Resolução (Resolution) RDC no 12, de 2 de Janeiro de 2001. Diário Oficial República Federativa do Brasil, Brasília, DF.

Britto, D. & Kronzucker, H. 2002 $NH4^+$ toxicity in higher plants: a critical review. *J. Plant Physiol.* **159**, 567–584.

Büyükkamaci, N. & Filibeli, A. 2002 Concentrated wastewater treatment studies using an anaerobic hybrid reactor. *Process Biochem.* **38**, 771–775.

Cáceres, R., Magrí, A. & Marfã, O. 2015 Nitrification of leachates from manure composting under field conditions and their use in horticulture. *Waste Manage.* **44**, 72–81.

Cichello, G. C. V., Ribeiro, R. & Tommaso, G. 2013 Characterization and kinetics of the anaerobic treatment of dairy wastewaters. *UNOPAR Científica Ciências Biológicas E da Saúde* **15** (1), 27–40.

Cohen, A., Breure, M., Van Andel, J. G. & Van Deursenn, A. 1982 Influence of phase separation on the anaerobic digestion of glucose – II. Stability, and Kinetic reponses to shock loadings. *Water Revista* **16**, 449–455.

Demirel, B., Yenigun, O. & Onay, T. T. 2005 Anaerobic treatment of dairy wastewaters: a review. *Process Biochem.* **40**, 2583–2595.

Dillalo, R. & Albertson, O. E. 1961 Volatile acids by direct titration. *J. WPCF* **33**, 356–365.

EMBRAPA – Empresa Brasileira de Pesquisa Agropecuária. Centro Nacional e Pesquisa de Solos 1999 *Sistema Brasileiro de Classificação de Solos (Brazilian System of Soil Classification).* Brasília, p. 412.

Farias, J. R. B., Bergamaschi, H. & Martins, S. R. 1994 Evapotranspiration inside plastic greenhouses. *Revista Brasileira de Agrometeorologia* **2** (1), 17–22.

Ferreira, D. F. 2011 Sisvar: a computer statistical analysis system. *Cienc. Agrotechnol.* **35**, 1039–1042.

Filgueira, F. 2008 *Novo Manual de Olericultura: Agrotecnologia Moderna na Produção E Comercialização de Hortaliças. (New Manual of Horticulture: Modern Agrotechnology in the Production and Marketing of Vegetables).* UFV, Viçosa, p. 402.

Fujihira, T., Seo, S., Yamaguchi, T., Hatamoto, M. & Tanikawa, D. 2018 High-rate anaerobic treatment system for solid/lipid-rich wastewater using anaerobic baffled reactor with scum recovery. *Bioresour. Technol.* **263**, 145–152.

Gomes, D. R. S., Papa, L. G., Cichello, G. C. V., Belancon, D., Pozzi, E. G., Balieiro, J. C. C., Monterrey-Quintero, E. S. & Tommaso, G. 2011 Effect of enzymatic pretreatment and increasing the organic loading rate of lipid-rich wastewater treated in a hybrid UASB reactor. *Desalination* **279**, 96–103.

Gomes, T. M., Rossi, F., Tommaso, G., Ribeiro, R., Macan, N. P. F. & Pereira, R. S. 2015 Treated dairy wastewater effect on the yield and quality of drip irrigated table beet. *Appl. Eng. Agric.* **31**, 255–260.

Gulhane, M., Pandit, P., Khardenavis, A., Singh, D. & Purohit, H. 2017 Study of microbial community plasticity for anaerobic digestion of vegetable waste in anaerobic baffled reactor. *Renew. Energ.* **101**, 59–66.

Guo, W., Nazim, H., Liang, Z. & Yang, D. 2016 Magnesium deficiency in plants: an urgent problem. *The Crop J.* **4**, 83–91.

Ji, G. D., Sun, T. H., Ni, J. R. & Tong, J. J. 2009 Anaerobic baffled reactor (ABR) for treating heavy oil produced water with high concentrations of salt and poor nutrient. *Bioresour. Technol.* **100**, 1108–1114.

Jürgensen, L., Ehimen, E. A., Born, J. & Holm-Nielsen, J. B. 2018 A combination anaerobic digestion scheme for biogas production from dairy effluent CSTR and ABR, and biogas upgrading. *Biomass Bioenerg.* **111**, 241–247.

Karadag, D., Köroglu, O. E., Ozkaya, B. & Cakmakci, M. 2015 A review on anaerobic biofilm reactors for the treatment of dairy industry wastewater. *Process Biochem.* **50**, 262–271.

Langenhoff, A. A. M., Intrachandra, N. & Stuckey, D. C. 2000 Treatment of dilute soluble and colloidal wastewater using an anaerobic baffled reactor: influence of hydraulic retention time. *Water Res.* **34** (4), 1307–1317.

Malavolta, E., Vitti, G. C. & Oliveira, S. A. 1997 *Avaliação do Estado Nutricional das Plantas: Princípios E Aplicações. (Assessment of the Nutritional Status of Plants: Principles and Applications)*, 2nd edn. Potafos, Piracicaba, p. 319.

Manariotis, I. D. & Grigoropoulos, S. G. 2002 Low-strength wastewater treatment using an anaerobic baffled reactor. *Water Environ. Res.* **74**, 170–176.

Marouelli, W. A., Silva, W. L. & Silva, H. R. 2008 *Sprinkler Irrigation in Vegetables: Water Quality, System Aspects and Practical Method of Management*, 2nd edn. Embrapa Informação Tecnológica, Brasília.

Matsumoto, E. M., Osako, M. S., Pinho, S. C., Tommaso, G., Gomes, T. M. & Ribeiro, R. 2012 Treatment of wastewater from dairy plants using anaerobic sequencing batch reactor (ASBR) following by aerobic sequencing batch reactor (SBR) aiming the removal of organic matter and nitrification. *Water Prac. Technol.* **7**, 2012.

Nasr, F. A., Doma, H. S. & Nassar, H. F. 2009 Treatment of domestic wastewater using an anaerobic baffled reactor followed by a duckweed pond for agricultural purposes. *Environmentalist* **29**, 270–279.

Nguyen, T. H., Watari, T., Hatamoto, M., Sutani, D., Setiadi, T. & Yamaguchi, T. 2020 Evaluation of a combined anaerobic baffled reactor–downflow hanging sponge biosystem for treatment of synthetic dyeing wastewater. *Environ. Tech. Innov.* **19**, 100913.

Oliveira, F. A., Carrilho, S. O., Medeiros, J. F., Maracajá, P. B. & Oliveira, M. K. T. 2011 Performance of lettuce cultivars under different salinity levels of the irrigation water. *Rev. Bras. de Engen. Agríc. E Amb.* **15**, 771–777.

Passeggi, M., López, I. & Borzacconi, L. 2012 Modified UASB reactor for dairy industry wastewater: performance indicators and comparison with the traditional approach. *J. Clean. Prod.* **26**, 90–94.

Pereira, B. F. F., He, Z. L., Silva, M. S., Herpin, U., Nogueira, S. F., Montes, C. R. & Melfi, A. J. 2011 Reclaimed wastewater: impact on soil – plant system under tropical conditions. *J. Hazard. Mater.* **192**, 54–61.

Putra, A. A., Watari, T., Maki, S., Hatamot, M. & Yamaguchi, T. 2020 Anaerobic baffled reactor to treat fishmeal wastewater with high organic content. *Environ. Tech. Innov.* **17**, 100586.

Raij, B. v., Cantarella, H., Quaggio, J. A. & Furlani, A. M. C. 1996 *Recomendações de Adubação E Calagem Para O Estado de São Paulo. (Fertilization and Liming Recommendations for the State of São Paulo)*. Instituto Agronômico, Campinas, p. 285.

Ripley, L. E., Boyle, W. C. & Converse, J. C. 1986 Improved alkalimetric monitoring for anaerobic digestion of high-strength wastes. *J. WPCF* **58**, 406–411.

Rosero-Henao, J. C., Bueno, B. E., De Souza, R., Ribeiro, R., De Oliveira A., L., Gomide, C. A., Gomes, T. M. & Tommaso, G. 2019 Potential benefits of near critical and supercritical pre-treatment of lignocellulosic biomass towards anaerobic digestion. *Waste Manage. Res.* **37**, 74–82.

Sayedin, F., Kermanshahi-pour, A. & He, S. 2018 Anaerobic digestion of thin stillage of corn ethanol plant in a novel anaerobic baffled reactor. *Waste. Manage.* **78**, 541–552.

Seborg, D. E., Edgar, T. F. & Mellichamp, D. A. 1989 *Process Dynamics and Control*. Wiley, New York.

Shahrivar, A. A., Rahman, M. M., Hagare, D. & Maheshwari, B. 2019 Variation in kikuyu grass yield in response to irrigation with secondary and advanced treated wastewaters. *Agric. Water Manage.* **222**, 375–385.

Silva, N., Junqueira, V. C. A., Silveira, N. F. A., Taniwaki, M. H. & Santos, R. F. S. 2010 *Food Microbiological Analysis Methods Manual*. Varela, São Paulo.

Singh, R., Gupta, R. K., Patil, R. T., Sharma, R. R., Asrey, R., Kumar, A. & Jangra, K. K. 2010 Sequential foliar application of vermicompost leachates improves marketable fruit yield and quality of strawberry (*Fragaria* x *ananassa* Duch). *Sci. Hortic.* **124**, 34–39.

Soh, Y. N. A., Kunacheva, C., Webster, R. D. & Stuckey, D. C. 2020 Composition and biotransformational changes in soluble microbial products (SMPs) along an anaerobic baffled reactor (ABR). *Chemosphere* **254**, 126775.

Stanchev, P., Vasilaki, V., Egas, D., Colon, J., Ponsa, S. & Katsou, E. 2020 Multilevel environmental assessment of the anaerobic treatment of dairy processing effluents in the context of circular economy. *J. Clean. Prod.* **261**, 121–139.

Tejada, M., Gonzalez, J. L., Hernandez, M. T. & Garcia, C. 2008 Agricultural use of leachates obtained from two different vermicomposting processes. *Bioresour. Technol.* **99**, 6228–6232.

Vanderzant, C. & Splittstoesser, D. F. 2001 *Compendium of Methods for the Microbiological Examination of Foods*, 4th edn. American Public Health Association, Washington, DC.

Wang, J., Huang, Y. & Zhao, X. 2004 Performance and characteristics of an anaerobic baffled reactor. *Bioresour. Technol.* **93**, 205–208.

Weiland, P. & Rossi, A. 1991 The start up, operation and monitoring of high-rate anerobic treatment systems: discusser's report. *Water Sci. Technol.* **24**, 255–277.

World Health Organization 2006 *Guidelines for the Safe use of Wastewater, Excreta and Greywater: Wastewater Use in Agriculture*. World Health Organization, Geneva.

Zandonadi, D. B., Santos, M. P., Medici, L. O. & Silva, J. 2014 Ação da matéria orgânica e suas frações sobre a fisiologia de hortaliças. (Action of organic matter and its fractions on the physiology of vegetables). *Hort. Brasileira* **32** (1), 14–20.

Zhender, A. J. B., Huser, B. A., Brock, T. D. & Wuhrmann, K. 1980 Characterization of an acetate-decarboxylating, non-hydrogen-oxidazing methane bacterium. *Arch. Microbiol.* **124**, 1–11.

Zhu, G., Zou, R., Jha, A. K., Huang, X., Liu, L. & Liu, C. 2015 Recent developments and future perspectives of anaerobic baffled bioreactor for wastewater treatment and energy recovery. *Crit. Rev. Env. Sci. Technol* **45**, 1243–1276.

First received 30 June 2020; accepted in revised form 2 October 2020. Available online 20 October 2020

Investigation of plasticizer production industry wastewater treatability using pressure-driven membrane process

Yasin Ozay, Oltan Canli, Bahar Ozbey Unal, Bulent Keskinler and Nadir Dizge

ABSTRACT

In this study, the treatability of plasticizer production industry wastewater was investigated using nanofiltration (NF) and reverse osmosis (RO) membranes. The effect of operating pressure, pH of the wastewater, and sequential treatment option on the permeate flux, COD, phthalate, and micropollutant removal efficiencies were examined. The steady-state permeate fluxes of NF270, NF90, and BW30 membranes were 47.1, 19.0, and 13.9 L/m^2/h for 15 bar, respectively. Sequential filtration using NF90 and BW30 membranes to protect the RO membrane was also tested. The initial and steady-state permeate fluxes were 30.4 and 18.9 L/m^2/h, respectively, for 15 bar. The effect of wastewater pH in the range 4.0–10.0 was also studied and maximum initial and steady-state permeate fluxes were obtained at pH = 10.0. The permeate quality of NF90 and NF90 + BW30 membranes operated at 15 bar was measured and they showed a high degree for phthalate removal from wastewater from 97.7% to 99.9%. Moreover, a high degree of micropollutants was also obtained from 88.4% to 99.9% for sequential filtration (NF90 + BW30). The COD reduction efficiencies were obtained at 15 bar as 23.3%, 81.5%, and 87.6% for individual NF270, NF90, and BW30 membranes, respectively. However, COD reduction efficiency was increased up to 90.8% when sequential filtration (NF90 + BW30) was applied.

Key words | membrane process, nanofiltration, phthalates, plasticizer production industry wastewater, reverse osmosis

Yasin Ozay
Nadir Dizge (corresponding author)
Department of Environmental Engineering,
Mersin University,
Mersin, 33343,
Turkey
E-mail: *ndizge@mersin.edu.tr*

Oltan Canli
Environment and Clean Production Institute, The Scientific and Technological Research Council of Turkey,
Marmara Research Center,
Kocaeli, 41470,
Turkey

Bahar Ozbey Unal
Institute of Earth and Marine Sciences,
Gebze Technical University,
Kocaeli, 41400,
Turkey

Bulent Keskinler
Department of Environmental Engineering,
Gebze Technical University,
Kocaeli, 41440,
Turkey

HIGHLIGHTS

- Plasticizer production industry wastewater was treated using a pressure-driven membrane process.
- Nanofiltration pre-treatment before reverse osmosis application improved the flux.
- The operation pressure and pH of feed water had a significant impact on permeate flux.

INTRODUCTION

Phthalic acid diesters, also commonly named as phthalates, are synthetic substances that are classified as potentially dangerous pollutants due to their wide availability in the environment (Bodzek *et al.* 2004). Plasticizers are mainly phthalates, which enhance the flexibility of a material. Most plasticizers have been used to bend polyvinyl chloride (PVC) for more than 60 years and some of them are also used for rubber products, paints, sealants, and biopolymer films (Staples *et al.* 1997). Phthalates are widely used in many industrialized countries, so they are known worldwide

doi: 10.2166/ws.2020.268

as food and environmental pollutants (Wormuth *et al.* 2006). Recent studies have shown that some phthalates disorder the human hormonal system and sexual growth. Further, phthalates are suspected to provoke respiratory and dermal diseases in children (Bang *et al.* 2011; He *et al.* 2015). Phthalic acid esters (PAE) are the most common monomeric plasticizers and they are uncoloured or yellowish oil-like liquids and almost odourless (Titow 1990; Abdel daiem *et al.* 2012). PAEs enter the environment as a result of their release during production, usage, and disposal, because they are not chemically bound to the polymeric matrix (Tran *et al.* 2016; Liu *et al.* 2018). Current research has mostly concentrated on the occurrence and risk rating of PAEs (Zheng *et al.* 2014). In some rivers and lakes, the concentrations of PAEs were reported from 0.54 to 90.4 mg/L in China and some African countries (Net *et al.* 2015; Benjamin *et al.* 2017). In addition, PAEs likely to be transferred to the food chain were detected in river sediments ranging from 2.3 to 80.1 mg/kg and in coastal sediments ranging from 5.2 to 78.1 mg/kg (Wang *et al.* 2014; Arfaeinia *et al.* 2019). PAE concentrations can reach up to 32.2 mg/L in landfill leachates and this concentration is respectively higher than those in municipal wastewaters elsewhere (Fudala-Ksiazek *et al.* 2017). A large amount of industrial or municipal wastewater is being treated every day in many countries but still some PAE compounds can be detected in treated water (Deblonde *et al.* 2011). It was reported that some treated wastewater samples in Saudi Arabia contained PAEs such as butyl benzyl phthalate (BBP), bis(2-ethylhexyl) phthalate (DEHP), dimethyl phthalate (DMP), diethyl phthalate (DEP), dibutyl phthalate (DBP) and dioctyl phthalate (DOP), and these compounds had the potential of inducing DNA damage (Al-Saleh *et al.* 2017).

Many results showed that plasticizers in different types of water (ground, drinking surface, and wastewater) need further treatment processes to save the food chain and environment. The removal of PAEs may be possible by certain technologies such as biodegradation, coagulation/flocculation, adsorption, advanced oxidation, and membrane processes (Abdel daiem *et al.* 2012; Fudala-Ksiazek *et al.* 2017). For example, Mesdaghinia *et al.* (2017) investigated the removal of PAEs with the adsorption process by using zeolite/Fe_3O_4 and reached a maximum removal efficiency of 95.4%. Fudala-Ksiazek *et al.* (2018) investigated the removal of phthalates from landfill leachate with the combination of nanofiltration (NF) and membrane bioreactor/ultrafiltration (MBR/UF) systems and results showed that PAE concentration was found under the limit of detection. Wei *et al.* (2016) evaluated the removal efficiency of trace PAEs from water with thin-film composite NF hollow fibre membranes and found a maximum rejection rate of 95.4% for diethylhexyl phthalate (DEHP). Zolfaghari *et al.* (2015) found 98% removal efficiency of di-(2-ethylhexyl) phthalate (DEHP) from synthetic municipal wastewater by using a submerged MBR and results showed that the removal efficiency for DEHP depended on the concentration of sludge. Yang *et al.* (2016) reported 99% removal efficiency for DEHP by a graphene-containing ceramic composite tubular membrane coupled with electrocoagulation and electrofiltration process.

It is very difficult to treat plasticizer production industry wastewater using physicochemical or biological treatment methods. Moreover, one-step treatment is not enough to achieve the desired discharge criteria. For this reason, membrane processes can be applied effectively due to both process intensification and modular structure. Additionally, water recovery can be also carried out by membrane processes. In this study, plasticizer production industry wastewater was treated using a pressure-driven membrane process. Two different nanofiltration (NF270 and NF90) and reverse osmosis (BW30) membranes, which were easily commercially available, were applied to COD and phthalate (diethyl phthalate (DEP) and di-(2-ethylhexyl) phthalate (DEHP)) removal from wastewater. The effect of transmembrane pressure and pH of wastewater were tested on the flux and permeate quality.

MATERIALS AND METHODS

The characterization of PAE producing process wastewater

Plasticizer production industry wastewater (PPIW) was kindly obtained from a firm located in Gebze, Turkey. The characterization of the wastewater is shown in Table 1. It can be clearly seen that the wastewater showed high chemical oxygen demand (COD) and acidic character. Chemical oxygen demand (COD) was measured according to the

Table 1 | Characterization of the plasticizer production industry wastewater

Parameter	Unit	Value
pH	–	4.0 ± 0.1
Conductivity	μS/cm	84 ± 5
COD	mg/L	8,460 ± 203
Diethyl phthalate	mg/L	0.32 ± 0.1
Di-(2-ethylhexyl) phthalate	mg/L	0.85 ± 0.1
2-Methyl-1-propanol	mg/L	19.12 ± 0.2
2-Butenal	mg/L	14.12 ± 0.2
2-Ethyl-1-hexanol	mg/L	8.32 ± 0.2
1,3-Dioxolane	mg/L	8.16 ± 0.2
1-Butanol	mg/L	5.95 ± 0.1
3-Methyl-3-buten-1-ol	mg/L	2.23 ± 0.1
Naphthalene	mg/L	2.06 ± 0.1
2-Methylphenol	mg/L	1.85 ± 0.1
Phenanthrene	mg/L	1.45 ± 0.1
2-Ethyl-N-(2-ethylhexyl)-1-hexanamine	mg/L	1.18 ± 0.1
2-Methyl-1-butanol	mg/L	1.15 ± 0.1
5-Methyl-3-hexen-2-one	mg/L	1.10 ± 0.1
Chlorobenzene	mg/L	1.00 ± 0.1

standard methods of 5220C (APHA 2005). A closed reflux titrimetric method was used for COD analysis. The conductivity and pH were measured using pH/Cond 340i Handheld Multimeters, WTW. The pH of the wastewater was adjusted with 1 N NaOH or HCl solution. Duplicate experiments were performed for all measurements. Standards were prepared in deionized (Millipore Direct-Q3UV) water displaying a resistance of not less than 18 MΩ cm.

The methodology of PAE determination included two steps. The extraction method procedure followed EPA 3511 (Padrón *et al.* 2014). The analytical GC-MS screening method allows the detection and semi-quantitative determination of a wide range of substances. Briefly, the steps include: (1) adding isotopically labeled internal standards to the migration water sample, (2) dichloromethane extraction of the samples, (3) evaporation of the extracts, (4) GC-MS analysis of the concentrated extract to identify organic substances from the wastewater at concentrations ≥2 mg/L and to semi-quantitatively determine their concentration.

Step I – Extraction of phthalates from water by micro-extraction technique: a mixture of isotopically labelled internal standards of different boiling points (d8-naphthalene, d10-phenanthrene, d5-chlorobenzene, d5-phenol, d10-p-xylene) was added to 10 mL water. A liquid/liquid extraction procedure was applied two times at pH 2 and 10, respectively. For each individual extraction step, 2 mL of dichloromethane (Merck, Darmstadt, Germany) were used. The four extracts were combined and the combined extract was dried using sodium sulfate. The extract was evaporated to obtain a final volume of 0.5 mL.

Step II – Gas chromatography–mass spectroscopy analyses (GCMS): all GC-MS measurements were performed with an Agilent 6890N gas chromatograph connected to an Agilent 5975C mass spectrometric detector (Agilent Technologies, Germany). The compounds were separated by Agilent HP-5MS capillary column (60 m × 0.25 mm ID × 0.25 mm film thickness) (Agilent Technologies, Germany). Helium (99.999%) was used as carrier gas at a constant flow rate of 1 Ml/min. The injector temperature was kept at 280 °C in splitless mode (4 min), and the oven temperature was programmed as follows: initial temperature 35 °C (hold 4 min), 8 °C/min to 300 °C (hold 20 min). The MS ionization was carried out in the electron impact ionization mode (El +, 70 eV) scanning from 35 to 550 amu at a scan speed of 1 scan/s in repetitive scan mode. The GC-MS interface and the ion source temperature were set at 280 °C and 230 °C, respectively. The first evidence of identification was derived from the comparison of mass spectra with the mass spectral database NIST (version 2.0d, Agilent Technologies, Germany) underlying the spectra after subtraction of the background.

Molecular structure and some chemical properties of pollutants detected in the wastewater are presented in Table 2.

Dead-end filtration studies

Membrane filtration experiments were carried out in the dead-end filtration system presented in Figure 1. The filtration system was a manufactured cylindrical, stainless steel module with a working volume of 250 mL (Sterlitech HP4750 Stirred Cell). The effective filtration area of the membrane was 14.6 cm^2. Inert nitrogen gas was used to enhance the required pressure. Permeate flow was monitored by weighing at every minute using computer software.

Table 2 | Properties of some pollutants detected in the wastewater

Name	Linear formula	Molecular weight (g/Mol)	Chemical structure
Diethyl phthalate	$C_{12}H_{14}O_4$	222.2	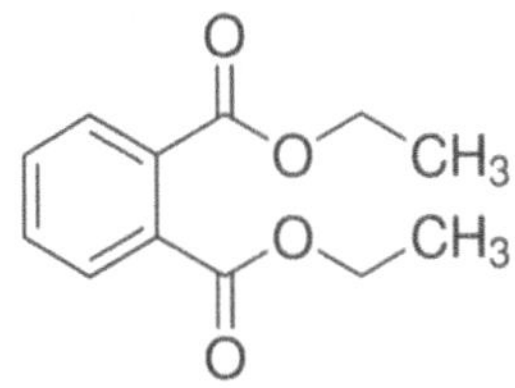
Di-(2-ethylhexyl) phthalate	$C_{24}H_{38}O_4$	390.56	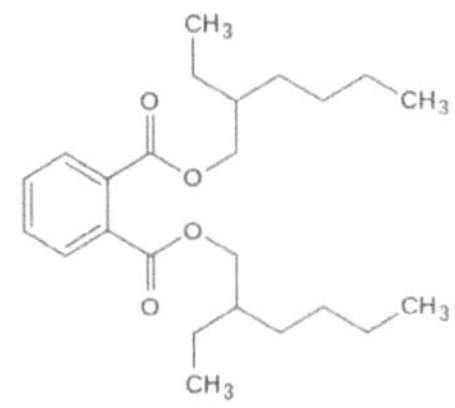
2-Methyl-1-propanol	$(CH_3)_2CHCH_2OH$	74.12	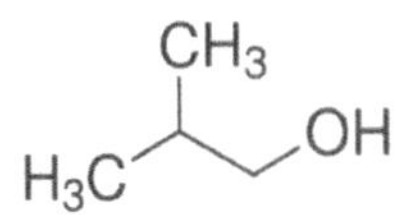
2-Butenal	C_4H_6O	70.09	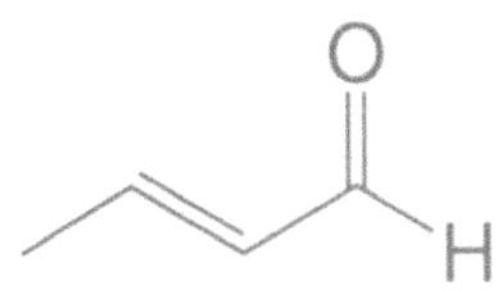
2-Ethyl-1-hexanol	$CH_3(CH_2)_3CH(C_2H_5)CH_2OH$	130.23	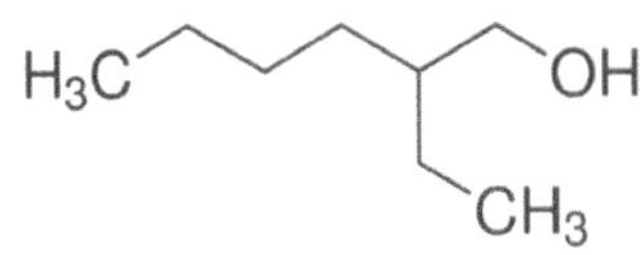
1,3-Dioxolane	$C_3H_6O_2$	74.08	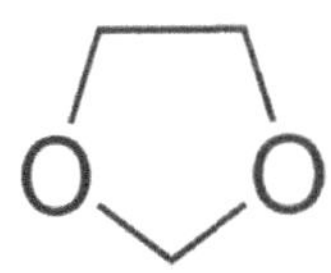
1-Butanol	$CH_3(CH_2)_3OH$	74.12	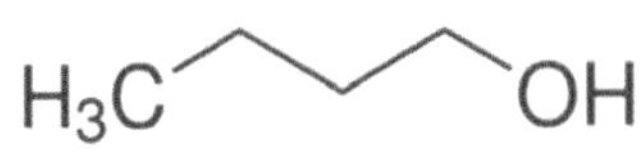
3-Methyl-3-buten-1-ol	$CH_2=C(CH_3)CH_2CH_2OH$	86.13	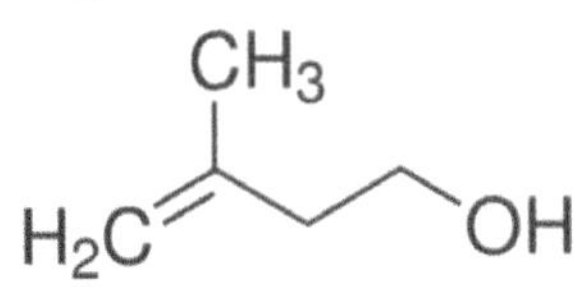
Naphthalene	$C_{10}H_8$	128.17	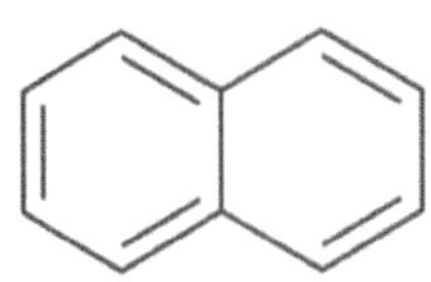
2-Methylphenol	C_7H_8O	108.14	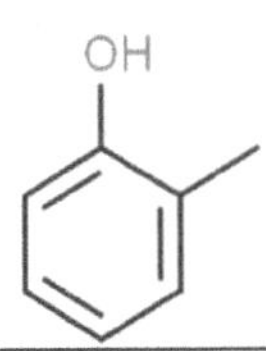

(continued)

Table 2 | continued

Name	Linear formula	Molecular weight (g/Mol)	Chemical structure
Phenanthrene	$C_{14}H_{10}$	178.23	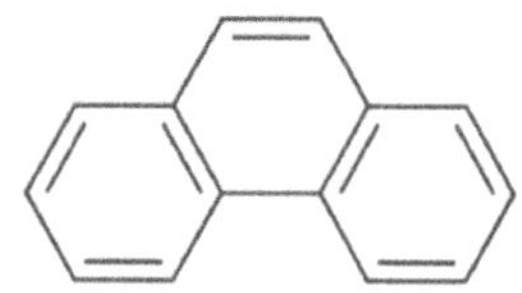
2-Ethyl-N-(2-ethylhexyl)-1-hexanamine	$C_{16}H_{35}N$	241.46	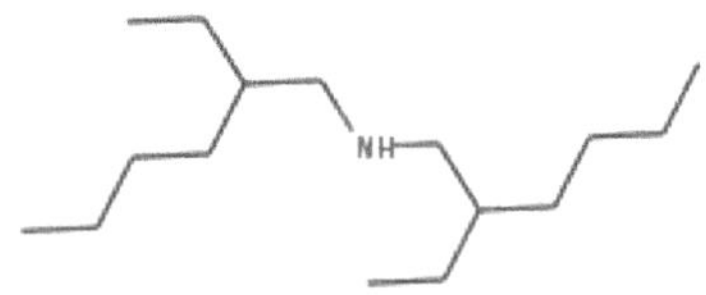
2-Methyl-1-butanol	$CH_3CH_2CH(CH_3)CH_2OH$	88.15	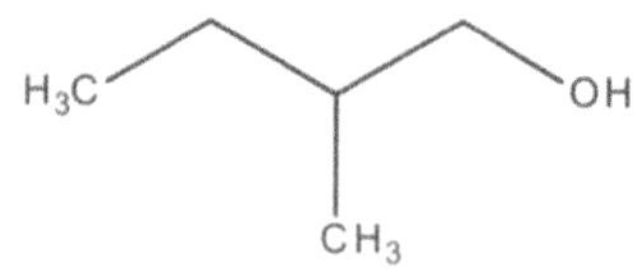
5-Methyl-3-hexen-2-one	$(CH_3)_2CHCH=CHCOCH_3$	112.17	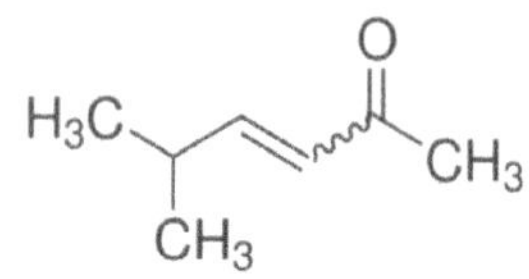
Chlorobenzene	C_6H_5Cl	112.55	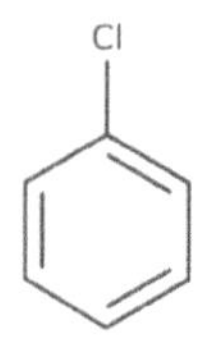

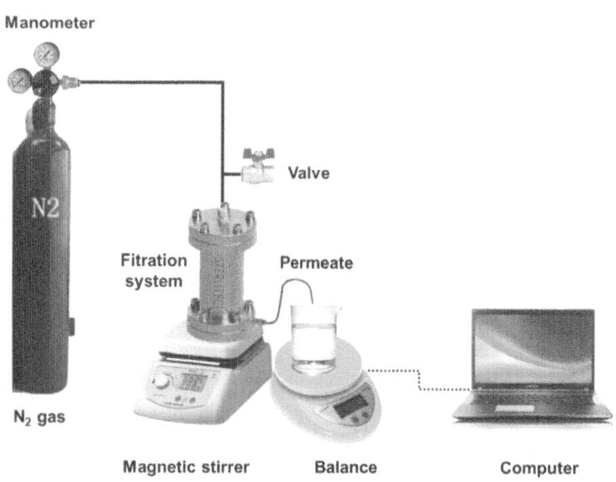

Figure 1 | Schematic diagram of the dead-end filtration set-up.

Membrane specifications

NF and RO membranes (Dow-Filmtec, USA) were used for the removal of COD and phthalates from the PPIW. The properties of the membranes are shown in Table 3. Before experiments, the membranes were immersed in distilled water overnight and they were compressed under a trans-membrane pressure (ΔP) of 30 bar for 1 h to avoid the compaction effect of the membrane.

Permeate flux and rejection measurements

The permeate flux was calculated by the amount of collected permeate through the membranes per unit time using Equation (1):

$$J_w = \frac{Q}{A \times \Delta t} \tag{1}$$

where Q is the volume of the permeate (L); A is the effective membrane area for filtration (m^2); and Δt is time (h).

Table 3 | Properties of NF and RO membranes used in filtration experiments

Membrane	Material	Molecular weight cutoff (Da)[a]	Stabilized salt rejection (%)	Maximum operating temperature (°C)	Maximum operating pressure (bar)	pH range
NF270	Polyamide	~200–400	>97.0	45	41	2–11
NF90	Polyamide	~200	>97.0	45	41	2–11
BW30	Polyamide	~100	99.5	45	41	2–11

[a]Information obtained from the manufacturer.

The removal efficiencies (R) were calculated using Equation (2):

$$\textit{Removal efficiency}\,(\%) = \frac{C_0 - C_t}{C_0} \times 100 \quad (2)$$

where C_0 is the initial COD or phthalate concentration and C_t is the COD or phthalate concentration measured at any time t.

Characterization of the membranes

Surface morphologies of the pristine and used membranes were investigated with a scanning electron microscope (Zeiss Supra 55, Germany) in high vacuum mode at 5.00 kV after coating with platinum–palladium using Ion SputterCoater Quorum Q150R ES.

A zetasizer (Malvern Zetasizer equipped with MPT-2 Titrator, Nano ZS) was used to measure the zeta potential of the wastewater. The measurements were carried out at 25 ± 1 °C in duplicate.

RESULTS AND DISCUSSION

The effect of transmembrane pressure on permeate flux

The effects of transmembrane pressure on permeate flux and cumulative permeate volume are given in Figure 2. The initial permeate flux of the NF270 membrane was improved from 64.1 to 122.5 L/m^2/h when the applied pressure was increased from 5 to 15 bar, respectively. The steady-state permeate flux increased from 29.5 to 47.1 L/m^2/h when the applied pressure increased from 5 to 15 bar after 120 min (Figure 2(a)). In addition the cumulative volumes were 133.5, 193.2, 232.1 mL for 5, 10, 15 bar, respectively, for 120 min filtration time (Figure 2(b)).

The NF90 membrane was also tested using the same pressures. The initial permeate flux increased from 20.9 to 39.0 L/m^2/h when the applied pressure increased from 5 to 15 bar for the NF90 membrane. The steady-state permeate flux increased from 13.1 to 19.0 L/m^2/h when the applied

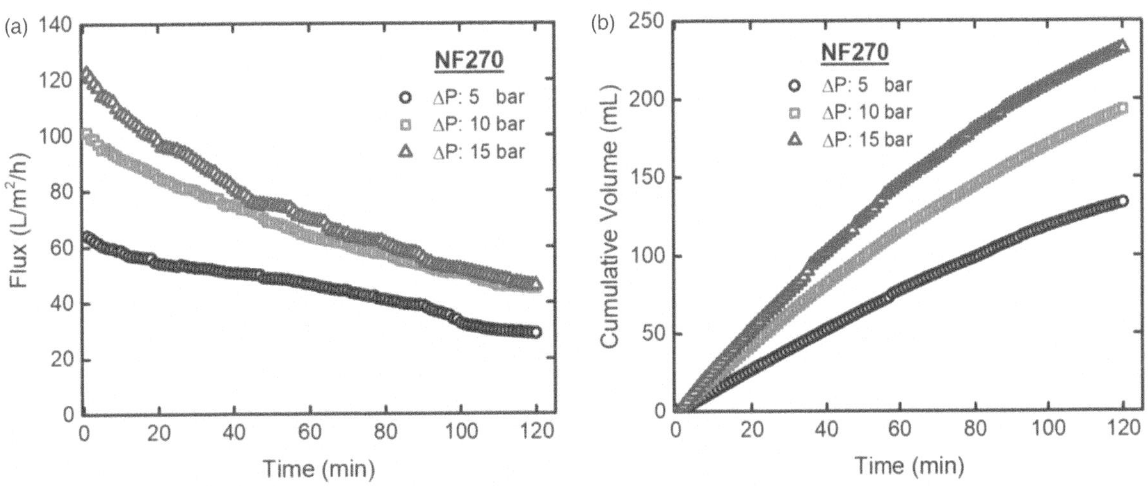

Figure 2 | Effect of transmembrane pressure on (a) permeate flux and (b) cumulative volume for NF270 membrane (wastewater pH = 4.0).

pressure increased from 5 to 15 bar after 120 min (Figure 3(a)). Moreover, the cumulative volumes were 40.6, 59.2, 69.1 mL for 5, 10, 15 bar, respectively, for 120 min filtration time (Figure 3(b)).

Figure 4 shows the permeate flux of the BW30 membrane for PPIW filtration at the operating pressures of 5, 10, 15 bar. The initial permeate flux increased from 7.9 to 19.7 L/m^2/h when the applied pressure increased from 5 to 15 bar. The steady-state permeate flux increased from 5.2 to 13.9 L/m^2/h when the applied pressure increased from 5 to 15 bar after 120 min (Figure 4(a)). Moreover, the cumulative volumes were 17.2, 28.6, 40.8 mL for 5, 10, 15 bar, respectively, for 120 min filtration time (Figure 4(b)).

The results showed that the NF270 membrane had a higher permeate water flux compared with the NF90 and BW30 membranes. This indicated that the permeability of the water molecules through the NF270 membrane pores was higher due to the larger MWCO value of the NF270 membrane (200–400 MWCO) than that of the BW30 membrane (~100 MWCO), as can be seen in Table 3. Also, the increment in the permeate flux for the higher operating pressures can be explained by a dissolution–diffusion model which is related to membrane transmembrane pressure. The salt flux can be explained by a dissolution–diffusion model defined as $J_s = -B\Delta C_s$, where J_s, B, and ΔC_s are the salt flux, salt permeability constant, and concentration difference of the salt across the

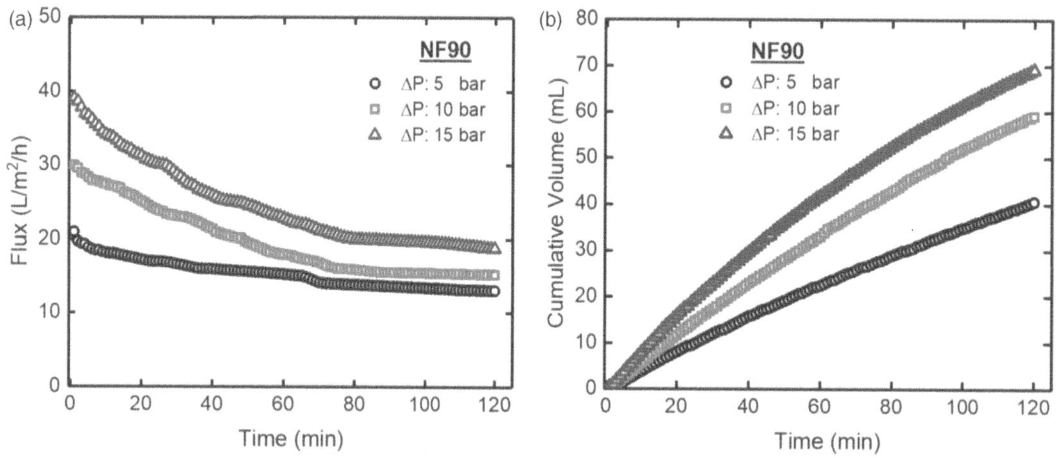

Figure 3 | Effect of transmembrane pressure on (a) permeate flux and (b) cumulative volume for NF90 membrane (wastewater pH = 4.0).

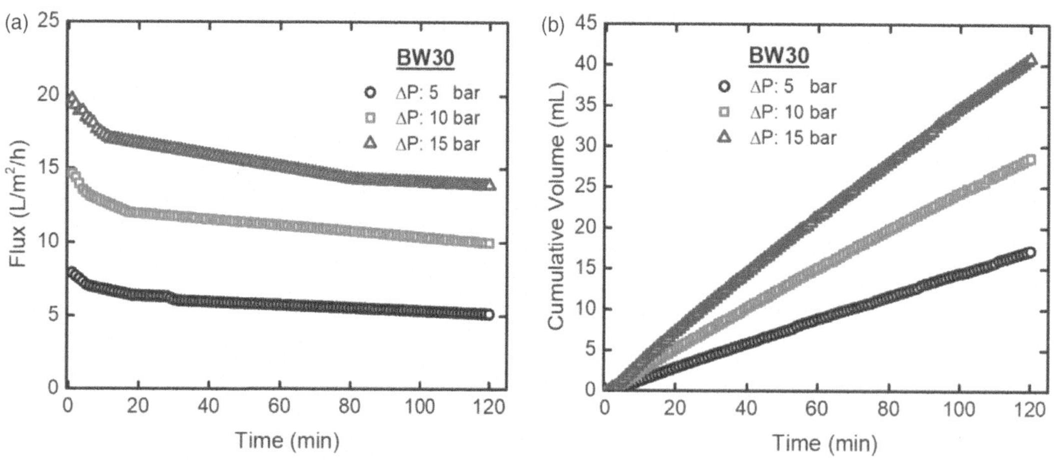

Figure 4 | Effect of transmembrane pressure on (a) permeate flux and (b) cumulative volume for BW30 membrane (wastewater pH = 4.0).

membrane, respectively. Therefore, transmembrane pressure improves the membrane flux (Bandini & Vezzani 2003). On the other hand, the NF270 and NF90 membranes illustrated more severe fouling phenomena than the BW30 membrane. Membrane fouling can occur by the plugging of membrane pores and/or the formation of a thin foulant layer on the membrane surface (Haan *et al.* 2018). The severe fouling attributed by PPIW was mainly due to it containing very high dissolved organic molecules including phthalates. PAEs might have caused an accumulation of a fouling layer on the NF270 and NF90 membranes' surface or into the membrane pores that caused a severe flux decline compared with the BW30 membrane (Bodzek *et al.* 2004; Haan *et al.* 2018). Moreover, the BW30 membrane had the least pore blockage phenomenon due to its extremely small MWCO value. Depending on the pressure, the solute concentration near the membrane surface increases the concentration polarization mechanism, increasing the cake layer formation as well as pore clogging on the membrane surface. Moreover, adsorption and cake adhesion mechanisms can be seen for membranes operated at high pressure (Haan *et al.* 2018). Therefore, more flux declines were observed in our study when the operating pressure was increased.

We also tested sequential filtration using NF90 and BW30 membranes to protect the RO membrane and to improve permeate water quality. The initial permeate flux increased from 16.0 to 30.4 $L/m^2/h$ when the applied pressure was increased from 5 to 15 bar. The steady-state permeate flux increased from 9.7 to 18.9 $L/m^2/h$ when the applied pressure was increased from 5 to 15 bar after 120 min filtration (Figure 5(a)). Moreover, the cumulative volumes were 30.6, 49.6, 57.9 mL for 5, 10, 15 bar, respectively, for 120 min filtration time (Figure 5(b)). The results show that sequential filtration (NF90 + BW30) enhanced higher permeate flux than the individual BW30 membrane. This might be explained as the NF90 membrane rejecting more dissolved organic matter as well as phthalates and protecting the BW30 membrane against fouling.

SEM images of the pristine and fouled membranes are seen in Figure 6. It can be clearly seen that much more pollutant was accumulated on the NF270 membrane and the pollutant formed a layer on the membrane surface. In addition, some open pores on the surface of the pristine NF90 membrane were closed by contaminants after filtration. Contaminants from wastewater were also detected on the BW30 membrane surface.

The effect of solution pH on permeate flux

The NF90 membrane was chosen for the pH studies since the NF90 membrane showed less flux decline compared with the NF270 membrane according to transmembrane pressure results. Moreover, the NF90 membrane provided better permeate water quality, as mentioned below in the section 'The effect of applied pressure and wastewater pH on permeate water quality'. In this section, the applicability of the NF90 membrane was also investigated as a pre-treatment stage of the BW30 membrane to increase its flux.

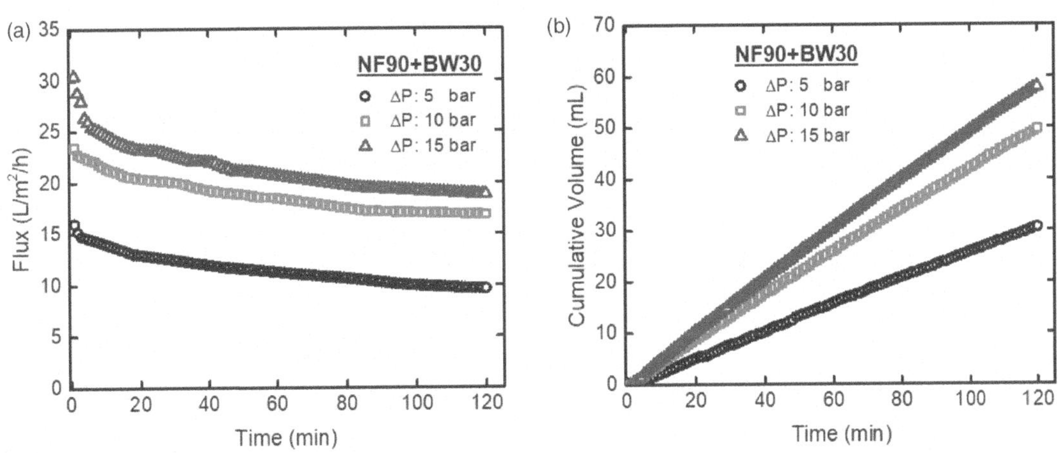

Figure 5 | Effect of transmembrane pressure on (a) permeate flux and (b) cumulative volume for sequential filtration of NF90 + BW30 membranes (wastewater pH = 4.0).

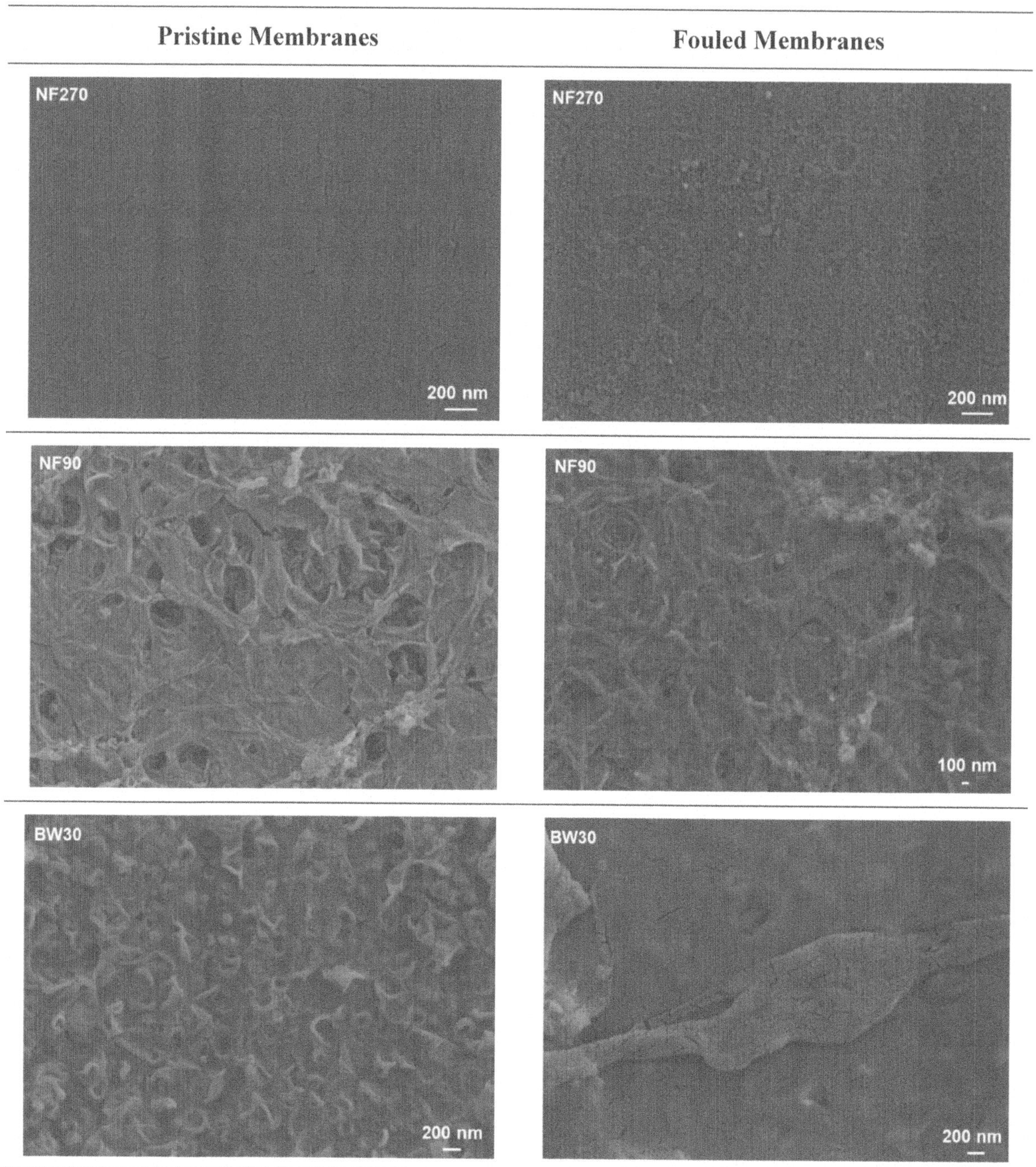

Figure 6 | SEM images of the pristine and fouled membranes (experimental conditions: $\Delta P = 15$ bar, wastewater pH = 4.0, filtration time = 120 min).

Therefore, the performance of individual NF90 and sequential NF90 + BW30 membranes was evaluated by measurement of the flux as a function of feed pH (4.0, 7.0, and 10.0) at 15 bar operating pressure. The initial permeate flux increased from 39.0 to 48.0 L/m^2/h when feed pH increased from 4.0 to 10.0 for the NF90 membrane. The steady-state permeate flux increased from 19.0 to 28.0 L/m^2/h when feed pH increased from 4.0 to 10.0 after 120 min filtration (Figure 7(a)). The results showed that when pH was increased to 10.0, there was a

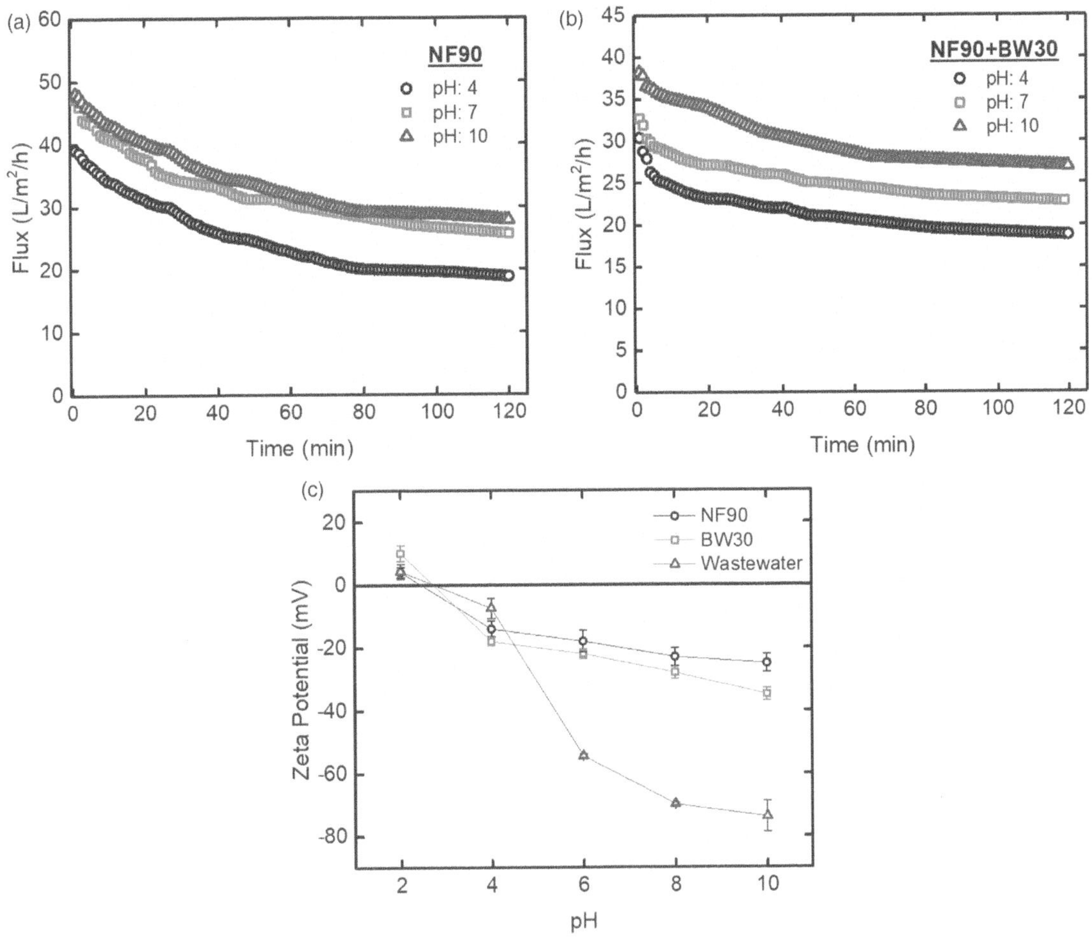

Figure 7 | Effect of solution pH on permeate flux for (a) NF90 membrane ($\Delta P = 15$ bar), (b) NF90 + BW30 membrane ($\Delta P = 15$ bar), (c) zeta potential of wastewater, NF90 and BW30 membranes.

contribution of electrostatic repulsion, which increased the retention of the charged molecules in the wastewater as well as permeate flux. On the other hand, feed pH affected BW30 performance. NF90 permeate was filtrated through the BW30 membrane for further treatment. The initial permeate flux increased from 30.4 to 38.2 L/m^2/h when feed pH increased from 4.0 to 10.0 for BW30 membrane. The steady-state permeate flux increased from 18.9 to 27.0 L/m^2/h when feed pH increased from 4.0 to 10.0 after 120 min filtration (Figure 7(b)). At pH 10.0, the electrostatic effects between substances with electrostatic charge in the wastewater and membrane surface were more expressive due to an increment in the negative zeta potential value of the membrane for the higher pH value. Figure 7(c) demonstrates the zeta potential of wastewater, NF90 and BW30 membranes for different pH values. The isoelectric point of the wastewater and BW30 membrane was about pH 3.0 while it was about pH 2.5 for the NF90 membrane. The zeta potentials of the wastewater and both membranes decreased with increased pH values. Similar results were obtained by Haan *et al.* (2018) and Wei *et al.* (2016) in which the relationship between flux and pH was explained by the electrostatic repulsion forces between the charged groups on the membrane and charged ions in the wastewater that prevented the adsorption of the pollutants on the membrane surface and reduced the membrane fouling.

As can be seen from Figure 8, the foulants on the membrane surface decreased with increasing pH values due to electrostatic repulsion forces, which is compatible with the results of Figure 7(b).

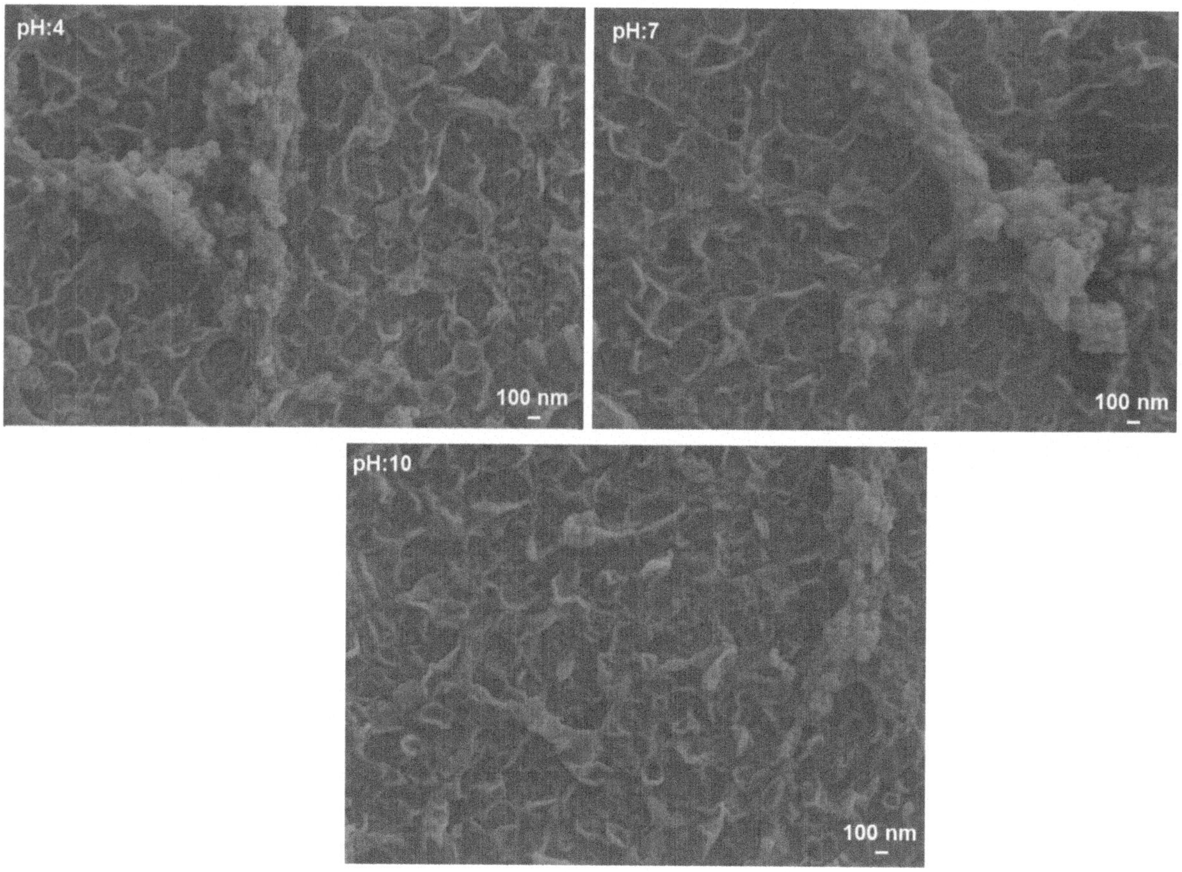

Figure 8 | SEM images of the BW30 membranes operating with different wastewater pH values (experimental conditions: NF90 membrane was applied as a pre-treatment stage of BW30 membrane, $\Delta P = 15$ bar, filtration time = 120 min).

The initial and steady-state fluxes of the membranes for different transmembrane pressure and wastewater pH are summarized in Table 4. The initial and steady-state fluxes of the membranes increased with the increasing of transmembrane pressure. The highest initial and steady-state fluxes were obtained by the NF270 membrane. Moreover, both initial and steady-state fluxes of the BW30 membrane increased when the NF90 membrane was applied as a pre-treatment stage before the BW30 membrane. For example, the initial and steady-state fluxes of the BW30 membrane were 19.7 and 13.9 L/m^2/h at 15 bar before NF90 pre-treatment. However, the initial and steady-state fluxes of the BW30 membrane increased up to 30.4 and 18.9 L/m^2/h at 15 bar and pH = 4.0 after NF90 pre-treatment. The wastewater pH also affected the fluxes and the initial and steady-state fluxes of the BW30 membrane increased up to 38.2 and 27.0 L/m^2/h at 15 bar and pH = 10.0 after NF90 pre-treatment. It can be concluded that pre-treatment is important for an RO membrane to obtain high flux.

The effect of applied pressure and wastewater pH on permeate water quality

Permeate water quality of samples measuring COD and phthalates was carried out by NF270, NF90, and BW30 membranes. The results are shown in Figure 9 for different applied pressures (5, 10, 15 bar) and wastewater pH (4.0, 7.0, 10.0). As shown in Figure 9(a), the NF270 membrane supplied the worst permeate water quality compared with the other membranes' permeate. Especially, the NF90 and BW30 membranes successfully reduced the initial COD concentration of the wastewater. The BW30 membrane showed a better performance in the treatment of plasticizer production industry wastewater. The maximum COD

Table 4 | Values of initial and steady-state flux for different operating parameters

Membrane type	Different ΔP (bar) (pH = 4.0)	Initial flux (J_0) (L/m²/h)	Steady-state flux (J) (L/m²/h)
NF270	5	64.1	29.5
	10	101.1	45.5
	15	122.5	47.1
NF90	5	20.9	13.1
	10	29.9	15.3
	15	39.0	19.0
BW30	5	7.9	5.2
	10	14.7	10.1
	15	19.7	13.9
NF90 + BW30	5	16.0	9.7
	10	23.4	16.9
	15	30.4	18.9
Membrane type	**Different pH (ΔP = 15 bar)**	**Initial flux (J_0) (L/m²/h)**	**Steady-state flux (J) (L/m²/h)**
NF90	4.0	39.0	19.0
	7.0	47.3	25.9
	10.0	48.0	28.0
NF90 + BW30	4.0	30.4	18.9
	7.0	34.4	22.8
	10.0	38.2	27.0

reduction efficiencies were obtained at 15 bar as 23.3%, 81.5%, 87.6% for NF270, NF90, BW30 membranes, respectively (Figure 9(a)). The reason was due to the smaller MWCO value of the BW30 membrane which prevented the solute contaminants going through the membrane into the permeate. The dissolved contaminant flux does not change too much, while the permeate water flux increases with increasing operating pressure. Therefore, the COD reduction efficiency also increased for the higher operating pressure (Wei *et al.* 2016). The COD reduction efficiencies were improved with increased pH values due to the electrostatic repulsion between the membrane and wastewater (Figure 9(b)).

Table 5 presents the removal efficiency obtained for particular components of the phthalates and micropollutants for the filtration of PPIW. The results indicated that individual NF90 and sequential NF90 + BW30 membranes supplied very effective removal of phthalates and micropollutants from wastewater. The rejection of 99.9% was obtained for diethyl and di-(2-ethylhexyl) phthalate when sequential NF90 + BW30 membranes were applied. The rejection of three phthalic acid esters (PAEs) by clean and fouled nanofiltration membranes was investigated by Zhu (2015). A rejection of 95.8% and 96.0% was obtained for dimethyl phthalate and diethyl phthalate, respectively, when an NF90 membrane was operated at 500 kPa transmembrane pressure, pH 7.5, 0.8 mg/L initial feed concentration, 10 mM NaCl.

Recently, PAEs have been commonly detected at higher concentrations in surface water and ground water. The removal of PAEs has drawn significant attention in the field of water treatment; however, it is difficult to remove PAEs using conventional water treatment technology. So far, little information has been provided on the removal of PAEs by NF and RO membranes. Thus, future research prospects should focus on the investigation of the relationship

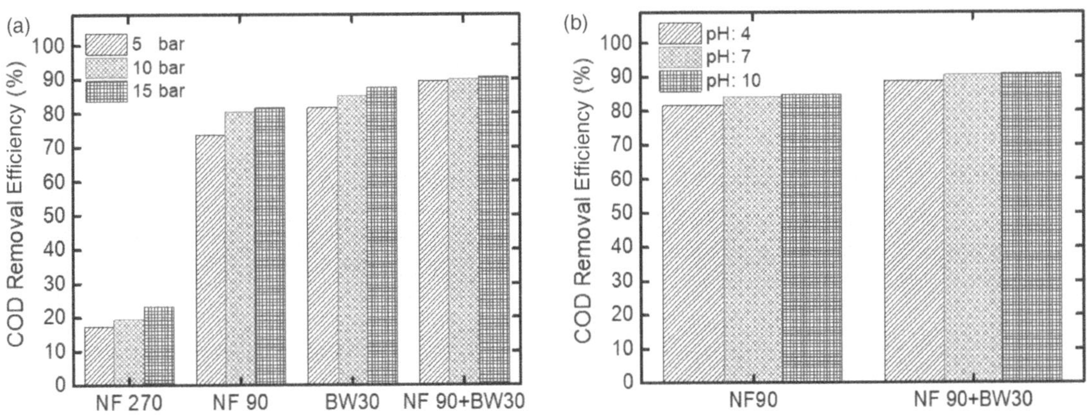

Figure 9 | The effect of (a) applied pressure (wastewater pH = 4.0) and (b) wastewater pH (ΔP = 15 bar) on COD reduction efficiency for different membranes.

Table 5 | Comparison of the removal efficiency (%) of phthalates and micropollutants during filtration (experimental conditions: $\Delta P = 15$ bar; wastewater pH = 10; filtration time = 120 min)

Name	NF90	NF90 + BW30
Diethyl phthalate	97.7	99.9
Di-(2-ethylhexyl) phthalate	98.9	99.9
2-Methyl-1-propanol	90.5	95.0
2-Butenal	82.6	88.5
2-Ethyl-1-hexanol	95.2	99.9
1,3-Dioxolane	82.5	88.4
1-Butanol	90.3	95.0
3-Methyl-3-buten-1-ol	91.2	95.8
Naphthalene	95.0	99.9
2-Methylphenol	94.3	99.9
Phenanthrene	96.5	99.9
2-Ethyl-N-(2-ethylhexyl)-1-hexanamine	98.1	99.9
2-Methyl-1-butanol	98.3	99.8
5-Methyl-3-hexen-2-one	94.5	99.9
Chlorobenzene	94.6	99.9

between the hydrophobicity of PAEs and rejection by different types of NF and RO membranes. Moreover, a pilot system should be operated for a long time to understand membrane behaviour versus different operating conditions.

CONCLUSIONS

In this study, the treatability of plasticizer production industry wastewater was examined by using commercial NF and RO membranes. The results indicated that the NF270 had the highest permeate flux value due to its larger pore size compared with NF90 and BW30 membranes. But, the initial permeate flux decline was much more for the NF270 membrane because of pore blockage. The operation pressure and pH of the feed water had a significant impact on permeate flux values. COD, phthalate and micropollutant removal efficiency was enhanced slightly with the adjustment of operating conditions for NF90 and NF90 + BW30 membranes. Furthermore, the use of sequential NF and RO membranes improved the initial and steady-state flux values of the BW30 membrane.

DATA AVAILABILITY STATEMENT

All relevant data are included in the paper or its Supplementary Information.

REFERENCES

Abdel daiem, M. M., Rivera-Utrilla, J., Ocampo-Pérez, R., Méndez-Díaz, J. D. & Sánchez-Polo, M. 2012 Environmental impact of phthalic acid esters and their removal from water and sediments by different technologies – a review. *Journal of Environmental Management* **109**, 164–178.

Al-Saleh, I., Elkhatib, R., Al-Rajoudi, T. & Al-Qudaihi, G. 2017 Assessing the concentration of phthalate esters (PAEs) and bisphenol A (BPA) and the genotoxic potential of treated wastewater (final effluent) in Saudi Arabia. *Science of The Total Environment* **578**, 440–451.

American Public Health Association (APHA) 2005 *Standard Methods for the Examination of Water and Wastewater*, 21st edn. American Public Health Association, Washington, DC, USA.

Arfaeinia, H., Fazlzadeh, M., Taghizadeh, F., Saeedi, R., Spitz, J. & Dobaradaran, S. 2019 Phthalate acid esters (PAEs) accumulation in coastal sediments from regions with different land use configuration along the Persian Gulf. *Ecotoxicology and Environmental Safety* **169**, 496–506.

Bandini, S. & Vezzani, D. 2003 Nanofiltration modeling: the role of dielectric exclusion in membrane characterization. *Chemical Engineering Science* **58**, 3303–3326.

Bang, D. Y., Lee, I. K. & Lee, B. M. 2011 Toxicological characterization of phthalic acid. *Toxicological Research* **27** (4), 191–203.

Benjamin, S., Masai, E., Kamimura, N., Takahashi, K., Anderson, R. C. & Faisal, P. A. 2017 Phthalates impact human health: epidemiological evidences and plausible mechanism of action. *Journal of Hazardous Materials* **340**, 360–383.

Bodzek, M., Dudziak, M. & Luks-Betlej, K. 2004 Application of membrane techniques to water purification. Removal of phthalates. *Desalination* **162**, 121–128.

Deblonde, T., Cossu-Leguille, C. & Hartemann, P. 2011 Emerging pollutants in wastewater: a review of the literature. *International Journal of Hygiene and Environmental Health* **214** (6), 442–448.

Fudala-Ksiazek, S., Pierpaoli, M. & Luczkiewicz, A. 2017 Fate and significance of phthalates and bisphenol A in liquid by-products generated during municipal solid waste mechanical–biological pre-treatment and disposal. *Waste Management* **64**, 28–38.

Fudala-Ksiazek, S., Pierpaoli, M. & Luczkiewicz, A. 2018 Efficiency of landfill leachate treatment in a MBR/UF system combined with NF, with a special focus on phthalates and bisphenol A removal. *Waste Management* **78**, 94–103.

Haan, T. Y., Shah, M., Chun, H. K. & Mohammad, A. W. 2018 A study on membrane technology for surface water treatment: synthesis, characterization and performance test. *Membrane Water Treatment* **9** (2), 69–77.

He, L., Gielen, G., Bolan, N. S., Zhang, X., Qin, H., Huang, H. & Wang, H. 2015 Contamination and remediation of phthalic acid esters in agricultural soils in China: a review. *Agronomy for Sustainable Development* **35**, 519–534.

Liu, Y. T., Li, T. & Peng, H. Y. 2018 Characteristics of phthalic acid esters pollution in urban surface runoff in Shanghai, China. *Huan Jing Ke Xue* **39** (11), 5007–5014 (in Chinese).

Mesdaghinia, A., Azari, A., Nodehi, R. N., Yaghmaeian, K., Bharti, A. K., Agarwal, S., Gupta, V. K. & Sharafi, K. 2017 Removal of phthalate esters (PAEs) by zeolite/Fe_3O_4: investigation on the magnetic adsorption separation, catalytic degradation and toxicity bioassay. *Journal of Molecular Liquids* **233**, 378–390.

Net, S., Sempéré, R., Delmont, A., Paluselli, A. & Ouddane, B. 2015 Occurrence, fate, behavior and ecotoxicological state of phthalates in different environmental matrices. *Environmental Science & Technology* **49** (7), 4019–4035.

Padrón, M. E. T., Afonso-Olivares, C., Sosa-Ferrera, Z. & Santana-Rodríguez, J. J. 2014 Microextraction techniques coupled to liquid chromatography with mass spectrometry for the determination of organic micropollutants in environmental water samples. *Molecules* **19**, 10320–10349.

Staples, C. A., Peterson, D. R., Parkerton, T. F. & Adams, W. J. 1997 The environmental fate of phthalate esters: a literature review. *Chemosphere* **35** (4), 667–749.

Titow, W. V. 1990 *PVC Plastics: Properties, Processing, and Applications*. Elsevier Applied Science, London, UK.

Tran, T. M., Minh, T. B., Kumosani, T. A. & Kannan, K. 2016 Occurrence of phthalate diesters (phthalates), *p*-hydroxybenzoic acid esters (parabens), bisphenol A diglycidyl ether (BADGE) and their derivatives in indoor dust from Vietnam: implications for exposure. *Chemosphere* **144**, 1553–1559.

Wang, J., Bo, L., Li, L., Wang, D., Chen, G., Christie, P. & Teng, Y. 2014 Occurrence of phthalate esters in river sediments in areas with different land use patterns. *Science of The Total Environment* **500–501**, 113–119.

Wei, X., Shi, Y., Fei, Y., Chen, J., Lv, B., Chen, Y., Zheng, H., Shen, J. & Zhu, L. 2016 Removal of trace phthalate esters from water by thin-film composite nanofiltration hollow fiber membranes. *Chemical Engineering Journal* **292**, 382–388.

Wormuth, M., Scheringer, M., Vollenweider, M. & Hungerbühler, K. 2006 What are the sources of exposure to eight frequently used phthalic acid esters in Europeans? *Risk Analysis* **26**, 803–824.

Yang, G. C. C., Chen, Y. C., Yang, H. X. & Yen, C. H. 2016 Performance and mechanisms for the removal of phthalates and pharmaceuticals from aqueous solution by graphene-containing ceramic composite tubular membrane coupled with the simultaneous electrocoagulation and electrofiltration process. *Chemosphere* **155**, 274–282.

Zheng, X., Zhang, B. T. & Teng, Y. 2014 Distribution of phthalate acid esters in lakes of Beijing and its relationship with anthropogenic activities. *Science of The Total Environment* **476–477**, 107–113.

Zhu, L. 2015 Rejection of organic micropollutants by clean and fouled nanofiltration membranes. *Journal of Chemistry* **2015**, 934318.

Zolfaghari, M., Drogui, P., Seyhi, B., Brar, S. K., Buelna, G., Dubé, R. & Klai, N. 2015 Investigation on removal pathways of di 2-ethyl hexyl phthalate from synthetic municipal wastewater using a submerged membrane bioreactor. *Journal of Environmental Sciences* **37**, 37–50.

First received 27 July 2020; accepted in revised form 5 October 2020. Available online 17 October 2020

Recovery of zinc from electroplating wastewater using green emulsion liquid membrane

M. Rajasimman, N. Rajamohan and S. Sujatha

ABSTRACT

In this research study, removal of zinc ions from the industrial wastewater was investigated using green emulsion liquid membrane technology. The liquid membrane was prepared by using waste cooking oil along with the surfactant, SPAN 80 and the internal phase, sulfuric acid. The extraction percentage of zinc increased with the increase in concentration of surfactant. The response surface methodology (RSM) analysis identified that the optimal variable values for the maximum extraction of zinc were: external pH – 3.8, surfactant concentration 4% (vol.), internal phase concentration – 1.61N, zinc concentration – 742 mg/L, external phase to emulsion volume ratio – 0.94 and carrier concentration – 8.9%. At the optimized conditions experiment was carried out and the maximum extraction was found to be 97.4%. The perturbation plot shows that the extraction of zinc was affected by variables in the following order of effect: zinc concentration > surfactant concentration > carrier concentration > external pH > external phase to emulsion volume ratio > internal phase concentration.

Key words | efficiency, emulsion, membrane, metal, recovery, zinc

M. Rajasimman
S. Sujatha
Department of Chemical Engineering,
Annamalai University,
Annamalai nagar 608002,
Tamil Nadu,
India

N. Rajamohan (corresponding author)
Chemical Engineering section, Faculty of Engineering,
Sohar University,
Sohar,
Oman
E-mail: *rnatarajan@su.edu.om*

HIGHLIGHTS

- The use of waste cooking oil for the extraction of zinc from industrial wastewater is a novel study.
- Optimization of operating variables using RSM for zinc recovery is a first of its kind.
- Studies on industrial wastewater for zinc recovery is a novel attempt.

GRAPHICAL ABSTRACT

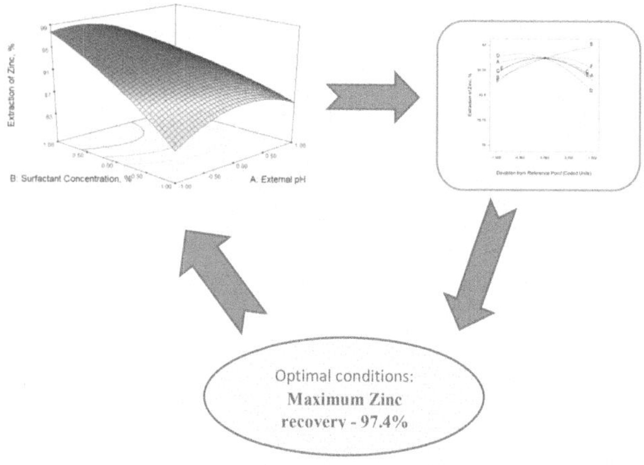

doi: 10.2166/ws.2020.294

INTRODUCTION

Pollution of water bodies due to metal contamination has gained significant attention recently due to the volume of the contaminated wastewater generated and recalcitrant nature of the metals. Use of metals has increased due to their increased use in automobile, infrastructure, insulation manufacturing and other specialty industries. Zinc, a metal used extensively galvanization applications, is a potential candidate used in corrosion protection through plating techniques. Zinc emanated from electroplating industries contributes a huge amount in the wastewater and is required to be treated or recovered. Due to the toxic nature of zinc, the disposal of untreated wastewater contaminated with zinc is unacceptable and leads to bioaccumulation of the metal in the ecosystem (Wang *et al.* 2019). Also, presence of zinc in water causes heath disorders like bone disorders, pancreatic dysfunctional, vertigo and other neurological damage when present in high concentrations (Hafshejani *et al.* 2015; Fernandez-Gonzalez *et al.* 2019).

Treatment of metal contaminated water has been investigated using several physico-chemical methods like adsorption, coagulation, precipitation and ion exchange. Many of these methods are considered to be inefficient in terms of low recovery rates at low metal concentrations, excessive chemical usage leading to secondary pollution, lack of eco-friendliness and disposal difficulties (Ma *et al.* 2018). Emulsion liquid membrane (ELM) is identified as a suitable recovery method for metals found in wastewater and other organic and inorganic components (Kislik 2010; Jusoh & Othman 2017). ELM, proposed as an advanced membrane process, operates through creation of a three phase system and enhanced mass transfer areas achieved by the water–oil emulsion globules. Emulsion stability is reported as one of the key factors contributing to extraction efficiency (Liu *et al.* 2017; Albaraka 2020). Design of experiments, a methodical approach for studying the effect of operating variables, is a key requirement for effective statistical analysis and development of a new approach. Response surface methodology (RSM) is a popular technique which helps identify the optimal conditions through a constructive set of desired experiments (Benyahia *et al.* 2014; Seifollahi & Rahbar-Kelishami 2019). Studies on removal of phenol using ELM has been proved to yield 83% removal efficiency under optimal conditions of treat ratio, agitation speed, extraction time and ionic liquid concentration (Rosly *et al.* 2019). Extraction of lead using ELM has reported a maximum recovery of 82.61% by Aliquat 336 at an acidic pH (5.5) from a feed phase of lead (II) nitrate of 207.2 ppm under 30 min of stirring at 210 rpm (Mesli & Belkhouche 2018). Application of ELM for recovery of amoxicillin was reported using RSM optimization technique and the extraction yield achieved was 99.8%. Removal of zinc has been studied using modified carbon nanotube membranes (Ali *et al.* 2019), cedar ash biosorption (Hafshejani *et al.* 2015), coconut shell-based activated carbon fiber adsorption (Shrestha *et al.* 2013), nano materials (Kocanova *et al.* 2017) and acid leaching (Wang *et al.* 2019). The novelty of this study is achieved by the utilization of used cooking oil as a carrier for the recovery of zinc using a green emulsion liquid membrane (GELM). The effects of operating parameters, namely, external pH, surfactant concentration, internal phase concentration, zinc concentration, external phase to emulsion volume ratio and carrier concentration, on the extraction of zinc were investigated and RSM was employed to optimize the operating parameters and study the interaction effects.

MATERIALS AND METHODS

Materials

The waste cooking oil used in this work was procured from a nearby restaurant. It is filtered through a white cotton cloth to remove solid impurities. Sulfuric acid (40%), the internal/stripping phase, and Span 80, the surfactant, were procured from Merck, Germany. Distilled water was used to perform the experiments. Wastewater was collected from a small scale electroplating industry and used.

Green emulsion liquid membrane (GELM) preparation

The extraction of zinc was carried out in 250 mL Erlenmeyer flask at room temperature in an agitator. GELM used in this work is water-in-oil emulsion. The internal phase solution was prepared using sulfuric acid in distilled water. The liquid membrane phase was prepared by dissolving a suitable amount of Span 80 and waste cooking oil in a 100 mL beaker. To form the GELM, the internal phase solution was added dropwise to the oil with the aid of a burette and stirred at 3,000 rpm in a magnetic stirrer for 12 min (Bjorkegren & Karimi 2012; Laki & Kargari 2016). A fresh emulsion was prepared for each experiment using the above said procedure.

Optimization of process variables

Box-Behnken Design (BBD) under RSM was employed to optimize the operating variables for the extraction of zinc from electroplating industry wastewater (Table 1). Batch experiments were conducted, to determine the effects of six variables, namely external pH, surfactant concentration, internal phase concentration, zinc concentration, external phase to emulsion volume ratio and carrier concentration on the extraction of zinc. Regression analysis was performed to estimate the response function as a second order polynomial. A second order polynomial model fitted to the experimental data is given by the following equation:

$$Y = \beta_0 + \sum_{i=1}^{k} \beta_i X_i + \sum_{i=1}^{k} \beta_{ii} X_i^2 + \sum_{i=1}^{k-1} \sum_{j=i+1}^{k} \beta_{ij} X_{ij} \tag{1}$$

where Y is the predicted response.

$\beta_i, \beta_{ii}, \beta_{ij}$ are coefficients estimated from regression and represent the linear, quadratic and interaction effect, respectively: β_0 is the intercept term: $X_i, X_j \ldots X_k$ are the input variables in coded form.

Table 1 | Process variables and their ranges for the extraction of zinc from electroplating wastewater

Variables	Range −1	0	+1
External pH	3	5	7
Surfactant (span 80) concentration	2	3	4
Internal phase concentration	0.5	1.5	2.5
Initial concentration of zinc	500	1,500	2,500
Emulsion to external phase volume ratio	0.8	1	1.2
Carrier concentration	4	8	12

Zinc extraction

The prepared emulsion was slowly dispersed in 150 mL of zinc containing electroplating wastewater with agitation speed in the range of 200–600 rpm. During the contact between wastewater and the emulsion, solute transport occurred through the membrane phase into the internal receiving phase, where it was concentrated. The mixture was then allowed to separate. The treated effluent from the emulsion was decanted. Sampling was started from the beginning at specified time intervals and the samples were kept stagnant for 20 min. The concentration of zinc in aqueous feed phase was determined by Atomic Absorption Spectroscopy (Perkin Elmer, USA). The performance was evaluated using the following equations:

$$\text{Extraction of zinc, } \% = \frac{C_i - C_f}{C_i} * 100 \tag{2}$$

$$\text{Recovery of zinc, } \% = \frac{C_{is}}{4 * TR * C_i} * 100 \tag{3}$$

where C_i – initial zinc concentration, mg/L

C_f – zinc concentration in the external feed phase after extraction, mg/L

C_{is} – zinc concentration in the stripping phase, mg/L

TR – treatment ratio.

RESULTS AND DISCUSSION

The impact of process variables on zinc extraction was predicted using the quadratic model given below:

$$\begin{aligned}\text{Zinc extraction, } \% = {} & 93.57 - 1.74^*A + 4.51^*B - 0.40^*C \\ & - 4.64^*D - 0.92^*E + 1.97^*F - 2.78^*A^*B \\ & - 2.39^*A^*C - 0.58^*A^*D + 1.36^*A^*E \\ & + 0.89^*A^*F + 0.10^*B^*C + 0.56^*B^*D \\ & - 1.80^*BE + 1.27^*BF + 0.04^*CD \\ & + 0.60^*CE - 0.76^*CF + 1.51^*DE \\ & - 0.74^*D^*F + 3.33^*EF - 3.45^*A^2 \\ & - 1.41^*B^2 - 4.56^*C^2 - 4.65^*D^2 \\ & - 5.05^*E^2 - 4.81^*F^2\end{aligned} \tag{4}$$

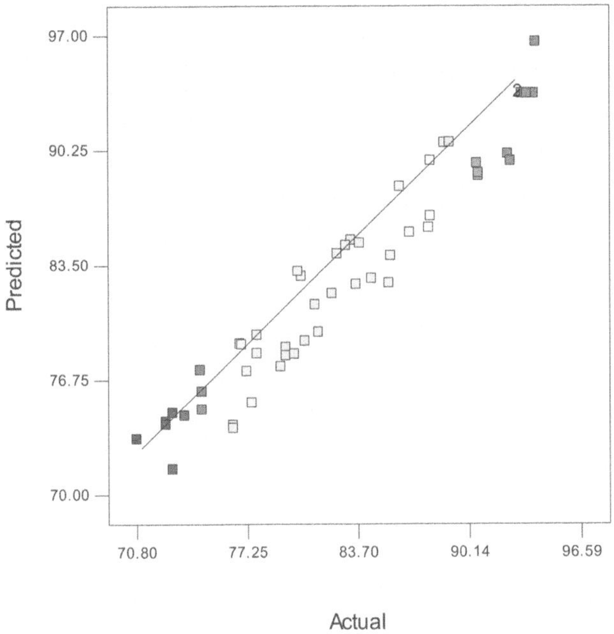

Figure 1 | Predicted vs experimental values of extraction of zinc.

where A - external pH, B - surfactant (span 80) concentration, C - internal phase concentration, D - initial concentration of zinc, E - emulsion to external phase volume ratio and F - carrier concentration.

Figure 1 shows a good match between the experimental and predicted zinc extraction ($R^2 = 0.9364$). Hence the above Equation (1) predicted the percentage of zinc extraction effectively. The ANOVA results obtained were tabulated in Table 2. The model F-value of 14.18 and *p*-value of less than 0.05 implies that the model was absolutely significant. From Table 2, it was observed that the terms A, B, D, F, AB, AC, BE, EF, A^2, C^2, D^2, E^2, F^2 were the influential parameters of the model.

Effect of pH

The external phase pH was reported to have significant effect on the extraction of metals in ELM. The extraction of zinc was studied in the pH range of 3.0–7.0. Figure 2 shows the effect of initial external phase pH on zinc extraction using emulsions. From this figure, it was inferred that, at a low pH 3, the extraction of zinc was found to be in the range of 80–85%. At pH around 3.8, maximum zinc recovery was obtained. From this figure, it was also found that the zinc extraction efficiency decreases with increase in pH above 3.8. In the pH range 3–4, effective extraction of zinc was observed and, above pH 5, there was no effect on zinc extraction from electroplating wastewater. In the pH range 3–4, extraction was high due to cation exchange reaction in which protons were released. At pH above 5, an increasing pH leads to formation of other species and similar observations were reported (Mellah & Benachour 2006; Fouad 2008). The optimal pH for maximum separation of zinc from wastewater was found to be pH 3.8.

Effect of surfactant concentration

The span 80 (surfactant) concentration sustains the stability of ELM. The effect of span 80 concentration on zinc extraction was depicted in Figure 2. When the surfactant concentration was increased from 2 to 4% (vol.), the extraction of zinc metal increased. At increasing surfactant concentration, viscosity of the emulsion membrane and emulsion stability increased. Mass transfer coefficient increase could have contributed to better removal of zinc. Further increasing the surfactant concentration resulted in high interfacial resistance to mass transfer and lower extraction efficiency was reported (Sulaiman *et al.* 2020). At lower span 80 concentration (2%), no emulsion was formed owing to a lack of the surfactant adsorbing the organic/aqueous interface (Chiha *et al.* 2010). The optimum surfactant concentration for the maximum extraction of zinc from electroplating wastewater was 4% (vol.).

Effect of internal phase concentration

The internal phase concentration affects the capacity of the emulsion to extract solute. The increase in acid concentration increases the emulsion capacity. The effect of internal phase stripping acid concentration on extraction of zinc was shown in Figure 3. The figure shows that there was an increase in extraction with an increase in internal phase acid concentration from 0.5 N to 1.61 N. The driving force in the ELM was difference in hydrogen ion chemical potentials between two aqueous phases. With increasing H_2SO_4 concentration from 0.5–1.61 N, increase in the

Table 2 | ANOVA table

Source	Sum of squares	df	Mean square	F value	*p*-value
Model	2,394.403	27	88.68159	14.17924	<0.0001
A	72.97594	1	72.97594	11.66808	0.0021
B	488.2526	1	488.2526	78.0664	<0.0001
C	3.856017	1	3.856017	0.616536	0.4394
D	516.8032	1	516.8032	82.63134	<0.0001
E	20.35042	1	20.35042	3.253815	0.0829
F	92.90535	1	92.90535	14.85458	0.0007
AB	61.88281	1	61.88281	9.894404	0.0041
AC	45.60125	1	45.60125	7.291155	0.0120
AD	5.463906	1	5.463906	0.873621	0.3586
AE	14.85125	1	14.85125	2.374557	0.1354
AF	6.30125	1	6.30125	1.007503	0.3248
BC	0.08	1	0.08	0.012791	0.9108
BD	2.475313	1	2.475313	0.395776	0.5348
BE	51.84	1	51.84	8.288665	0.0079
BF	13.005	1	13.005	2.079361	0.1612
CD	0.0128	1	0.0128	0.002047	0.9643
CE	2.88	1	2.88	0.460481	0.5034
CF	9.333025	1	9.333025	1.492251	0.2328
DE	18.30125	1	18.30125	2.926175	0.0991
DF	4.3808	1	4.3808	0.700443	0.4103
EF	88.445	1	88.445	14.14142	0.0009
A^2	122.7019	1	122.7019	19.61872	0.0002
B^2	20.30827	1	20.30827	3.247077	0.0832
C^2	214.1246	1	214.1246	34.23625	<0.0001
D^2	222.5357	1	222.5357	35.5811	<0.0001
E^2	262.196	1	262.196	41.92236	<0.0001
F^2	238.2325	1	238.2325	38.09084	<0.0001
Residual	162.6124	26	6.254324		
Lack of Fit	162.2991	21	7.728529	123.3276	<0.0001
Pure Error	0.313333	5	0.062667		
Cor Total	2,557.015	53			

Std. Dev. – 2.50; R^2 – 0.9364; Adj R^2 – 0.8704; C.V. % – 3.02; Pred R^2 – 0.6682.
PRESS – 848.38; Adeq Precision –13.939.

extraction of zinc was observed as the capacity of receiving phase increased. It was expected that emulsion viscosity increases by increasing amount of H_2SO_4 in the internal phase thus decreasing the difference of densities. The increasing viscosity of membrane resulted in increasing droplet size (Chaouchi & Hamdaoui 2014).

Effect of zinc concentration

The effect of zinc concentration was studied between 500 mg/L to 2,500 mg/L. At low zinc concentration, most of the solute diffusing within the emulsion globule was stripped by the internal phase droplets that were situated

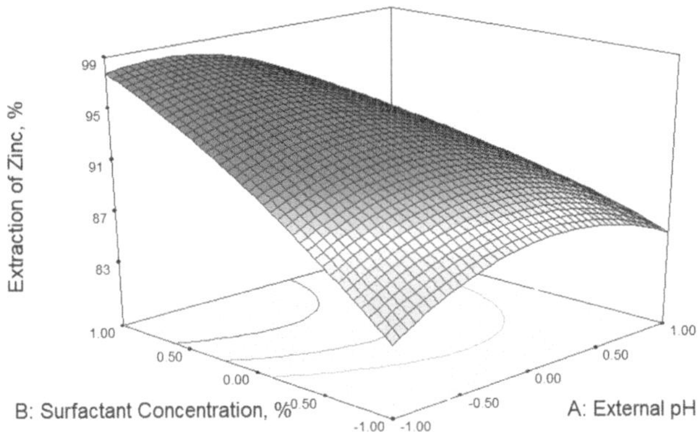

Figure 2 | Effect of external pH and surfactant concentration on zinc extraction.

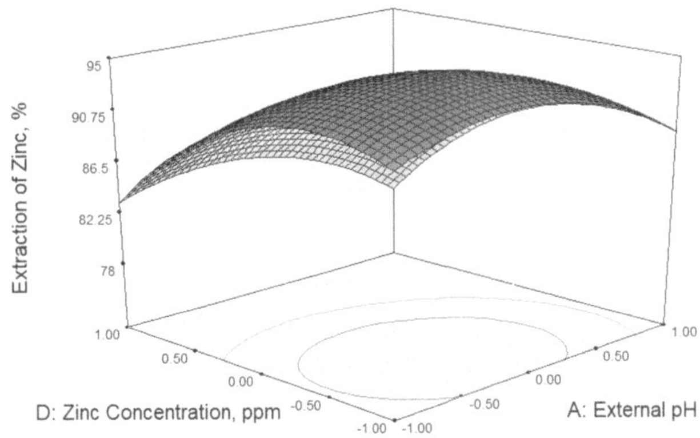

Figure 4 | Effect of external pH and zinc concentration on zinc extraction.

in the peripheral regions of the emulsion globule. When the zinc concentration increases, the peripheral droplets get rapidly exhausted, necessitating the solute to permeate deeper within the globule prior to get stripped. Therefore, an increase in zinc concentration, upto 742 mg/L, corresponds to an increase in diffusional path lengths. Increased path lengths lead to a decline in the extraction of zinc. This was clearly observed in Figure 4. Further increase in the feed concentration of zinc led to decline in the extraction. This could be due to the increase in mass transfer resistance in the emulsion globule (Daas & Hamdaoui 2010).

Effect of emulsion to external phase volume ratio

The ratio of emulsion phase to the feed phase, known as treat ratio (TR), was a measure of emulsion holdup in the system. Increase in treat ratio results in an increase in the extraction capacity of the emulsion. It also results in an increase in the amount of carrier and overall surface area for mass transfer in the system. An increase in emulsion phase to the feed phase was expected to increase both the rate and extent of extraction (Sengupta *et al.* 2006). Figure 5 shows the effect of emulsion phase to the feed phase on extraction behavior. When the emulsion phase holdup was low, the zinc extraction was low. Increase in the ratio up to 0.94 substantially increased the extraction of zinc but further increase in the ratio leads to decrease the zinc extraction. Increase in emulsion holdup results in a plethora of complex effects by increasing the amount of extractant in the system and resulted in formation of larger globules. It shifts the globule size distribution to the higher end of the spectrum. Larger globule sizes lead to a decrease in external mass transfer areas that, in some cases, become rate-limiting.

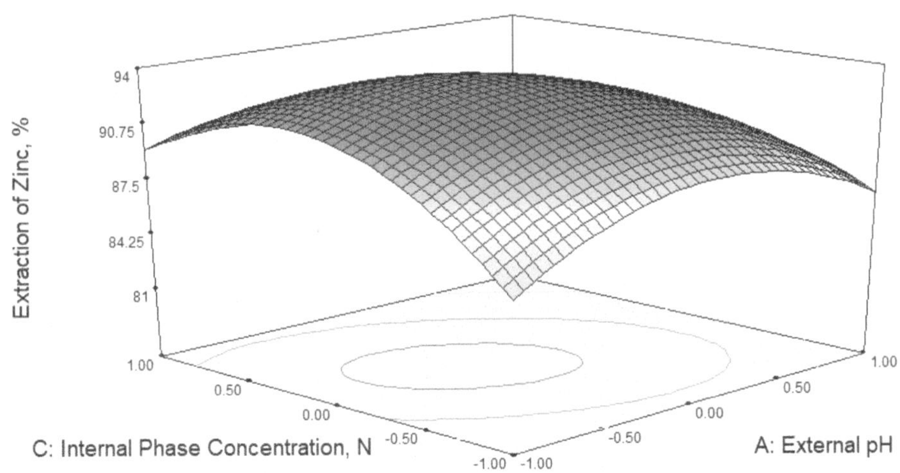

Figure 3 | Effect of external pH and internal phase concentration on zinc extraction.

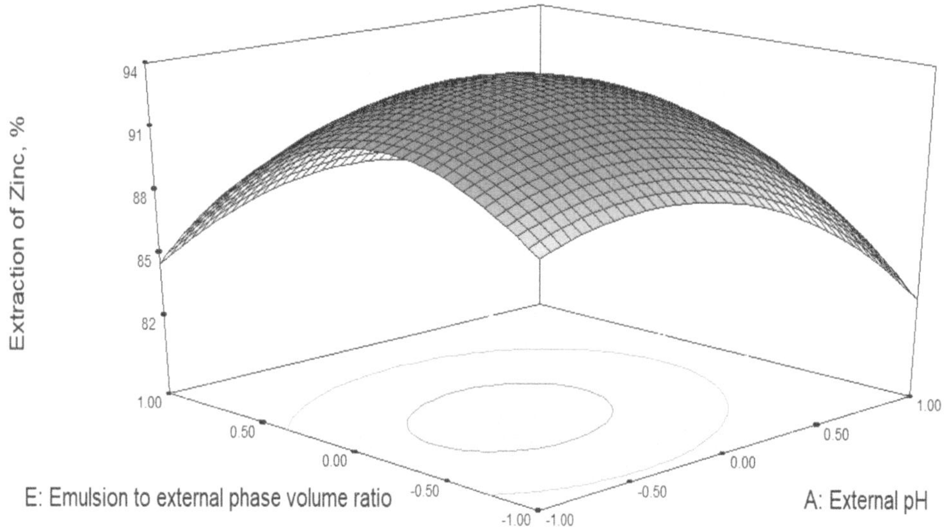

Figure 5 | Effect of external pH and emulsion to feed phase concentration on zinc extraction.

Large globules also cause an increase the effective diffusion path lengths within the globule that results in a decline in extraction rates. Increase in emulsion holdup also enhances globule interactions causing globule breakage, globule coalescence, redispersion, etc., which result in release of the encapsulated zinc back to the feed phase (Sengupta *et al.* 2006). Hence, increase in treat ratio brings forward a net effect which was a complex interaction of all these effects and it was not easy to decipher any single cause as dominant. In this case, apparently it appears that increase in extractant concentration was the prime reason for the enhancement of extraction rates.

Effect of carrier concentration

In ELM, the carrier concentration is one of the important variables to be optimized. Figure 6 shows the effect of carrier concentration on extraction behavior. Increase in

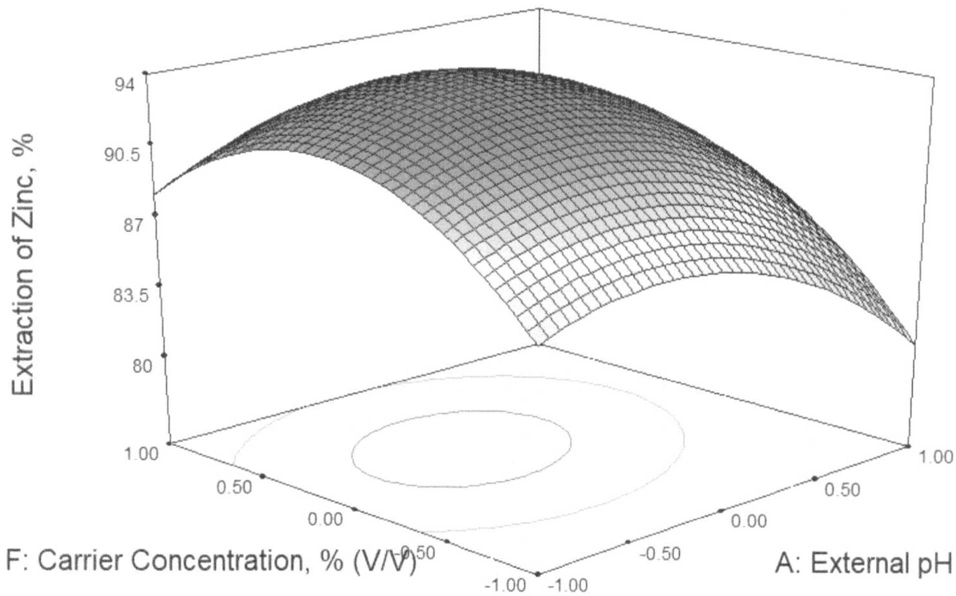

Figure 6 | Effect of external pH and carrier concentration on zinc extraction.

the carrier concentration in the oil membrane phase of the emulsion from 4% (v/v) to 8.9% (v/v) led to enhancement of zinc removal from feed phase. Increase in the carrier concentration in the range under investigation also augments the rate of extraction in another fashion, utilizing the surface active nature of the extractant. Decline in carrier concentration causes an increase in interfacial tension between the emulsion and feed phase, which leads to an increase in emulsion globule size, thereby lowering the external surface area and, as a consequence, lowering the extraction rates as well. Some extractant have a tendency to aggregate at high concentrations that tends to reduce its metal binding capacity. Decline in extraction rates were observed at these carrier concentrations deviating from the usual trend of increase in extraction rates with increase in carrier concentrations.

Optimum conditions for maximum zinc extraction

In Figures 2–6, the elliptical nature of the contours and bending degree of the axes show the interaction between individual variables in affecting zinc extraction. Figure 7 is the perturbation plot which shows the comparative effect of the variables on the response. A steep curvature in curves D and B shows that the zinc extraction was very sensitive to surfactant concentration, and zinc concentration. The comparatively semi-flat F and A curves show less sensitivity of carrier concentration and pH on the zinc extraction when compared to surfactant concentration, and zinc concentration. The internal phase concentration and external phase to emulsion volume ratio has less impact in the extraction process when compared to other variables. The perturbation plot (Figure 7) shows that the extraction of zinc was affected by variables in the following order of effect: zinc concentration > surfactant concentration > carrier concentration > external pH > external phase to emulsion volume ratio > internal phase concentration.

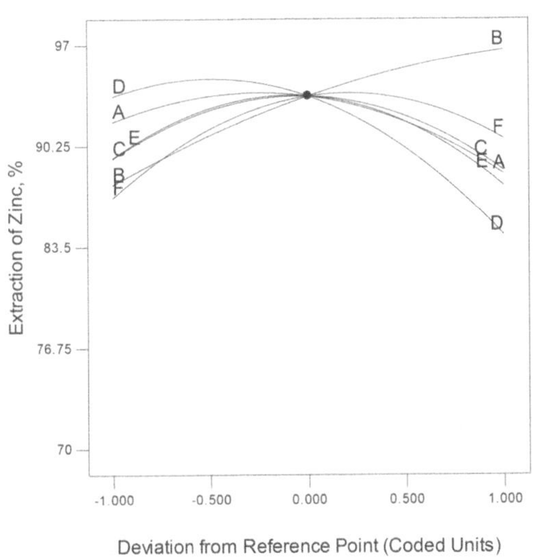

Figure 7 | Perturbation plot showing effect of variables on zinc extraction from electroplating wastewater.

The RSM analysis identified that the optimal variable values for the maximum extraction of zinc were: external pH – 3.8, surfactant concentration 4% (vol.), internal phase concentration – 1.61 N, zinc concentration – 742 mg/L, external phase to emulsion volume ratio – 0.94 and carrier concentration – 8.9%. At the optimized conditions experiment was carried out and the maximum extraction was found to be 97.4%. From Table 3, at unoptimized conditions, the maximum zinc extraction 94% (Run no. 36). But after finding the exact optimum conditions using RSM, the percentage of zinc extraction was increased by 3.5%. At these conditions, the recovery of zinc from the stripping phase was found to be 78%. The outcome of this work proved that efficient recovery of zinc from electroplating wastewater was feasible using GELM technology. In addition, this study has identified the optimal conditions required for the recovery of metal from electroplating wastewater.

CONCLUSIONS

The zinc extraction from electroplating wastewater was investigated using waste cooking oil based emulsion liquid membrane as green technology and appreciable zinc extraction efficiency was achieved. The process parameters including external pH, surfactant concentration, internal phase concentration, zinc concentration, external phase to emulsion volume ratio and carrier concentration were optimized using RSM method. The interaction effects between the variables were well represented by the contour plots. In the order of effect, zinc concentration is identified to

Table 3 | BBD design and results of extraction of zinc from electroplating wastewater using GELM

S. No.	A	B	C	D	E	F	Extraction of zinc	
							Experimental	Predicted
1	1	−1	0	1	0	0	72.90	74.80
2	0	1	−1	0	−1	0	89.00	90.68
3	0	0	−1	−1	0	1	86.12	88.09
4	0	−1	−1	0	−1	0	80.00	78.26
5	0	1	1	0	1	0	83.00	84.64
6	0	−1	0	0	1	1	84.50	82.67
7	1	−1	0	−1	0	0	87.90	86.36
8	−1	0	0	−1	−1	0	92.40	90.00
9	0	1	0	0	−1	1	90.60	89.45
10	0	0	0	0	0	0	93.70	93.57
11	−1	0	1	0	0	1	83.80	84.79
12	1	0	0	1	1	0	77.50	75.40
13	0	0	0	0	0	0	93.50	93.57
14	0	1	0	0	1	−1	79.20	77.52
15	1	0	0	−1	−1	0	83.30	84.96
16	1	1	0	1	0	0	77.80	79.37
17	0	1	1	0	−1	0	90.70	88.88
18	1	0	0	1	−1	0	72.90	71.49
19	0	0	−1	1	0	1	77.20	77.25
20	−1	−1	0	1	0	0	76.40	73.89
21	−1	0	0	−1	1	0	85.50	82.42
22	−1	0	−1	0	0	1	83.60	82.34
23	0	−1	−1	0	1	0	76.90	78.82
24	0	0	−1	1	0	−1	70.80	73.27
25	1	0	1	0	0	−1	72.50	74.11
26	0	0	−1	−1	0	−1	81.20	81.15
27	0	1	0	0	1	1	88.70	90.66
28	−1	0	0	1	−1	0	76.80	78.87
29	0	0	0	0	0	0	93.70	93.57
30	0	0	1	−1	0	1	87.80	85.68
31	0	0	1	−1	0	−1	82.20	81.80
32	0	0	0	0	0	0	93.40	93.57
33	1	0	0	−1	1	0	80.40	82.82
34	0	0	1	1	0	1	74.60	75.03
35	1	0	−1	0	0	−1	79.50	78.16
36	−1	1	0	−1	0	0	94.00	96.59
37	−1	0	1	0	0	−1	82.50	84.15
38	0	0	0	0	0	0	93.20	93.57

(continued)

Table 3 | continued

S. No.	A	B	C	D	E	F	Extraction of zinc	
							Experimental	Predicted
39	0	0	1	1	0	−1	76.40	74.08
40	−1	1	0	1	0	0	92.55	89.59
41	1	0	1	0	0	1	77.80	78.30
42	0	1	0	0	−1	−1	87.90	89.61
43	0	−1	1	0	−1	0	74.60	76.06
44	0	−1	1	0	1	0	80.60	79.02
45	0	−1	0	0	−1	−1	81.40	79.54
46	1	0	−1	0	0	1	86.70	85.40
47	0	−1	0	0	1	−1	73.60	74.65
48	−1	−1	0	−1	0	0	80.20	83.12
49	0	1	−1	0	1	0	85.60	84.04
50	0	−1	0	0	−1	1	72.50	74.28
51	−1	0	−1	0	0	−1	79.50	78.65
52	−1	0	0	1	1	0	74.50	77.33
53	1	1	0	−1	0	0	90.70	88.71
54	0	0	0	0	0	0	93.90	93.57

influence the most. The optimum condition for maximum removal is: external pH – 3.8, surfactant concentration 4% (vol.), internal phase concentration – 1.61 N, zinc concentration – 742 mg/L, external phase to emulsion volume ratio – 0.94 and carrier concentration – 8.9%. Under optimal conditions, the maximum extraction was found to be 97.4% and the recovery of zinc from the stripping phase was 78%. The successful recovery of zinc using a waste resource based ELM confirmed the sustainable application.

DISCLOSURE STATEMENT

The authors declare that they have no known competing financial interests in relation to this published work.

DATA AVAILABILITY STATEMENT

All relevant data are included in the paper or its Supplementary Information.

REFERENCES

Albaraka, Z. 2020 Carrier-mediated liquid membrane systems for lead (II) ion separations. *Chem. Papers* **74**, 77–88. https://doi.org/10.1007/s11696-019-00868-w.

Ali, S., Aziz Ur Rehman, S., Shah, I. A., Farid, M. U., An, A. K. & Huang, H. 2019 Removal of zinc from water and wastewater effluents by hydroxylated and carboxylated carbon nanotube membranes: behaviors and mechanisms of dynamic filtration. *J. Hazard. Mater.* **365**, 64–73. https://doi/10.1016/j.jhazmat.2018.10.089.

Benyahia, N., Belkhouche, N. & Jonsson, J. A. 2014 A comparative study of experimental optimization and response surface methodology of Bi(III) extraction by emulsion organophosphorus liquid membrane. *J. Environ. Chem. Eng.* **2**, 1756–1766. https://doi.org/10.1016/j.jece.2014.07.003.

Bjorkegren, S. & Karimi, R. F. 2012 *A Study of the Heavy Metal Extraction Process Using Emulsion Liquid Membranes*. Chalmers University of Technology, Göteborg, Sweden.

Chaouchi, S. & Hamdaoui, O. 2014 Extraction of priority pollutant 4-nitrophenol from water by emulsion liquid membrane: emulsion stability, effect of operational conditions and membrane reuse. *J. Disper. Sci. Technol.* **35** (9), 1278–1288. https://doi.org/10.1080/01932691.2013. 844704.

Chiha, M., Ahmedchekkat, F. & Petrier, C. 2010 Study on ultrasonically assisted emulsification and recovery of copper(II) from wastewater using an emulsion liquid membrane process. *Ultrason. Sonochem.* **17**, 318–325. https://doi.org/10.1016/j.ultsonch.2009.09.001.

Daas, A. & Hamdaoui, O. 2010 Extraction of bisphenol A from aqueous solutions by emulsion liquid membrane. *J. Membrane Sci.* **348** (2010), 360–368. https://doi.org/10.1016/j.memsci.2009.11.026.

Fernandez-Gonzalez, R., Martin-Lara, M. A., Moreno, J. A., Blazquez, G. & Calero, M. 2019 Effective removal of zinc from industrial plating wastewater using hydrolyzed olive cake: scale-up and preparation of Zinc-based biochar. *J. Cleaner Prod.* **227**, 634–644. https://doi.org/10.1016/j.jclepro.2019.04.195.

Fouad, E. A. 2008 Zinc and copper separation through an emulsion liquid membrane containing di-(2-ethylhexyl) phosphoric acid as a carrier. *Chem. Eng. Technol.* **31**, 370–376. https://doi.org/10.1002/ceat.20070043.

Hafshejani, L. D., Nasab, S. B., Mafi Gholami, R., Moradzadeh, M., Izadpanah, Z., Hafshejani, S. B. & Bhatnagar, A. 2015 Removal of zinc and lead from aqueous solution by nanostructured cedar leaf ash as biosorbent. *J. Mol. Liq.* **211**, 448–456. http://dx.doi.org/10.1016/j.molliq.2015.07.044.

Jusoh, N. & Othman, N. 2017 Stability of palm oil-based emulsion liquid membrane for succinic acid extraction from aqueous solution. *Appl. Membr. Sci. Technol.* **19**, 1–17. https://doi.org/10.11113/amst.v19i1.19.

Kislik, V. S. 2010 *Liquid Membranes: Principles and Applications in Chemical Separations and Wastewater Treatment*, 1st edn. Elsevier, Amsterdam, The Netherlands.

Kocanova, V., Cuhorka, J., Dusek, L. & Mikulasek, P. 2017 Application of nano filtration for removal of zinc from industrial wastewater. *Desalin. Water Treat.* **75**, 342–347. https://doi:10.5004/dwt.2017.20453.

Laki, S. & Kargari, A. 2016 Extraction of silver ions from aqueous solutions by emulsion liquid membrane. *J. Memb. Sci. Res.* **2** (1), 33–40. https://doi:10.22079/JMSR.2016.15876.

Liu, H., Zhang, Y., Huang, J., Liu, T., Xue, N. & Shi, Q. 2017 Optimization of vanadium (IV) extraction from stone coal leaching solution by emulsion liquid membrane using response surface methodology. *Chem. Eng. Res. Des.* **123**, 111–119. https://doi.org/10.1016/j.cherd.2017.05.001.

Ma, H., Kokkilic, O., Langlois, R., Song, X., Qin, Y. & Waters, K. E. 2018 Selective separation of copper and nickel ions from aqueous solutions containing calcium by emulsion liquid membranes using central composite design. *Can. J. Chem. Eng.* **9999**, 1–13. https://doi.10.1002/cjce.23418.

Mellah, A. & Benachour, D. 2006 The solvent extraction of zinc and cadmium from phosphoric acid solution by di-2-ethyl hexyl phosphoric acid in kerosene diluent. *Chem. Eng. Process* **45**, 684–690. https://doi.org/10.1016/j.cep.2006.02.004.

Mesli, M. & Belkhouche, N. E. 2018 Emulsion ionic liquid membrane for recovery process of lead. Comparative study of experimental and response surface design. *Chem. Eng. Res. Des.* **129**, 160–169. https://doi.org/10.1016/j.cherd.2017.11.011.

Rosly, M. B., Jusoha, N., Othmana, N., Abdul Rahman, H., Mohamed Noah, N. F. & Raja Sulaiman, R. N. 2019 Effect and optimization parameters of phenol removal in emulsion liquid membrane process via fractional-factorial design. *Chem. Eng. Res. Des.* **145**, 268–278. https://doi.org/10.1016/j.cherd.2019.03.007.

Seifollahi, Z. & Rahbar-Kelishami, A. 2019 Amoxicillin extraction from aqueous solution by emulsion liquid membranes using response surface methodology. *Chem. Eng. Technol.* **42**, 156–166. https://doi.org/10.1002/ceat.201800089.

Sengupta, B., Sengupta, R. & Subrahmanyam, N. 2006 Process intensification of copper extraction using emulsion liquid membranes: experimental search for optimal conditions. *Hydrometallurgy* **84**, 43–53. https://doi.org/10.1016/j.hydromet.2006.04.002.

Shrestha, S., Son, G., Lee, S. H. & Lee, T. G. 2013 Isotherm and thermodynamic studies of Zn (II) adsorption on lignite and coconut shell-based activated carbon fiber. *Chemosphere* **92**, 1053–1061. https://doi.org/10.1016/j.chemosphere.2013.02.068.

Sulaiman, R. N. R., Abdul Rahman, H., Othman, N., Rosly, M. B., Jusoh, N. & Mohd Noah, N. F. 2020 Extraction of reactive dye via synergistic Aliquat 336/D2EHPA using emulsion liquid membrane system. *Korean J. Chemical Eng.* **37**, 141–150. https://doi.org/10.1007/s11814-019-0418-4.

Wang, J., Wang, Z., Zhang, Z. & Zhang, G. 2019 Removal of zinc from basic oxygen steelmaking filter cake by selective leaching with butyric acid. *J. Clean. Prod.* **209**, 1–9. https://doi.org/10.1016/j.jclepro.2018.10.253.

First received 13 August 2020; accepted in revised form 22 October 2020. Available online 4 November 2020

Real-time monitoring as an adaptive strategy towards green treatment of textile effluent using biosorbent from *Acalypha indica*

C. Sivapragasam, V. Aruna Janani, A. Andappan, B. Archana, M. Vasudevan and N. Natarajan

ABSTRACT

Performance of green treatment systems such as adsorption to treat textile effluents often suffers lack of longevity and efficiency due to the presence of complex compounds of varying reactivity. There is scope for improving the operational efficiency of such processes using real-time monitoring systems. The present study aimed to evaluate the performance of an activated biosorbent prepared from the leaves of *Acalypha indica* for treating textile industry effluent by simulating process control with real-time monitoring. Batch experiments were performed with synthetic and real-time dye effluents to identify the optimum conditions (pH = 3.0, dosage = 1.0 g/L; time = 1 h) for the highest adsorption capacity (6 mg g^{-1} and 2 mg g^{-1}). The evaluation of physical parameters suggested best fit for Freundlich isotherm model and pseudo-second-order kinetic model. The LabVIEW-based simulation control system enabled close monitoring of pH and temperature during the process. Based on the inputs, an alteration of initial pH has resulted in substantial reduction in chemical oxygen demand (COD) (73.91%), turbidity (52.43%) and total dissolved solids (TDS) (19.43%). The average incremental increase was highest for COD (45.80 ± 0.06%) compared to TDS (10.13 ± 0.06%) and turbidity (−1.74 ± 0.03%) for varying dosage (3 g to 11 g). The proposed framework for incorporating a process-control-based monitoring system can help to achieve better performance.

Key words | *Acalypha indica*, adsorption models, biosorption, LabVIEW, real-time monitoring, textile effluent

C. Sivapragasam (corresponding author)
B. Archana
Department of Civil Engineering,
Kalasalingam Academy of Research and Education,
Krishnankoil 626128,
Tamil Nadu,
India
E-mail: *sivapragasam25@gmail.com*

V. Aruna Janani
A. Andappan
Department of Chemical Engineering,
Kalasalingam Academy of Research and Education,
Krishnankoil 626128,
Tamil Nadu,
India

M. Vasudevan
Department of Civil Engineering,
Bannari Amman Institute of Technology,
Sathyamangalam 638401,
Tamil Nadu,
India

N. Natarajan
Department of Civil Engineering,
Dr Mahalingam College of Engineering and Technology,
Pollachi 642003,
Tamil Nadu,
India

HIGHLIGHTS

- Activated carbon produced from the leaves of *Acalypha indica* as a biosorbent.
- Adsorption isotherm provides best fit with Freundlich model.
- Adsorption kinetics indicate that the textile effluent follows pseudo-second-order kinetics.
- LabVIEW has been used for the measurement of pH and temperature of the effluent.
- Adjustment of the initial pH led to substantial reduction in the COD, TDS and turbidity.

doi: 10.2166/ws.2021.033

INTRODUCTION

Water pollution is a perilous issue threatening our very existence despite extraordinary advances in technological innovations. There is an open call for water conservation either by reducing the usage economically or by enhancing reuse/recycle to meet the growing demands (Zhang *et al.* 2016; Kausar *et al.* 2018). Among the water consuming industries, textile industries and pulp and paper industries are of major concern not just due to their enormous volumetric rate but more importantly due to their hazardous nature of their ingredients. Depending on the nature of dyeing operations, there are several categories of dyes such as direct, reactive, vat, azo and sulphide dyes which are regularly used in the textile industries (Saravanan *et al.* 2012a; Kausar *et al.* 2018). For example, direct dyes are water soluble while azo dyes consist of insoluble azo groups (Kanwal *et al.* 2017). Hence, a common strategy for the removal of dyes is still a major issue in the treatment of the textile effluents (Xu *et al.* 2017).

Being the second largest producer and a major consumer (63%) of textile products, the Indian textile industry is regulated by a few statutory laws such as Environment Act (1986), Environment Protection Rules (1986), Water Act (1974), Hazardous Waste Rules (2008), and the Prevention and Control of Pollution Act (1981). Upon realizing the limitations for these legislations in creating a positive impact on the environment, further guidelines were issued by the National Green Tribunal (2014) and the Central Pollution Control Board (2015) to implement zero liquid discharge. The target industries include pharmaceuticals, paper and pulp, tanneries, and refineries and dyeing and textiles. This has resulted in a never-before strive for the industries to achieve complete recycling of their effluents by installing additional treatments units such as adsorption, evaporation and incineration.

Having sufficiently exposed the problems with textile effluents, it is important to compare the available solution strategies for the elimination of dye compounds from water (Long *et al.* 2017). It is inevitable to accept the associated advantages and disadvantages of the available treatment techniques (physical methods such as adsorption and membrane separation, and chemical processes such as coagulation, chemical oxidation, froth-flotation, oxidation, ozonation and solvent extraction) as it is difficult to establish the absolute superiority of one process over other (Troster *et al.* 2002; Saravanan *et al.* 2012b; Ang *et al.* 2015; Crini & Lichtfouse 2018). Another important aspect is the number of stages in the downstream process line in textile dyeing before reaching the final product (such as desizing, bleaching, dyeing, printing and finishing) where the quantity and strength of the effluent vary considerably (Srebrenkoska *et al.* 2014). Hence, from a resource recovery point of view, hybrid solutions are suggested to achieve simultaneous separation of dyes and salts along with purification of water (Lin *et al.* 2015; Kumar & Saravanan 2017). However, the selective reactivity and huge economic requirements can pose sufficient hindrance to the stakeholders to rethink about modifying some of the existing sustainable solutions such as adsorption treatment.

Recent studies have indicated that adsorption is still a preferred technique for the remediation of various compounds in the textile effluents due to their promising results in sustainability measurements (Vasudevan *et al.* 2016; Padmanabhan *et al.* 2017; Barakan *et al.* 2020; Linggi *et al.* 2020; Mohammadi *et al.* 2020). In view of the characteristic porosity contributing to a high surface area and notable adsorption capacity, activated carbon (AC) derived from natural biomass is found to be efficient in the removal of several kinds of dyes, heavy metals, and gases (Saravanan *et al.* 2012a; Kausar *et al.* 2018). Biosorbents prepared from natural materials and wasted agricultural products are found to be effective in removing water-soluble synthetic dyes such as Alcian Blue, Methylene Blue, Malachite Green, Violet 6 and Basic Red 18 (Khorramfar *et al.* 2010; Gupta *et al.* 2011; Mallampati *et al.* 2015; Jendia *et al.* 2020). Similarly, heavy metals are also removed from the aqueous solution by specific biosorbents under selected experimental conditions (Acharya *et al.* 2009; Boudrahem *et al.* 2011; Ranjith *et al.* 2011; Vasudevan *et al.* 2016; Padmanabhan *et al.* 2017; Natarajan & Al-Qasmi 2018; Shyla *et al.* 2018; Poonam & Kumar 2020; Ramos-Vargas *et al.* 2020).

The synthesis of low-cost biosorbents has assumed high importance owing to the economic benefit, while the treatment efficiency is on par with many advanced techniques

(Peng *et al.* 2017; Cechinel *et al.* 2014). In this study, the biosorptive capacity of an indigenous medicinal plant, *Acalypha indica* was studied for treating textile industry effluents. Biosorbents prepared from the leaves of *Acalypha indica* were found to be effective in removal of Reactive Red 198 (Praveena *et al.* 2011), Methylene Blue (Ramathilagam & Ananthakumar 2016) and Methyl Orange (Janani *et al.* 2019) from aqueous solutions. Suneetha *et al.* (2015) used the cost-efficient AC obtained from the stems of *Acalypha indica* plant for the elimination of fluoride from the polluted water. Although this species has been used in several studies for the elimination of water-soluble dyes, the application of the same species for treatment of a real-time textile effluent is not well understood.

In many cases, it is not feasible to undergo a forensic investigation to characterize these pollutants in detail due to various economic, technical and institutional limitations. Nonetheless, basic physico-chemical analysis of textile effluents still proves to provide many practically useful insights to adopt conservative and sustainable remediation strategies. The compositional heterogeneity of the textile effluents makes it cumbersome to predict the variability in sorptive mechanisms, as the expected physico-chemical parameters undergo incoherent variations during the adsorption process. These problems can be quite severe during large-scale applications where even a slight variation in pH or temperature can cause significant reduction in the treatment efficiency, causing further economic liability. In this situation, a real-time process monitoring scheme can help the production engineers to intervene during the process and, in some cases, can make suitable modifications to improve the performance of the system (Schreiner *et al.* 2006).

The graphical programming environment called Laboratory Virtual Engineering Workbench (LabVIEW) is widely used to simulate the temporal variability of the selected environmental parameters by real-time monitoring and simultaneous recording of the data. In a previous study, Sivapragasam *et al.* (2017) used the LabVIEW platform for real-time estimation of the water footprint for hydropower generation. However, it is important to customize these process-control systems based on the real-time monitoring data inputs. Therefore, the objectives of the study were: (i) to prepare and characterize AC from *Acalypha indica*, (ii) to compare the removal efficiencies and performance mechanisms while using a real-time textile effluent instead of synthetic dye solution, and (iii) to evaluate the variations in pH and temperature as process-control parameters by real-time monitoring using the LabVIEW environment. This work is novel compared with other similar studies on adsorption in the sense that a conceptual design and implementation of LabVIEW has been proposed for the adsorption process control for the first time.

MATERIALS AND METHODS

Sampling and analysis of textile effluent

The effluent was collected from a textile industry in Rajapalayam Taluk of Tamil Nadu, and was processed water after bleaching prior to any preliminary or primary treatment in the industry. The physico-chemical characteristics of the real-time textile industry effluent (TIE) samples collected are given in Table 1. The industry had already attempted a few popular treatment methods like coagulation, precipitation and ion exchange processes for treating their effluent. However, considering the overall economic and environmental impacts, they are interested in designing an adsorption-based treatment system for the removal of toxic dyes from their effluents.

Preparation and characterization of AC

In order to prepare the biosorbent, about 100 leaves of *Acalypha indica* were collected from local gardens in Rajapalayam Taluk of Tamil Nadu and washed with distilled water and subsequently dried at 70 °C for a time period of 48 h. The dried leaves were powdered and the particles were sieved using a British Standard Sieve mesh size 60

Table 1 | Characteristics of textile industry effluent collected

Parameter	Concentration
Biological oxygen demand (BOD)	11.2 mg l^{-1}
Chemical oxygen demand (COD)	2,000 mg l^{-1}
Dissolved oxygen (DO)	6.7 mg l^{-1}
pH	9.9
Total dissolved solids (TDS)	1,330 mg l^{-1}
Turbidity	124.7 NTU

(250 μ) and stored in an airtight container. The dried powder was mixed with 1 M hydrochloric acid (HCl) at an impregnation ratio of 1:1 (*w/v*) for 24 hours for activation at room temperature, followed by oven drying at 70 °C for another 24 hours. This chemical activation was followed by thermal activation to promote carbonization by keeping the dried materials inside a stainless steel box and heating in a muffle furnace at a temperature of 250 °C for 20 minutes. After subsequent washing with distilled water and drying, the AC was made ready for use.

In order to ascertain the recovery of used AC from the effluent after adsorption, an experiment was performed to prepare magnetized activated carbon (MAC) using chemical precipitation. For this, 3.9 grams of ferrous sulphate ($FeSO_4.7H_2O$) was mixed with 7.8 grams of ferric chloride (Fe_2Cl_3) and was heated at 40 °C in a hot air oven for 30 min. To this mixture, 3.3 grams of prepared carbon was added, followed by drop-wise addition of 100 ml of 5 M sodium hydroxide (NaOH) solution with continuous stirring. The resulting precipitate of ferric oxide (Fe_2O_3) was separated by filtration and taken for reuse after washing and drying at 60 °C (Altıntıg *et al.* 2017).

In order to characterize the surface features of the adsorbent, various instrumentation techniques were employed such as Brunauer–Emmett–Teller (BET) analysis for calculating the surface area, scanning electron microscope (SEM) image analysis (EVO18, M/s Carl Zeiss, Germany) and X-ray diffractometer analysis (D8 Advanced ECO, M/s Bruker, India) for determining the surface morphology and Fourier transform infrared (FTIR) spectroscope (RTracer-100, M/s Shimadzu, Japan) for identifying the presence of functional groups on the surface of the prepared AC adsorbent. The FTIR analysis was performed in the region from 4,000 cm^{-1} to 400 cm^{-1}.

Batch adsorption studies

In order to understand the feasibility of adopting adsorption as an effective remediation technology for TIE, a few batch experiments were performed in the laboratory. As a reasonable assumption, Malachite Green (MG) was used as the representative compound for preparing a synthetic dye solution (SDS) with varying initial concentrations 250, 500, 750 and 1,000 mg l^{-1} in order to match with the physico-chemical characteristics of the collected sample (Table 1). For this, a standard graph was plotted between the selected range of initial concentrations and the corresponding absorbance values obtained with a UV-visible spectrophotometer (Model-1800, M/s Shimadzu, Japan) at a wavelength of 325 nm. The wavelength was adjusted to this maximum value in order to match the normal peak absorbance range reported for azodyes which are very commonly present in textile effluents. Based on the trial experiments, the initial concentrations for the TIE samples were varied from 30, 50, 70, 90 and 110 mg l^{-1} in order to estimate the suitable dosage conditions.

A series of batch experiments was conducted by varying the adsorbent dosage, contact time and pH of the solution to estimate the maximum adsorption capacity for both SDS and TIE samples. While testing with the SDS, the adsorbent dosage was varied in the range 0.05, 0.1, 0.15 and 0.2 g per 100 ml of prepared sample, whereas the dosage range for treating the TIE samples was kept higher (1, 3, 5, 7, 9 and 11 g) for the same volume. The pH of the sample was modified using either NaOH or HCl to make the trials for the range of 3.0, 5.0, 7.0 and 9.0. All the experiments were conducted in 200 ml Erlenmeyer flasks with continuous agitation by placement in a rotary shaker at 200 rpm to maintain a good contact between the adsorbent and the dye compounds. The equilibrium experiments were performed for the selected optimum time (1 h) by keeping the optimum dosage (as mentioned in the Results section) and varying the initial concentration from 250 mg l^{-1} to 1,000 mg l^{-1}. For finding the temporal variation of the prevailing concentration in the agitated medium during the kinetic studies, 2 ml samples were collected at equal intervals of 10 minutes up to a period of 4 h. The collected samples were centrifuged at 10,000 rpm for 10 minutes and the supernatant was taken for measurement. In a similar way, the collected effluent samples were also treated with the prepared biosorbent by maintaining the optimum dosage and pH conditions for maximum adsorption.

The adsorption capacity of the prepared biosorbent was determined using the following equation (Cazettaa *et al.* 2011):

$$q_e = \frac{(C_0 - C_t)V}{W} \tag{1}$$

where q_e is the adsorption capacity (mg g^{-1}), C_o and C_t are the initial and equilibrium concentrations of the sample (mg g^{-1}) respectively, V is volume of the sample (ml) and W is mass of adsorbent (g).

Monitoring of environmental parameters

As the nature and extent of physico-chemical modifications happening in the system during the adsorption process are practically very difficult to monitor merely by visual means, we used the LabVIEW platform to simulate the temporal variability of the selected environmental parameters. In the present study, a ds18b20 temperature sensor and Robocraze pH sensor were used to measure the temperature and pH of the effluent respectively as a function of time. The sensor was connected to the Data Acquisition (DAQ) unit of LabVIEW which has a triggering and data storing set up to acquire and store the sensor data in a very short period of time. The program was set to record the temperature and pH readings every 5 minutes. The DAQ was initialized by giving an input while the output data were stored in a specified file path. The measured output readings are in voltage which can be converted to temperature and pH readings using Equations (2) and (3):

$$T = (V - 0.72) \times 100 \tag{2}$$

$$pH = 3.5 \times V + 0.05 \tag{3}$$

where T is the temperature (°C) and V is the voltage reading (V).

The additive and multiplicative terms in Equations (2) and (3) represent the calibration constants, indicating the possible range of correct data with offsets for the measured values. The LabVIEW platform developed for the temperature and pH measurements is shown below (Figure 1). A conceptual design framework of real-time process-control systems using LabVIEW is also presented in this study in a later section.

RESULTS AND DISCUSSION

Adsorbent characterization studies

Surface morphological features

The firsthand evidence of adsorptive removal of MG using AC was observed by comparing the SEM images before and after adsorption (Figure 2). On a peripheral screening approach, the presence of bright spots on the AC grains (Figure 2(b)) indicated the active mass transfer of MG molecules from the aqueous phase to the solid (sorbed) phase. Unlike a uniform surface coating, the discontinuous surface-filling phenomenon provided evidence for the prevailing non-linear mass transfer mechanism which will be further elucidated using adsorption kinetic models. This inference is in corroboration with the conclusions reported by Arami *et al.* (2005) who have reported a pseudo-second-order (PSO) kinetic model to represent the rapid adsorption in a continuous flow/mixing process. The BET surface area analysis showed that the AC is essentially micro-porous and possesses larger surface area (in the order of 1,500 m^2 g^{-1}) when compared to the plain leaves without activation (in the order of 70 m^2 g^{-1}).

Significance of pre-treatment in adsorbent preparation

As the surface functional groups play a major role in the adsorption process, activation of adsorbents generally follow some selective pre-treatment processes to enhance their sorptive capacity. Literature shows that compared to granular activated carbon (GAC), fine-sized MAC is more hydrophilic in nature due to the interaction between separated bond spins and conduction electron spins (Nakayama *et al.* 1993). Generally, magnetization is shown to improve the presence of oxygen on the adsorbent surface, thus leading to the formation of more C–O and M–O bonds as observed in FTIR spectra (Rekos *et al.* 2016; Anyika *et al.* 2017). Presence of more hydroxyl groups can also impact higher adsorption due to the abundance of hydrogen bonds (Moosavi *et al.* 2020). However, MAC is usually preferred for removal of heaving metals form aqueous solution due to their easiness in separation after adsorption (Adibmehr & Faghihian 2018). In addition, it also showed

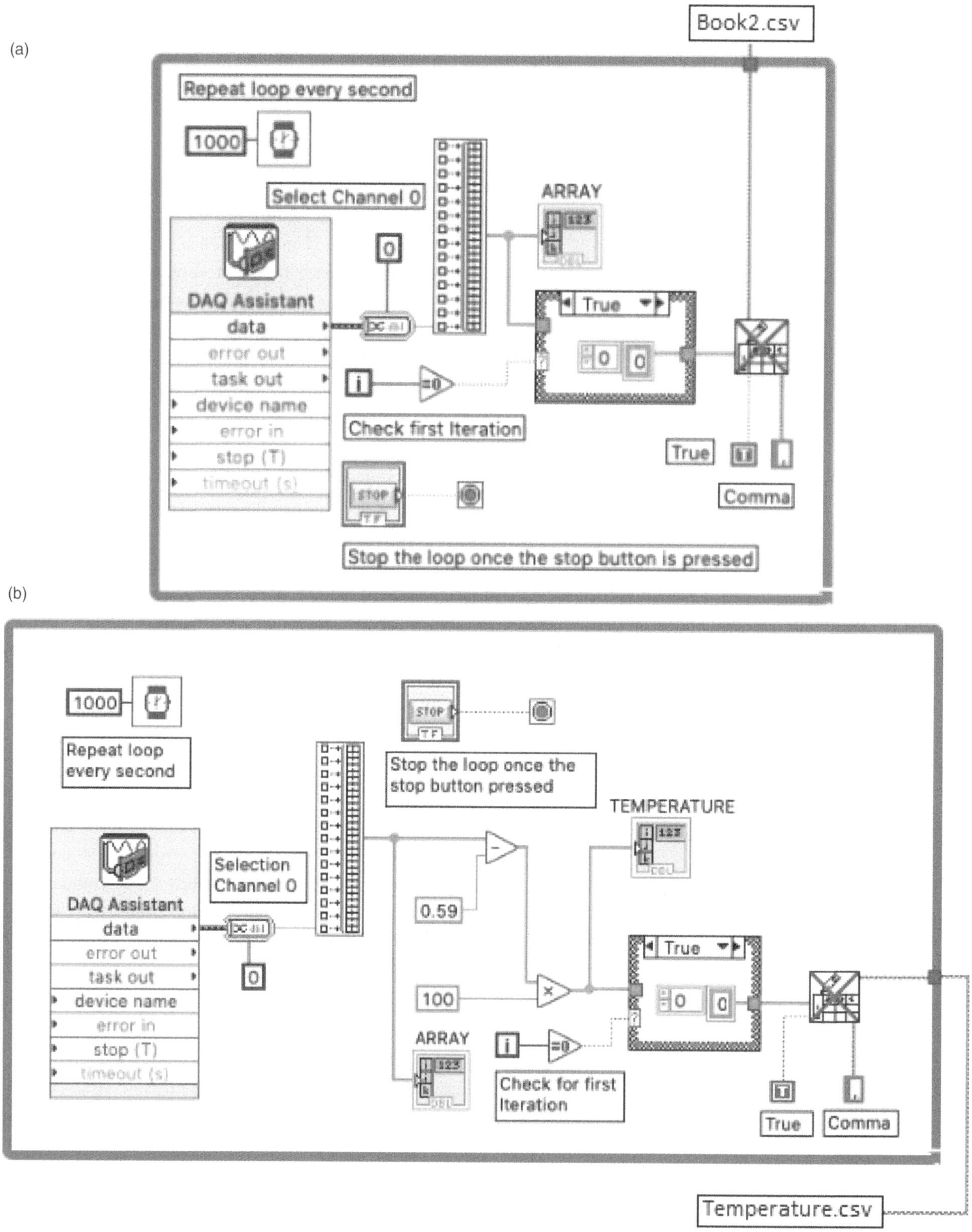

Figure 1 | The LabVIEW configuration environment for the continuous monitoring of (a) pH and (b) temperature.

Figure 2 | SEM images showing the change in surface roughness of the prepared AC (a) before and (b) after adsorption.

improved adsorption capacity under highly acidic conditions (pH 1.8–2.4) (Shirun *et al.* 1997). Yang *et al.* (2008) observed high adsorption capacity for Methylene Blue by rice husk-based AC after magnetization with 23% (w/w) Fe_3O_4.

In order to elucidate the specific reactive combination for the adsorbent and the dye compounds, a preliminary comparison was made between the MAC and normal AC in terms of their adsorptive capacities. Figure 3(a) depicts the comparison of the observed adsorption capacity for MAC and normal AC. The normal AC is found to be more effective in dye removal (in terms of q_{eq}) when compared to the magnetized dye for the given set of adsorption conditions (1 g adsorbent for 30 mg l^{-1} synthetic dye solution). It was observed that ferrous sulphate used in the magnetization process had added additional colour to the

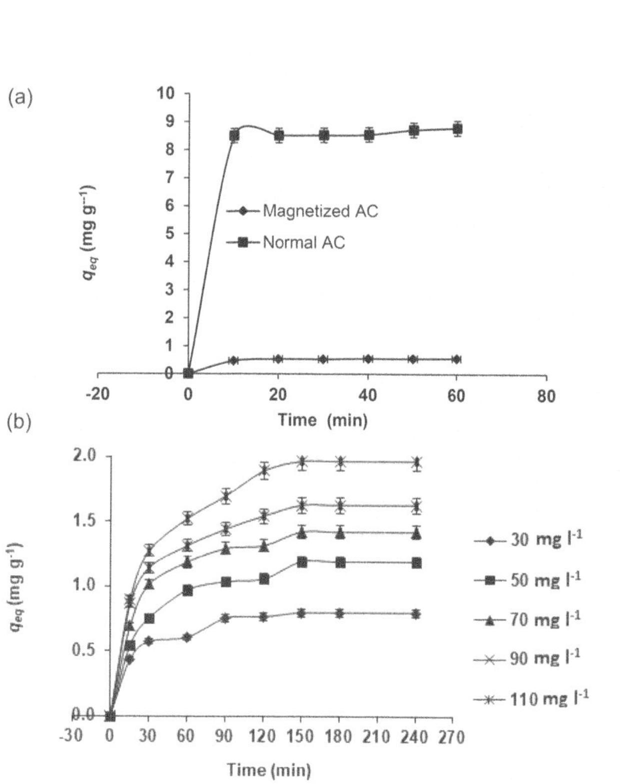

Figure 3 | Comparison of the variation in adsorption capacity in response to the variations in (a) adsorbent pre-treatment and (b) initial dye concentration.

solution due to the formation of magnetic iron oxides thereby reducing the preference of the target dye compounds to occupy the adsorbent surface. It can also be inferred that the magnetized carbon may not be preferred for the treatment of mixed dye effluent due to the formation of additional free radicals that can impart more reactivity to the toxic dyes in aqueous environment (Praveena *et al.* 2011; Suneetha *et al.* 2015). In addition, the reduced activity of GAC may also be attributed to the non-metallic nature of the pollutant (MG dye) and the irregular pore structure formed during the activation (Moosavi *et al.* 2020). Due to the reduced adsorption capacity, it may lead to additional expense for treatment for removal of coloured dyes. The temporal variations in the adsorptive removal efficiency showed promising results for the TIE samples when the initial concentration was varied from 30 mg l^{-1} to 110 mg l^{-1}. Figure 3(b) showed that higher adoption capacity can be obtained for higher initial concentrations indicating a responsive time delay in reaching the equilibrium at higher concentrations. Hence it can be understood that adsorption mechanism is not limited by the surface features of the adsorbent within the selected range of initial concentrations.

A detailed analysis of this result was further carried out with a two-factor ANOVA without repetition to understand the significance of the variations in some of the experimental conditions on the expected outcome. The analysis considered contact time and initial concentration as independent variables and maximum adsorption capacity as the dependent variable for both SDS and TIE samples (Table 2). The sum of squares (*SS*) values indicated the amount of variation caused by the systemic difference within the selected range of independent parameters (i.e. contact time and initial concentration) separately (Maxwell *et al.* 2017). While considering both samples, the SS values for contact time (1,175.73 and 3,377.76) were lower compared to the values for initial concentration (3,271.34 and 25,007.03). This shows that the relative change in adsorption capacity with incremental increase in contact time is less significant. In other words, the adsorption process followed a fairly uniform increasing trend for the increase in contact time. However, the change in initial concentration had drastically changed the adsorption capacity at every time step due to the increased physico-chemical interactions, resulting in higher values of internal variance.

The degree of freedom (*df*) and mean square (*MS*) are the consequent parameters based on the variance, in order to compute the *F*-values separately (Maxwell *et al.* 2017). As the *F*-values for the initial concentrations are higher compared to their *F*-critical values, we may infer the significant influence of variation in initial concentration on adsorption compared

Table 2 | Results of ANOVA study for various contact time and initial concentrations for synthetic dye solution (SDS) and real textile industry effluent (TIE)

Sample	Source of variation	Sum of squares (*SS*)	Degree of freedom (*df*)	Mean square (*MS*)	*F*-value	*p*-value	*F*-crit
SDS sample	Contact time	1,175.73	6	195.95	2.57	0.0457	2.51
	Initial concentration	3,271.34	4	817.83	10.72	3.94e-05	2.77
TIE sample	Contact time	3,377.76	6	562.96	1.09	0.3897	2.42
	Initial concentration	25,007.03	5	5,001.41	9.70	1.37e-05	2.53

to the variation in contact time (Mandel 1961; Alin & Kurt 2006). This can also be inferred using the p-values, as the higher p-values (great than 0.05) for the time interval indicated that the variation in time interval had no significant influence on the adsorption capacity compared to the initial concentrations. It was also observed that the variation in MS of initial concentrations was high for the results with TIE samples compared to SDS samples, indicating the possibility of progressive adsorption due to the low range of initial concentrations used in the experiments. This analysis provides the statistical evidence for the cause of difference in errors while comparing the differences in adsorption capacities.

Influence of pH and adsorbent dosage on adsorption capacity

The performance of the adsorbent was further evaluated for its suitability with the environment in terms of optimum conditions of pH and adsorbent dosage. The adsorbent was found to have strong surface reactivity in an acidic environment as observed from the variation in adsorption capacity (q_e) with respect to pH (Figure 4(a)). The adsorption capacity of the prepared biosorbent was found to be at maximum when the pH of the solution was 3.05 owing to the strong electrostatic attraction between anionic dye molecules and the cationic adsorbent. It is also important to note that the formation of non-polar functional groups such as the methyl group ($R–CH_3$) may prevail at higher pH values that can reduce the adsorption capacity (Aljeboree *et al.* 2014). Hence from the results obtained, a pH of 3.0 was selected as the optimum for the rest of the experiments. Figure 4(b) represents the variation of q_e when the adsorbent dosage was varied from 0.05 to 11 g. The uptake was found to be maximum (6.6 mg g^{-1}) at an adsorbent dosage of 5 g and thereafter decreased. The steep escalation in the adsorption capacity during the initial increase in the dosage can be attributed to presence of a significant number of active sites owing to the large surface area of the adsorbent. However, the later decrease in the adsorption capacity may be due to the inefficient filling of active sites, and possibly due to the agglomeration of biomass.

The optimum conditions obtained for SDS samples were further compared with those for TIE samples having a different range of initial concentrations. The optimum pH for TIE

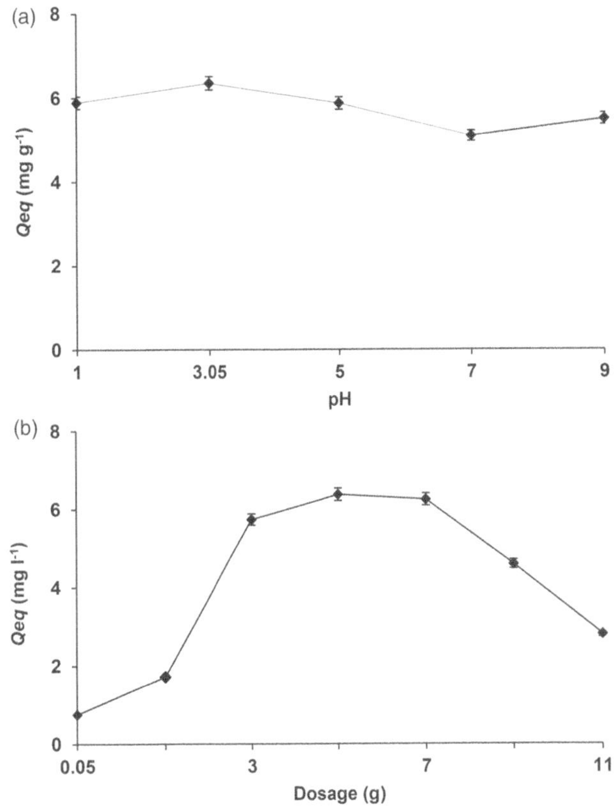

Figure 4 | Impact of (a) solution pH and (b) adsorption dosage on adsorption capacity of synthetic dye solution (SDS) samples.

samples was found to be at a low value of 2.0 compared to SDS samples showing the higher preference of acidic pH for the adsorbent (Figure 5(a)). It is also inferred from the results that the process produced a significant shift towards acidic nature during the process of adsorption and the change in pH was found to produce a direct linear response with the adsorbent dosage. This is because the increase in adsorbent dosage causes aggregation of adsorbent and consequently, a decrease in the available adsorption sites and hence, the adsorption intensity (Ramathilagam & Ananthakumar 2016). It was also revealed from the studies that the adsorption was efficient at an optimal dosage of 5 g per 100 ml of the TIE samples (Figure 5(b)). Hence it is quite appropriate to anticipate that a relatively small quantity of adsorbent would be sufficient for effectively removing the colour of the textile effluent within the equilibrium time of 60 minutes.

The significance of variation among the pH and adsorbent dosage on the adsorption capacity was further evaluated based on the two-way ANOVA test without repetition (Table 3).

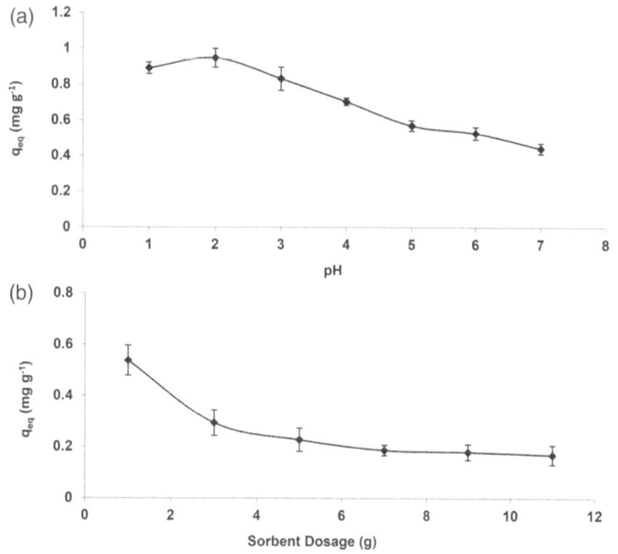

Figure 5 | Impact of (a) solution pH and (b) adsorption dosage on adsorption capacity of real textile industry effluent (TIE) samples.

Table 3 | Results of ANOVA study for the significance of variations in pH and adsorbent dosage for synthetic dye solution (SDS) and textile industry effluent (TIE) samples

Source of variation	Sum of squares (*SS*)	Degree of freedom (*df*)	Mean square (*MS*)	*F*-value	*p*-value	*F*-crit
pH	9.65	4	2.41	0.61	0.666	3.84
Type of sample	75.95	2	37.97	9.63	0.007	4.46
Dosage	41.27	5	8.25	1.53	0.264	3.32
Type of sample	100.41	2	50.21	9.31	0.005	4.10

The results indicated that the *p*-value was higher than the assigned level of significance (0.05) for the variations in pH and dosage, showing their lesser impact compared to the variation in type of sample. This was also confirmed with the comparison of F-values with respect to the F-critical values (Maxwell *et al.* 2017). Furthermore, the sum of squares (*SS*) was found to be highest for type of sample (75.95 and 100.41) having only 2 degrees of freedom (*df*) compared to the dosage (41.27) and pH (9.65) with degrees of freedom of 5 and 4 respectively. The higher degrees of freedom correspond to the availability of more number of trials (or the number of samples) as in the case of varying pH and dosage. This shows that the removal efficiency of the prepared adsorbent is not at the same level for SDS and TIE samples owing to the different range of chemicals present, as well as the different range of initial concentrations attempted. This was in accordance with the previous results that the impact of variation of environmental conditions (pH and dosage) varies with the effluent being treated (Kanwal *et al.* 2017).

Mechanistic modeling of biosorption for different samples

The basic purpose of biosorption is to encourage direct attachment of dissolved organic/metallic compounds onto the surface of the solid phase material (biosorbent) where they can passively concentrate and be separated out from the system at a later time. The physical force of attraction holds the molecules on the adsorbent surface as the pore-filling process primarily depends on the size, shape and surface affinity of the compounds and the reduction in adsorption energy. Although many mechanistic models are available to explain various physico-chemical interactions happening at the pore scale, the exact mechanism of non-metabolic biosorption is still not fully understood. The results from the batch experiments are further evaluated to fit with the best isotherm and kinetic models in order to ascertain more inference about the prevailing mechanism of contaminant removal.

Isotherm models

Mechanistic models were used to compare the adsorption behavior of the prepared AC for the selected range of concentrations of SDS and TIE samples. A plot of C_e *vs* C_e/q_e was drawn for the Langmuir isotherm and ln*(qe)* for the Freundlich isotherm. The values of correlation coefficients and regression factors indicated that the results fitted well with the Freundlich model ($R^2 = 0.9998$ and 0.9677) and the Langmuir model ($R^2 = 0.9821$ and 0.803) for the SDS sample and for the TIE samples respectively (Supplementary Figures S1 and S2). This has a direct implication on the adsorption mechanism as the Freundlich model suggests an increasing rate of adsorption due to the presence of multi-layers in the adsorbent surface (Iqbal & Ashiq 2007). The details of the isotherm constants are provided in Table 4.

Table 4 | Adsorption parameters obtained from isotherm studies

Type of sample	Langmuir model			Freundlich model		
SDS	$Q_0 = 0.1905$	b = 8.85	$R^2 = 0.9821$	K = 9.5162	$n = 3.285$	$R^2 = 0.9998$
TIE	$Q_0 = 196.08$	b = 0.01	$R^2 = 0.803$	K = 5.026	$n = 1.506$	$R^2 = 0.9677$

Kinetic models

A comparison of the adsorption rate is essential to differentiate the performance of samples with varying concentrations and chemical constituents. The basic comparison between the pseudo-first-order (PFO) and PSO models indicated that the present results fitted more with PSO both for SDS and TIE samples (Supplementary Figures S3 and S4). Higher values of q_e (8.643 mg g^{-1}) for PSO indicated faster adsorption to the multiple layers as evident with the better fit for the Freundlich isotherm (Table 5). The higher value for TIE samples may be attributed to the varying characteristics of adsorbing chemicals present in the real-time effluent (Lin *et al.* 2015).

Decontamination efficiency on real-time effluent

The specific treatment efficiencies of various physico-chemical parameters are obtained from the experiments with varying adsorbent dosage. It was observed that the removal efficiency of COD, TDS and turbidity increased with increase in adsorbent dosage (Figure 6). This was attributed to the increased surface area of the adsorbent promoting intensive mass transfer. However, it is to be noted that the optimum performance of the adsorbent may not be achieved for the selected environmental conditions at a higher dosage. This could be verified based on the expected mass transfer kinetics as per the above results. In fact, this inference may serve as a feasible solution strategy based on the techno-economic conditions as well. In case of TDS and turbidity removal, the performance was quite apparent about 50–60% even at a higher dosage. As most of the chemical ingredients are possibly unknown in the real-time effluents, the COD removal efficiency has a significant role in the overall selection of the adsorbent. Similarly, the actual kinetics of adsorption may be significantly different for different dye compounds; hence, it was not possible to assess the real-time removal efficiency over a pre-defined period of time.

Real-time process monitoring

Sensitivity studies

The uncertainty in understanding the actual rate of adsorption and the details of chemical interactions during the process can be addressed to some extent by incorporating a real-time monitoring platform. In the present study, the real-time monitoring results for temperature and pH showed that there was an interesting shift in the mass transfer during the adoption. When the pH of the solution was initially fixed at 3 (based on the previous results for optimum conditions), the LabVIEW-empowered simulator could detect an increasing temporal variation in pH to a value of 6.0 after a period of 1 h (Figure 7(a)). This shows the specific affinity of the prepared AC adsorbent within

Table 5 | Adsorption parameters obtained from kinetic studies

Synthetic dye solution (SDS) (C = 250 mg l^{-1})						Textile industry effluent (TIE) (C = 110 mg l^{-1})					
PFO			PSO			PFO			PSO		
k_1	q_e	R^2	k_2	q_e	R^2	k_1	q_e	R^2	k_2	q_e	R^2
0.0214	1.7939	0.4217	0.6660	8.6430	0.9998	0.027	91.16	0.9491	0.0063	105.26	0.9923

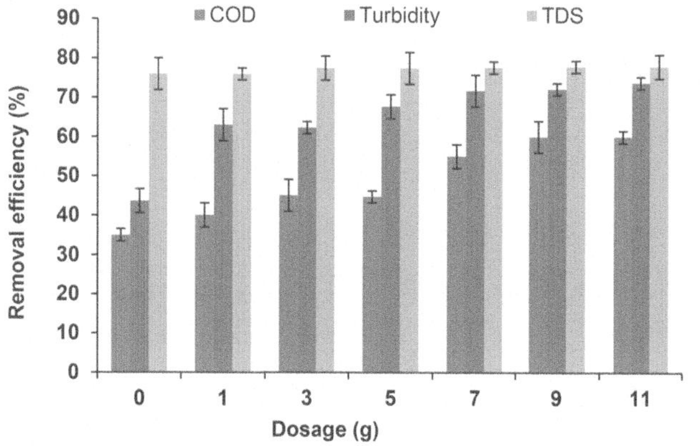

Figure 6 | Removal efficiencies of COD, TDS and turbidity from the TIE sample using the prepared AC adsorbent.

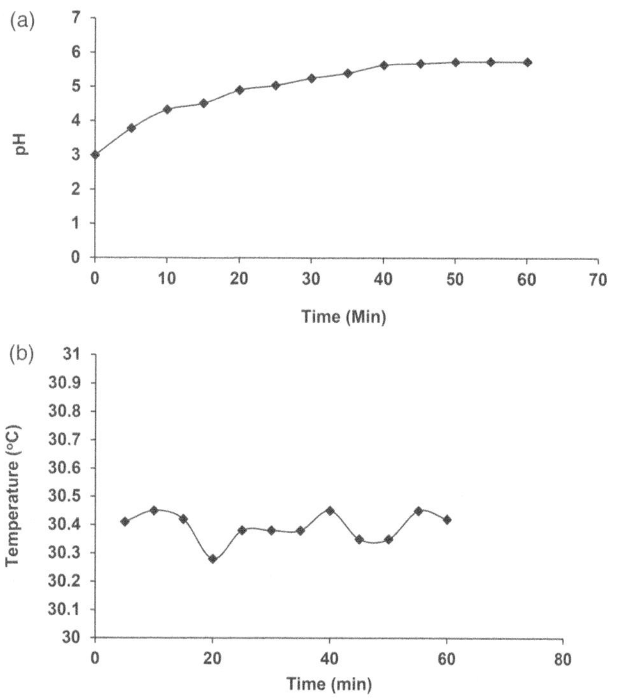

Figure 7 | The sensitivity of (a) pH and (b) temperature during the real-time monitoring using the LabVIEW environment.

the acidic range of pH. The variation in temperature of the solution showed minute but alternative variations during the adsorption process (Figure 7(b)). This variation was quite significant to understand the thermodynamics of adsorption as temperature serves as the direct indication of energy exchange during the mass transfer. However, in order to further elucidate about the adsorption mechanism, it is important to incorporate the temperature variations in the classical adsorption thermodynamic models.

It is quite possible that the simulator can be trained towards an expected range of performance while operating under varying concentrations of sample solutions and adsorbent dosages. Therefore, the performance of the present simulator for the TIE sample was compared for two different scenarios, (i) initial pH as it is and (ii) initial pH fixed at 3.0. The simulation results showed that alteration of initial pH had a significant effect on the removal efficiencies of COD, compared to turbidity and TDS (Table 6). This is a clear indication of the sensitivity of the system to express the real-time chemical changes happening during the adsorption. The smaller variations in the removal efficiencies of turbidity and TDS may be attributed to the net occurrence of finely adsorbed media present in the solution causing some residual impact on the colour and visibility. As explained earlier, the improved performance at higher adsorbent dosages shall be considered only as an indication of the increased mass transfer rather than deciding on the optimum adsorption conditions.

Conceptual framework

The conceptual framework of the proposed LabVIEW control system is illustrated here with the help of a simple

Table 6 | Variation in removal efficiencies of turbidity, TDS and COD as observed before and after alteration of initial pH

	Turbidity		TDS		COD	
Dosage (g)	**Before alteration of pH**	**After alteration of pH**	**Before alteration of pH**	**After alteration of pH**	**Before alteration of pH**	**After alteration of pH**
3	21.93%	14.94%	0%	0%	5.77%	60.87%
5	33.06%	32.24%	0%	17.14%	14.85%	63.04%
7	49.80%	48.57%	6.86%	17.43%	25.00%	69.57%
9	50.58%	51.93%	7.81%	19.43%	33.33%	73.91%
11	53.43%	52.43%	8.13%	19.43%	33.33%	73.91%

flowchart (Figure 8). This framework ensures that automatic control actions will adjust and maintain the parameters at the set point so that it will not interfere with the efficacy of the process. The optimized values for the parameters will be fed in the setup and the time interval can be set at which this program should take the readings. For example, in case of pH monitoring, this will enable addition of a base or an acid depending on the changes to the effluent until the pH reaches the set point. One specific outstanding feature of such a system is that the process control can also be achieved through a remotely operated mechanism by which the physical presence of the operator may be minimized. Signals or alarms will be produced if the value of the parameters exceeds the prescribed limits. This enables the operator to detect the unusual behavior in the system that needs to be addressed.

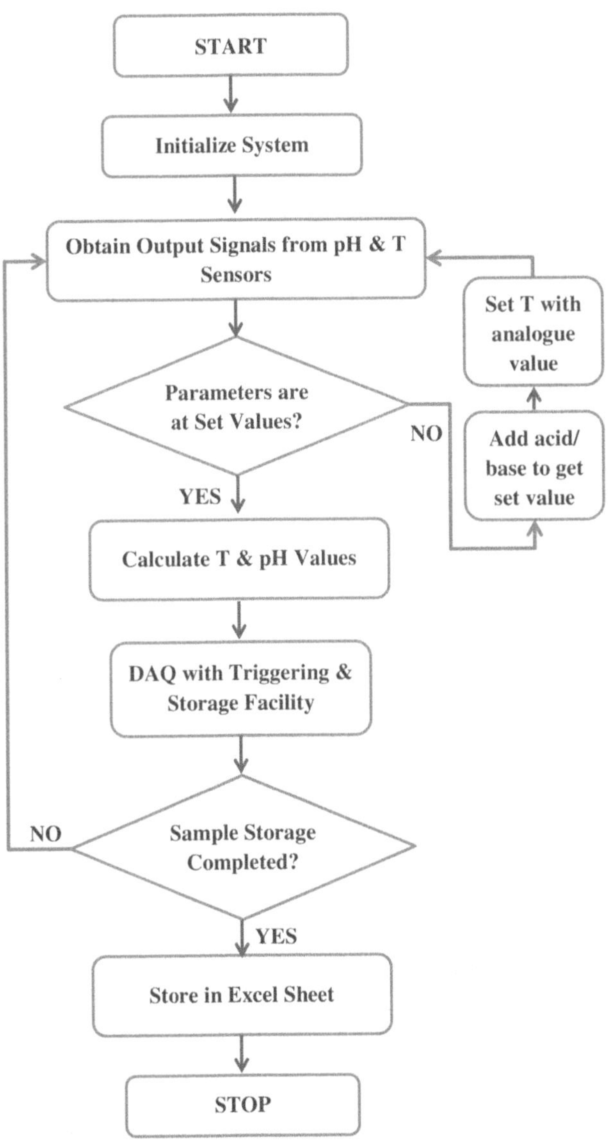

Figure 8 | Process flow sheet depicting the LabVIEW program.

Scope and challenges in real-time applications

The results of biosorption using *Acalypha indica* for treating a mixture of complex dye compounds suggest the possibility of developing a low-cost technology for industrial effluents before their final disposal to the environment. There is enormous scope in optimizing the entire process flow to accommodate the intervention at any stage of textile processing to withdraw, treat and reuse the effluents in a continuous manner. The suggested process-control framework can be elaborated by incorporating many other parameters (both systemic and environmental) of critical influence. While considering the complex interactions between the financial risk and process reliability of large-scale adsorption systems (as seen in a packed tower or agitated media), there are certainly advantages on the real-time applications of biosorption using low-cost natural materials.

The process-control method presented in this study is simple and convenient for scaling up; however, it can be improved with an additional optimization module. There should be an in-built protocol for selecting additional parameters which may vary in their number and preference from case to case. One critical factor to consider is the regeneration capacity of the biosorption system under regular feed of heterogeneous dye chemicals. When the pH and dosage are not regularly monitored, there may be chances of adsorbent becoming under-utilized due to the agglomeration of partially coagulated particles, which may even influence the hydraulic mixing regime in real-time adsorption towers. Therefore, the real-time monitoring can help the system maintain conditions specifically suitable for adsorption without causing unfavorable coagulation behavior. The accuracy and service life of the sensors and other major equipments, controlling the feed, mixing and separation are also to be carefully evaluated. The separation of the treated effluent and the used adsorbent (for further reuse) can further deploy necessary mechanisms for the recovery of valuable dyes and associated salts. It has to be further developed as a viable option towards achieving circular economy-based sustainable treatment of textile effluents.

As our results propose, the industry-scale application of biosorption using *Acalypha indica* could eventually result in mass-scale cultivation of the crop which is a traditional medicinal herb and listed in the Pharmacopoeia of India as an expectorant to treat asthma and pneumonia. As we understand from the literature, the current regulatory policy does not, however, hinder the gross cultivation or alternate industrial application for the herbs unless it violates the existing rules such as Biological Diversity Act (2002) and the Drugs and Cosmetics Act (1940, amended in 2005). The current regulations concerning the cultivation and marketing of medical plants, especially from forest area are governed by the Scheduled Tribes and Other Traditional Forest Dwellers (Recognition of Forest Rights) Act (2006) and a notification regarding the submission of consumption data of Raw Material used by ASU Drugs Manufactures (Link 1; Link 2). As far as the authors understand, there are no regulatory restrictions in promoting the industrial usage of *Acalypha indica* in India (Sahoo & Manchikanti 2013; Katoch *et al.* 2017).

In order to encourage investments and production in the technology sector, the Government of India has announced attractive booster plans in the latest financial budget 2020. According to the referred sources, there is a proposal to withdraw the dividend distribution tax (DDT) to prevent the cascading effect on taxation (Link 3). Furthermore, many start-up initiatives are promoted with attractive early-life funding and seed funding along with income tax benefits. These policies are also applicable to the textile industries operating with common effluent treatment plants in India. Considering the suitability of agro-climatic conditions, the parent plant of the adsorbent (*Acalypha indica*) can be cultivated in industrial scale in most of the South-East Asian countries. Based on the current economic conditions and strategic industrial policies, most of these countries are on a par with India in managing large numbers of isolated industries utilizing the dye chemicals. Hence it is envisaged that the results of the present study will support the futuristic industrial operation trends in those countries.

CONCLUSION

Real-time monitoring and process-control-based biosorption system for treating real-time textile effluents has been studied using LabVIEW. The selected biosorbent prepared from *Acalypha indica* indicated the satisfactory removal of colour and concentration of dyes from the original textile effluent within a time period of 60 minutes, under an optimal adsorbent dosage of 0.1 gram per 100 ml of diluted sample. The primary investigation on the adsorption mechanism (Freundlich isotherm and PSO kinetics) suggested the possibility for enhanced capacity upon surface modification and controlled interactions. The conceptual framework for a real-time monitoring–controlling system was demonstrated by measuring pH and temperature for varying initial conditions of chemical concentration and adsorbent dosage to result in substantial reduction in COD (73.91%), TDS (19.43%) and turbidity (52.43%). The study also discusses the techno-economic avenues for scaling up the process for real-time industrial applications.

DATA AVAILABILITY STATEMENT

All relevant data are included in the paper or its Supplementary Information.

REFERENCES

Acharya, J., Sahu, J. N., Sahoo, B. K., Mohanty, C. R. & Meikap, B. C. 2009 Removal of chromium (VI) from wastewater by activated carbon developed from Tamarind wood activated with zinc chloride. *Chemical Engineering Journal* **150**, 25–39.

Adibmehr, M. & Faghihian, H. 2018 Magnetization and functionalization of activated carbon prepared by oak shell biowaste for removal of Pb_{2+} from aqueous solutions. *Chemical Engineering Communications* **205** (4), 519–532.

Alin, A. & Kurt, S. 2006 Testing non-additivity (interaction) in two-way ANOVA tables with no replication. *Statistical Methods in Medical Research* **15**, 63–85.

Aljeboree, A. M., Alshirifi, A. N. & Alkaim, A. F. 2014 Kinetics and equilibrium study for the adsorption of textile dyes on coconut shell activated carbon. *Arabian Journal of Chemistry* **10**, 3381–3393.

Altıntıg, E., Altundag, H., Tuzen, M. & Sari, A. 2017 Effective removal of methylene blue from aqueous solution using magnetic loaded activated carbon as novel adsorbent.

Chemical Engineering Research and Design **122**, 151–163.

Ang, W. L., Mohammad, A. W., Hilal, N. & Leo, C. P. 2015 A review on the applicability of integrated/hybrid membrane processes in water treatment and desalination plants. *Desalination* **363**, 2–18.

Anyika, C., Asri, N. A. M., Majid, Z. A., Yahya, A. & Jaafar, J. 2017 Synthesis and characterization of magnetic activated carbon developed from palm kernel shells. *Nanotechnology for Environmental Engineering* **2** (1), 16.

Arami, M., Limaee, N. Y., Mahmoodi, N. M. & Tabrizi, N. S. 2005 Removal of dyes from colored textile wastewater by orange peel adsorbent: equilibrium and kinetic studies. *Journal of Colloid and Interface Science* **288**, 371–376.

Barakan, S., Aghazadeh, V., Beyragh, S. A. & Mohammadi, S. 2020 Thermodynamic, kinetic and equilibrium isotherm studies of As(V) adsorption by Fe(III)-impregnated bentonite. *Environment, Development and Sustainability* **22**, 5273–5295.

Boudrahem, F., Benissad, F. A. & Soualah, A. 2011 Adsorption of lead(II) from aqueous solution by using leaves of date trees as an adsorbent. *Journal of Chemical & Engineering Data* **56**, 1804–1812.

Cazettaa, A. L., Alexandro, M. M., Vargasa, E., Nogamia, M., Kunitaa, M. H., Guilherme, M. R., Martins, A. C., Silva, T. L., Moraes, J. C. G. & Almeida, V. C. 2011 NaOH-activated carbon of high surface area produced from coconut shell: kinetics and equilibrium studies from the methylene blue adsorption. *Chemical Engineering Journal* **174**, 117–125.

Cechinel, M. A. P., Souza, S. G. & de Souza, A. A. U. 2014 Study of lead (II) adsorption onto activated carbon originating from cow bone. *Journal of Cleaner Production* **65**, 342–349.

Crini, G. & Lichtfouse, E. 2018 Wastewater treatment: an overview. In: *Green Adsorbents for Pollutant Removal, Environmental Chemistry for a Sustainable World* (G. Crini & E. Lichtfouse, eds), 18. Springer Nature, pp. 1–22.

Gupta, N., Kushwaha, A. K. & Chattopadhyaya, M. C. 2011 Application of potato (*Solanum tuberosum*) plant wastes for the removal of methylene blue and malachite green dye from aqueous solution. *Arabian Journal of Chemistry* **9** (1), 707–716.

Iqbal, M. J. & Ashiq, M. N. 2007 Adsorption of dyes from aqueous solutions on activated charcoal. *Journal of Hazardous Materials* **139**, 57–66.

Janani, A., Saravanan, S., Rajesh, S., Sivakumar, P. & Anirbid, S. 2019 Batch and column studies on the removal of methyl orange by *Acalypha indica* biomass using gravitational search algorithm as an optimisation tool. *Desalination and Water Treatment* **147**, 385–397.

Jendia, A. H., Hamzah, S., Abuhabib, A. A. & El-Ashgar, N. M. 2020 Removal of nitrate from ground water by eggshell bio-waste. *Water Supply*. doi.org/10.2166/ws.2020.151.

Kanwal, A., Bhatti, H. N., Iqbal, M. & Noreen, S. 2017 Basic dye adsorption onto clay/$MnFe_2O_4$ composite: a mechanistic study. *Water Environment Research* **89**, 301–311.

Katoch, D., Sharma, J. S., Banerjee, S., Biswas, R., Das, B., Goswami, D., Harwansh, R. K., Katiyar, C. K. & Mukherjee, P. K. 2017 Government policies and initiatives for development of Ayurveda. *Journal of Ethnopharmacology* **197**, 5–31. doi:10.1016/j.jep.2016.08.018.

Kausar, A., Iqbal, M., Javed, A., Aftab, K., Bhatti, H. N. & Nouren, S. 2018 Dye adsorption using clay and modified clay: a review. *Journal of Molecular Liquids* **265**, 395–407.

Khorramfar, S., Mahmoodi, N. M., Arami, M. & Gharanjig, K. 2010 Equilibrium and kinetics studies of the cationic dyes removal capability of a novel bio sorbent *Tamarindus indica* from textile wastewater. *Coloration Technology* **126**, 261–268.

Kumar, P. S. & Saravanan, A. 2017 Sustainable wastewater treatments in textile sector. In: *Sust Fibres Textiles* (A. R. Horrocks & S. C. Anand, eds). Woodhead Publishing, United Kingdom, pp. 323–346.

Lin, J., Ye, W., Huang, J., Ricard, B., Baltaru, M. C., Greydanus, B., Balta, S., Shen, J., Vlad, M., Sotto, A. & Luis, P. 2015 Toward resource recovery from textile wastewater: dye extraction, water and base/acid regeneration using a hybrid NF-BMED process. *ACS Sustainable Chemistry & Engineering* **3** (9), 1993–2001.

Linggi, L., Berlin, M., Malik, M., Roy, S., Natarajan, N. & Vasudevan, M. 2020 Feasibility of adsorptive removal of NH_4^+ and NO_3^- species from clayey aquifer using special soils. *Environment, Development and Sustainability*. doi.org/10.1007/s10668-020-00889-6.

Link 1: Acts and Rules. National Medicinal Plants Board, Ministry of AYUSH, Government of India. Available from: https://www.nmpb.nic.in/acts-rights-rules (accessed on 8 December 2020).

Link 2: Why does India Need a Medicinal Plant Policy? Seema Bhatt. Available from: https://www.devalt.org/newsletter/jan98/of_2.htm (accessed on 8 December 2020).

Link 3: Budget 2020: Incentives for startups, DDT abolition are big positives for tech sector. Available from: https://www.cnbctv18.com/technology/budget-2020-focus-on-startups-ddt-abolition-are-big-positives-for-tech-sector-5207681.htm (accessed on 8 December 2020).

Long, D. D., Wang, Q. & Han, J. R. 2017 Biosorption of copper (II) from aqueous solutions by sclerotiogenic *Aspergillus oryzae* G15. *Water Environment Research* **89**, 703–713.

Mallampati, R., Xuanjun, L., Adin, A. & Valiyaveettil, S. 2015 Fruit peels as efficient renewable adsorbents for removal of dissolved heavy metals and dyes from water. *ACS Sustainable Chemistry and Engineering* **3**, 117–1124.

Mandel, J. 1961 Non-additivity in two-way analysis of variance. *Journal of the American Statistical Association* **56**, 878–888.

Maxwell, S. E., Delaney, H. D. & Kelley, K. 2017 *Designing Experiments and Analyzing Data: A Model Comparison Perspective*. Routledge, Taylor and Francis Group, USA.

Mohammadi, N., Mousazadeh, B. & Hamoule, T. 2020 Synthesis and characterization of NH_2-SiO_2@Cu-MOF as a high-performance adsorbent for Pb ion removal from water environment. *Environment, Development Sustainability*. doi.org/10.1007/s10668-020-00646-9.

Moosavi, S., Lai, C. W., Gan, S., Zamiri, G., Akbarzadeh Pivehzhani, O. & Johan, M. R. 2020 Application of efficient magnetic particles and activated carbon for dye removal from wastewater. *ACS Omega* **5** (33), 20684–20697.

Nakayama, A., Suzuki, K., Enoki, T., Di Vittorio, S. L., Dresselhaus, M. S., Koga, K., Endo, M. & Shindo, N. 1993 Magnetic properties of activated carbon fibers. *Synthetic Metals* **57** (1), 3736–3741.

Natarajan, R. & Al-Qasmi, F. 2018 Biosorption of fluoride using *Anogeissus dhofarica* – effect of process variables and kinetic studies. *Water Supply* **18** (5), 1803–1809.

Padmanabhan, K., Vasudevan, M., Ramakrishnan, S. & Natarajan, N. 2017 Feasibility study on decontamination of aged spent wash using activated carbon from *Limonia acidissima* shell. *Desalination and Water Treatment* **82**, 369–378. http://dx.doi.org/10.5004/dwt.2017.21054.

Peng, W., Li, H., Liu, Y. & Song, S. 2017 A review on heavy metal ions adsorption from water by graphene oxide and its composites. *Journal of Molecular Liquids* **230**, 496–504.

Poonam & Kumar, N. 2020 Experimental and kinetic study of removal of lead (Pb^{+2}) from battery effluent using sweet lemon (*Citrus limetta*) peel biochar adsorbent. *Environment, Development and Sustainability* **22**, 4379–4344.

Praveena, N., Saravanan, P., Dharmendra Kumar, M., Gandhi, N. N. & Renganathan, S. 2011 Biosorption of reactive Red 198 from an aqueous solution using *Acalypha indica*. *Asia-Pacific Journal of Chemical Engineering* **7** (5), 761–768.

Ramathilagam, G. & Ananthakumar, K. 2016 Adsorptive removal of methylene blue using low cost and efficient biosorbent: equilibrium, kinetics and thermodynamics studies. *International Journal of Environmental Pollution* **6**, 440–453.

Ramos-Vargas, S., Huirache-Acuna, R., Rutiaga-Quinones, J. G. & Cortes-Martinez, R. 2020 Effective lead removal from aqueous solutions using cellulose nanofibers obtained from water hyacinth. *Water Supply*. https://doi.org/10.2166/ws.2020.173

Ranjith, L., Shukla, S. P., Vennila, A., Purushothaman, C. S., Lakshmi, M. S., Aruna, S. & Padmanabhan, A. K. 2011 An assessment on spirulina platensis as a biosorbent for arsenic removal. *Water Supply* **11**, 370–378.

Rekos, K., Kampouraki, Z. C., Samanidou, V. & Deliyanni, E. 2016 Magnetic graphene oxide-polystyrene and magnetic activated carbon-polystyrene nanocomposites as sorbents for bisphenol A. *EGUGA*. EPSC2016-10045.

Sahoo, N. & Manchikanti, P. 2013 Herbal drug regulation and commercialization: an Indian industry perspective. *The Journal of Alternative and Complementary Medicine* **19** (12), 957–963. doi: 10.1089/acm.2012.0275.

Saravanan, P., Sivakumar, P., Geoprincy, G., Nagendra, G. N. & Renganathan, S. 2012a Biosorption of acid green 1 using dried *Rhodoturula glutinis* biomass. *Indian Journal of Environmental Protection* **32**, 207–214.

Saravanan, P., Sivakumar, P., Suganya, T., Gandhi, N. N. & Renganathan, S. 2012b Bioaccumulation of reactive red 11 using live yeast *Rhodotorula glutinis*. *Indian Journal of Environmental Protection* **32**, 249–255.

Schreiner, C., Branzila, M., Trandabat, A. & Ciobanu, R. C. 2006 Air quality and pollution mapping system using remote measurements and GPS technology. *Global NEST Journal* **8** (3), 315–323.

Shirun, Z., Shouxin, L., Guizhen, F., Ying, Z. & Zhifen, C. 1997 Preparation of magnetic activated carbon. *Journal of Forestry Research* **8** (4), 250–253.

Shyla, S. S., Rajayyan, M. K. & Malairajan, S. 2018 Removal of lead (II) ions from synthetic wastewater using lantana camara leaves bio carbon. *International Journal of Current Advanced Research* **7** (4), 12144–12150. http://dx.doi.org/10.24327/ijcar.2018.12150.2130.

Sivapragasam, C., Rajasekaran, M. P. & Vinotha, M. 2017 A conceptual framework for real time estimation of WFP for small hydroelectric power plant. *Water Policy* **19** (6), 1049–1062. doi.org/10.2166/wp.2017.289.

Srebrenkoska, V., Zhezhova, S., Risteski, S. & Golomeova, S. 2014 *Methods for Waste Waters Treatment in Textile Industry*. Available from: https://core.ac.uk/download/pdf/35338582.pdf (accessed on 8 December 2020).

Suneetha, M., Sundar, S. N. & Ravindhranath, K. 2015 Extraction of fluoride from polluted waters using low-cost active carbon derived from stems of *Acalypha indica* plant. *Asian Journal of Water, Environment and Pollution* **12** (3), 33–49.

Troster, I., Fryda, M., Herrmann, D., Schäfer, L., Hänni, W., Perret, A., Blaschke, M., Kraft, A. & Stadelmann, M. 2002 Electrochemical advanced oxidation process for water treatment using DiaChem® electrodes. *Diamond and Related Materials* **11** (3–6), 640–645.

Vasudevan, M., Ajithkumar, P. S., Singh, R. P. & Natarajan, N. 2016 Mass transfer kinetics using two-site interface model for removal of Cr(VI) from aqueous solution with cassava peel and rubber tree bark as adsorbents. *Environmental Engineering Research* **21** (2), 152–163. http://dx.doi.org/10.4491/eer.2015.152.

Xu, X., Tian, M., Qu, L. & Zhu, S. 2017 Graphene oxide/chitosan/polyvinyl-alcohol composite sponge as effective adsorbent for dyes. *Water Environment Research* **89**, 555–563.

Yang, N., Zhu, S., Zhang, D. & Xu, S. 2008 Synthesis and properties of magnetic Fe_3O_4-activated carbon nanocomposite particles for dye removal. *Materials Letters* **62** (4–5), 645–647.

Zhang, L., Zeng, Y. & Cheng, Z. 2016 Removal of heavy metal ions using chitosan and modified chitosan: a review. *Journal of Molecular Liquids* **214**, 175–191.

First received 30 August 2020; accepted in revised form 19 January 2021. Available online 2 February 2021

Resource-based agricultural non-point source pollution control using biological agents

Delong Kan and Pengfei Duan

ABSTRACT

Non-point source (NPS) pollution is a major cause of the deterioration of surface water quality. Effectively controlling NPS pollution, especially that from rural areas, is critical in improving the quality and safety of water resources. However, most of the approaches employed for NPS control are low-efficiency or impractical owing to ignorance of the willingness of participants, as well as the high cost. To improve the effectiveness and efficiency, this study newly proposes a resource-based approach for agricultural NPS pollution control via using biological agents. In this approach, the domestic organic waste and livestock manure are economically utilized to produce organic fertilizers and plant protection agents to replace harmful chemical fertilizers and pesticides, respectively. The approach has been applied in the Danjiangkou catchment area in Nanyang City, and the results show that: (1) the most severe pollutants, total P and total N concentrations have been effectively reduced and their values in 2025 are predicted to be the same level as that before significant deterioration; (2) with unpowered sewage treatment and spontaneous participation of farmers, the treatment cost is significantly reduced; (3) the resource-based treatment of domestic waste and livestock manure supports the development of an efficient and sustainable agricultural economy.

Key words | agriculture, biological agent, non-point source pollution, resource-based, South-North Water transfer project

Delong Kan
Pengfei Duan (corresponding author)
Henan Key Laboratory of Ecological Security for Water Source Region of Mid-line of South-to-North Diversion Project, Collaborative Innovation Center of Water Security for Water Source Region of Mid-line of South-to-North Diversion Project of Henan Province,
Nanyang Normal University,
Nanyang, 473061,
China
E-mail: *dpf19830905@nynu.edu.cn*

HIGHLIGHTS

- The resource-based approach is cost-saving and efficient for agricultural NPS pollution control.
- The domestic sewage is treated with micro-power or no power.
- The domestic organic waste and livestock manure are economically utilized.
- A spontaneous collaboration arises between villagers, enterprises, and governments.

INTRODUCTION

Non-point source (NPS) pollution deteriorates the surface water quality of rivers and lakes. In contrast with point source pollution, which can be easily monitored and controlled, NPS pollution remains underestimated and poorly controlled (Wu *et al.* 2017). The contribution of NPS pollution to the overall environmental pollution has gradually increased, and NPS pollution is becoming the primary source of surface and groundwater pollution (Ongley *et al.* 2010; Wu *et al.* 2018). According to previous research, 30–50% of the Earth's total surface area is under the influence of NPS pollution (Xin *et al.* 2017); furthermore, agricultural NPS pollution, which is harmful to the aquatic environment, is considered the most prevalent form of NPS pollution (Fan *et al.* 2019; Yi *et al.* 2020). In the Clean Water Act of the United States of America (USA), legislated in 1972, NPS pollution is defined as 'any source of water

doi: 10.2166/ws.2020.291

pollution that does not meet the legal definition of 'point source' in section 502(14) of the Clean Water Act' (United States 1972). Agricultural NPS pollution, which is mainly caused by fertilizers, pesticides, sewage sludge, and livestock manure, includes nutrients such as N and P, as well as those in pesticides, and other organic or inorganic pollutants (Li *et al.* 2010; Li *et al.* 2019). Agricultural NPS pollution is an important cause of water quality deterioration in surface waters (Liu *et al.* 2020). Owing to the prevalence, extensive dispersion, and uncertain transport paths of NPS pollutants, the mechanism of agricultural NPS pollution is not well-known; however, it does involve a long latent period, which is significantly harmful (Zeng 2019). Agricultural NPS pollution is considered as a world-wide environmental issue (Yang *et al.* 2017), and it hinders the sustainable rural development in China (Liu *et al.* 2017). Therefore, the development of an effective approach for controlling NPS pollution is crucial.

Research on NPS pollution was first conducted in the 1960s in several developed countries, such as the USA, Britain, and Japan, and global research interest expanded from the 1970s. In the 1980s, research on agricultural NPS pollution underwent a period of vigorous development. By the 1990s, new pollutants had become the focus of research (Tong *et al.* 2011; Adu & Kumarasamy 2018). And then numerical researches on NPS were presented, including the pollution source identification, modeling and estimation, control and management and so on (Xiang *et al.* 2017; Yang *et al.* 2017). By using ArcHydro (ArcGIS extension model) and a long-term hydrological and NPS pollution model, Wilson & Weng (2010) evaluated the impact of urban land-use change on lake water quality in Calumet, Chicago. Burow *et al.* (2010) analyzed 5101 samples collected via the National Water Quality Assessment program of the US Geological Survey in 51 research areas from 1991 to 2003, and found that the nitrate concentration was highest in shallow groundwater in the agricultural subsoil. Currently, numerical NPS pollution models are available, and detailed reviews can be found in studies conducted by Yang *et al.* (2017) and Adu & Kumarasamy (2018). Among these models, the Soil and Water Assessment Tool (SWAT), agricultural non-point source pollution model AnnAGNPS, and areal non-point source watershed environment response simulation (ANSWER) have been commonly applied (Liu *et al.* 2016).

In China, research on NPS pollution started in the 1980s. Thereafter, studies on urban NPS pollution were conducted in Beijing, Shanghai, Hangzhou, Suzhou, Changsha, Nanjing, and Chengdu, among other cities (Li *et al.* 2010). Simultaneously, research on rural NPS pollution was successively conducted in the Yuqiao Reservoir, Dianchi Lake, Taihu Lake, Chaohu Lake, Jinjiang River Basin, and Dongjiang River Basin (Chen *et al.* 2010; Cai *et al.* 2017; Xue *et al.* 2019). Since the 1990s, macro characteristics and statistical models (black-box or empirical models) of pesticide and fertilizer pollution have played an important role in agricultural NPS pollution research (Stephenson *et al.* 2010). Currently, the combination of the NPS pollution model with 3S (GIS, GPS, and RS) technology (Emili & Greene 2013; Yaghi & Salim 2017) and the water quality models of watersheds have become new research hotspots in water quality management (Zhang *et al.* 2015). However, research on NPS pollution in China is still limited, mainly concentrated on the assessment of NPS pollution loads, simulation of pollutant transport, and data management with GIS. The deficiencies regarding public policy and administration with low participation of households are the most challenging (Wu & Ge 2019; Zhang *et al.* 2020). There is still much work to fill the gap between research and its applications in this field.

Owing to the unique context of NPS pollution in China, the theories and experience from developed countries cannot be simply transplanted (Ongley *et al.* 2010). Agricultural NPS pollution control should utilize various sustainable prevention, treatment, and control technologies, and be informed by the natural conditions and pollution characteristics of the concerned catchment (Yang *et al.* 2013; Wu & Ge 2019; Yi *et al.* 2020; Zhou *et al.* 2020). However, owing to the high-cost and low reward for farmers in China, the agricultural NPS control is low-efficiently pushed forward only by the government. Besides, the agricultural harvest heavily relying on fertilizers and pesticides (Wu & Ge 2019) also hinders the NPS pollution control. Therefore, this study proposes a practical approach for agricultural NPS pollution control using the biological agents for waste utilization. In this resource-based approach, the domestic sewage disposal is conducted with micro-power or without power, and the domestic organic waste, sewage sludge, and livestock manure are economically utilized to produce biological agents to replace the harmful chemical fertilizers and pesticides that contribute significantly to NPS pollution. Therefore, the best contribution of this research is advancing the NPS pollution control towards being more low-cost, efficient and environment-friendly: (1) the treatment of domestic waste and livestock manure is turned into the production of biological agents; (2) the produced biological agents reduce the application of harmful chemical fertilizers and pesticides, which contribute a lot to the NPS pollution; (3) the biological agents bring income to the farmers, promoting their willingness to participate in NPS pollution control. For agricultural NPS pollution, the farmers' willingness is a knotty problem of long standing.

The proposed approach has been applied in Xichuan County and Xixia County since 2010 when the water quality significantly deteriorated, and the results have successfully demonstrated its value with regard to NPS pollution control. This paper is organized as follows. The methodology for agricultural NPS pollution control using biological agents is first introduced, as well as the SWAT model for pollution prediction. The study area and the issue of NPS pollution are then described. The application results for to the catchment of the Danjiangkou Reservoir in Nanyang City are then analyzed. Finally, conclusions and suggestions are given.

METHODOLOGY

Framework

As shown in Figure 1, NPS pollution in rural areas usually involves domestic sewage and waste from villages in addition to pollutants from farmland. Livestock manure, which is released into rivers, also contributes to NPS pollution. The NPS pollutants converge into lakes or reservoirs through different pathways, such as streamflow, overland flow, and groundwater flow. NPS pollution is a complex issue owing to the diversity of pollution sources and transport pathways (Adu & Kumarasamy 2018).

The agricultural NPS pollution framework depicted in Figure 2 employs a series of high–efficiency biological agents for waste, fecal sludge, and soil treatment. It adopts the mode of 'the government leading and coordinating market operation and farmer participation', thereby allowing for the resource-based treatment of domestic waste and livestock manure, the low–energy and sustainable treatment of sewage, and the development of an efficient eco-agricultural circular economy. The produced organic fertilizers with zero agricultural residues can effectively replace chemical fertilizers, while the plant protection agents can effectively replace chemical pesticides. Further, these high-quality agricultural products can increase farmers' income while ensuring agricultural efficiency and food safety.

Biological agents

As shown in Figure 2, the whole framework of NPS pollution control runs on the base of biological agents, which unite all parts of NPS pollution control via a stable circulating system. In this research, the biological agents are some

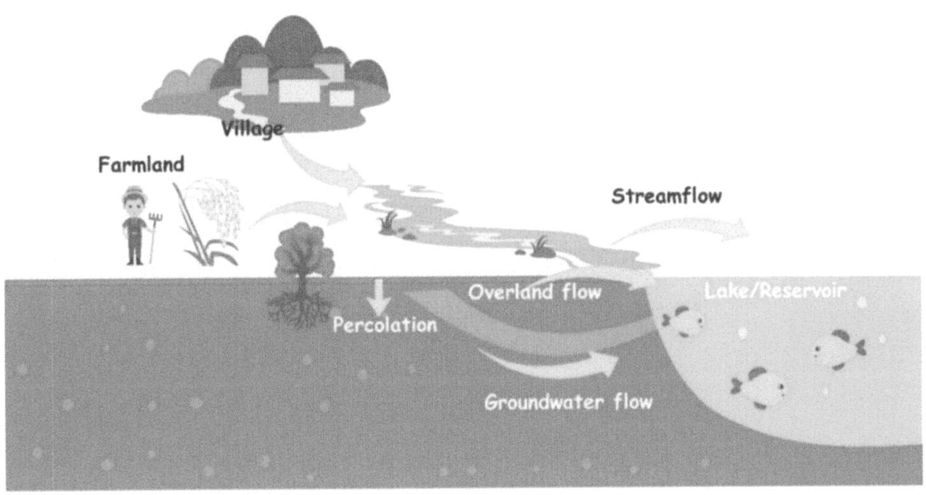

Figure 1 | Agricultural NPS pollution sources and pathways.

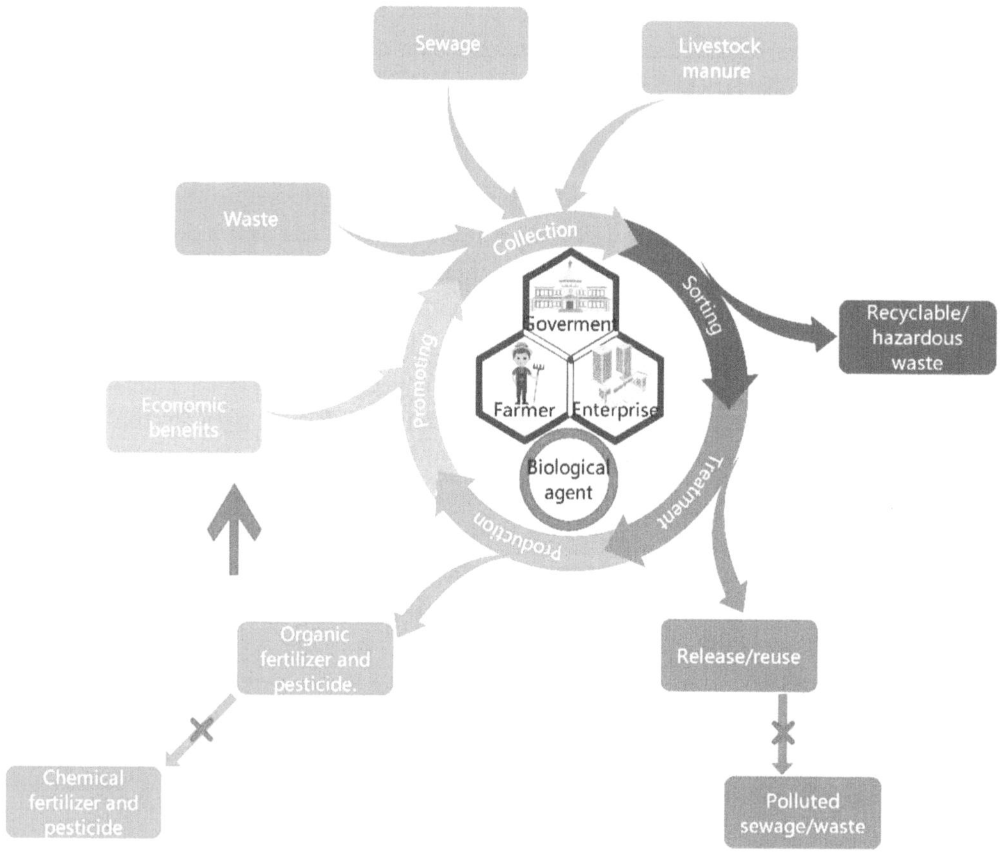

Figure 2 | Agricultural NPS pollution control based on biological agents.

organic matters that serve for environmental protection, mainly including the deodorizers and antidotes, eco-agricultural synergists, plant protection agents and soil purification agents, and water purification agents, as shown in Figure 3. The biological agents are produced from the domestic waste and livestock manure, and can be used in the domestic

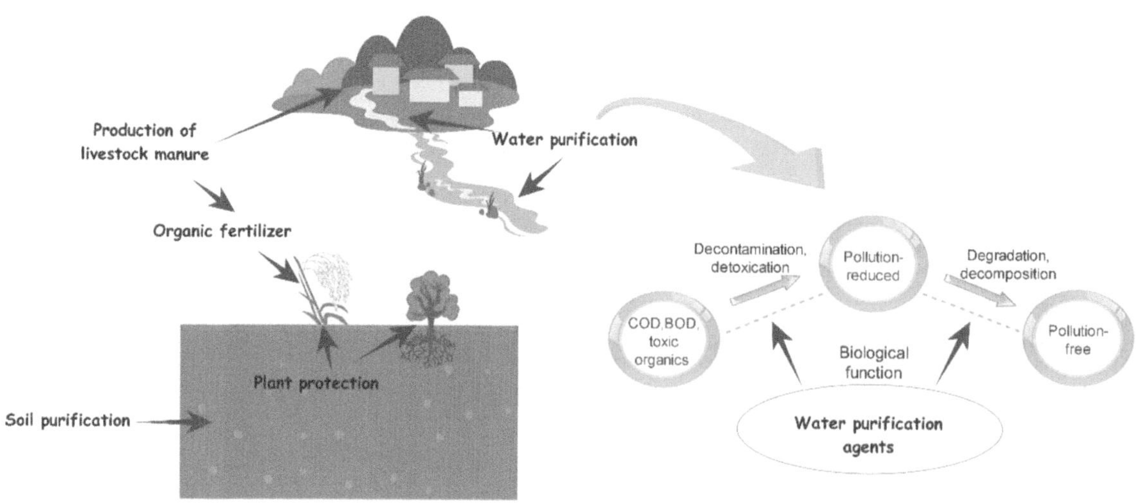

Figure 3 | Function of biological agents.

sewage treatment to improve the efficiency, and used in farming to reduce the application of chemical fertilizer and pesticide.

Water purification agents (Figure 3), as an example, can effectively degrade the pollutants in domestic sewage mainly with their biological function. After the addition of the biological agents, the microbes gain or increase the ability to decompose the nutrients and chemical fertilizers. The water purification agents can quickly degrade the hazardous substances, thereby reducing the chemical oxygen demand (COD), biological oxygen demand, and total suspended solids, among others. The sludge can also be purified with organic pollutants reduced to harmless fish food.

Components

The proposed approach is composed of three parts, namely the production of biological agents, treatment of waste and livestock manure, and domestic sewage treatment.

Production of biological agents

The biological agents are produced from the domestic waste and livestock manure as a critical component of our proposed approach. A production flow chart of biological agents is shown in Figure 4. Different biological agents are derived from different fermentation bacteria but share the same production line.

Treatment of domestic waste and livestock manure

(i) Collection: The household garbage is stored at a fixed point, and the household cleaner is responsible for the collection of garbage. The garbage is transported to the garbage transfer station (concentration point) in the village. The waste disposal enterprise will arrange for garbage trucks to collect and transport the garbage to the pretreatment plant, thereby forming an industrial operation mode, as shown in Figure 5.

(ii) Management of livestock manure: the management of livestock manure is conducted by households, with separate storage and enterprise collection. The treatment of livestock manure includes dry manure cleaning, manure composting, and manure fertilizer returning.

(iii) Garbage disposal: domestic organic waste and livestock manure are recycled into organic fertilizers after pretreatment (including sorting, crushing, and deodorization), composting, production, and other processes.

Treatment of domestic sewage

The popular technology of ponding treatment (Liu *et al.* 2019) involves a biological wastewater treatment process based on natural purification with the joint action of bacteria and algae on organic pollutants. Following this technology, villages, towns, communities, and immigrant villages are equipped with a sewage pipe network, and a three-stage treatment mode is proposed for the unpowered and biological treatment of domestic sewage, as shown in Figure 6.

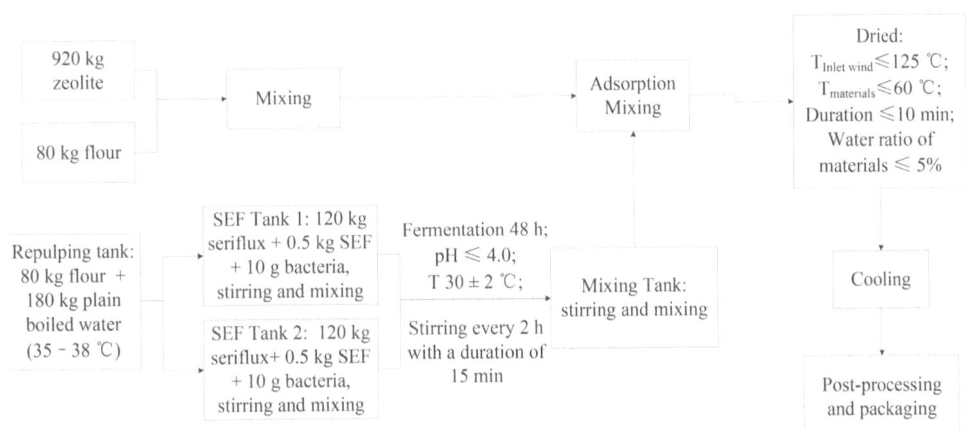

Figure 4 | The production flow chart of biological agents.

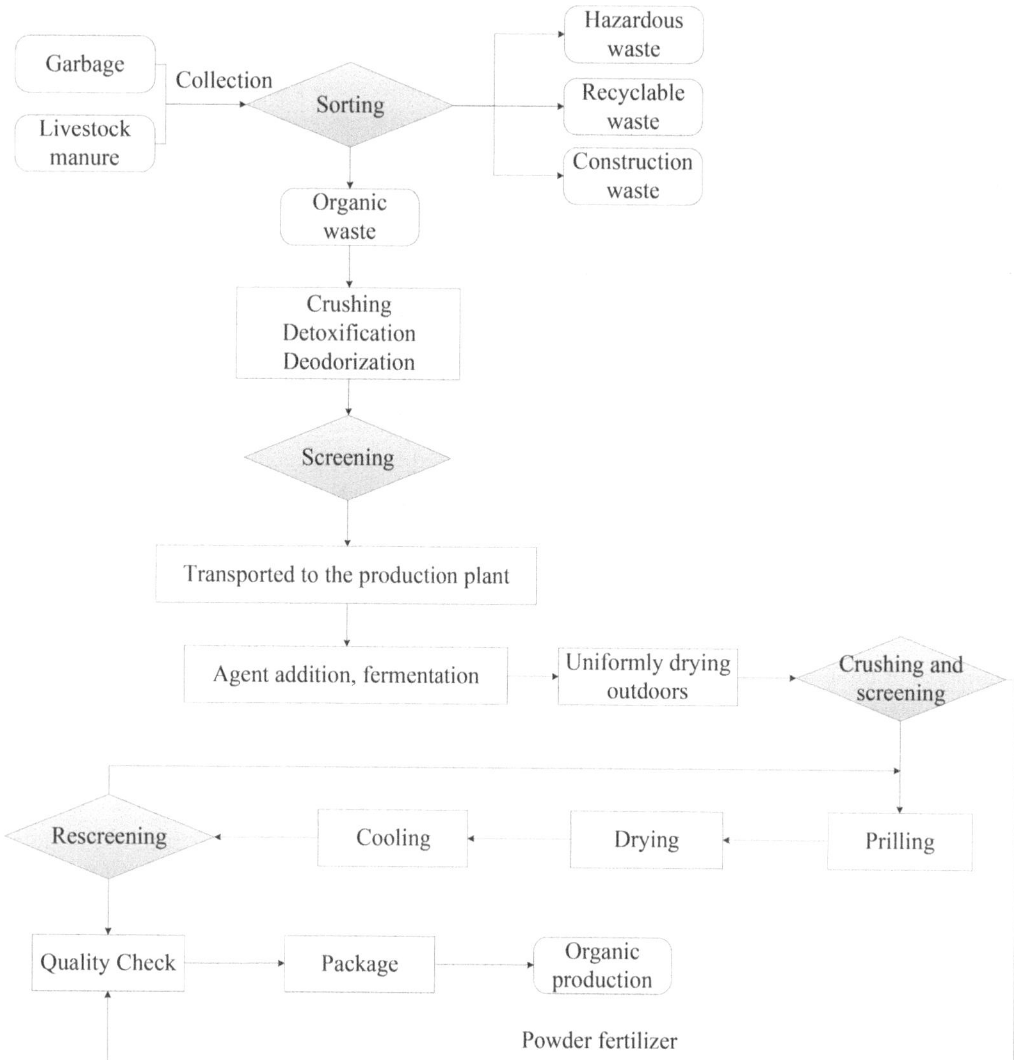

Figure 5 | Treatment flowchart of domestic waste and livestock manure.

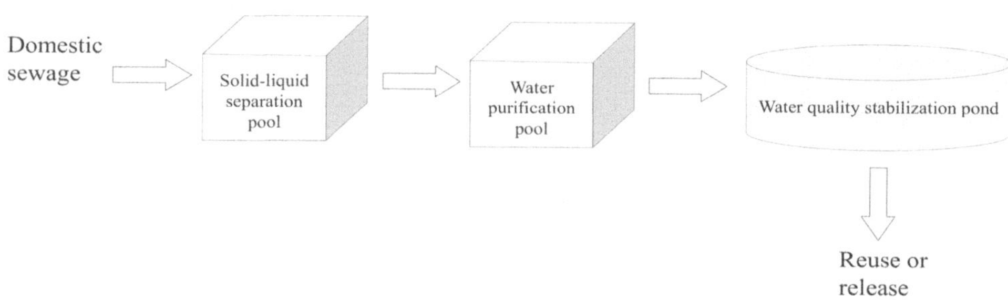

Figure 6 | Domestic sewage treatment process.

SWAT model for assessment

Due to the limited observation in time and space, the effectiveness of NPS pollution control for those cases when and where pollutants were monitored is assessed using the SWAT model, which is an effective tool in the simulation of pollutant distribution and identification (Ouyang *et al.* 2008; Liu *et al.* 2016). Considering nitrogen as an example, the SWAT can simulate the migration and transformation of nitrogen in different forms, as shown in Equation (1) (for open channel) and Figure 7.

$$\frac{\partial N}{\partial t} = \frac{\partial\left(A_x D_L \frac{\partial N}{\partial x}\right)}{A_x \partial x} - \frac{\partial(A_x u N)}{A_x \partial x} + \frac{dN}{dt} + \frac{S}{V} \quad (1)$$

where N denotes the concentration of any one form of nitrogen (mg/L), A_x is the cross-sectional area (m^2), D_L is the dispersion coefficient (m^2/s), S denotes the external pollution sources or sinks (g/s), u is the flow velocity (m/s), V is the water volume of calculated river reach (m^3), t is time (s), and x is distance (m).

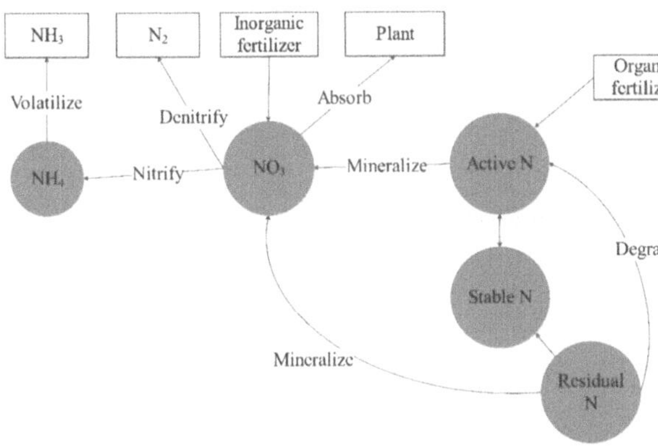

Figure 7 | Simulation of nitrogen cycle in SWAT (Neitsch *et al.* 2011).

STUDY AREA

Overview of the study area

The proposed approach for NPS pollution control was applied to the Danjiangkou catchment area (the source of the Middle Route of the South–North Water Transfer Project, SNWTP) in Nanyang City. Four counties were included; that is, Xichuan County, Xixia County, Neixiang County, and Dengzhou City, as shown in Figure 8. The total area of this study catchment is 6,361.93 km^2, including 67,400 hm^2 of arable land. Most of the arable land in this area is distributed in narrow regions along the rivers with low elevations and decentralized settlements. The distribution of precipitation is uneven, and the precipitation from May to October accounts for 80% of the total annual precipitation.

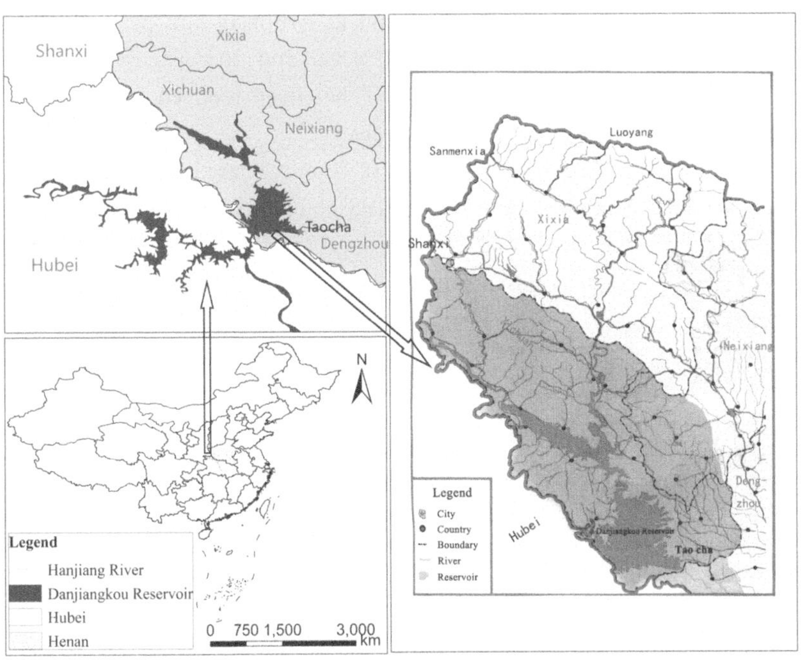

Figure 8 | The study area.

In addition to being an important catchment area, Nanyang City is the only junction and core area of the point (head of the canal), line (main canal), and face (reservoir) of the SNWTP. The catchment area in Nanyang City covers 31 towns, 596 administrative villages, and 950,000 people, and accounts for 82% of the total area of the water resource protection zone in Henan Province. The total length of the Middle Route of the SNWTP is 1,267 km from the Taocha Canal head in Nanyang City to Tuancheng Lake in Beijing City. The length of the section in Nanyang City is 185.5 km, which accounts for 1/7 of the total length. Owing to its sensitive regional location, Nanyang City is responsible for ensuring the quality of water sent to Beijing City and for the sustainable management of water resources.

Agricultural non-point source pollution data

According to the field surveys, the COD and the amount of N in the form of ammonia related to the domestic sewage in the Danjiangkou Reservoir and its upstream catchment is 7,448 t/year and 860 t/year, respectively. The output of rural domestic waste and livestock manure is 327,000 t/year and 480,000 t/year, respectively; the annual output of N and P from 67,400 hm^2 arable land is 141,000 t/year, with an annual average volume of applied chemical fertilizers and pesticides up to 140 kg/mu and 1 kg/mu, respectively. Agricultural NPS pollution, from domestic sewage, waste, livestock manure, chemical fertilizers, and pesticides, has become the main factor affecting the water quality in the study area. Table 1 shows the annual average volume of NPS pollution from chemical fertilizers and pesticides in the study area since 2006.

RESULTS

Improvement in environmental protection

Domestic waste and manure treatment results

The employed waste and manure treatment system in the study area can produce 210,000 t/y of organic fertilizers, in which the organic matter content is greater than 45%. The resource-based treatment not only deals with the domestic waste and livestock manure, but also significantly mitigates the pollution from chemical fertilizers and pesticides. Organic fertilizers produced from the resource-based treatment system can effectively reduce COD by 35,000 t/y and ammonia N by 845 t/y, and can save almost 100 million yuan/y (in RMB) for the investment in fertilizers. Compared with conventional fertilizers, organic fertilizers can significantly reduce the soil nitrate content (Wang *et al.* 2018), and the average N loss can be reduced by 30–50% owing to the lower N mineralization rate of organic fertilizers. For example, in Jiuzhong Town in Xichuan County, the application of chemical fertilizers and pesticides has been reduced by 4,000 t/y and 30 t/y since 2010, respectively, and the loss of N and P has been reduced by 216 t/y and 72 t/y, respectively. In Wuliqiao Town in Xixia County, the application of chemical fertilizers and pesticides has been reduced by 4,000 t/year and 20 t/year, respectively, and the loss of N and P has been reduced by 192 t/year and 60 t/year, respectively.

Following the idea of resource-based treatment in this research, parts of the NPS pollutants have the potential to be beneficial if they are placed in the right position. In this research, the organic waste and livestock manure are

Table 1 | Pollution from chemical fertilizers and pesticides in the study area since 2006 (tons/year)

	Chemical fertilizers					Pesticides	
County	Consumption	N	N loss	P	P loss	Consumption	Loss
Xichuan County	48,000	15,000	3,300	6,790	2,038	90	60
Xixia County	32,000	10,000	2,000	8,000	2,400	122	90
Neixiang County	9,360	2,920	584	716	215	42	28
Dengzhou City	2,895	356	71	136	40	13	9
Total	92,255	28,276	5,955	15,642	4,693	267	187

collected and reused on the base of biological agents. Compared with the traditional household biogas digester (Rajendran *et al.* 2013), the resource-based treatment is more advantageous in both efficiency and sustainability.

In addition, the high-quality agricultural products with zero agricultural residues can improve farmers' income, agricultural efficiency, and food safety. The saving costs in chemical fertilizers have greatly encouraged farmers to participate in the NPS pollution control. The resource-based treatment of domestic waste and livestock manure supports the development of an efficient and sustainable agricultural economy.

Domestic sewage treatment results

Domestic sewage treatment utilizes centralized collection, decentralized treatment, and micro-power, thereby allowing for the effective collection and treatment of domestic sewage in dispersed rural settlements. With the help of biological agents, the average removal rate of domestic sewage using the three-stage ponding system is shown in Table 2.

The sewage effluent enters a solid-liquid separation tank at the front end of the sewage treatment system and then enters the water quality purification tank. In the early stage, deodorants and antidotes are added to activate the genes of bacteria for microbial deodorization, and then a special water purification agent is added to form a treatment layer. Most of the organic matter in the wastewater is removed, and the upper part of the clear liquid is collected in the water quality stabilization tank, which is equipped with a biological matrix.

Table 2 | Average removal rate of domestic sewage using three-stage ponding system

Processing stage		COD (mg/L)	BOD (mg/L)	Ammonia N (mg/L)
Solid–liquid separation tank	Inlet concentration	360	160	25
	Outlet concentration	200	80	20
	Removal rate (%)	45	50	20
Water purification pool	Inlet concentration	200	80	20
	Outlet concentration	140	56	20
	Removal rate (%)	30	30	0
Water stabilization tank	Inlet concentration	140	56	20
	Outlet concentration	70	20	12
	Removal rate (%)	50	60	40

After three-stage removal, the quality of domestic sewage can be effectively improved, with the removal rate reaching 80.6%, 87.5%, and 52.0%, respectively for the COD, BOD and Ammonia N. In this no-power treatment system, three different kinds of biological agents are used sequentially in the three stages. Most of the organic matter in the wastewater is removed after the purification pool, while the stabilization tank further clears the wastewater with a relatively longer standing time before reuse. Further, the removal rate of COD and BOD is much better than for Ammonia N owing to the effects of water purification agents, which present almost no impact on the Ammonia N in the water purification pool. It is plausible that the demonstrated domestic sewage treatment is practical although the high removal rate may not be achieved when it is applied to other regions.

Pollution control assessment

The Danjiangkou Reservoir is the source of the Middle Route of SNWTP, and its water quality is significantly affected by the NPS pollution in Nanyang City. Owing to the implementation of the resource-based approach, there has been a reduction of TN and TP in the Danjiangkou Reservoir, as shown in Figure 9.

There was a significant deterioration around 2008 while both the TN and TP doubled in concentration (Figure 9). Fortunately, after the NPS pollution control was applied, the deterioration was successfully stopped, and the indicators of TN and TP were sustained at the Level three (1.0 mg/L) and Level two (0.025 mg/L) respectively, according to the Environmental Quality Standards for Surface Water (GB 3838-2002, China). Further, the water quality was gradually restored with a reduction tendency for both TP and TN since 2011.

In addition, the predicted TN and TP from SWAT are also shown in Figure 9. According to the prediction results, the water quality at Taocha in 2025 is predicted to recover to the same level as that before significant deterioration, with the TN and TP reaching Level two (0.5 mg/L) and Level one (0.01 mg/L), respectively. The simulation results from SWAT provide more evidence to the assessment of our proposed approach. Due to the long lag time, the SWAT prediction with some numerical error is used in the

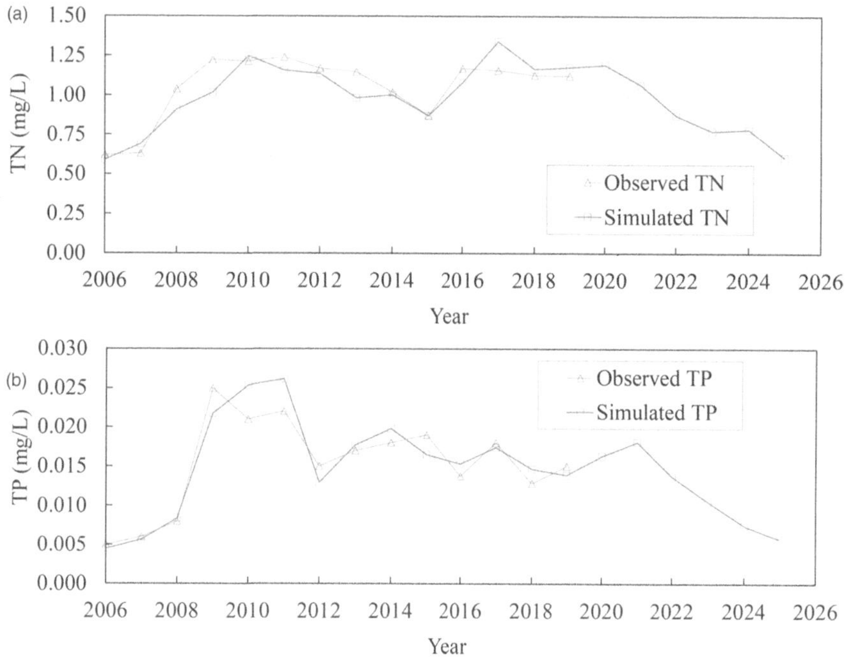

Figure 9 | Observed and simulated TN and TP in Danjiangkou Reservoir at Taocha.

assessment. However, the evident reduction tendency is convincing for the effectiveness of NPS pollution control.

Reduction in treatment costs

With the resource-based treatment, the cost of management and the maintenance of mechanical equipment is significantly reduced. The operation mode allows for the maximum utilization of garbage resources using safe, sustainable, and cost-effective technology. The cost of domestic sewage treatment with no power is less than 0.1 yuan/m^3, which is far lower than traditional treatment costs (about 1.0 yuan/m^3).

In practice, the management mode of 'household storage, group collection, village concentration, township supervision, and enterprise transportation' is adopted for the treatment of domestic waste; the mode of 'household storage, enterprise transportation and disposal ' is adopted for the treatment of livestock manure; and the maintenance of sewage treatment is the responsibility of qualified operation enterprises. In terms of the chemical fertilizers and pesticides, first, the operation enterprise and farmers reduce the use of chemical fertilizers and pesticides with replacement by organic fertilizers. Second, the use of organic fertilizers and plant protection agents is encouraged with a subsidy of 0.45–0.75 yuan/m^2 of farmland for farmers. In these operation modes, the farmers are no longer passive followers or 'spectators', but active participants. Finally, with the collaboration of farmers, enterprises, and government in NPS pollution control, the cost to the government is also significantly reduced, with a reduction ratio larger than 50%.

CONCLUSIONS

Agricultural NPS pollutants are major causes of the deterioration of surface water quality. They present a complex challenge owing to their prevalence, extensive dispersion, low concentration, and dynamic pollution pathways. NPS pollution is primarily from livestock and poultry farming, crop cultivation, and villages. NPS pollution is generally underestimated and poorly controlled, thereby resulting in long-term damage to the environment. This study proposes a resource-based approach for NPS pollution control using environmental biological agents, aiming at improving the efficiency of agricultural NPS pollution. In the proposed approach, the systematical operation mode of 'government leading and coordinating market operation and farmer participation' is employed to realize the resource-based treatment of domestic waste and livestock manure, unpowered and biological treatment of sewage, and prevention and control of agricultural NPS pollution. After almost one decade of implementation of the proposed approach in Nanyang City, the NPS pollution of Danjiangkou catchment has been successfully controlled. The optimal utilization of garbage resources and sustainable production of biological agents also reduced the operating costs. The application of chemical fertilizers and pesticides has been reduced significantly, along with a reduction in the loss of N and P. Through the collaboration of farmers, enterprises and the government, the cost to the government has been significantly reduced, thereby enhancing the effectiveness and efficiency.

Although the proposed approach provides useful strategies for the control of agricultural NPS pollution, the large scale of livestock and farming with chemical fertilizers and pesticides, as well as the overdispersed housing, significantly hinder the NPS control. In Figure 9, although the proposed approach has initially been implemented since 2010, the water quality improvement was too slight until 2015. To address this issue and speed up the water quality improvement, the environmental protection institutions in the townships and villages should be improved. A corporate and market-oriented operation of rural pollution control should be implemented, and farmers should be further encouraged to actively participate. In terms of NPS pollution from livestock manure, the strategy of pollution control at the source position, such as clean breeding, reducing the volume of manure pollution, and facilitating follow-up treatment and utilization, is essential. Process control and pollution reduction during transport should also be improved, including manure pollution collection and storage, dry manure cleaning, and manure resource utilization.

ACKNOWLEDGEMENTS

This work was supported by the Scientific and Technological Research Project of Henan Provincial Science and Technology Department under Grant number 182102110239 and the National Natural Science Foundation of China under Grant number 31501260. We would like to thank Editage (www.editage.cn) for English language editing. We would like to thank the review work by the editors and the two anonymous reviewers. We would like to thank the help from Jiabiao Wang in the revision process.

DATA AVAILABILITY STATEMENT

All relevant data are included in the paper or its Supplementary Information.

REFERENCES

Adu, J. T. & Kumarasamy, M. V. 2018 Assessing non-point source pollution models: a review. *Polish Journal of Environmental Studies* **27** (5), 1913–1922.

Burow, K. R., Nolan, B. T., Rupert, M. G. & Dubrovsky, N. M. 2010 Nitrate in groundwater of the United States, 1991–2003. *Environmental Science & Technology* **44** (13), 4988–4997.

Cai, S., Shi, H., Pan, X., Liu, F., Cui, Y. & Xie, H. 2017 Integrating ecological restoration of agricultural non-point source pollution in Poyang Lake basin in China. *Water* **9** (10), 745.

Chen, D., Lu, J., Wang, H., Shen, Y. & Kimberley, M. O. 2010 Seasonal variations of nitrogen and phosphorus retention in an agricultural drainage river in east China. *Environmental Science and Pollution Research* **17** (2), 312–320.

Emili, L. A. & Greene, R. P. 2013 Modeling agricultural nonpoint source pollution using a geographic information system approach. *Environmental Management* **51** (1), 70–95.

Fan, L., Yuan, Y., Ying, Z., Lam, S. K., Liu, L., Zhang, X., Liu, H. & Gu, B. 2019 Decreasing farm number benefits the mitigation of agricultural non-point source pollution in China. *Environmental Science and Pollution Research* **26** (1), 464–472.

Li, J., Lu, W., Zeng, X., Yuan, J. & Yu, F. 2010 Analysis of spatial–temporal distributions of nitrate-N concentration in Shitoukoumen catchment in northeast China. *Environmental Monitoring and Assessment* **169** (1–4), 335–345.

Li, X., Liu, W., Yan, Y., Fan, G. & Zhao, M. 2019 Rural households' willingness to accept compensation standards for controlling agricultural non-point source pollution: a case study of the Qinba water source area in northwest China. *Water* **11** (6), 1251.

Liu, R., Xu, F., Zhang, P., Yu, W. & Men, C. 2016 Identifying non-point source critical source areas based on multi-factors at a basin scale with SWAT. *Journal of Hydrology* **533**, 379–388.

Liu, T., Liu, H. & Yin, F. 2017 Review on the research progress of agricultural non-point source pollution. *Science Technology Innovation* **36**, 11–12.

Liu, J., Qin, X. L., She, N., Wu, L. Y. & Bu, Z. W. 2019 Innovative design of the biological wastewater treatment system. In *IOP Conference Series: Earth and Environmental Science October*. Vol. 344, No. 1, p. 012144. IOP Publishing, Bristol, UK.

Liu, Y., Li, H., Cui, G. & Cao, Y. 2020 Water quality attribution and simulation of non-point source pollution load flux in the Hulan River basin. *Scientific Reports* **10** (1), 1–15.

Neitsch, S. L., Arnold, J. G., Kiniry, J. R. & Williams, J. R. 2011 *Soil and Water Assessment Tool Theoretical Documentation Version 2009*. Texas Water Resources Institute. Texas AgriLife Research and USDA Agriculural Research Service, Temple, Texas, USA.

Ongley, E. D., Xiaolan, Z. & Tao, Y. 2010 Current status of agricultural and rural non-point source pollution assessment in China. *Environmental Pollution* **158** (5), 1159–1168.

Ouyang, W., Hao, F. H. & Wang, X. Y. 2008 Regional non-point source organic pollution modeling and critical area identification for watershed best environmental management. *Water, Air, and Soil Pollution* **187**, 251–261.

Rajendran, K., Aslanzadeh, S., Johansson, F. & Taherzadeh, M. J. 2013 Experimental and economical evaluation of a novel biogas digester. *Energy Conversion and Management* **74**, 183–191.

Stephenson, K., Aultman, S., Metcalfe, T. & Miller, A. 2010 An evaluation of nutrient nonpoint offset trading in Virginia: a role for agricultural nonpoint sources? *Water Resources Research* **46** (4), 1–11.

Tong, Y., Deng, Z. & Gang, D. D. 2011 Nonpoint source pollution. *Water Environment Research* **83** (10), 1683–1703.

United States 1972 *Federal Water Pollution Control Act Amendments of 1972*. Pub.L. 92–500, October 18. Available from: https://www.britannica.com/topic/Clean-Water-Act.

Wang, Y., Zhu, Y., Zhang, S. & Wang, Y. 2018 What could promote farmers to replace chemical fertilizers with organic fertilizers? *Journal of Cleaner Production* **199**, 882–890.

Wilson, C. & Weng, Q. 2010 Assessing surface water quality and its relation with urban land cover changes in the Lake Calumet Area, Greater Chicago. *Environmental Management* **45** (5), 1096–1111.

Wu, H. & Ge, Y. 2019 Excessive application of fertilizer, agricultural non-point source pollution, and farmers Policy Choice. *Sustainability* **11** (4), 1165.

Wu, Y., Liu, J., Shen, R. & Fu, B. 2017 Mitigation of nonpoint source pollution in rural areas: from control to synergies of multi ecosystem services. *Science of the Total Environment* **607**, 1376–1380.

Wu, S., Liu, H., Liu, S., Wang, Y., Gu, B., Jin, S., Lei, Q., Zhai, L. & Wang, H. 2018 Review of current situation of agricultural non-point source pollution and its prevention and control technologies. *Strategic Study of Chinese Academy of Engineering* **20** (05), 31–38.

Xiang, C., Wang, Y. & Liu, H. 2017 A scientometrics review on nonpoint source pollution research. *Ecological Engineering* **99**, 400–408.

Xin, Z., Jintian, C., Yuqi, L. & Lei, W. 2017 Geo-cognitive computing method for identifying 'source-sink' landscape patterns of river basin non-point source pollution. *International Journal of Agricultural and Biological Engineering* **10** (5), 55–68.

Xue, H., Sayama, T., Takara, K., He, B., Huang, G. & Duan, W. 2019 Non-point source pollution estimation in the Pingqiao River Basin, China, using a spatial hydrograph-separation approach. *Hydrological Sciences Journal* **64** (8), 962–973.

Yaghi, A. & Salim, H. 2017 Integration of Rs/GIS for surface water pollution risk modeling. Case study: Al-Abrash Syrian Coastal Basin. International Archives of the Photogrammetry. *Remote Sensing and Spatial Information Sciences* **42** (2/W7), 949–954.

Yang, L., Shi, W., Xue, L., Song, X., Wang, S. & Chang, Z. 2013 Reduce-retain-reuse-restore technology for the controlling the agricultural non-point source pollution in countryside in China: general countermeasures and technologies. *Journal of Agro-Environment Science* **1**, 6–13.

Yang, B., Huang, K., Sun, D. & Zhang, Y. 2017 Mapping the scientific research on non-point source pollution: a bibliometric analysis. *Environmental Science and Pollution Research* **24** (5), 4352–4366.

Yi, X., Lin, D., Li, J., Zeng, J., Wang, D. & Yang, F. 2020 Ecological treatment technology for agricultural non-point source pollution in remote rural areas of China. *Environmental Science and Pollution Research* **27**, 1–13.

Zeng, J. 2019 Research on ecological compensation mechanism for preventing agricultural non-point source pollution. In *Proceedings of the 9th International Conference on Social Science and Education Research (SSER).*

Zhang, J. L., Li, Y. P., Wang, C. X. & Huang, G. H. 2015 An inexact simulation-based stochastic optimization method for identifying effluent trading strategies of agricultural nonpoint sources. *Agricultural Water Management* **152**, 72–90.

Zhang, T., Yang, Y., Ni, J. & Xie, D. 2020 Construction of an integrated technology system for control agricultural non-point source pollution in the Three Gorges Reservoir Areas. *Agriculture, Ecosystems & Environment* **295**, 106919.

Zhou, L., Liu, Y., Wang, Y. & Hu, X. 2020 Assessment and analysis of agricultural non-point source pollution loads in China: 1978–2017. *Journal of Environmental Management* **263**, 110400.

First received 10 July 2020; accepted in revised form 17 October 2020. Available online 30 October 2020

Effective removal of fly ash by *Penicillium chrysogenum* and determination of direct fly ash toxicity with *Daphnia magna*

Burcu Ertit Taştan

ABSTRACT

This study demonstrates the removal of fly ash with *Penicillium chrysogenum*, a newly isolated species of fungus, and acute toxicity assessment with *Daphnia magna*. In the study, two different removal mechanisms were compared, both bio-removal and bio-sorption. Six different ash and three different biomass concentrations were used simultaneously. Although other fungal species in the literature failed at such a high concentration of fly ash, *P. chrysogenum* was able to tolerate it even at 10% concentration. The highest bio-removal yield was recorded as 100% at 0.5% fly ash concentration. Maximum bio-sorption yield was 95.27% after 24th hour. The evaluation results of fly ash bio-toxicity by *D. magna* showed that the no observed effect level (NOEL) was 0.2 mg/L and the low observed effect level (LOEL) was 0.5 mg/L. The element analysis, determined by ED-XRF, clarified that Ca, Si, Fe and S were the common elements in this ash. This is the first study in the literature where fly ash removal was carried out using *P. chrysogenum* for both bio-removal and bio-sorption and needs to be developed in the future.

Key words | ash bio-removal, bio-sorption, bio-toxicity, *Daphnia magna*, fly ash, *Penicillium chrysogenum*

Burcu Ertit Taştan
Health Services Vocational School,
Gazi University,
06830, Gölbaşı, Ankara,
Turkey
E-mail: *burcuertit@gazi.edu.tr*

HIGHLIGHTS

- Fly ash removal by *P. chrysogenum* was investigated.
- The fungus achieved the highest removal yields.
- Highest pulp densities of fly ash were tolerated by the fungus in the literature.
- Highest biosorption yield was determined in the literature as 95.27% after 24 hours.
- Evaluation of fly ash bio-toxicity by *D. magna* was revealed for the first time.

doi: 10.2166/ws.2020.303

GRAPHICAL ABSTRACT

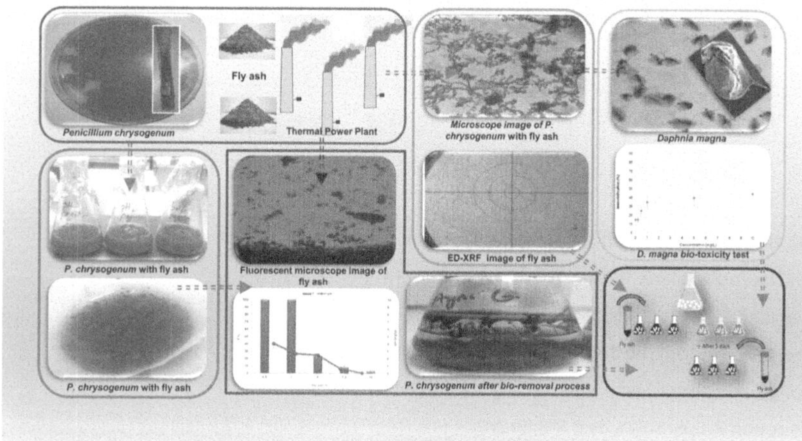

INTRODUCTION

Fly ash is the largest byproduct of coal plants. Approximately 80% of the total coal ash is fly ash and 20% is bottom ash (Querol *et al.* 1995). Fly ash is produced at 1,200–1,700 °C from a variety of organic and inorganic compounds (Vassilev & Vassileva 2005). The geological factors and operating conditions are important parameters that affect the chemical characterization of fly ash (Siddique 2010). Also, the pH of fly ash varies from 4.5 to 13.25 depending on the sulfur and CaO content of the coal (Riehl *et al.* 2010). SiO_2, Al_2O_3, Fe_2O_3 and CaO are the main components of the fly ash.

Fly ash is toxic to human health due to its small particles (Borm 1997; Ahmaruzzaman 2010). Money spent on health problems caused by thermal power plants using coal was calculated as 55 trillion Turkish lira (Querol *et al.* 1995). The fly ash produced worldwide is estimated to be around 500 million tonnes (Ahmaruzzaman 2010). Turkey is one of the countries in the world with the highest coal consumption.

Cost effective biological methods are used to reduce hazardous materials. One of the most effective microorganisms used in biological treatments are fungi, which are very effective microorganisms in heavy metal accumulation (Leung *et al.* 2010; Taştan *et al.* 2010). Therefore, they are frequently preferred in biotechnological applications.

The use of fungi in the bio-removal process of heavy metals from fly ash is one of the most popular research areas in recent years (Xu & Ting 2009). *Penicillium chrysogenum*, a species of fungus mostly found in damp environments, is prefered in treatment studies due to its ease of culture, availability, low cost and adaptation to adverse conditions (Aranciaga *et al.* 2012; Ferreira-Guedes & Leitao 2018). However, according to our best knowledge there is no study on the bio-removal and bio-sorption of fly ash by *Penicillium* genus. Previous studies have shown that *Aspergillus* species fail at high fly ash concentrations. The aim of the present study is to reveal the potential usability of *P. chrysogenum* at high fly ash concentrations with a low cost culture media, minimum operation time and minimum biomass concentrations.

There are quite a limited number of studies in the literature investigating fly ash toxicity with *Daphnia magna* (Blinova *et al.* 2012; Rodrigues *et al.* 2020), and they the acute toxicity of fly ash for *D. magna*. Millions of tonnes of fly ash are produced by factories every year, and this is a very serious global environmental problem. This study adds much new information to the literature to provide a solution to this global problem.

MATERIALS AND METHODS

Fly ash

Incineration fly ashes of different sizes (116, 232, 248, 265, 348 and 809 μm) were obtained from power plants in the Central Anatolia Region in Turkey. Fly ashes were quantified by preparing 10,000 ppm stock solutions continuously mixed with a magnetic stirrer, after which they were autoclaved at 121 °C for 15 min before use (Taştan 2017).

Culture conditions

The aim was to use a low cost culture medium in this study. Therefore, fungal biomass was inoculated into 250 mL Erlenmeyer flasks containing sterile fly ash in 100 mL T6 nutrimedia (Maysa, Turkey Rev: NMT6.01.0113) as a low cost industrial medium (Taştan 2017). The pH of the growth medium was adjusted to 6.0 by adding diluted (0.01 M) and concentrated (1 M) sulfuric acid or sodium hydroxide solutions. The flasks were incubated at 25 ± 2 °C on a rotary shaker (VWR 5000 Model Orbital Shaker) at 100 rpm for 7 days. In order to quantify the initial fungal biomass concentration, 0.5 g/L dry weight of fungal biomass was inoculated in 100 mL culture medium.

In some treatment studies, a more effective removal process was tested by using two types of culture stages (Wang *et al.* 2009). In the first type of culture stage, the fungus was inoculated into culture media at the start time ($t = 0$ day) and at the same time sterile fly ash was added to the media ($t = 0$ day), then they were incubated for 7 days. In the second type of culture stage, first the fungus was pre-incubated for 5 days ($t = 5$ days) without fly ash and then sterile fly ash was added on the fifth day ($t = 5$ days) and incubated for another 2 days. A schematic of the experimental setup is shown in Figure 1.

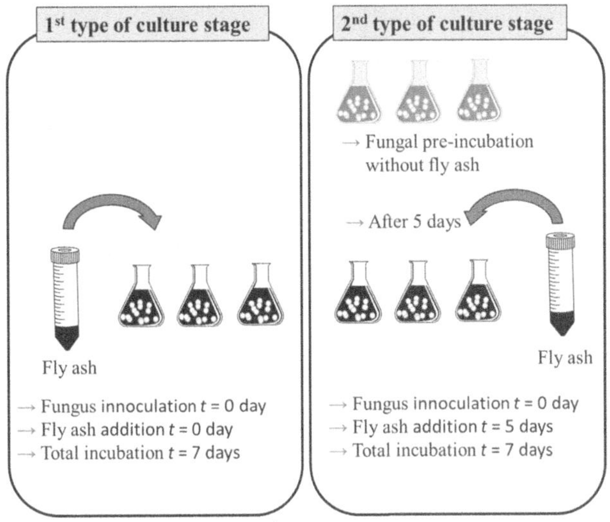

Figure 1 | Schematic of the experimental setup.

Isolation of microorganism

Cheese from Erzurum (Turkey) was selected as the inoculum and it was spread on potato dextrose agar (PDA) in Petri plates and incubated at 30 ± 1 °C. After 7 days of incubation, fungal colonies were isolated and then purified by streaking the cells repeatedly on the PDA. The pure cultures were kept at 4 °C and were transferred to fresh PDA every 3 months (Taştan 2017).

Identification of new isolate with PCR and sequencing

Whole cells from an exponentially growing culture of the isolate were used for 5.8S rRNA gene amplification. The 5.8S rRNA region was amplified with primers as designed by Bokulich & Mills (2013), forward ITS1 5′-TCC GTA GGT GAA CCT GCG G-3′, reverse ITS2 5′- GCT GCG TTC TTC ATC GAT GC-3′. Polymer Polymerase chain reaction (PCR) was carried out in 50 μL reactions and the reaction mix included 0.3 μM of each primer, 0.2 mM of each dNTP, 2.0 mM $MgCl_2$ and 100 ng of template DNA. Taq DNA polymerase was used in the amplification. Amplification by PCR technique was carried out by an initial denaturation at 95 °C for 4 min, followed by 35 cycles of denaturation at 94 °C for 30 seconds, annealing at 57 °C for 30 seconds, an extension at 72 °C for 1 min, and a final extension of 72 °C for 5 min. Applied Biosystems Genetic Analyzer 3130 was used in sequencing.

Bio-removal process

Four different culture media compositions were prepared for the fungal ash bio-removal process. (i) T6 nutrimedia (12 g/L),

(ii) T6 nutrimedia (12 g/L) + corn syrup (5 mL/L) (Vanikoy, Turkey), (iii) T6 nutrimedia (12 g/L) + whey powder (2.5 g/L) (Maybi, Turkey) and (iv) T6 nutrimedia (12 g/L) + corn syrup (5 mL/L) + whey powder (2.5 g/L). To assess the effect of incineration fly ashes of different sizes (116, 232, 248, 265, 348 and 809 μm) on fungal ash bio-removal, the fungus was cultivated separately and simultaneously in the first type of culture stage and the second type of culture stage.

The effect of pH on fungal ash bio-removal was investigated at pH 4.0, 5.0, 6.0, 7.0 and 8.0 at 1% fly ash concentration in the second type of culture stage. To detect the effect of fly ash concentration, 0.5, 1, 5, 7.5 and 10 g fly ash was included in 100 mL culture media in the second type of culture stage. The concentration of fly ash was expressed as 0.5, 1, 5, 7.5 and 10%. The effect of increasing fungal biomass concentrations was investigated at 1% concentration of selected optimum fly ash in the second type of culture stage. The initial fungal biomass concentrations corresponded to the dry weights $X = 0.25$ g/L, $X = 0.5$ g/L and $X = 1$ g/L.

Bio-sorption process

The fungal biomass was centrifuged at 5,000 rpm for 10 min at the end of the logarithmic growth period. After washing the pellet twice with dH_2O it was autoclaved (121 °C, 15 min, ALP CLG-40M, Japan). The wet cells were converted to dry weight using the wet/dry weight standardization method ($R^2 = 98.5\%$). Bio-sorption studies were carried out in 100 mL aqueous solution in 250 mL Erlenmeyer flasks as described by Laçin *et al.* (2015) at 5, 15 and 30 min, and 1, 2, 4, 6, and 24 hours with 1% fly ash concentration. The fungal ash bio-sorption yield (Y %) was calculated using Equation (1) as described below. An ash-free control group was used as the blank.

Acute immobilization and bio-toxicity assessment

EC_{50} is the concentration estimated to immobilize 50% of *D. magna* (daphnids) within a stated exposure period. Immobilization is taken to mean those animals that are not able to swim within 15 seconds, after gentle agitation of the test vessel (OECD 2004). To detect the accute immobilization of fly ash, daphnids were used as the test species (Balusamy *et al.* 2015). The daphnids were cultivated at a stable temperature of 20 ± 1 °C in 16:8 hours light:dark periods. Increasing concentrations of fly ash at 0, 0.2, 1, 5 and 10 mg/L were tested in the test medium. Bio-toxicity studies were conducted as three replicates and five daphnids were used for each concentration and for the control group. After the 24, 48 and 72 hours, daphnids in each group were tested for immobilization.

Analytical methods

In the experiments 3 mL samples were taken daily for bio-removal experiments and taken hourly for bio-sorption experiments during the incubation period. Erlenmeyer flasks not containing fly ash were used as the control media.

The fungus to be used during calculations was passed through a series of steps as follows:

i. After the incubation period was completed, the fungi were filtered through coarse filter paper.
ii. The fungi were washed to allow cells to separate from medium residues.
iii. The fungi containing ash and the fungi without ash (control media) were weighed separately according to their purpose as wet or dry weight and tared.

Fungal ash bio-removal and bio-sorption yields (Y %) and maximum specific fly ash uptake (q_m) was calculated using Equations (1) and (2);

$$Y\,(\%) = \frac{(C_0 - C_t)}{C_0} \times 100 \quad (1)$$

$$q_m = \frac{(C_0 - C_t)}{X} \quad (2)$$

In these equations, q_m is the maximum amount of fly ash taken up per unit dry weight of fungal cells (mg/g), X is dry biomass (g/L), C_0 is the initial fly ash concentration (mg/L) and C_t is the final fly ash concentration (mg/L), respectively (Taştan *et al.* 2010, 2012).

The biomass was calculated by determining the dried cell bulk. The cell bulk was calculated by measuring the centrifuged pellets, which were dried at 80 °C overnight (Taştan *et al.* 2016). The dried cell bulk was expressed as the ash-free control group (X_c) and the ash-containing group (X_s). The fly ash concentration was measured as 0.5, 1, 5, 7.5 and 10 g fly ash in 100 mL culture media and expressed as 0.5, 1, 5, 7.5 and 10%.

An energy dispersive X-ray fluorescence spectrometer (ED-XRF, Shimadzu) was used for element analyses. Schimadzu reference materials (standards) were used for calibrating the ED-XRF instrument. The samples were dried at 110 °C for 12 hours before the analyses and detected with a silicon drift detector (Taştan 2017).

All of the tests were carried out three times. All data were calculated with corresponding standard error as formulated and described by Kenney & Keeping (1962), Equation (3); σ is the square root of the estimated error variance of the quantity.

$$SE = \sqrt{\sigma^2} \tag{3}$$

RESULTS AND DISCUSSION

Effect of growth medium on fungal ash bio-removal

As a result of PCR and sequencing, the isolated new strain was identified as *P. chrysogenum*. The effect of different media compositions were tested for increasing the performance of fungal fly ash uptake and summarized in Table 1. The bio-removal yield obtained in T6 nutrimedia was higher than the yield of T6 nutrimedia + corn syrup media, but it was lower than the yields that were obtained in T6 nutrimedia + whey powder. In this step, dry biomass amounts of the ash-free control group (X_c) and ash-containing sample groups (X_s) were compared. The lowest yield was obtained in T6 nutrimedia + corn syrup + whey powder media, while the lowest X_s was obtained in T6 nutrimedia. Therefore T6 nutrimedia + whey powder media was more efficient in fly ash removal process and also the X_s value (2.65 g/L) recorded in this medium was only 21.6% lower than of X_c (3.38 g/L). By contrast, the maximum X_c was obtained in T6 nutrimedia + corn syrup media, while 30.8% decrease was observed in X_s amount of fungus. The highest difference between X_s and X_c amounts was observed in T6 nutrimedia as 34.6%. In a summary, the highest bio-removal yield, highest X_s amount and the minimum difference between X_s and X_c amounts were obtained in T6 nutrimedia + whey powder media.

In the removal of various contaminants, it was found that different culture media had a direct effect on fungal removal efficiency (Marcial *et al.* 2006; Taştan *et al.* 2016). The use of this inexpensive and commercial T6 nutrimedia culture medium in the literature is very recent. Studies have shown that the medium is effective in ash removal by different types of fungi (Taştan 2017). In this study, *P. chrysogenum* was able to tolerate ash up to 10% concentration using T6 nutrimedia culture medium as can be seen in Table 1. In a study using the same medium, *Fusarium oxysporum* and *P. glabrum* removed 6.36 and 18.65% fly ash at 7.5% pulp concentration, respectively (Taştan 2017). In another study, standard sucrose medium was used as the culture medium and *A. niger* did not show any growth at 6% ash concentration (Xu & Ting 2009). The results show that this commercial medium is a highly productive medium that will contribute to the literature.

Effect of different fly ashes on fungal ash bio-removal

The fungus was cultivated in the two independent types of bio-removal process with six different fly ashes (Table 2). In the first type of culture stage, fungus was first cultivated

Table 1 | Effects of different media on bio-removal yield (*Y* %) of fly ash by *P. chrysogenum* (X_c; dry biomass amounts of ash-free control group, X_s; and ash-containing sample groups)

Culture media	*Y* %	X_S (g/L)	X_C (g/L)
T6 nutrimedia	76.79 ± 1.11	2.06 ± 0.28	3.15 ± 0.71
T6 nutrimedia + corn syrup	71.60 ± 0.41	2.47 ± 0.37	3.57 ± 0.40
T6 nutrimedia + whey powder	83.64 ± 0.55	2.65 ± 0.32	3.38 ± 0.22
T6 nutrimedia + corn syrup + whey powder	42.25 ± 0.28	2.57 ± 0.11	3.33 ± 0.15

Table 2 | Fungal ash removal yields (*Y* %) with different types of culture stages

	Sizes of fly ash (μm)					
Type of culture stage	**116**	**232**	**248**	**265**	**348**	**809**
First	78.06 ± 1.74	46.49 ± 6.01	82.32 ± 0.67	14.75 ± 3.77	62.67 ± 1.44	45.25 ± 0.52
Second	100 ± 0	63.52 ± 0.74	62.62 ± 4.37	49.81 ± 1.56	51.40 ± 4.34	22.97 ± 3.75

with incineration fly ashes of different sizes (116, 232, 248, 265, 348 and 809 μm) at 1% fly ash concentration. According to results, the maximum bio-removal yield, 82.32%, was obtained with fly ash of 248 μm size and the minimum bio-removal yield, 45.25%, was obtained with fly ash of 809 μm size, due to the larger size of this fly ash. In the second type of culture stage, fungus was introduced with fly ash after 5 days of pre-incubation. According to these results, the maximum bio-removal yield was obtained with fly ash of 116 μm size. The minimum bio-removal yield, 100%, was obtained again in fly ash of 809 μm size. The second type of culture stage allowed the rapid transformation of fungal spores into micelles and removed the fly ash with higher yields in a shorter time (2 days). By contrast, while the ash size was 26.3 μm, *A. niger* showed an improvement at 0–5% fly ash pulp density, but did not show an improvement at 6% (Xu & Ting 2009). In another study, more effective treatment was obtained by using two types of culture stages (Wang *et al.* 2009), and 45% decrease in lag phase was realized by pre-washing before the fungal bioleaching process and thus the bioleaching process was accelerated by 30%. When all results are taken into consideration, 116 μm size of fly ash was selected for further testing in the second type of culture stage ($t = 5$ days).

Effect of pH on fungal ash bio-removal

The effect of pH on fungal ash removal yield by *P. chrysogenum* is presented in Table 3. Although the same removal yields were observed at pH 4.0, 5.0 and 6.0, yields decreased up to 55.61% at pH 8.0. The maximum X (dry weight g/L) values were obtained as 3.08 g/L at pH 6.0 despite these equal removal yields. Therefore, the optimum pH was selected as 6.0 (Table 3).

In another study, two types of culture stages were used. In the first one, fungus and ash were incubated at the same time, in the second, fly ash was added after the fungus developed for 2 days. At the end of day 15 in the first type of culture stage, the pH decreased to about 6.0 and in the second type of culture stage pH decreased to about 4.0 (Xu *et al.* 2014), and these results support this study.

Effect of increasing fly ash concentrations on fungal ash bio-removal

The results of increasing fly ash concentrations from 0.5 to 10% are summarized in Figure 2. As seen in the Figure 2, *P. chrysogenum* tolerated the fly ash well up to high concentrations. The maximum fungal ash removal yields were obtained at the lowest two concentrations of fly ash (0.5 and 1%). The fungus was able to completely remove low concentrations, and it was also able to tolerate high concentrations up to 10%. In another study the bioleaching kinetics of *A. niger* in the presence of fly ash at 1–6% concentrations in a bioleaching process was investigated. According to the results, fungal growth was not observed at higher than 6% of fly ash concentrations (Xu & Ting 2009).

Table 3 | Selection of pH in fungal fly ash bio-removal (*Y* %)

	pH				
	4.0	**5.0**	**6.0**	**7.0**	**8.0**
Y %	100 ± 0	100 ± 0	100 ± 0	94.06 ± 4.10	55.61 ± 7.67
X_S (g/L)	2.36 ± 0.06	2.88 ± 0.19	3.08 ± 0.12	2.47 ± 0.56	2.00 ± 0.23

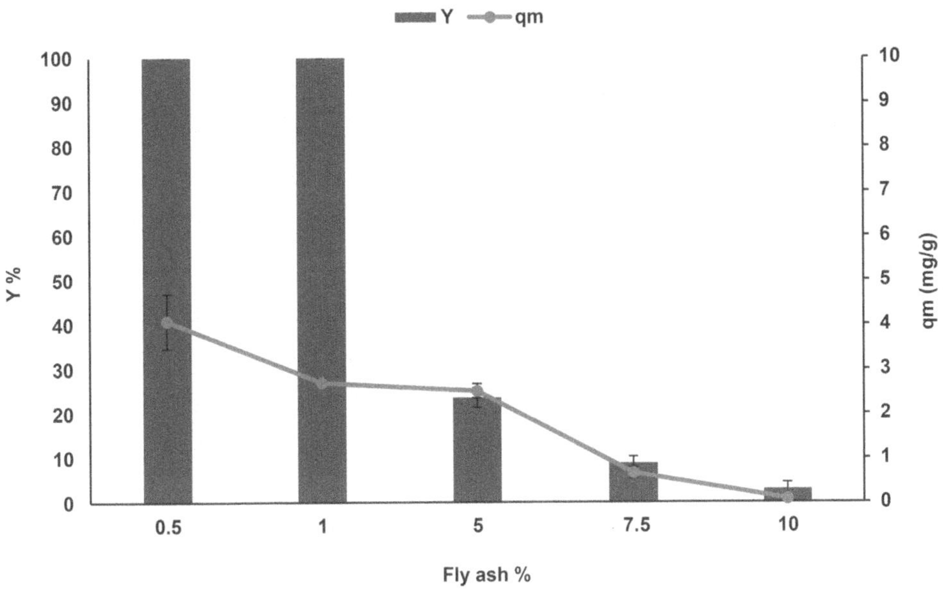

Figure 2 | Effect of increasing fly ash concentrations on the fungal fly ash bio-removal yields (Y%).

In a study by Xu & Ting (2004), different parameters were investigated on the bioleaching of fly ash by *A. niger*. Results showed that fly ash concentration was more important than the spore concentration and the time of addition of fly ash. In another study the scanning electron microscope (SEM) results of *A. niger* emphasized that the fungal hyphae were impacted negatively by increasing fly ash concentrations (Xu *et al.* 2014).

According to Figure 2, the maximum specific fly ash uptake (q_m) was 4.08 mg/g at 0.5% fly ash concentration. By contrast, the fungus was able to uptake fly ash at 0.08 mg/g at the highest fly ash concentration (10%). The mechanism of fungal cellular uptake was expressed in this stage of experiments. Therefore bio-sorption experiments were also conducted. Further studies are needed to better understand the mechanism, such as genetic, enzymatic and molecular analyses and advanced microscopic studies using transmission electron microscopy (TEM) and SEM.

Shortening the time factor in bio-removal studies affects profits at industrial scales. In the present study *P. chrysogenum* grew in 0.5–10% concentrations of fly ash with the help of adding fly ash after an incubation period, meaning that fungal growth was not inhibited by increasing concentrations of fly ash. Similar to the results of the present study, Xu & Ting (2004) added fly ash after 2 days of fungal incubation and studied 1–8% concentrations. They noted that the fungus *A. niger* was able to grow in the presence of 4% fly ash.

Table 4 | Effect of initial biomass concentrations on the fungal fly ash bio-removal yields (Y%)

	Initial fungal biomass concentrations X_0 (g/L)		
	0.25	**0.5**	**1**
6 hours	97.42 ± 0.51	98.71 ± 0.32	100 ± 0
24 hours	100 ± 0	100 ± 0	100 ± 0
48 hours	100 ± 0	100 ± 0	100 ± 0

Effect of increasing biomass concentrations on fungal ash bio-removal

To find a suitable biomass concentration of *P. chrysogenum*, 0.25, 0.5, 1 g/L initial biomass concentrations were tested.

Table 5 | Bio-sorption yields of fly ash by *P. chrysogenum*

Time	1 min	5 min	15 min	30 min	1 hour	2 hours	4 hours	6 hours	24 hours
Y %	50.35 ± 3.16	51.48 ± 2.6	57.93 ± 0.02	64.04 ± 2.07	57.88 ± 1.45	59.62 ± 4.76	64.64 ± 3.58	66.78 ± 0.12	95.27 ± 3.60

As seen in Table 4, the minimum fungal ash removal yield, 97.42%, was obtained at 0.5 g/L initial biomass concentration in the sixth hour of incubation. When the initial biomass concentration increased to 1 g/L, the yield increased to 100% in the sixth hour. All of the fungal fly ash removal yields increased to 100% after 24 hours at all biomass concentrations. As can be seen from the results, the fungus can be recorded as a very active species for ash removal. Regardless of the initial fungal biomass concentrations within the studied ranges, ash removal yields reached to 100% after 24–48 hours. This led to the result that *P. chrysogenum* is a highly effective biological material that can be used for ash removal in further studies. To the best of our knowledge, investigating the effect of increasing fungal biomass concentrations on ash removal has not been conducted before in the literature. However, ash removal studies in the current literature showed that increasing biomass is associated with increased removal efficiency (Taştan *et al.* 2012; Taştan & Dönmez 2015).

Bio-sorption experiments

To better understand the mechanism of fly ash removal by fungi, the effect of bio-sorption on fungal ash uptake were tested. The data obtained after 24 hours of incubation with autoclaved (dead) fungus are summarized in Table 5. The maximum bio-sorption yield was 95.27% after 24 hours. Moreover half of fly ash was biosorbed in the first hour of the incubation. Therefore bio-sorption of fly ash by *P. chrysogenum* biomass can be recommended as fast way to remove fly ash, a harmful pollutant, from the environment.

In another study, the bio-sorption of heavy metals was carried out by *A. niger* biomass in bioleaching of fly ash. The fungal biomass was in contact with a heavy metal solution extracted from fly ash. This research showed that the bio-sorption of metal ions decreased clearly in the presence of fly ash compared to that in the absence of fly ash (Yang *et al.* 2009). It was determined that the hours-long bio-sorption process was more effective in ash removal than the days-long fungal cellular uptake mechanism.

Element analyses

The element analysis results conducted by ED-XRF are shown in Table 6. In the samples, Ca, Fe and Si were the

Table 6 | Elemental analyses of fly ash, growing media and biomass of *P. chrysogenum* with fly ash analyzed by ED-XRF

	Ca%	Fe%	Si%	S%	K%	Ti%	Sr%	Mn%	Ni%	Cr%	Cu%	Zr%	V%	Zn%	Rb%	P%	Mo%
Fly ash	38.49 ± 0.16	16.52 ± 0.10	35.40 ± 0.28	5.85 ± 0.03	1.67 ± 0.02	1.18 ± 0.01	0.31 ± 0.00	0.22 ± 0.01	0.04 ± 0.00	0.05 ± 0.00	0.06 ± 0.00	0.07 ± 0.01	0.08 ± 0.00	0.02 ± 0.01	0.03 ± 0.00	–	0.02 ± 0.00
Growing media	77.99 ± 0.09	0.16 ± 0.01	0.78 ± 0.01	0.72 ± 0.02	2.74 ± 0.02	–	0.11 ± 0.01	0.02 ± 0.00	–	–	0.07 ± 0.00	0.02 ± 0.00	–	0.14 ± 0.00	–	17.25 ± 0.02	–
Biomass of *P. chrysogenum* with fly ash	39.12 ± 0.23	17.74 ± 0.15	30.24 ± 0.23	2.53 ± 0.18	3.19 ± 0.01	1.21 ± 0.02	0.32 ± 0.01	0.25 ± 0.00	0.04 ± 0.00	0.05 ± 0.01	0.06 ± 0.00	0.07 ± 0.00	0.08 ± 0.00	0.04 ± 0.00	0.03 ± 0.00	5.01 ± 0.42	–

common elements in the fly ash used in this study. The Ca concentrations were higher than the other major elements in the all samples. The reason why the Ca amounts were higher in the biomass of *P. chrysogenum* was due to the high Ca levels observed in growth medium (77.99%). As stated before, the Ca concentration probably was related to the organic matter in the coal samples (Sutcu & Karayigit 2015). Therefore all the samples had high Ca concentrations. The average yield of Si in fly ash was 35.40%. Fe, S, K and Ti were the other remediated elements in fly ash, with the yields of 16.52, 5.85, 1.67 and 1.18%, respectively (Table 6). Elemental analysis results also show that this application can be useful in bio-removal of fly ash. Therefore the isolated fungi, *P. chrysogenum*, could be suggested as efficient bio-material in the process of remediation of metals from fly ash when it is scaled up.

Bayat (2002) examined the removal of Cr and Cd from two fly ashes. The adsorption of Cd was found to be higher than Cr. In another study, some properties of various fly ashes in Turkey were detected and high Ca levels were found (Bayat 1998). In the present study, the reason for high fungal tolerance to increasing fly ash concentrations may be to the high Ca levels in fly ash and in the growth medium.

Acute immobilization test

The EC_{50} concentration was found to be 10 mg/L. However, no LD_{50} value was determined by *D. magna* after 72 hours of fly ash exposure time. The no observed effect level (NOEL) was found at 0.2 mg/L and low observed effect level (LOEL) was found at 0.5 mg/L fly ash (Figure 3).

There are only a few studies in the literature on the toxicity tests of fly ash using crustaceans. In a study, the toxicity evaluation of municipal solid waste incineration fly ash was evaluated with the role of cadmium by a clone of *D. pulex*. They explained that the leachate was more than 20 times more toxic than expected from its cadmium concentration for survival of *D. pulex* (Kaneko 1996). There are some studies about evaluation of the toxicity of lanthanum oxide (Balusamy *et al.* 2015), acid dyes (Dehghani *et al.* 2016), graphene oxide (Zhang *et al.* 2019), and pesticides (Taştan *et al.* 2017). But according to our best knowledge there is a limited number of studies in the literature investigating the direct effects of fly ash toxicity. In one of the limited number of studies conducted, it was found that the EC_{50} value of *D. magna* in fly ash toxicity was 30.80% (Rodrigues *et al.* 2020), and in another, the inhibition was between 10 and 65% in different fly ash water eluates (Blinova *et al.* 2012).

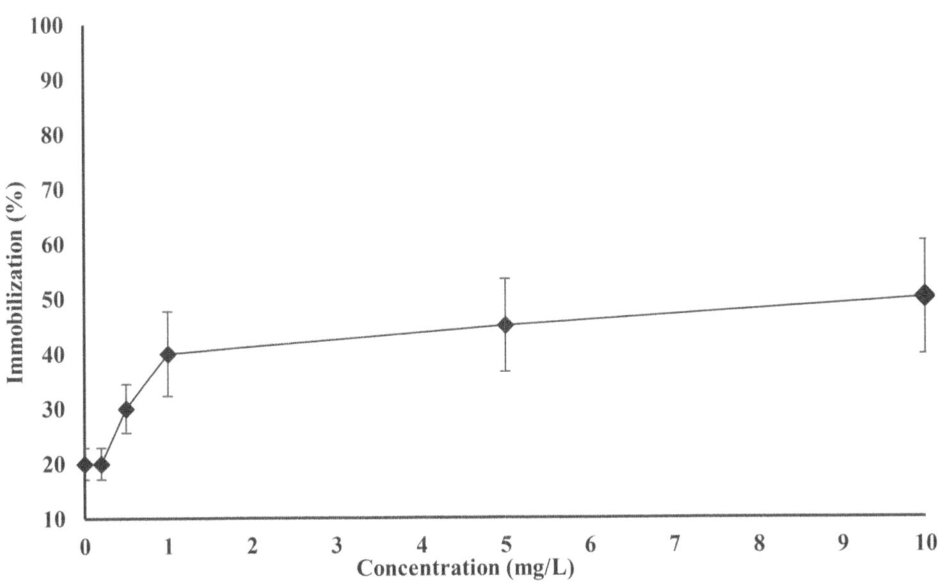

Figure 3 | Effect of increasing fly ash concentrations on the mobility of *D. magna* after 72 hours of exposure.

Therefore the present study is the first to reveal the bio-toxicity evaluation of fly ash with above mentioned requirements.

CONCLUSIONS

This study highlights the potential usage of *P. chrysogenum* in the bio-removal of fly ash. Six different sizes of fly ash, two different culture stage techniques and three different fungal biomasses were used. In the first type of culture stage, the maximum removal rate was 82.32%, and in the second type of culture stage the maximum removal rate was 100%. The second type of culture stage allowed the rapid transformation of fungal spores into micelles. Thus, the fungal biomass developed rapidly. Fungal biomass could tolerate 10% fly ash concentration, which to the best of our knowledge, is the highest ash concentration in the literature. ED-XRF results helped to show the fate of metals in fly ash. In this study, in which both bio-sorption and bio-removal mechanisms were investigated, bio-sorption was found to be a highly effective mechanism, as well as bio-removal. *P. chrysogenum* could serve as an effective bio-agent for bio-removal and bio-sorption of fly ash from aqueous media. The present method was found to be advantageous as compared to other reported bio-removal studies, as it allowed the fungus to tolerate high concentrations of fly ash.

ACKNOWLEDGEMENTS

Financial support is gratefully acknowledged from the Scientific and Technological Research Council of Turkey (TÜBİTAK) (Grant number 115Y077). The author also thanks Şevda Akyürek for technical support.

DATA AVAILABILITY STATEMENT

All relevant data are included in the paper or its Supplementary Information.

REFERENCES

Ahmaruzzaman, M. 2010 A review on the utilization of fly ash. *Progress in Energy and Combustion Science*. https://doi.org/10.1016/j.pecs.2009.11.003.

Aranciaga, N., Durruty, I., Gonzalez, J. F. & Wolski, E. A. 2012 Aerobic biotransformation of 2, 4, 6–trichlorophenol by *Penicillium chrysogenum* in aqueous batch culture: degradation and residual phytotoxicity. *Water SA*. https://doi.org/10.4314/wsa.v38i5.5.

Bayat, O. 1998 Characterisation of Turkish fly ashes. *Fuel*. https://doi.org/10.1016/S0016-2361(97)00274-3.

Balusamy, B., Taştan, B. E., Ergen, S. F., Uyar, T. & Tekinay, T. 2015 Toxicity of lanthanum oxide (La_2O_3) nanoparticles in aquatic environments. *Environmental Sciences: Processes and Impacts*. https://doi.org/10.1039/c5em00035a.

Bayat, B. 2002 Comparative study of adsorption properties of Turkish fly ashes: II. The case of chromium (VI) and cadmium (II). *Journal of Hazardous Materials*. https://doi.org/10.1016/S0304-3894(02)00141-3.

Blinova, I., Bityukova, L., Kasemets, K., Ivask, A., Käkinen, A., Kurvet, I., Bondarenko, O., Kanarbik, L., Sihtmäe, M., Aruoja, V., Schvede, H. & Kahru, A. 2012 Environmental hazard of oil shale combustion fly ash. *Journal of Hazardous Materials*. http://dx.doi.org/10.1016/j.jhazmat.2012.05.095.

Bokulich, N. A. & Mills, D. A. 2013 Improved selection of internal transcribed spacer-specific primers enables quantitative, ultra-high-throughput profiling of fungal communities. *Applied and Environmental Microbiology*. https://doi.org/10.1128/AEM.03870-12.

Borm, P. J. A. 1997 Toxicity and occupational health hazards of coal fly ash (CFA). A review of data and comparison to coal mine dust. *Annals of Occupational Hygiene*. https://doi.org/10.1016/S0003-4878(97)00026-4.

Dehghani, M. H., Mahdavi, P., Tyagi, I., Agarwal, S. & Gupta, V. K. 2016 Investigating the toxicity of acid dyes from textile effluent under UV/ZNO process using *Daphnia magna*. *Desalination and Water Treatment*. https://doi.org/10.1080/19443994.2016.1141327.

Ferreira-Guedes, S. & Leitao, A. L. 2018 Simultaneous removal of dihydroxybenzenes and toxicity reduction by *Penicillium chrysogenum* var. *halophenolicum* under saline conditions. *Ecotoxicology and Environmental Safety*. https://doi.org/10.1016/j.ecoenv.2017.12.046.

Kaneko, H. 1996 Evaluation of municipal waste incinerator fly ash toxicity and the role of cadmium by two aquatic toxicity tests. *Waste Management*. https://doi.org/10.1016/S0956-053X(96)00097-9.

Kenney, J. F. & Keeping, E. S. 1962 Linear Regression and Correlation. In *Mathematics of Statistics*. Part 1, 3rd edition. Van Nostrand, Princeton, NJ, pp. 252–285.

Laçin, B., Taştan, B. E. & Dönmez, G. 2015 Detection of boron removal capacities of different microorganisms in wastewater and effective removal process. *Water Science and*

Technology **72** (10), 1832–1839. https://doi.org/10.2166/wst.2015.409.

Leung, H. M., Wu, F. Y., Cheung, K. C., Ye, Z. H. & Wong, M. H. 2010 Synergistic effects of arbuscular mycorrhizal fungi and phosphate rock on heavy metal uptake and accumulation by an arsenic hyperaccumulator. *Journal of Hazardous Materials*. https://doi.org/10.1016/j.jhazmat.2010.05.042.

Marcial, J., Barrios-Gonzalez, J. & Tomasini, A. 2006 Effect of medium composition on pentachlorophenol removal by *Amylomyces rouxii* in solid-state culture. *Process Biochemistry*. https://doi.org/10.1016/j.procbio.2005.07.010.

OECD 2004 *Guideline 202: Daphnia* sp., *Acute Immobilisation Test.* OECD guidelines for testing of chemicals.

Querol, X., Fernández-Turiel, J. & López-Soler, A. 1995 Trace elements in coal and their behaviour during combustion in a large power station. *Fuel*. https://doi.org/10.1016/0016-2361(95)93464-O.

Riehl, A., Elsass, F., Duplay, J., Huber, F. & Trautmann, M. 2010 Changes in soil properties in a fluvisol (calcaric) amended with coal fly ash. *Geoderma*. https://doi.org/10.1016/j.geoderma.2009.11.025.

Rodrigues, P., Silvestre, J. D., Flores-Colen, I., Viegas, C. A., Ahmed, H. H., Kurda, R. & de Brito, J. 2020 Evaluation of the ecotoxicological potential of fly ash and recycled concrete aggregates use in concrete. *Applied Sciences*. http://dx.doi.org/10.3390/app10010351.

Siddique, R. 2010 Utilization of coal combustion by-products in sustainable construction materials. *Resources, Conservation and Recycling*. https://doi.org/10.1016/j.resconrec.2010.06.011.

Sutcu, E. C. & Karayigit, A. I. 2015 Mineral matter, major and trace element content of the Afşin-Elbistan coals, Kahramanmaraş, Turkey. *International Journal of Coal Geology*. https://doi.org/10.1016/j.coal.2015.04.007.

Taştan, B. E. 2017 Clean up fly ash from coal burning plants by new isolated fungi *Fusarium oxysporum* and *Penicillium glabrum*. *Journal of Environmental Management*. https://doi.org/10.1016/j.jenvman.2017.05.062.

Taştan, B. E. & Dönmez, G. 2015 Biodegradation of pesticide triclosan by A. versicolor in simulated wastewater and semi-synthetic media. *Pesticide Biochemistry and Physiology*. http://dx.doi.org/10.1016/j.pestbp.2014.11.002.

Taştan, B. E., Ertuğrul, S. & Dönmez, G. 2010 Effective bioremoval of reactive dye and heavy metals by *Aspergillus versicolor*. *Bioresource Technology*. https://doi.org/10.1016/j.biortech.2009.08.099.

Taştan, B. E., Duygu, E. & Dönmez, G. 2012 Boron bioremoval by a newly isolated *Chlorella* sp and its stimulation by growth stimulators. *Water Research*. https://doi.org/10.1016/j.watres.2011.10.045.

Taştan, B. E., Özdemir, C. & Tekinay, T. 2016 Effects of different culture media on biodegradation of triclosan by *Rhodotorula mucilaginosa* and *Penicillium* sp. *Water Science and Technology*. https://doi.org/10.2166/wst.2016.221.

Taştan, B. E., Tekinay, T., Çelik, H. S., Özdemir, C. & Cakir, D. N. 2017 Toxicity assessment of pesticide triclosan by aquatic organisms and degradation studies. *Regulatory Toxicology and Pharmacology*. https://doi.org/10.1016/j.yrtph.2017.10.030.

Vassilev, S. V. & Vassileva, C. G. 2005 Methods for characterization of composition of fly ashes from coal-fired power stations: a critical overview. *Energy and Fuels*. https://doi.org/10.1021/ef049694d.

Wang, Q., Yang, J., Wang, Q. & Wu, T. 2009 Effects of water-washing pretreatment on bioleaching of heavy metals from municipal solid waste incinerator fly ash. *Journal of Hazardous Materials*. https://doi.org/10.1016/j.jhazmat.2008.05.125.

Xu, T. J. & Ting, Y. P. 2004 Optimisation on bioleaching of incinerator fly ash by *Aspergillus niger* – Use of central composite design. *Enzyme and Microbial Technology*. https://doi.org/10.1016/j.enzmictec.2004.07.003.

Xu, T. J. & Ting, Y. P. 2009 Fungal bioleaching of incineration fly ash: metal extraction and modeling growth kinetics. *Enzyme and Microbial Technology*. https://doi.org/10.1016/j.enzmictec.2009.01.006.

Xu, T. J., Ramanathan, T. & Ting, Y. P. 2014 Bioleaching of incineration fly ash by *Aspergillus niger* – precipitation of metallic salt crystals and morphological alteration of the fungus. *Biotechnology Reports*. https://doi.org/10.1016/j.btre.2014.05.009.

Yang, J., Wang, Q., Luo, Q., Wang, Q. & Wu, T. 2009 Bio-sorption behavior of heavy metals in bioleaching process of MSWI fly ash by *Aspergillus niger*. *Biochemical Engineering Journal*. https://doi.org/10.1016/j.bej.2009.05.022.

Zhang, Y., Meng, T., Shi, L., Guo, X., Si, X., Yang, R. & Quan, X. 2019 The effects of humic acid on the toxicity of graphene oxide to *Scenedesmus obliquus* and *Daphnia magna*. *Science of the Total Environment*. https://doi.org/10.1016/j.scitotenv.2018.08.280.

First received 13 June 2020; accepted in revised form 26 October 2020. Available online 9 November 2020

Mesoporous structure regulation of activated carbons and effects on the synergistic efficacy of synchronous adsorption and bio-degradation in biological activated carbon process

Xu-Jin Gong, Yu-Qi Dong and Wei-Guang Li

ABSTRACT

Mesoporous activated carbon MCGL-4 was tailored for simultaneous enhancement of adsorption and bio-degradation by multistage depth-activation (MDA). Synergistic efficacy of synchronous adsorption and bio-degradation was evaluated in pilot-scale bio-enhanced activated carbon (BEAC) system. Results identified that MCGL-4 obtains synchronously well-developed meso- (0.7605 cm^3/g), micro- (0.2655 cm^3/g) and macro-porous (0.143 cm^3/g) structures. Higher volume during 20.4–208.2 Å (0.6848 cm^3/g) ensured higher adsorption capacities for natural organic matters (NOM). The initial immobilized biomass and stabilities on MCGL-4 were also significantly promoted. Rapid small-scale column tests system (RSSCTs) tests showed that adsorption capacities for humic-like organics were 67,725.32 mg·DOC/(kg·carbon) at 39.50 $m^3 \cdot H_2O$/(kg·carbon). In BEAC system, MCGL-4 achieved higher removal efficiency for fulvic acid, humic acid and aromatic organic matters than commercial carbons. At 39.50 $m^3 \cdot H_2O$/(kg·carbon), cumulative uptake of organic-pollutants achieved by MCGL-4 was 94,850.51 mg·DOC/(kg·carbon). The proportion occupied by bio-degradation were 31,674.70 mg·DOC/(kg·carbon). It also confirmed that bio-degradation ability was much higher than commercial carbons after mesoporous structures regulation by MDA process.

Key words | activated carbon, adsorption, bio-degradation, bio-enhanced activated carbon (BEAC), natural organic matters, pore-structure regulation

Xu-Jin Gong (corresponding author)
Yu-Qi Dong
School of Energy and Civil Engineering,
Harbin University of Commerce,
Harbin 150090,
China
E-mail: *xujin_gong@126.com*

Wei-Guang Li
School of Environment,
Harbin Institute of Technology,
Harbin 150090,
China

HIGHLIGHTS

- MCGL-4 was tailored for simultaneous bio-enhancement and adsorption by multistage depth-activation.
- MCGL-4 is a promising adsorbent for effective removal of natural organic matters from water.
- MCGL-4 was more conductive to simultaneous biodegradation and adsorption of organic-pollutants.
- Mesopores (20.4 ~ 208.2 Å) played important role in adsorption of humic-like organics.

NOMENCLATURE

KBV	Volume of water treated by per kilogram of activated carbon ($m^3 \cdot H_2O$/(kg·carbon))
S_{BET}	BET surface area (m^2/g)
S_{micro}	*t*-Plot micropore area (m^2/g)
V_{total}	Total pore volume (cm^3/g)
V_{micro}	*t*-Plot micro-porous volume (cm^3/g)
V_{mes}	Calculated mesoporous volume: $V_{mes} = V_{total} - V_{micro}$ (cm^3/g)

doi: 10.2166/ws.2020.293

$V_{treated}$ Volume of water treated by carbons (m^3)
V_{IN} Adsorption capacities for iodine(mg/g)
V_{MB} Adsorption capacities for methylene blue (mg/g)
V_{HA} Adsorption capacities for humic acid (mg/g)
N Days of operation (d)
Q_{in} Flow rate (L/d)
W_{AC} Weight of carbon in column (kg)
QC Cumulative uptake of DOC (mg·DOC/(kg·carbon))
Δt_i Time interval between two sampling dates (d)
DOC^i_{in} Influent DOC at time i (mg/L)
DOC^i_{out} Effluent DOC at time i (mg/L)

INTRODUCTION

Advanced water treatment using activated carbon has been applied worldwide because of the high efficiency and economic performance, such as carbon adsorption (Li *et al.* 2012; Gong *et al.* 2013), ozone-biological activated carbon (O_3-BAC) (Gao *et al.* 2010) and bio-enhanced activated carbon (BEAC) (Zhang *et al.* 2011; Al-Amrani *et al.* 2012). Bio-enhancement in BEAC process ensured higher purification efficiency and service-life (Piai *et al.* 2020). The use of BEAC/BAC for removing natural organic matters and toxic xenobiotics is advisable as the final stage of water purification (Smolin *et al.* 2020). However, tailoring or selecting excellent carbon for BEAC process remains a challenge. The method for the assessment of the contribution of adsorption and bio-degradation in dynamic BEAC/BAC process is also extremely complex (Piai *et al.* 2020; Smolin *et al.* 2020; Ti *et al.* 2020; Zhiteneva *et al.* 2020).

High adsorption capacities for natural organic matter (NOM) and trace organic-pollutants (TrOPs) require synchronously well-developed micro- and meso-porosity distribution of carbons (Gong *et al.* 2020). Studies suggested that meso-pores (2–50 nm) and secondary micro-pores (1–2 nm) principally controlled adsorption of NOM from water (Velten *et al.* 2011). NOM suffered severely competitive adsorption with trace organic-pollutants (TrOPs) on carbon surfaces with primarily small micro-pores (<1.0 nm) (Quinlivan *et al.* 2005). On the other hand, synchronous adsorption and bio-degradation is the core mechanism of organic-pollutants removal in BEAC process (Velten *et al.* 2011; Gibert *et al.* 2012). And it can be significantly affected by pore structure distribution of meso- and macros-pores (Aktaş & Çeçen 2007).

Therefore, tailoring an efficient carbon with a perfect matching between adsorption and bio-degradation (both biomass and bio-activity levels on carbon surfaces) was a significantly important, and pore structure regulation is the promising way to improve adsorption and bio-degradation efficiency simultaneously (Sulaymon *et al.* 2010). Studies suggested that carbons with simultaneously well-developed meso- and micro-porous structures may achieve better matching between adsorption and bio-degradation (Singh *et al.* 2012). However, micro-porous carbons were still commonly employed in drinking water treatment plants (DWTPs) due to the industrialization restriction in China. Research and development of new-type mesoporous carbons was financially supported by the Ministry of Housing and Urban-Rural Development of China in the present work. A series of innovative mesoporous carbons were successfully prepared. However, it is still a challenging work to quantify the effects of pore structure regulation on synergy-effect between bio-degradation and adsorption. Therefore, quantification of adsorption and bio-degradation on new carbons in BEAC/BAC process was also conducted in present work.

MATERIALS AND METHODS

Activated carbon preparation

Three new carbons were prepared by multi-step procedure in Figure 1, including precursor-blending, re-agglomeration, potassium hydroxide (KOH) impregnation, carbonization and multistage depth-activation (MDA) (Li *et al.* 2014; Gong *et al.* 2015a, 2015b). Take carbon MCGL-4 as an example, the detailed preparation parameters were described as the following: coal-blending (35% Taixi coal +30% Shen Fu coal) and coconut shell (30%) were first mixed and crushed to 100–200 mesh. The mixture was agglomerated using refined tar (5%) as agglomerate. The agglomerated material was crushed and screened again to 2–10 mm. Impregnation was conducted before carbonization using 10% KOH (100 rpm, 60 min). Carbonization was consequently conducted at 600 °C in N_2 protection environment for 30 min. MDA was conducted in

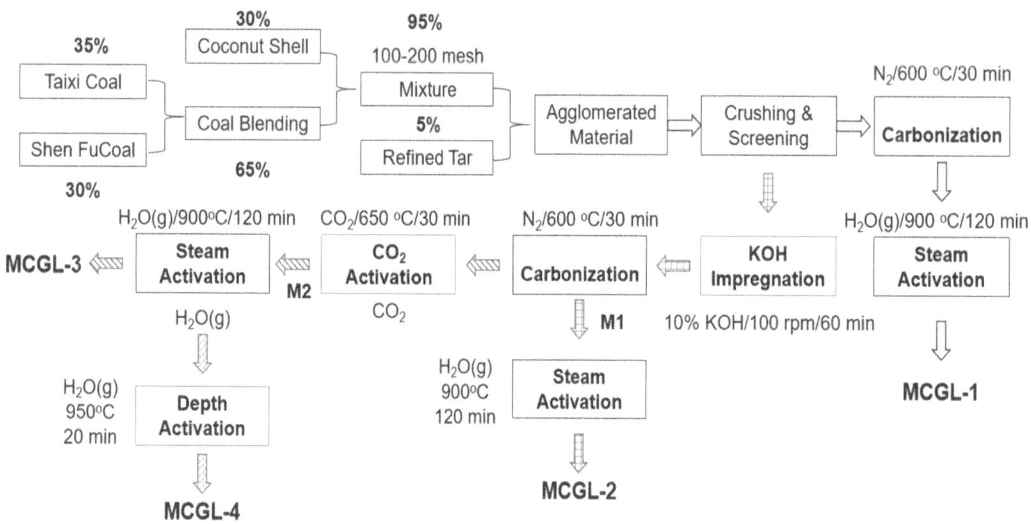

Figure 1 | Preparation procedure of new carbons.

activation apparatus, and MDA can be divided into three stages: CO_2 activation (650 °C/30 min), steam activation (900 °C/120 min) and depth-steam activation (950 °C/20 min).

Activated carbon characterization

Nitrogen isotherms were measured by ASAP2020 Sorptometer at 77 K and the specific surface area (S_{BET}) was calculated by Brunauer–Emmett–Teller (BET) model (Praetorius & Voigt 2015; Spiesshofer *et al.* 2015; Wang 2016). Pore size distribution/pore volume distribution (PSDs/PVDs) were calculated by *t*-plot method (Gong *et al.* 2013). Especially, micro-porous volume (V_{micro}) was also identified by Horvath-Kawazoe (H-K) model. Mesoporous volume (BJH-V_{mes}) was calculated by Barrett-Joyner-Halenda (BJH) model (Choi *et al.* 2015; Haluszka *et al.* 2015). Adsorption capacities for iodine (V_{IN}), methylene blue (V_{MB}) and humic acid(V_{HA}) were determined according to the National Standards of China (GB/T 7702.7-2008, 7702.6-2008 and 7701.2-2008).

Commercial carbons

Three kinds of commercial carbons in Table 1 were also employed as control group, which were commonly used in a water purification plant in China.

Table 1 | Parameters of commercial carbons employed in present work

		Commercial carbons		
Parameters	**Unit**	**SX-10**	**F400**	**ZJ15**
Specific surface area (S_{BET})	m^2/g	921	1,169	924
External surface area (S_{Ext})	m^2/g	256.31	–	301.39
Microporous surface area (S_{Micro})	m^2/g	556.09	939.5	519.17
Total pore volume (V_{total})	cm^3/g	0.6536	0.8963	0.3714
Microporous volume (V_{micro})	cm^3/g	0.2390	0.4716	0.2541
Mesoporous volume(V_{mes})	cm^3/g	0.2927	0.3170	0.042
Macroporous volume (V_{mar})	cm^3/g	0.1219	0.1077	0.0753
Average pore diameter (D)	Å	26.98	28.94	18.26

Experimental apparatus

Testing apparatus of functional bacteria immobilization

As shown in Figure 2(a), carbons (300 mL) were fixed into conical bottles after steam sterilization, and the empty bed contact time (EBCT) was 30 min. Functional bacteria were enriched from steady BEAC process, mainly contains *Acinetobacter Harbinensis sp.*, three strains of *Pseudomonas sp.* and *Bacillus subtilis.* After fermenting, functional bacteria (10^9 CFU/L) were circularly immobilized onto carbons (24 h). Biomass (phospholipid contents) on different carbons was determined at different times.

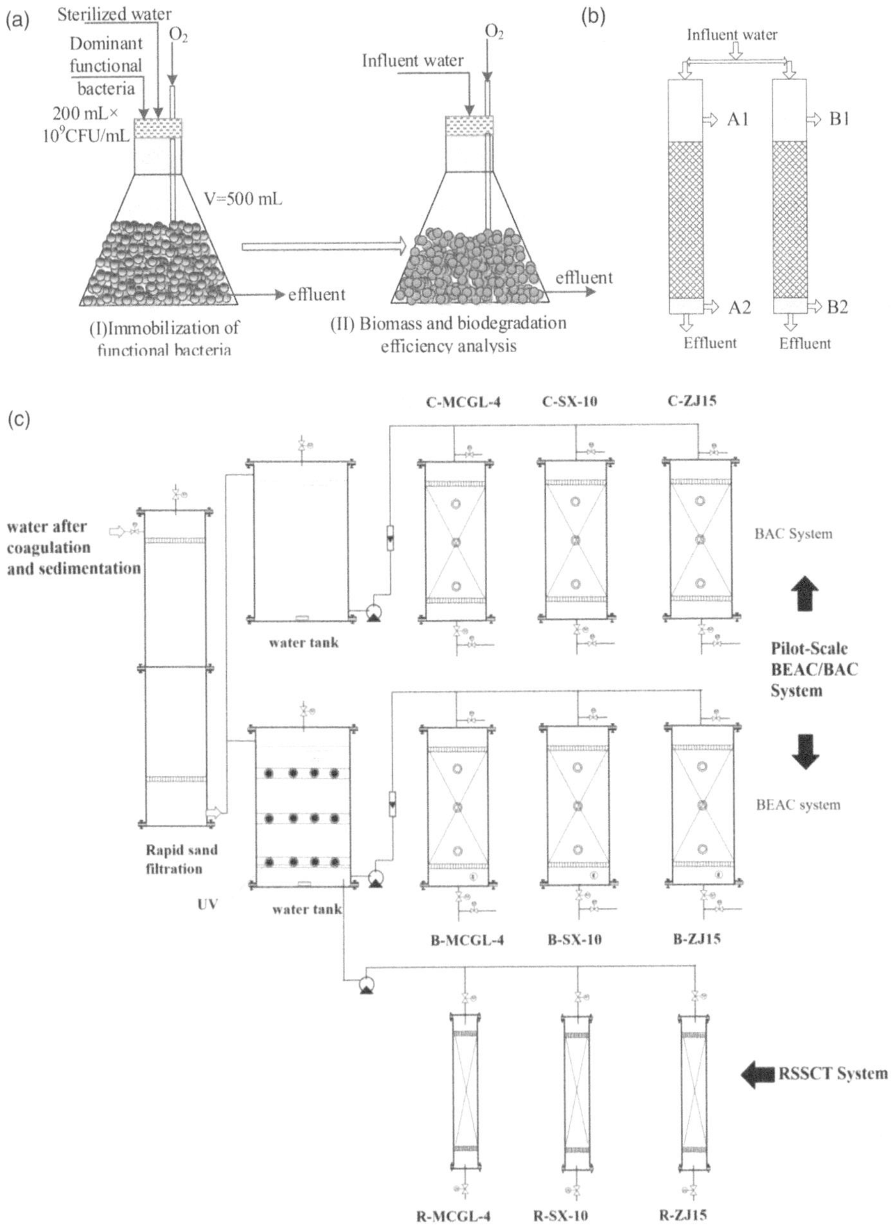

Figure 2 | Experimental set-up. (a) Evaluation set-up for initially immobilized biomass, (b) evaluation set-up for dissolved oxygen utilization, (c) pilot-scale BEAC/BAC system and RSSCT tests system.

Utilization rates of dissolved oxygen (DO) during biodegradation by different carbons were determined using apparatus in Figure 2(b) in both adsorption system (column A) and bio-enhancement system (column B with functional bacteria immobilization). The influent raw water was sterilized by ultraviolet and fully aerated. Dissolved organic carbon (DOC) and NH_4^+-N were adjusted to 4.0 ± 0.25 mg/L and 0.50 ± 0.16 mg/L, respectively, using sodium acetate and ammonium chloride. After stable operation for a week, changes of DOC, DO and NH_4^+-N were determined for 5 days, using HQ30D portable DO analyzer and TOC analyzer (TOC-VCPH).

Herein, bio-degradation can be ignored in column A and the DOC and NH_4^+-N removal can be attributed to adsorption, rather than bio-degradation. And DO consumption in column A can be defined as ΔDO_{CS} (surface affinity for DO by adsorption process). Therefore, utilization efficiency of DO which was caused by surface bio-degradation in column B can be calculated using as following Equations (1)–(8):

No.	Equations	Units	Annotation
(1)	$\Delta DO_{CS} = DO_{A1} - DO_{A2}$	mg/L	Surface affinity for DO by adsorption in column A
(2)	$\Delta DOC_{Bio} = \Delta DOC_B - \Delta DOC_A$	mg/L	DOC removal by bio-degradation in column B
(3)	$\Delta N_{Bio} = \Delta(NH_4^+ - N)_B - \Delta(NH_4^+ - N)_A$	mg/L	NH_4^+-N removal by bio-degradation in column B
(4)	$\Delta DO_C = 2.66\text{mg} \cdot O_2/(\text{mg} \cdot DOC) \times \Delta DOC_{Bio}$	mg/L	Theoretical amount of DO required for DOC degradation
(5)	$\Delta DO_N = 4.33\text{mg} \cdot O_2/(\text{mg} \cdot NH_4^+ - N) \times \Delta N_{Bio}$	mg/L	Theoretical amount of DO required for NH_4^+-N degradation
(6)	$\Delta DO_{TH} = \Delta DO_C + \Delta DO_N + \Delta DO_{CS}$	mg/L	Theoretical consumption of DO in column B
(7)	$\Delta DO_R = DO_{B1} - DO_{B2}$	mg/L	Actual consumption of DO in column B
(8)	$PADO = \frac{\Delta DO_C + \Delta DO_N}{\Delta DO_R} \times 100\%$	%	Actual effective utilization rate of DO by bio-degradation

Pilot-scale column testing system (BAC/BEAC process)

As shown in Figure 2(c), pilot-scale BAC/BEAC system (1.0 $m^3 \cdot h^{-1}$) and rapid small-scale column tests system (RSSCTs) were established, treating water from Songhua River. Influent water was pre-treated by coagulation, inclined-tube sedimentation and rapid sand filtration. Pilot-scale BEAC system contains three parallel columns (φ50 mm × 2.10 m), packed with 1.20 m carbon bed and operated in up-flow mode (184 L/d). After enrichment and domestication, cyclic immobilization of the dominant functional in BEAC columns was conducted according to the methods described by Zhang (Zhang *et al.* 2011). Ultraviolet sterilization (100 $W \cdot m^{-2}$) was employed in raw water tank to minimize the bacteria interference. Difference between BAC and BEAC system was the formation of bio-degradation. Bio-degradation formed naturally during operation. Therefore, there was no ultraviolet sterilization in the raw tank. Operation parameters and influent water qualities of BEAC and BAC system are summarized in Table 2.

Rapid small-scale column tests (RSSCTs) system

RSSCTs system was established according to the dispersed flow pore surface diffusion model using EBCT and hydraulic loading to describe the adsorption process (Greenwald *et al.* 2015). RSSCTs can obtain breakthrough curves in a fraction of the time with a fraction of water (Sperlich *et al.* 2005). It was consequently employed to quickly quantify adsorption capacities and parameters are summarized in Table 2. Carbon samples were pre-treated as follows: (1) carbons were thoroughly washed with deionized water until all dust was removed and the pH of the wash water was steady; (2) well-washed carbons were dried at 150 °C for 4 h; (3) carbon water slurry was heated to boiling for 30 min and loaded into the column after cooling to room temperature. The feed water qualities were listed in Table 2.

Sampling and analysis

Removal efficiencies of DOC and UV_{254} were determined in pilot-scale and RSSCT system. Three-dimensional

Table 2 | Operational parameters and feed water qualities of PCT and RSSCT system

Parameters	Pilot system	RSSCT system
Granule size – (mm)	0.60–2.50 (1.50)	0.21
Empty bed contact time (EBCT) (min)	20	0.40
Flow rate (m^3/d)	0.185	0.064
Filtering velocity (m/h)	3.9	27.86
Height (m)	2.10	0.35
Carbon bed depth (m)	1.20	0.24
Column diameter (cm)	5	1.1
Temperature (°C)	15.45 ± 7.93	17.10 ± 3.59
DOC(mg/L)	3.28 ± 0.47	3.20 ± 0.15
UV_{254} (cm^{-1})	0.078 ± 0.015	0.074 ± 0.012
Turbidity (NTU)	0.38 ± 0.16	0.35 ± 0.11

fluorescence spectra (3D-EEM) of water samples were scanned with 5 nm increments by a fluorescence spectrophotometer (FP-6500) (Gong *et al.* 2013). Gas chromatography-mass spectrometry analysis (GC-MS) was performed using chromatography-mass spectrometer (6890GC-5973/5975MSD).

Cumulative uptake of micro-pollutants

The concentration levels of micro-pollutants (DOC) were reported as functions of the operational time and water volume fed to carbon columns divided by mass of GAC (*KBV*). The *KBV* can be described by the following equation:

$$KBV = \frac{V_{\text{Treated}}}{M_{AC}} = \frac{N \cdot Q_{\text{in}}}{1000 \cdot M_{AC}} \tag{9}$$

The *KBV* of pilot-scale system reached up to 39.50 $m^3 \cdot H_2O/(kg{\cdot}carbon)$ and it was equivalent to 246 days of operation. Performance of micro-pollutants removal from water by carbons in BEAC/BAC and RSSCTs system was also represented by the cumulative uptake DOC (*QC*) as a function of *KBV*.

$$QC = \frac{Q_{\text{in}}}{2M_{AC}} \cdot \sum_{i=1}^{n} \Delta t_i \cdot \begin{bmatrix} (\text{DOC}_{\text{Mn}}^{\text{in},i-1} + \text{DOC}_{\text{Mn}}^{\text{in},i}) \\ -(\text{DOC}_{\text{Mn}}^{\text{out},i-1} + \text{DOC}_{\text{Mn}}^{\text{out},i}) \end{bmatrix} \tag{10}$$

RESULTS

Pore structure distribution

Results of N_2 adsorption/desorption isotherms suggested that MCGL-4/3 obtained remarkable type-IV isotherms with capacities of 815/733 cm^3/g. However, N_2 adsorption capacities achieved by MCGL-2/1 were under 520 cm^3/g. Physico-chemical properties based on N_2 adsorption/desorption isotherms are summarized in Figure 3 and Table 3. As shown in Table 3, total porous volumes (V_{total}) of MCGL-4 and MCGL-3 reached to 1.1688 and 1.0495 cm^3/g, respectively, higher than MCGL-2 (0.8029 cm^3/g) and MCGL-1

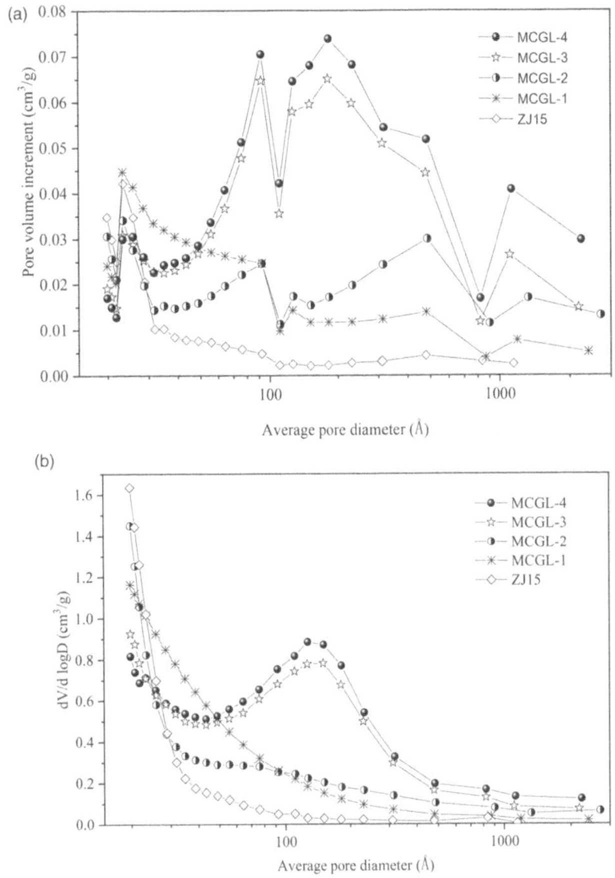

Figure 3 | Physico-chemical properties of new carbons. (a) Distribution of pore size and pore volumes, (b) dV/dlog*D*.

(0.7041 cm^3/g). Mesoporous volume (V_{mes}) of MCGL-4 increased to 0.7605 cm^3/g with highest mesoporosity ($V_{\text{mes}}/V_{\text{total}} = 65.11\%$), followed by MCGL-3 (0.6740 cm^3/g, 64.22%). Figure 3(b) further indicates that MCGL-4 had higher pore volume increment (>50 Å) with significant changes in peak value during 50~300 Å. V_{mes} of MCGL-4 was mainly contributed by pore structures during 20.4~52.5 Å (0.2405 cm^3/g), 52.5~208.2 Å (0.4443 cm^3/g) and 208.2~408 Å (0.1225 cm^3/g). MCGL-3 achieved highest micro-porous volume ($V_{\text{micro}} = 0.2900$ cm^3/g) and micro-porous surface area ($S_{\text{Micro}} = 638.42$ m^2/g), followed by MCGL-4 ($V_{\text{micro}} = 0.2655$ cm^3/g and $S_{\text{Micro}} = 588$ m^2/g).

Results also suggested that MCGL-3 obtained highest iodine adsorption value (1,185 ± 5.6 mg/g), followed by MCGL-2 (1,062 ± 8.9 mg/g), MCGL-4 (1,110 ± 6.8 mg/g) and MCGL-1 (1,127 ± 7.4 mg/g). However, MCGL-4

Table 3 | Physico-chemical properties of new carbons

Index	Unit	Activated carbons			
		MCGL-4	MCGL-3	MCGL-2	MCGL-1
S_{BET}	m^2/g	1,265	1,316	1,183	1,090
S_{Ext}	m^2/g	655.41	677.99	705.91	744.94
S_{Micro}	m^2/g	588	638.42	477.51	363.62
V_{total}	cm^3/g	1.1688	1.0495	0.8029	0.7041
V_{micro}	cm^3/g	0.2655	0.2900	0.2368	0.1594
V_{mes}	cm^3/g	0.7605	0.6740	0.3986	0.3502
V_{mar}	cm^3/g	0.1629	0.0855	0.1675	0.1945
D	Å	36.84	32.93	26.12	25.85
Pore volume distributions					
20.4~52.5Å	cm^3/g	0.2405	0.2364	0.2035	0.1648
52.5~208.2Å	cm^3/g	0.4443	0.3979	0.1748	0.1200
208.2~408Å	cm^3/g	0.1225	0.1105	0.0440	0.0335
408~943Å	cm^3/g	0.0685	0.0561	0.0413	0.0224
943~3,000Å	cm^3/g	0.0705	0.0412	0.0300	0.0188

achieved highest properties on methylene blue (251 ± 3.2 mg/g), higher than MCGL-3 (234 ± 4.1 mg/g), MCGL-2 (205 ± 4.3 mg/g) and MCGL-1 (196 ± 1.4 mg/g).

Adsorption capacities of NOM

Adsorption property of humic acid (V_{HA}) was employed to evaluate NOM adsorption. Results suggested that V_{HA} of commercial carbons SX-10, F400 and ZJ15 were 0.626, 0.797 and 0.303 mg/g, respectively. After pore structures regulation, values of V_{HA} of MCGL-4 and MCGL-3 increased to 1.033 and 0.908 mg/g, respectively, higher than MCGL-2 (0.562 mg/g) and MCGL-1 (0.692 mg/g). It also identified that V_{HA} has fine linear correlation with the V_{mes} ($R^2 = 0.9541$) and V_{total} ($R^2 = 0.9293$) based on 7 carbons used in present work. And the corresponding pore volume (20.4 ~ 208.2 Å) has the highest linear correlation coefficient with V_{HA} ($R^2 = 0.9830$), followed by pores during 208.2 ~ 408 Å ($R^2 = 0.9741$), 52.5 ~ 208.2 Å ($R^2 = 0.9617$) and 20.4 ~ 52.5 Å ($R^2 = 0.8827$). Therefore, increase of mesoporous volume (MCGL-3 and MCGL-4), especially during the range of 20.4 ~ 208.2 Å, may well facilitate the adsorption capacities for NOM. Molecular weight distribution of NOM can influence its diffusion rate and adsorbability. Ultrafiltration membrane classification of molecular weight of humic acid was consequently conducted. And relationships between pore distributions and adsorption capacities of humic acid with different ranges of molecular weight are illustrated in Figure S1 (supplementary materials).

As shown in Figure S1(a)–(c), pore volumes during 52.5 to 408 Å was well correlated with adsorption capacity of humic acid ($R^2 > 0.96$), when average molecular weight were below 1,600 *Da*. As mentioned above, pore volumes during 52.5~208.2 Å were much more higher than 208.2~408 Å. Therefore, mesoporous volume during 52.5~208.2 Å may mainly provide surface adsorption site. When average molecular weight increased to 2,700~5,100 *Da* (Figure S1(d)–(f), adsorption capacities also increased with the increase of mesoporous volume during 52.5~208.2 Å. However, the linear correlation was not significant. Well linear correlation ($R^2 > 0.96$) was observed during 408~903 Å, it suggested macroporous structure may also play an important role.

Immobilization properties for functional bacteria

Another goal of present work is to improve the properties of bio-enhancement on carbon surface. Studies suggested that initially immobilized biomass of functional bacteria in BEAC process can significantly shorten the biofilm formation, and consequently improve bio-degradation of organic-pollutants. As shown in Figure S2(a), biomass (phospholipid content) on carbons increased over time of cyclic loading. After 24 hours, MCGL-4 obtained 9.13 mmol-P/g of phospholipid content, higher than other 6 carbons. After surface back-washing (5 minutes), phospholipid content on MCGL-4 decreased to 8.06 mmol-P/g. However, it was still higher than other carbons. It indicated that immobilized biomass and stabilities on MCGL-4 were significantly promoted. As shown in Figure S2(b), after 240 hours of continuous culture, MCGL-4 achieved higher phospholipid content (30.36 mmol-P/g), with average growth rate of 2.213 mmol-P/g/d.

Water purification performance of MCGL-4

As mentioned above, it suggested that new carbon MCGL-4 has advantages in NOM adsorption, immobilization capability and biodegradability in BEAC process. Therefore,

MCGL-4 was selected as research target in subsequent studies treating source water from Songhua River.

Changings of DOC

Purification performances of pilot-scale systems were determined under 246 days of operation (KBV = 39.50 $m^3 \cdot H_2O/(kg \cdot carbon)$). Changing curves of DOC in BEAC/BAC system are shown in Figure 4(a).

In pilot-scale BAC system (Figure 4(a)), C-MCGL-4 achieved lower average effluent DOC (1.32 ± 0.70 mg/L) than C-SX-10 (1.82 ± 0.79 mg/L) and C-ZJ15 (2.50 ± 0.72 mg/L). Effluent DOC of C-ZJ15 exceed 3.0 mg/L during KBV ranges of 19.2–39.50 $m^3 \cdot H_2O/(kg \cdot carbon)$. However, it remained stable lower level in C-MCGL-4 (2.03 ± 0.15 mg/L) and C-SX-10 (2.58 ± 0.25 mg/L).

In the pilot-scale BEAC system, it indicated that B-MCGL-4 achieved the highest DOC removal efficiency (76% ± 15%) with lowest average value (0.98 ± 0.56 mg/L). Effluent DOC reached a steady state (1.57 ± 0.15 mg/L) after 120 days of operation (19–39.50 $m^3 \cdot H_2O/(kg \cdot carbon)$). However, immobilization did not significantly facilitate the DOC removal in B-ZJ15 (2.34 ± 0.65 mg/L). Removal efficiency of DOC achieved by B-SX-10 increased 12% in comparison with C-SX-10.

3D-EEM And GC-MS spectrometry

The excitation (EX) and emission (EM) boundaries of 3D-EEM were classified into five regions by Chen and Liu *et al.* (Chen *et al.* 2003; Liu *et al.* 2015). Regions-I and -II (EX < 250 nm, EM < 380 nm) associates with aromatic proteins, Region-III (EX < 250 nm, EM > 380 nm) represents fulvic acid-like compounds, Region-IV (EX > 250 nm, EM < 380 nm) represents soluble microbial by-product-like compounds, and Region V (EX > 250 nm, EM > 380 nm) represents humic-like organic compounds. Results of 3D-EEM suggested that the major organic compounds in source water were humic acid-like organics, with two prominent fluorescence peaks (EX/EM = 263 nm/438 nm, EX/EM = 308 nm/423 nm) in Region V. After sand filter purification, the fluorescence intensity decreased slightly.

Figure 4(b) and 4(c) illustrate the average fluorescence intensity during KBV ranges of 5~30 $m^3 \cdot H_2O/(kg \cdot carbon)$.

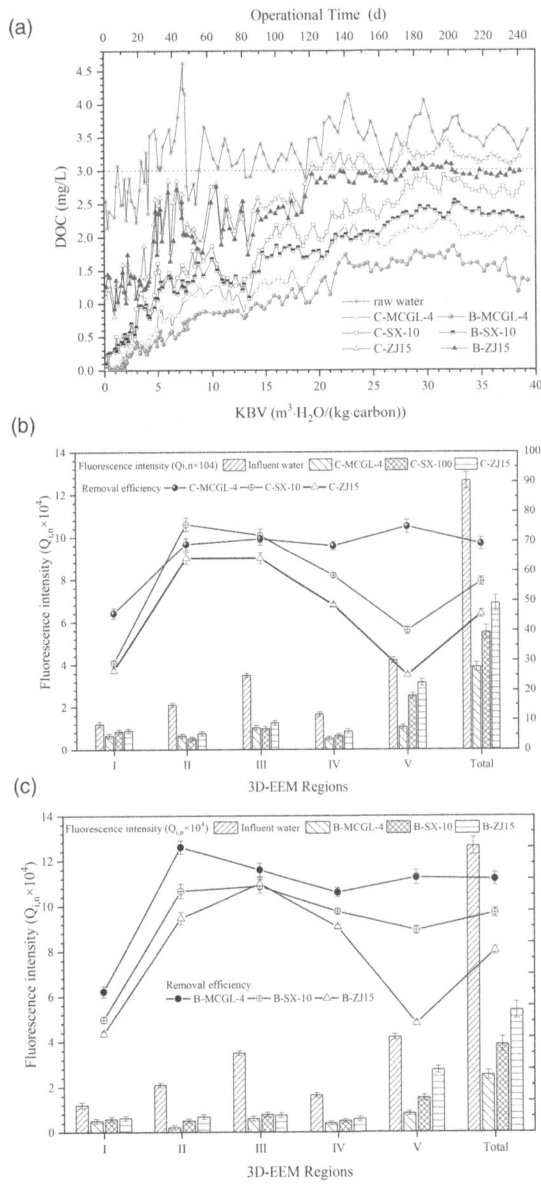

Figure 4 | Water purification performances of carbons in pilot-scale BAC and BEAC system. (a) Changing curves of DOC, (b) change of fluorescence intensity in BAC system, (c) change of fluorescence intensity in BEAC system.

The total intensity of influent water of pilot-scale system was 1.27×10^5, with higher proportion in Region III (fulvic acid-like compounds, 3.5×10^4, 27.56%) and V (humic-like organic compounds, 4.2×10^4, 33.08%). Therefore, removal of fulvic acid- and humic acid-like organics were the core objectives. The total fluorescence intensities (removal efficacy) of C-MCGL-4、C-SX-10 and C-ZJ15 decreased to 3.89×10^4 (69.25%), 5.51×10^4 (56.44%) and 6.88×10^4 (45.62%), respectively. C-MCGL-4 removed 75% of humic acid-like organics (Region V), followed by 68.49% of fulvic acid (Region IV). It indicated that fulvic acid- and humic acid-like organics can be well removed by new carbon MCGL-4 in BAC system.

In pilot-scale BEAC system, the total fluorescence intensity decreased by 34.96% after bio-enhancement of B-MCGL-4, in comparison with C-MCGL-4. And descend ranges of fluorescence intensities achieved by B-MCGL-4 were 23.08% (Region I), 67.69%(II), 41.18%(III), 23.08%(IV) and 21.90%(V). Therefore, B-MCGL-4 achieved highest removal efficiency for aromatic proteins (Region II and I), followed by fulvic acid-like compounds (Region-III). Moreover, bio-enhancement also improved the removal of soluble microbial by-product-like compounds (Region IV) humic-like organic compounds (Region V) in B-MCGL-4 system.

GC-MS was also used to characterize the purification performances of carbons (at $20\,m^3 \cdot H_2O/(kg \cdot carbon)$). Results suggested that about 101 kinds of organic-pollutants were detected in influent water of pilot-scale system, mainly contains alkanes (30%), benzene compounds (19.7%), naphthalenes (11.1%), olefins (6.2%), and organic ester compounds (4.9%) (Wang *et al.* 2012, 2018). For C-MCGL-4, about 65 kinds of organics were detected, and peak abundance and area decreased sharply during the retention time ranges of 6–12 min and 17–30 min (mainly alkanes, benzene and naphthalene compounds). In BEAC system (B-MCGL-4), peak abundance and area decreased during retention time range of 5–20 min in comparison with C-MCGL-4. The amount of alkanes, benzene and naphthalene compounds decreased by 15.26%, 16.37% and 8.29%, respectively.

Biological proliferation on carbon surface in BEAC system

Biological proliferation on carbon surface in B-MCGL-4 was also determined. The biomass on B-MCGL-4, B-SX-10 and B-ZJ15 reached to 45.6, 34.5 and 28.6 mmol-P/gat 39.50 $m^3 \cdot H_2O/(kg \cdot carbon)$.

DISCUSSION

Pore structure regulation mechanisms

The normal distributions of pore structures of 47 kinds of commercial activated carbons (Figure S3) were summarized. New carbon MCGL-4 showed higher levels of V_{total} and V_{mes} than most of carbons, and it also has average level of V_{micro}. It identified that synchronous well-developed mesoporous and micro-porous structures are the most distinguishing characteristics of MCGL-4.

Catalytic carbonization or activation is a promising way to regulate pore structures, and KOH was usually used to promote condensation polymerization and aromatization, so as to form rich micro-porous structures. Martins (Martins *et al.* 2015) pointed that proper amount of KOH impregnation was beneficial for development of micro-porous structure, since KOH can promote the changings of pyrolysis, formatting of initial pore structures and crystallite graphite structures. KOH impregnation was introduced in preparation of MCGL-2, and its V_{total} ($0.8029\,cm^3/g$) and V_{micro} ($0.2368\,cm^3/g$) increased by 14.03% and 48.56%, respectively, compared to MCGL-1. Moreover, mesoporous volume of MCGL-2 ($0.3986\,cm^3/g$) increased by 13.82% in comparison with MCGL-1. That is to say, higher microprous volumes can be attributed to KOH impregnation process, by which mesoporous volumes increased during the collapse of micro-pores by steam activation.

CO_2 activation was employed after carbonization in MCGL-3. Compared to MCGL-2, V_{total} and V_{mes} of MCGL-3 increased by $0.2466\,cm^3/g$ and $0.2754\,cm^3/g$, respectively. Studies suggested that (Gong *et al.* 2009, 2013) activating agent and time have significant effects on surface physiochemical changes. Especially, surface activation using CO_2 as activator was proved to be more likely to form abundant micro-pores due to its linear molecular structure. The increase of micro-pores attributed by CO_2 activation may offer the potential mesoporous foundation during the subsequent activation. Moreover, pore structure distribution of the intermediate product (M1) after CO_2 activation was also determined. Results suggested that about $0.2024\,cm^3/g$ of microprous volume increased in comparison with MCGL-2. It also interprets the effects of CO_2 activation.

Depth activation at 950 °C was employed in MCGL-4, and it can cause higher ablation rate of carbons. Studies also suggested micro-porous carbon can be commonly obtained with ablation rate below 50%. Mesoporous or macro-porous carbons can be obtained when ablation rate was higher than 70%. Well-developed micro-porous and mesoporous structures may be obtained during ranges of ablation rate of 50–70%. Higher ablation rate of C-GL-4 (68.5% ± 2.30%) was confirmed because of the employment of MDA. In comparison with MCGL-3 (with ablation rate of 55.7% ± 1.80%), depth activation at 980 °C causes collapse of micro-pores, by which V_{total}, V_{mes} and V_{mar} of MCGL-4 increase by 0.1279 cm^3/g, 0.0789 cm^3/g and 0.0774 cm^3/g. However, V_{micro} of MCGL-4 decreased by 0.0284 cm^3/g because of the collapse of micro-pores during depth activation.

One objective of present work aims at improving adsorption kinetics and capacities for NOM by pore structure regulation. Results of 3D-EEM and V_{HA} identified that higher mesoporosity (65.11%) and BJH-V_{mes} (0.7605 cm^3/g) of MCGL-4 resulted in accelerating kinetics and capacity for humic-like organics (Kalkan *et al.* 2011; Lu *et al.* 2014; Gong *et al.* 2015a; Greenwald *et al.* 2015). Figure S4 suggested that adsorption of humic acids by MCGL-4/3 followed pseudo-second-order kinetics ($R^2 > 0.998$) when average molecular weight ranged from 2,700 to 5,100 *Da*. It indicated that a key step of velocity control was the adsorption reaction stage, rather than pore diffusion stage. However, adsorption process achieved by MCGL-2/1 followed Weber-Morris kinetic model ($R^2 > 0.998$), and it indicated that adsorption process was mainly influenced by the diffusion of adsorbents in the pores of the adsorbents. The adsorption kinetics results confirmed that mesoporous regulation during 20.4~408 Å promoted adsorption capacities and kinetics.

Comprehensive quantitative indicators of carbons

Selecting or tailoring an excellent carbon for bio-enhancement is a complex process, especially when the pilot-scale or production experimental data were scarce. Lots of factors can influence the carbon selection such as raw water quality, biomass, bio-activities and parameters of filtration, etc. Therefore, methods which can be used to prejudge properties of carbons based on physical and chemical characteristics are of extraordinary value. Under relatively steady operation mode, the key factors of BEAC process were biomass and bio-activities, especially the initial immobilization capacity of biomass (BI). It determined the extent and efficacy of bio-degradation in initial and stably BEAC process. In order to evaluate the BI by different carbons with different PSD/PVD distribution, 'Comprehensive Quantitative Indicators (*CQI*)' was proposed.

Figure 5(a) firstly calculated the linear correlation coefficients (R^2) between initially immobilized biomass (*BI*, 24 h) and parameters of seven carbons used in present work. Linear fitting results identified that orders of correlation coefficient ($R^2 > 0.800$) were as followed: $V_{mes}(0.9827) > V_{HA}(0.9692) > D(0.948) > V_{total}(0.9323) > V_{MB}(0.8896) >$ *oxygen content* $(0.8221) > V_{micro}(0.0011)$. It confirmed that functional bacteria immobilization was significantly influenced by mesoporous structures, adsorption capacities of NOM and surface oxygen content. While V_{micro} had little correlation, and it also explained the deficiency of micro-porous carbons.

An Ideal Model of Carbons for BEAC process (IMC_{BEAC}) was proposed in the present work. Parameters of IMC_{BEAC} were assigned as follows: $S_{BET} \geq 1{,}400$ m^2/g, $V_{total} \geq 1.90$ cm^3/g, $V_{micro} \geq 0.60$ cm^3/g, $V_{mes} \geq 1.00$ cm^3/g, $V_{mar} \geq 0.25$ cm^3/g, $D \geq 40$ Å, $V_{IN} \geq 1{,}400$ mg/g, $V_{MB} \geq 350$ mg/g, $V_{HA} \geq 2.0$ mg/g, *strength* ≥ 90% and density ≥ 500 g/L. Based on the IMC_{BEAC}, Comprehensive Quantitative Indicators (*CQI*) was proposed consequently to evaluate the initial immobilization capacity of biomass (*BI*) by different carbons. It was identified using Equation (11).

$$CQI = \sum_{i=1}^{n} CQI_i = \sum_{i=1}^{n} \left(W_i \times \frac{V_i}{V_{i-IMC}} \right) \quad (11)$$

Here, V_i was parameters of carbons used in present work (Tables 1 and 3); W_i was the linear correlation coefficients (R^2) in Figure 5(a), $V_{i\text{-}IMC}$ was parameters of IMC_{BEAC} as mentioned above.

As shown in Table 4, the calculated *CQI* of IMC_{BEAC} was 6.97, followed by MCGL-4(5.36)>MCGL-3(4.95)> MCGL-2(4.12)>F400(3.78) = MCGL-1(3.78)>SX-10(3.53)> ZJ15(2.51). Linear correlation analysis between *CQI*

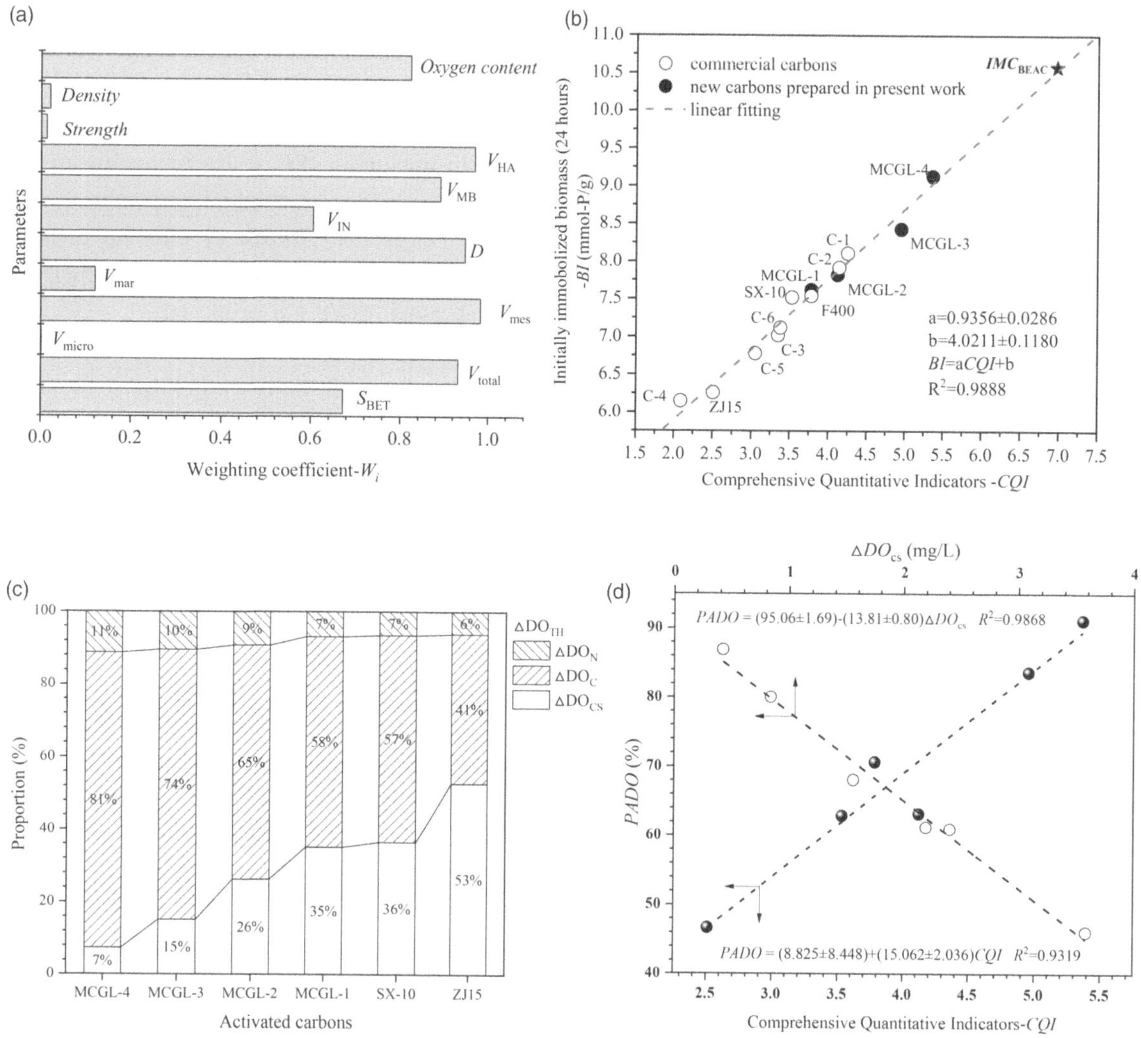

Figure 5 | Comprehensive quantitative Indicators. (a) W_i, (b) linear fitting of *CQI* and *BI*, (c) parameters of bioactivity, (d) linear fitting between *PADO* and *CQI*.

and *BI* was conducted in Figure 5(b), based on 12 tested carbons. Results suggested that initial immobilization capacity of biomass (*BI*) can be well reflected by *CQI* of carbons. According to equations ($BI = 0.9231CQI + 4.0649(R^2 = 0.9776)$) obtained from Figure 5(b), calculated *BI* of IMC_{BEAC} was 10.499 mmol-P/g. This phenomenon suggested a feasible method to rapidly evaluate the surface immobilization ability of functional bacteria. *CQI* proved to be a promising parameter ($R^2 = 0.9319$) to prejudge *BI* of different carbons. Results of pilot-scale BEAC and BAC system also identified that the selected new carbon MCGL-4 obtained higher efficacy for organics removal and bio-degradation efficiency.

DO consumption and utilization efficiency during bio-degradation

Yapsakli *et al.* (Yapsakli & Cecen 2010; Yapsakli *et al.* 2010) and Aktas & Cecen (Aktas & Cecen 2010) suggested that bio-degradation was closely related to DO consumption and utilization efficiency. Surface affinity for DO was affected by surface oxidation degree. The amount of DO which can be available for bio-degradation decreased with the increase in surface affinity for DO. Surface oxygen content was also proved to be an important factor that can affect the immobilization of functional bacteria ($R^2 = 0.8221$). Therefore, DO consumption and

Table 4 | Comprehensive quantitative indicators (CQI, dimensionless unit)

i	Parameters	W_i	Activated carbons IMC	MCGL-4	MCGL-3	MCGL-2	MCGL-1	SX-10	ZJ15	F400	C-1	C-2	C-3	C-4	C-5	C-6
1	Strength	0.0124	0.01	0.01	0.01	0.01	0.01	0.01	0.01	0.01	0.01	0.01	0.01	0.01	0.01	0.01
2	Density	0.0202	0.02	0.02	0.02	0.02	0.02	0.02	0.02	0.02	0.02	0.02	0.02	0.02	0.02	0.02
3	Oxygen content	0.8221	0.82	0.82	0.74	0.69	0.59	0.54	0.34	0.41	0.60	0.65	0.51	0.35	0.46	0.43
4	S_{BET}	0.6721	0.67	0.60	0.63	0.57	0.52	0.44	0.44	0.56	0.48	0.51	0.44	0.58	0.49	0.60
5	V_{total}	0.9323	0.93	0.58	0.51	0.39	0.35	0.32	0.18	0.44	0.41	0.36	0.27	0.24	0.33	0.27
6	V_{micro}	0.0011	–	–	–	–	–	–	–	–	–	–	–	–	–	–
7	V_{mes}	0.9827	0.98	0.74	0.66	0.39	0.34	0.29	0.04	0.31	0.39	0.34	0.20	0.09	0.15	0.22
8	V_{mar}	0.1207	0.12	0.08	0.04	0.08	0.09	0.06	0.04	0.05	0.03	0.04	0.03	0.01	0.04	0.04
9	D	0.948	0.95	0.90	0.78	0.62	0.61	0.64	0.43	0.69	0.79	0.75	0.66	0.24	0.46	0.58
10	V_{IN}	0.6043	0.60	0.48	0.51	0.49	0.46	0.42	0.41	0.48	0.49	0.44	0.41	0.48	0.47	0.44
11	V_{MB}	0.8896	0.89	0.64	0.59	0.52	0.50	0.48	0.44	0.47	0.61	0.64	0.50	0.02	0.39	0.51
12	V_{HA}	0.9692	0.97	0.50	0.44	0.34	0.27	0.30	0.15	0.34	0.42	0.37	0.29	0.04	0.23	0.27
CQI			6.97	5.36	4.95	3.78	4.12	3.53	2.51	3.78	4.25	4.14	3.35	2.09	3.05	3.38

utilization efficiency during bio-degradation were determined in Figure 5 and Table 5 according to methods described earlier.

As shown in Table 5, considerable differences were observed in surface affinity for DO during adsorption process (ΔDO_{CS}) in column-A. And it indicated that ΔDO_{CS} were negatively correlated with thermal ablation rate of carbons. Fitting results also showed that linear correlation coefficient between ΔDO_{CS} and surface oxygen content was 0.9725. The bio-degradation efficiency of DOC and NH_4^+-N in column B increased with the decrease of ΔDO_{CS}. The theoretical amounts of DO required in bio-degradation of DOC (ΔDO_C) and NH_4^+-N (ΔDO_N) in column B (MCGL-4) were 4.62 ± 0.12 mg/L and 0.64 ± 0.15 mg/L, respectively. The

Table 5 | Parameters relating to DO consumption and utilization during biodegradation

Parameters	unit	Activated carbons MCGL-4	MCGL-3	MCGL-2	MCGL-1	SX-10	ZJ15
DO	mg/L	7.76 ± 0.27	7.65 ± 0.34	7.82 ± 0.26	7.66 ± 0.31	7.69 ± 0.21	7.88 ± 0.13
ΔDO_{CS}	mg/L	0.42 ± 0.07	0.84 ± 0.05	1.56 ± 0.06	2.20 ± 0.06	2.41 ± 0.13	3.59 ± 0.06
ΔDOC_{Bio}	mg/L	1.74 ± 0.06	1.55 ± 0.09	1.45 ± 0.04	1.37 ± 0.12	1.42 ± 0.03	1.05 ± 0.04
ΔN_{Bio}	mg/L	0.15 ± 0.03	0.13 ± 0.03	0.12 ± 0.04	0.10 ± 0.05	0.10 ± 0.04	0.09 ± 0.03
ΔDO_C	mg/L	4.62 ± 0.10	4.13 ± 0.06	3.84 ± 0.11	3.64 ± 0.09	3.77 ± 0.08	2.80 ± 0.11
ΔDO_N	mg/L	0.64 ± 0.15	0.58 ± 0.12	0.55 ± 0.14	0.43 ± 0.06	0.43 ± 0.09	0.42 ± 0.13
ΔDO_{TH}	mg/L	5.69 ± 0.18	5.55 ± 0.13	5.95 ± 0.16	6.28 ± 0.11	6.62 ± 0.12	6.80 ± 0.12
ΔDO_R	mg/L	5.77 ± 0.17	5.62 ± 0.14	6.08 ± 0.17	6.46 ± 0.13	6.69 ± 0.17	6.89 ± 0.10
$PADO$	%	91.17 ± 0.59	83.67 ± 2.81	70.65 ± 2.28	63.16 ± 3.11	62.89 ± 2.39	46.66 ± 3.49
Ablation rate	%	68.5 ± 2.30	55.70 ± 1.80	49.30 ± 2.70	43.5 ± 3.20	40.15 ± 3.3	33.63 ± 1.8
Oxygen content	%	15.96	9.02	8.45	7.22	6.57	4.12

total theoretical consumption of DO in column B (MCGL-4) ($\Delta DO_{TH} = \Delta DO_C + \Delta DO_N + \Delta DO_{CS}$) was 5.69 ± 0.18 mg/L. As shown in Figure 5(c) (in column B, MCGL-4), percentage occupied by ΔDO_{CS}, ΔDO_C and ΔDO_N were 7, 81 and 11%, respectively. Actual effective utilization rate of DO (*PADO*) by bio-degradation achieved by MCGL-4 in column B reached to 91.17%. In comparison with MCGL-4, micro-porous carbon ZJ15, SX-10 and F400 obtained higher percentage which was occupied by ΔDO_{CS}, rather than bio-degradation of DOC and NH_4^+-N. Therefore, *PADO* was significantly influenced by ΔDO_{CS} ($R^2 = 0.9868$). Moreover, relationship between *PADO* and *CQI* was also determined in Figure 5(d). The fitting results showed that *CQI* was also a promising parameter ($R^2 = 0.9319$) to prejudge *PADO* (bio-activity) of different carbons.

Synergistic efficacy of synchronous adsorption-bio-degradation enhanced by pore regulation

Gibert (Gibert *et al.* 2012), Velten (Velten *et al.* 2011) and Aktaş & Çeçen (Aktaş & Çeçen 2007) pointed out that simultaneous adsorption and bio-degradation were the predominant factors contributing to organic-pollutants removal in BEAC process. The synergistic efficacy of adsorption and bio-degradation can be significantly affected by pore structure regulation (Wang *et al.* 2007; Sulaymon *et al.* 2010). However, it is difficult to quantify the relative contribution rate of the two mechanisms at different operational stages due to lack of methodologies (Oleinikova *et al.* 2018; Schupp *et al.* 2018; Li *et al.* 2019). Cumulative uptake of DOC (*QC*) and RSSCT tests were introduced in present work to quantify synergistic efficacy between adsorption and bio-degradation.

Cumulative uptake of DOC (*QC*)

The changing curves of *QC* calculated from data in Figure 4(a) are illustrated in Figure 6(a) and 6(b). It indicated that *QC* values of pilot-scale BEAC/BAC increased with the extent of operational time/*KBV*. In BAC system (Figure 6(a)), *QC* of C-MCGL-4 reached to 79,614.35 mg·DOC/(kg·carbon) when KBV was 39.50 $m^3 \cdot H_2O$/(kg·carbon) (246 days). While *QC* values of C-SX-10 and C-ZJ15 were 56,539.40 and

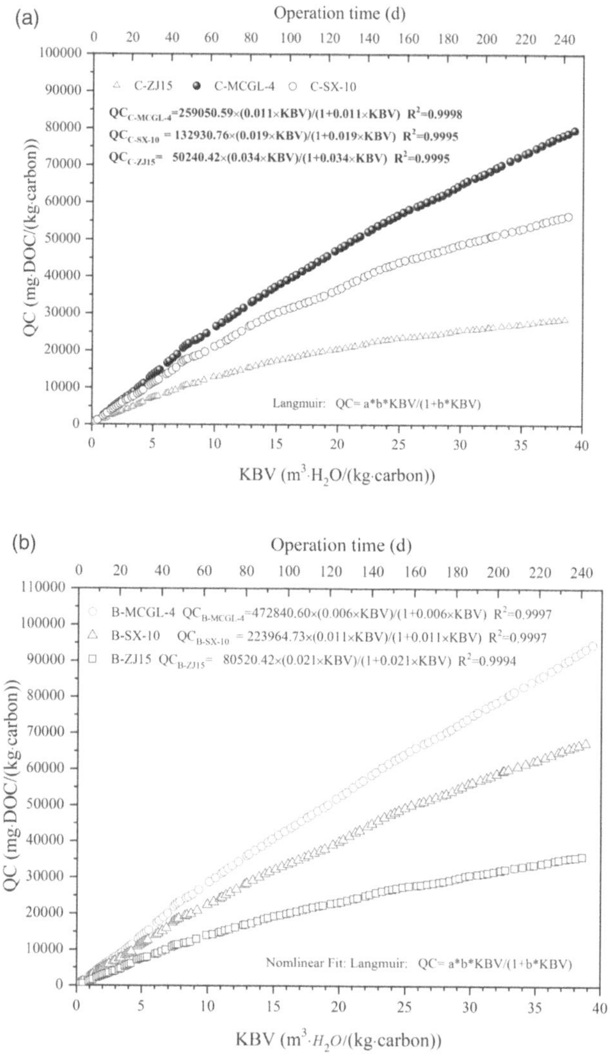

Figure 6 | Cumulative uptake of DOC and related parameters. (a) Pilot-scale BAC system; (b) Pilot-scale BEAC system.

28,584.34 mg·DOC/(kg·carbon), respectively. In BEAC system (Figure 6(b)), *QC* obviously increased from the initial stage due to immobilization. *QC* values of B-MCGL-4, B-SX-10 and B-ZJ15 increased to 94,655.50, 67,366.73 and 35,941.86 mg·DOC/(kg·carbon), respectively, at 39.50 $m^3 \cdot H_2O$/(kg·carbon). Well nonlinear fitting ($R^2 > 0.999$) in Figure 6(a) and 6(b) also suggested that *QC* obtained from BAC/BEAC system can be well fitted by nonlinear equation as a function of *KBV*.

Although it is difficult to directly quantify the relationship between adsorption and bio-degradation, it is clear that effects of bio-degradation caused by immobilization of functional bacteria in BEAC system can be first evaluated based on data from BAC system. Variable α and β were consequently employed as a function of KBV. Variable α is described by the following Equation (12):

$$\alpha_{C-\mathrm{i}} = \frac{QC_{B-\mathrm{i}}}{QC_{C-\mathrm{i}}} \tag{12}$$

Here, parameter i represents carbon MCGL-4, SX-10 or ZJ15. Results suggests $\alpha_{\text{C-MCGL-4}}$ was always higher than 1.0 even at the initial stage of operation; however, it required 3.78 $m^3 \cdot H_2O$/(kg·carbon) for $\alpha_{\text{C-SX-10}}$ and 3.09 $m^3 \cdot H_2O$/(kg·carbon) for $\alpha_{\text{C-ZJ15}}$, respectively. It indicated that simultaneous bio-degradation and adsorption occurred in B-MCGL-4 after bacteria immobilization, while B-SX-10 and B-ZJ15 still required a short operational time to achieve effective bio-degradation. The $\alpha_{\text{C-MCGL-4}}$, $\alpha_{\text{C-SX-10}}$ and $\alpha_{\text{C-ZJ15}}$ were 1.184, 1.192 and 1.257, respectively, at 39.50 $m^3 \cdot H_2O$/(kg·carbon). However, increment value $\Delta QC_{\text{MCGL-4}}$ $(QC_{B-MCGL-4} - QC_{C-MCGL-4})$ reached to 15,041.16 mg·DOC/(kg·carbon)), and it was obviously higher than $\Delta QC_{\text{SX-10}}(QC_{B-SX-10} - QC_{C-SX-10} =$ 10,827.33 mg·DOC/(kg·carbon)$^-$) and ΔQC_{ZJ15} $(QC_{B-ZJ15} - QC_{C-ZJ15} = 7{,}357.52$ mg·DOC/(kg·carbon)).

Synergy effect between bio-degradation and adsorption

RSSCTs have been widely employed as an established way to accurately predict removal of numerous trace organic chemicals via adsorption to GAC by utilizing pilot-scale columns and thus minimize removal via bio-degradation (Anumol *et al.* 2015). Therefore, cumulative uptake of DOC calculated from the RSSCTs system was compared to results from a BEAC/BAC filter operated in parallel to determine the relative contribution of biodegradation (illustrated in Figure 7(a)) (Zhiteneva *et al.* 2020).

Results suggested that RSSCT-calculated QC can be well nonlinearly represented by RSSCT-calculated KBV ($R^2 > 0.997$). R-MCGL-4 achieved highest capacity (67,725.32 mg·DOC/(kg·carbon)) at 61.40 $m^3 \cdot H_2O$/(kg·carbon), and 63,175.79 mg·DOC/(kg·carbon)) at 39.50 $m^3 \cdot H_2O$/(kg·carbon).

Under the assumption above, $\Delta RQC_{\text{B-MCGL-4}}$, $\Delta RQC_{\text{B-SX-10}}$ and $\Delta RQC_{\text{C-MCGL-4}}$ were defined in Equations (13)–(15).

$$\Delta RQC_{B-MCGL-4} = (QC_{B-MCGL-4} - QC_{R-MCGL-4})_{39.5KBV} \tag{13}$$

$$\Delta RQC_{B-SX-10} = (QC_{B-SX-10} - QC_{R-SX-10})_{39.5KBV} \tag{14}$$

$$\Delta RQC_{C-MCGL-4} = (QC_{C-MCGL-4} - QC_{R-MCGL-4})_{39.5KBV} \tag{15}$$

When RSSCT-calculated KBV were 39.50 $m^3 \cdot H_2O$/(kg·carbon)), the calculated $\Delta RQC_{\text{B-MCGL-4}}$, $\Delta RQC_{\text{B-SX-10}}$

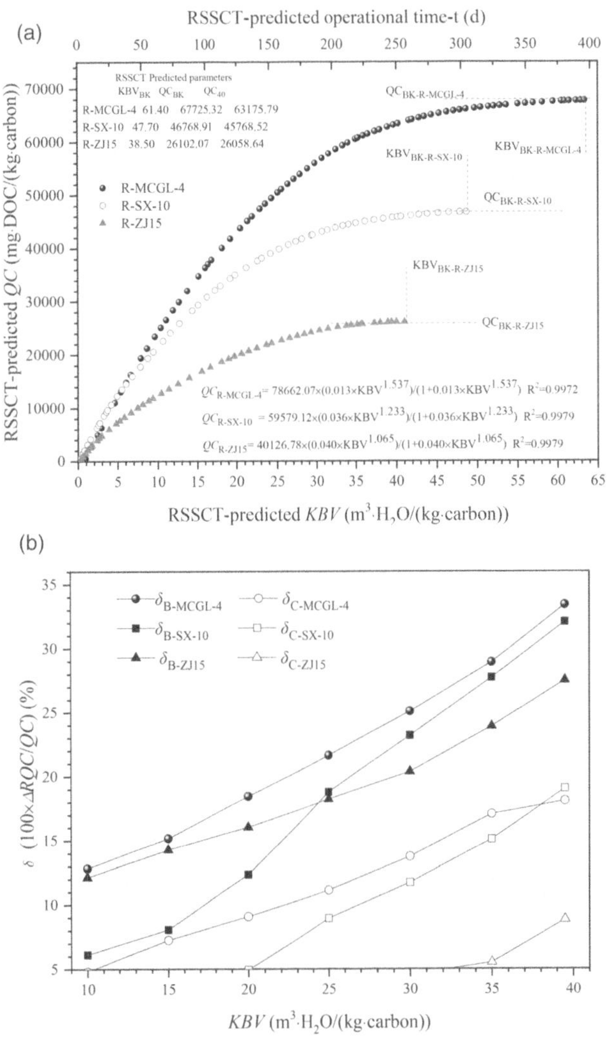

Figure 7 | Variation of the RSSCT-predicted parameters. (a) Variation of RSSCT-predicted QC, (b) variation of δ.

and $\Delta RQC_{C\text{-}MCGL\text{-}4}$ were 31,674.72, 21,598.21 and 16,963.54 mg·DOC/(kg·carbon), respectively. It identified that B-MCGL-4 achieved highest bio-degradation abilities. Parameter δ ($100^*\Delta RQC/QC$) can be consequently used to represent the contributions of bio-degradation in either BEAC or BAC process for DOC removal.

$$\delta_{B-i}(\%) = 100 \times \frac{\Delta RQC_{B-i}}{QC_{B-i}} \quad (i \text{ was MCGL-4, SX-10 or ZJ15}) \quad \delta_{C-i}(\%) = 100 \times \frac{\Delta RQC_{C-i}}{QC_{C-i}} \tag{16}$$

As shown in Figure 7(b), $\delta_{B\text{-}MCGL\text{-}4}$ was always higher than $\delta_{B\text{-}SX\text{-}10}$ and $\delta_{B\text{-}ZJ15}$ in BEAC system. The $\delta_{B\text{-}MCGL\text{-}4}$ increased from 12.83% to 33.39% with the extent of KBV, while $\delta_{B\text{-}SX\text{-}10}$ and $\delta_{B\text{-}ZJ15}$ ranged during 6.13~32.06% and 12.14~27.49%, respectively. Moreover, $\delta_{B\text{-}SX\text{-}10}$ was lower than $\delta_{B\text{-}ZJ15}$ during KBV ranges of 10–24 $m^3 \cdot H_2O$/(kg·carbon), it may be attributed to higher adsorption efficiency for DOC by carbon SX-10. However, biological activities were slight ($\delta < 5\%$) in C-SX-10 (KBV < 20 $m^3 \cdot H_2O$/(kg·carbon)) and C-ZJ15 (KBV < 33 $m^3 \cdot H_2O$/(kg·carbon)). δ_c obtained from C-MCGL-4, C-SX-10 and C-ZJ15 increased to 18.08%, 17.05% and 8.84%, respectively, at 39.50 $m^3 \cdot H_2O$/(kg·carbon). Therefore, bio-degradation in B-MCGL-4 accounted for the highest percentages in DOC removal than B-SX-10, B-ZJ15 and BAC process.

CONCLUSIONS

New carbon MCGL-4, tailored by KOH impregnation and MDA, obtained synchronously well-developed meso- (0.7605 cm^3/g), micro- (0.2655 cm^3/g) and macro-porous (0.143 cm^3/g) structures. Mesoporous regulation of MCGL-4 resulted in the promotion of both NOM adsorption and initially immobilized biomass/bio-activity. Mesoporous structure during 20.4~208.2 Å played the most important role in the promotion of adsorption kinetics and capacity for humic-like organics, followed by 208.2~408 Å. Comprehensive Quantitative Indicators of Carbons (*CQI*) was proved to be a promising parameter to prejudge the initial immobilized biomass and bio-activity of different carbons. A new methodology is proposed to quantify the relative contribution rate of adsorption and bio-degradation based on cumulative uptake of DOC (*QC*) in RSSCTs tests system and pilot-scale BEAC system. *QC* achieved by B-MCGL-4 reached to 94,850.51 mg·DOC/(kg·carbon) at 39.50 $m^3 \cdot H_2O$/(kg·carbon). Based on the RSSCTs-calculated cumulative adsorption of DOC, bio-degradation abilities achieved by MCGL-4 in BEAC system reached up to 31,674.70 mg·DOC/(kg·carbon)).

ACKNOWLEDGEMENTS

This study was supported by the Natural Science Foundation of China (NSFC) (Funding No. 51708162, 51608149) and Young Innovative Talents Training Program of Heilongjiang Regular Undergraduate Institutions of Higher Learning (Funding No. UNPYSCT-2018131). The authors also thank Harbin Institute of Technology (HIT) and Shanxi Xinhua Chemical Co., Ltd, in the preparation phases of activated carbons.

COMPLIANCE WITH ETHICAL STANDARDS

The manuscript is the original work of authors and it has not been previously submitted to *Water Supply* or other journals for simultaneous consideration. The manuscript has not been published previously. This study is not split up into several parts to increase the number of submissions and submitted to various journals or to one journal over time. No data have been fabricated or manipulated (including images) to support conclusions. No data, text, or theories by others are presented as if they were the authors' own. All authors mutually agree for its submission to *Water Supply*. The publication is approved by all authors and tacitly or explicitly by the responsible authorities where the work was carried out.

DATA AVAILABILITY STATEMENT

All relevant data are included in the paper or its Supplementary Information.

REFERENCES

Aktaş, Ö. & Çeçen, F. 2007 Bioregeneration of activated carbon: a review. *Int. Biodeterior. Biodegrad.* **59** (4), 257–272.

Aktas, O. & Cecen, F. 2010 Adsorption and cometabolic bioregeneration in activated carbon treatment of 2-nitrophenol. *J. Hazard. Mater.* **177** (1–3), 956–961.

Al-Amrani, W. A., Lim, P. E., Seng, C. E. & Ngah, W. S. 2012 Bioregeneration of mono-amine modified silica and granular activated carbon loaded with Acid Orange 7 in batch system. *Bioresour. Technol.* **118**, 633–637.

Anumol, T., Sgroi, M., Park, M., Roccaro, P. & Snyder, S. A. 2015 Predicting trace organic compound breakthrough in granular activated carbon using fluorescence and UV absorbance as surrogates. *Water Res.* **76** (0), 76–87.

Chen, W., Westerhoff, P., Leenheer, J. A. & Booksh, K. 2003 Fluorescence excitation-emission matrix regional integration to quantify spectra for dissolved organic matter. *Environ. Sci. Technol.* **37** (24), 5701–5710.

Choi, S. W., Kim, S. & Jung, Y. 2015 Dynamic heterogeneity in crossover spin facilitated model of supercooled liquid and fractional Stokes-Einstein relation. *J. Chem. Phys.* **142** (24), 244506.

Gao, Y. N., Li, W. G., Zhang, D. Y. & Wang, G. Z. 2010 Bio-enhanced activated carbon filter with immobilized microorganisms for removing organic pollutants in the Songhua River. *Water Sci. Technol.* **62** (12), 2819–2828.

Gibert, O., Lefevre, B., Fernandez, M., Bernat, X., Paraira, M., Calderer, M. & Martinez-Llado, X. 2012 Characterising biofilm development on granular activated carbon used for drinking water production. *Water Res.* **47**, 1101–1110.

Gong, G.-z., Xie, Q., Zheng, Y.-f., Ye, S.-f. & Chen, Y.-f. 2009 Regulation of pore size distribution in coal-based activated carbon. *New Carbon Materials* **24** (2), 141–146.

Gong, X., Li, W., Wang, K. & Hu, J. 2013 Study of the adsorption of Cr(VI) by tannic acid immobilised powdered activated carbon from micro-polluted water in the presence of dissolved humic acid. *Bioresour. Technol.* **141**, 145–151.

Gong, X.-J., Li, W.-G., Wang, G.-Z., Zhang, D.-Y., Fan, W.-B. & Yin, Z.-D. 2015a Characterization and performance evaluation of an innovative mesoporous activated carbon used for drinking water purification in comparison with commercial carbons. *Environ. Sci. Pollut. Res.* **22** (17), 13291–13304.

Gong, X.-J., Li, W.-G., Zhang, D.-Y., Fan, W.-B. & Zhang, X.-R. 2015b Adsorption of arsenic from micro-polluted water by an innovative coal-based mesoporous activated carbon in the presence of co-existing ions. *Int. Biodeterior Biodegradation* **102**, 256–264.

Gong, X.-J., Li, Y.-S., Dong, Y.-Q. & Li, W.-G. 2020 Arsenic adsorption by innovative iron/calcium in-situ-impregnated mesoporous activated carbons from low-temperature water and effects of the presence of humic acids. *Chemosphere* **250**, 126275.

Greenwald, M. J., Redding, A. M. & Cannon, F. S. 2015 A rapid kinetic dye test to predict the adsorption of 2-methylisoborneol onto granular activated carbons and to identify the influence of pore volume distributions. *Water Res.* **68** (0), 784–792.

Haluszka, D., Lorincz, K., Molnar, G., Tamas, G., Kolonics, A., Szipocs, R., Karpati, S. & Wikonkal, N. M. 2015 In vivo second-harmonic generation and ex vivo coherent anti-stokes raman scattering microscopy to study the effect of obesity to fibroblast cell function using an Yb-fiber laser-based CARS extension unit. *Microsc. Res. Tech.* **78** (9), 823–830.

Kalkan, Ç., Yapsakli, K., Mertoglu, B., Tufan, D. & Saatci, A. 2011 Evaluation of biological activated carbon (BAC) process in wastewater treatment secondary effluent for reclamation purposes. *Desalination* **265** (1–3), 266–273.

Li, W., Gong, X., Li, X., Zhang, D. & Gong, H. 2012 Removal of Cr(VI) from low-temperature micro-polluted surface water by tannic acid immobilized powdered activated carbon. *Bioresour. Technol.* **113**, 106–113.

Li, W. G., Gong, X. J., Wang, K., Zhang, X. R. & Fan, W. B. 2014 Adsorption characteristics of arsenic from micro-polluted water by an innovative coal-based mesoporous activated carbon. *Bioresour. Technol.* **165**, 166–173.

Li, Q., Lu, H., Yin, Y., Qin, Y., Tang, A., Liu, H. & Liu, Y. 2019 Synergic effect of adsorption and biodegradation enhance cyanide removal by immobilized Alcaligenes sp. strain DN25. *J. Hazard. Mater.* **364**, 367–375.

Liu, X., Yang, Q., Chen, W., Mo, L., Chen, S., Kang, J. & Song, X. 2015 A ratiometric fluorescent probe for rapid, sensitive and selective detection of sulfur dioxide with large Stokes shifts by single wavelength excitation. *Org. Biomol. Chem.* **13** (32), 8663–8668.

Lu, X., Jiang, J., Sun, K., Wang, J. & Zhang, Y. 2014 Influence of the pore structure and surface chemical properties of activated carbon on the adsorption of mercury from aqueous solutions. *Mar. Pollut. Bull.* **78** (1–2), 69–76.

Martins, A. C., Pezoti, O., Cazetta, A. L., Bedin, K. C., Yamazaki, D. A. S., Bandoch, G. F. G., Asefa, T., Visentainer, J. V. & Almeida, V. C. 2015 Removal of tetracycline by NaOH-activated carbon produced from macadamia nut shells: kinetic and equilibrium studies. *Chem. Eng. J.* **260** (0), 291–299.

Oleinikova, O. V., Shirokova, L. S., Drozdova, O. Y., Lapitskiy, S. A. & Pokrovsky, O. S. 2018 Low biodegradability of dissolved organic matter and trace metals from subarctic waters. *Sci. Total Environ.* **618**, 174–187.

Piai, L., Blokland, M., van der Wal, A. & Langenhoff, A. 2020 Biodegradation and adsorption of micropollutants by biological activated carbon from a drinking water production plant. *J. Hazard. Mater.* **388**, 122028.

Praetorius, S. & Voigt, A. 2015 A Navier-Stokes phase-field crystal model for colloidal suspensions. *J. Chem. Phys.* **142** (15), 154904.

Quinlivan, P. A., Li, L. & Knappe, D. R. 2005 Effects of activated carbon characteristics on the simultaneous adsorption of aqueous organic micropollutants and natural organic matter. *Water Res.* **39** (8), 1663–1673.

Schupp, T., Austin, T., Eadsforth, C. V., Bossuyt, B., Shen, S. M. & West, R. J. 2018 A review of the environmental degradation, ecotoxicity, and bioaccumulation potential of the low molecular weight polyether polyol substances. *Rev. Environ. Contam. Toxicol.* **244**, 53–111.

Singh, S. K., Townsend, T. G., Mazyck, D. & Boyer, T. H. 2012 Equilibrium and intra-particle diffusion of stabilized landfill leachate onto micro- and meso-porous activated carbon. *Water Res.* **46** (2), 491–499.

Smolin, S., Kozyatnyk, I. & Klymenko, N. 2020 New approach for the assessment of the contribution of adsorption, biodegradation and self-bioregeneration in the dynamic process of biologically active carbon functioning. *Chemosphere* **248**, 126022.

Sperlich, A., Werner, A., Genz, A., Amy, G., Worch, E. & Jekel, M. 2005 Breakthrough behavior of granular ferric hydroxide (GFH) fixed-bed adsorption filters: modeling and experimental approaches. *Water Res.* **39** (6), 1190–1198.

Spiesshofer, J., Heinrich, J., Lehmann, R., Efken, C., Fox, H., Bitter, T., Korber, B., Horstkotte, D. & Oldenburg, O. 2015 Respiratory effects of adaptive servoventilation therapy in patients with heart failure and Cheyne-Stokes respiration compared to healthy volunteers. *Respiration* **89** (5), 374–382.

Sulaymon, A. H., Ali, A. F. M. & Al-Naseri, S. K. 2010 Mathematical models application for natural organic matter adsorption onto activated carbon. *Desalination Water Treat.* **24** (1–3), 93–100.

Ti, Q., Gu, C., Cai, J., Fan, X., Zhang, Y., Bian, Y., Sun, C. & Jiang, X. 2020 Understanding the role of bacterial cellular adsorption, accumulation and bioavailability regulation by biosurfactant in affecting biodegradation efficacy of polybrominated diphenyl ethers. *J. Hazard. Mater.* **393**, 122382.

Velten, S., Boller, M., Koster, O., Helbing, J., Weilenmann, H. U. & Hammes, F. 2011 Development of biomass in a drinking water granular active carbon (GAC) filter. *Water Res.* **45** (19), 6347–6354.

Wang, X. 2016 Population balance modeling of precipitated calcium carbonate (PCC) flocculation induced by cationic starches. *Int. J. Miner. Process* **148**, 9–14.

Wang, H., Ho, L., Lewis, D. M., Brookes, J. D. & Newcombe, G. 2007 Discriminating and assessing adsorption and biodegradation removal mechanisms during granular activated carbon filtration of microcystin toxins. *Water Res.* **41** (18), 4262–4270.

Wang, H., Jahandar Lashaki, M., Fayaz, M., Hashisho, Z., Philips, J. H., Anderson, J. E. & Nichols, M. 2012 Adsorption and desorption of mixtures of organic vapors on beaded activated carbon. *Environ. Sci. Technol.* **46** (15), 8341–8350.

Wang, H., Qu, B., Liu, H., Ding, J. & Ren, N. 2018 Analysis of organochlorine pesticides in surface water of the Songhua river using magnetoliposomes as adsorbents coupled with GC-MS/MS detection. *Sci. Total Environ.* **618**, 70–79.

Yapsakli, K. & Cecen, F. 2010 Effect of type of granular activated carbon on DOC biodegradation in biological activated carbon filters. *Process Biochem.* **45** (3), 355–362.

Yapsakli, K., Mertoglu, B. & Çeçen, F. 2010 Identification of nitrifiers and nitrification performance in drinking water biological activated carbon (BAC) filtration. *Process Biochem.* **45** (9), 1543–1549.

Zhang, D., Li, W., Zhang, S., Liu, M., Zhao, X. & Zhang, X. 2011 Bacterial community and function of biological activated carbon filter in drinking water treatment. *Biomed. Environ. Sci.* **24** (2), 122–131.

Zhiteneva, V., Ziemendorf, E., Sperlich, A., Drewes, J. E. & Hubner, U. 2020 Differentiating between adsorption and biodegradation mechanisms while removing trace organic chemicals (TOrCs) in biological activated carbon (BAC) filters. *Sci. Total Environ.* **743**, 140567.

First received 25 May 2020; accepted in revised form 22 October 2020. Available online 4 November 2020

Experimental study on the performance of modified flocculant (PEI-T) for removal of turbidity and heavy metals from aqueous solution

Wang Jin-xi and Rajendra Prasad Singh

ABSTRACT

The current study was focused on the evaluation of the performance of a modified flocculant (PEI-T) for the removal of turbidity and heavy metals from wastewater. Thioglycolic acid was introduced into the molecular chain of cationic flocculant polyethyleneimine (PEI) through an amidation reaction to form a new flocculant PEI-T. Results revealed that: (a) PEI-T and PEI had similar turbidity removal effect, but PEI-T had a good function of catching heavy metal copper and mercury; (b) when Cu^{2+} and turbidity-causing substances coexist in a water sample, Cu^{2+} and turbidity-causing substances promote the removal of each other and the residual concentration of Cu^{2+} and turbidity decrease further; and (c) the presence of Ca^{2+} and Mg^{2+} can promote the removal of copper ions by PEI-T.

Key words | heavy metals removal, modified flocculant, PEI, TGA, turbidity

Wang Jin-xi (corresponding author)
School of Chemical Engineering,
Lanzhou University of Arts and Science,
Lanzhou 730000,
China
E-mail: *wjxwjx1999@163.com*

Rajendra Prasad Singh
School of Civil Engineering,
Southeast University,
Nanjing, 211189,
China

HIGHLIGHTS

- Modified flocculant (PEI-T) was evaluated for removal of turbidity and heavy metals from wastewater.
- Thioglycolic acid was introduced into the molecular chain of cationic flocculant polyethyleneimine (PEI) to form a new flocculant, PEI-T.
- These flocculants had similar turbidity removal effect.
- Cu^{2+} and turbidity-causing substances promote their removal when Cu^{2+} and turbidity-causing substances coexist.

INTRODUCTION

Flocculants are efficient water treatment chemicals, which are widely used in the actual water treatment process. Flocculation is a very important process (Mutairi *et al.* 2004; Chang 2011; Yang *et al.* 2012) in solid-liquid separation of water treatment to produce the flocculation phenomenon through collection of solid particles in suspension. Compared with other water treatment methods, flocculation has the advantages of being more effective, and offering easier operation and low cost (Vilarrasa-García *et al.* 2017; Santos *et al.* 2018; Song *et al.* 2020). Due to the rapid industrialization, urbanization, environmental protection requirements, and enhanced environmental awareness, green treatment of wastewater has become a research hotspot in the field of environmental protection (Duan *et al.* 2019; Li *et al.* 2019). The study of natural non-toxic polymers has received increasing attention from researchers. At present, many researchers are engaged in the research and development of new flocculants for more effective, economic and sustainable treatment of wastewaters.

Metal ions in water cannot be degraded easily (Namasivayam & Ranganathan 1998; Schwarzenbach *et al.* 2016). Due to their toxicity and bioaccumulation in the food chain, it is of great significance to effectively reduce

doi: 10.2166/ws.2020.306

the concentration of metal ions in water (Özverdi & Erdem 2006). Selective removal of metal ions from simulated wastewater by sulfonated calix[4]arene intercalated with layered double hydroxide (SC4A-NO_3-LDH) was reported in a previous study (Reda & Zhang 2019). The authors reported that SC4A-NO3-LDH exhibited high adsorption performance towards metal ions with selectivity order of Fe(III) > Cu(II) > Pb(II) > Ag(I) for heavy metal ions, and Eu(III) > Nd(III) > La(III) for rare earth metal ions. Kinetics of Fe^{3+}, Cu^{2+} and Eu^{3+} were reported to follow the pseudo-second-order reaction, suggesting chemisorption.

Copper-containing wastewater mainly comes from metal mining and smelting, metal processing, machinery manufacturing, electroplating, circuit board production and other industries. With the advancement of industrial processes and the intensification of human activities, a large number of copper-containing wastewaters are being discharged by industrial enterprises and this will lead to the deterioration of the water environment, and affect human health, drinking water safety and seriously damage the aquatic ecosystem. When irrigated with wastewater containing copper, copper can accumulate in soil and crops, causing harm. Therefore, copper-containing wastewater must be treated effectively. The treatment methods for copper-containing wastewater mainly include chemical precipitation, electrolysis, adsorption, ion exchange, membrane separation and so on. The conventional treatment of copper is chemical precipitation with lime or sulfide, but this kind of treatment method requires a large amount of reagent, needs to adjust the pH value, and the amount of sludge is large. Moreover, the complexing agent in the wastewater will interfere with the precipitation of copper ions, so it needs to be removed by pretreatment (Meng & Hu 2000). The harm from mercury-containing wastewater is well known. In the current study, the traditional flocculant was modified by adding groups able to capture heavy metals, so the traditional flocculant could effectively capture heavy metals in water while removing turbidity.

Flocculation can easily and effectively remove 80–95% of suspended solids and 65–95% of colloidal substances (Fang *et al.* 2018; Ma *et al.* 2019). The organic polymer flocculants mainly adsorb colloidal particles in water to the molecular chain of the flocculants through the adsorption bridging effect to form flocs (Fang *et al.* 2018; Zhang *et al.* 2018; Wang *et al.* 2019). Compared with inorganic flocculants, organic polymer flocculation has the advantages of good stability, lower dosage, wide application range, strong adsorption and bridging ability, fast flocculation speed, small impact of pH value and temperature, and less sludge. Its flocculation effect is affected by molecular weight, charge density, and dosage, mixing time, flocculant stability and other factors (Bolto & Gregory 2007; Lu *et al.* 2014; Fang *et al.* 2018; Wang *et al.* 2019).

Polyethyleneimine (PEI) is a traditional polymer flocculant, which can be used to remove colloidal substances and fine suspended solids in water but has little effect on dissolved heavy metal ions in water (Fosso-Kankeu *et al.* 2016). In this study, sulfhydryl (-SH), which has a strong coordination effect on heavy metals, was introduced into the traditional flocculant polyethyleneimine through reaction, so the traditional flocculant can not only remove turbidity, but also capture heavy metals. In consideration of the challenges and cost control of the access reaction, thioglycolic acid TGA ($HOOCCH_2SH$) containing -SH was added into the long chain of polyethyleneimine (PEI) by amidation reaction to form modified polyethylene thioacetamide (PEI-T). This expanded the function of the traditional flocculant, and can effectively capture and remove heavy metals in the water while removing turbidity. By adding the modified flocculants, not only can the turbid substances in water be removed, but also the heavy metals in the water can be collected and removed, which enlarges the scope of application of the flocculant and reduces the process of water treatment.

There are many primary and secondary amines in the molecular chain of the polymer flocculant polyethyleneimine, which have high reaction activity. However, amidation is a very common chemical reaction. In this study, amines and carboxylic acids were amidated by appropriate means, and sulfhydryl groups were connected into the long chain of polyethyleneimine. The reaction formula is as follows (Equation (1)):

(1)

There are many types of amidation reaction. The EDC has a linear structure, which has been widely used in the condensation of the carboxyl group and primary amine. The use of EDC as an activator of amidation makes the reaction easy to control, has fewer side reactions and the product is easy to purify. Therefore, the current study was focused on the evaluation of the performance of a modified flocculant (PEI-T) for removal of turbidity and heavy metals and the activator EDC.HCl.

MATERIALS AND DETERMINATION

Modification experiments

In the reaction vessel, a certain amount of PEI solution, catalyst and TGA solution were added, and the reaction was stirred at room temperature. In order to determine the best conditions and change the influencing factors (EDC, nFEI: nTGA, PEI, pH, time), the PEI-T was characterized by the removal effect of Cu^{2+}, and the subsequent reaction was carried out after the optimization. Finally, the best conditions for preparing PEI-T were obtained at the following conditions: room temperature, nPEI: nTGA = 1:2, PEI concentration was 4%, time $t = 12$ h, catalyst EDC = 6 ml, initial pH = 2.5.

Flocculation experiments

A six-line stirrer was taken as the main experimental equipment, 400 ml copper or turbidity water samples were prepared, different test conditions were adapted, rapid agitation ($120r < min^{-1}$) was taken for 2 min, slow agitation ($40R < min^{-1}$) was taken for 10 min, and then allowed to stand for 15 min, a pipette was used to suck up the supernatant 2 cm from the liquid level to determine the content of copper or turbidity. The experiment was repeated 3 times, and the final average value was taken, and lastly the relative error of the experiment was ensured as $\leq 5\%$.

Experiment methods

The residual concentration of Cu^{2+} and Hg^{+} was measured by an atomic absorption spectrophotometer, and the turbidity was measured by turbidity meter (APHA 2005).

Data analysis

Outcomes of the present study were statistically analyzed by ANOVA and Origin 8.1 was used for preparing the figures.

RESULTS AND ANALYSIS

Characterization of PEI-T and description of product properties

PEI-T was precipitated and washed with acetone several times. After filtration and vacuum drying (50–), both PEI-T and PEI were analyzed by infrared spectroscopy respectively with potassium bromide tablets. The functional groups of PEI-T were determined by comparison. Figure 1 shows the infrared spectra of PEI and PEI-T, respectively. The upper spectrum is for PEI-T and the lower spectrum is for PEI. It can be seen from Figure 1 that PEI is transformed into PEI-T with a wave number of 1,667.14 cm^{-1}, indicating that PEI-T contains a tertiary amide group and an absorption peak with wave number of 2,562.96 cm^{-1} is generated, indicating that the PEI-T molecule already contains thiol group.

The sulfur content of PEI-T 2 g was measured by an elemental analyzer. The results of elemental analysis show that the content of sulfur in PEI-T is 0.198%. FTIR and elemental analysis showed that TGA was successfully attached to PEI. The product PEI-T was milky white, liquid, pH = 3.12, density 1.16 g/ml, solid content 4.31%.

Performance of turbidity removal before and after modification

Distilled water was added, the pH was adjusted to 5, and different amounts of PEI and modified flocculant PEI-T

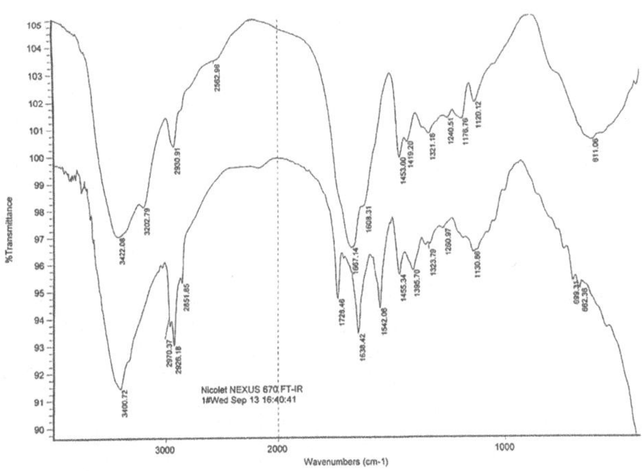

Figure 1 | The IR spectra of PEI-T and PEI.

were added. These results are presented in Figure 2. Results revealed the slight difference between modified PEI-T and the original flocculant PEI in the turbidity removal effect (Figure 2). As a flocculant, they almost have the same effect due to having the same turbidity removal mechanism. PEI is a cationic organic polymer flocculant. The molecule contains a large number of primary amine groups (-NH^2), tertiary amine groups (-NH-) and quaternary amine groups. It could adsorb H^+ with positive charge, and mainly promoted flocculation and sedimentation of colloidal substances in water by reducing the turbidity of the colloidal substances through 'adsorption neutralization' and 'adsorption bridge' (Wang *et al.* 2017). The mechanism of PEI-T turbidity removal is the same as that of PEI.

In addition to its turbidity function, PEI-T could also capture and remove heavy metals. It can be seen from Figure 3 that with the increase in the pH value, the removal rate of turbidity by PEI-T decreased. In general, the pH value had a certain effect on the turbidity removal of PEI-T, but it was small. For the modified flocculant PEI-T, there was an equilibrium of ionization and hydrolysis in water (Equation (2)). Results revealed that when the pH is low, the equilibrium is shifted to the right, and the cationic flocculant fully exerted adsorption and electric neutralization. When the pH is high, the equilibrium shifts to the left, and the negative charge is the key form in -S-, and the adsorption neutralization can be inhibited. These results are consistent with the findings of Augustine *et al.* (2019).

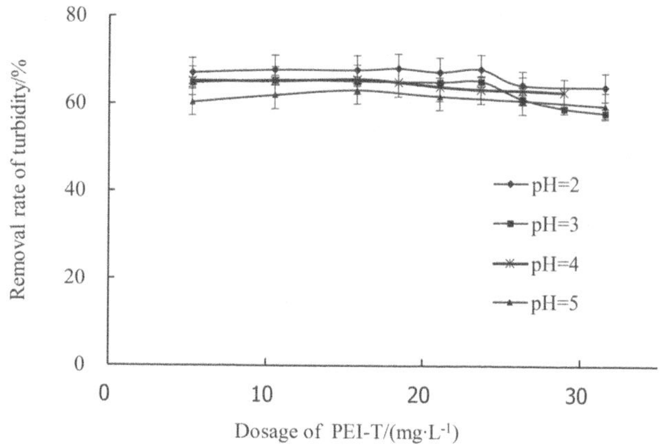

Figure 3 | Turbidity removal performance of PEI-T at different pH.

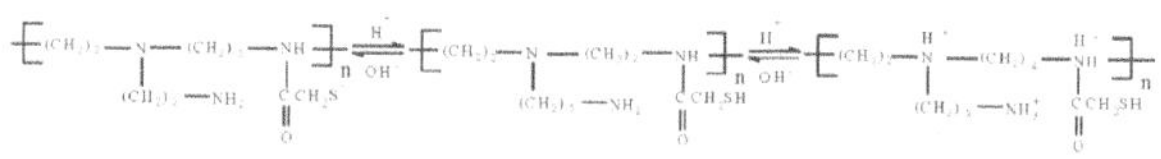

(2)

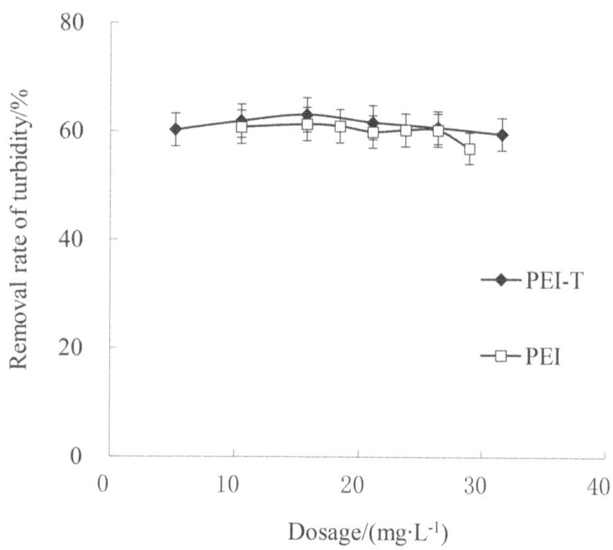

Figure 2 | Comparison of turbidity removal performance between PEI-T and PEI.

Performance of modified flocculant for Cu removal

Effect of dosage of medicament on Cu^{2+} removal

At pH 5, PEI-T was added to water samples with different concentrations of Cu^{2+}, and the residual Cu^{2+} concentration in the water was measured after flocculation and precipitation. The experimental results are shown in Figure 4. This experiment did not compare the copper removal effect of PEI, because PEI has no effect on copper.

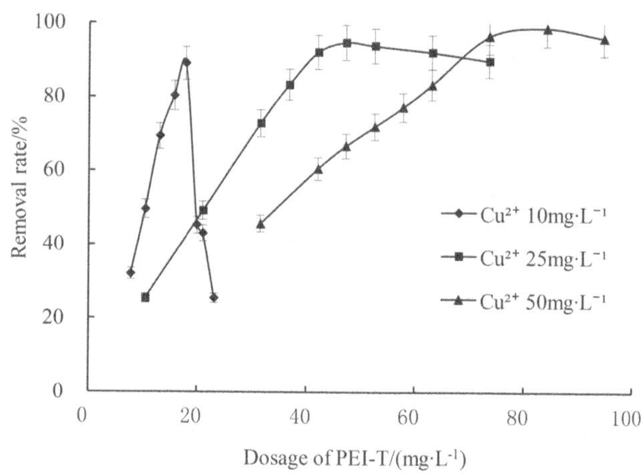

Figure 4 | Effect of PEI-T dosage on Cu^{2+} removal rate.

It can be seen from Figure 4 that there were optimal dosages for water samples with different concentrations of Cu^{2+}. Before the optimal point, the removal rate of Cu^{2+} decreased with the increase in the dosage. When the optimal point was reached, the removal rate of Cu^{2+} was highest. With continued increase in the amount of PEI-T, the Cu^{2+} removal rate began to decline. This was because with the increase in the addition of PEI-T, the Cu^{2+} removal rate increased to the maximum. After adding an excessive amount of PEI-T, the -SH in the PEI-T molecule that did not participate in the reaction dissociates with a negative charge, making it a negatively charged particle. The excess negative charge on the particle causes the chelate to generate electrostatic repulsion. Collision weakens and the flocculation effect decreases, which is why the Cu^{2+} removal rate started to decrease. Figure 4 also revealed that the optimal dosage increased with the increasing Cu^{2+} concentration in the water sample. With the increase of Cu^{2+} concentration, the -SH on the PEI-T molecule complexed with Cu in the form of coordination and covalent bonds to form a stable insoluble heavy metal complex with a cross-linked spatial network structure to precipitate PEI-T-Cu; there is also a redox reaction, and the reaction formulas are as presented in Equations (3) and (4).

$$2RCOCH_2SH + Cu^{2+} \rightarrow RCOCH_2S - Cu(II) - SCH_2COR + 2H^+ \qquad (3)$$

$$4RCOCH_2SH + Cu^{2+} \rightarrow RCOCH_2S - Cu(I) - SCH_2COR + RCOCH_2S\text{-}SCH_2COR \qquad (4)$$

The lower the Cu^{2+} concentration, the closer the PEI-T to Cu^{2+} reaction ratio is to 2: 1 ($mg \cdot L^{-1}$), and the larger the initial Cu^{2+} concentration, the smaller the reaction ratio between PEI-T and Cu^{2+} (<2). At the same time, $-NH_2$ on the parent chain and the nitrogen atom on the equivalent group will coordinate with H^+ in the solution, and will also undergo an ion exchange reaction with Cu^{2+}. This process could promote the removal rate of copper (Equation (5)).

$$\underset{H^+}{\overset{H^+}{R\ddot{N}}}COCH_2SH + Cu^{2+} \longrightarrow \overset{Cu_{2+}}{R\dot{N}}COCH_2SH + 2H^+ \qquad (5)$$

Effect of pH on PEI-T removal of Cu^{2+}

A water sample containing 25 $mg \cdot L^{-1}$ Cu^{2+} was taken to start the experiment. The pH value of the raw water was adjusted, and the effect of pH value on the removal of Cu^{2+} by PEI-T was studied. The experimental pH range was set between 3.0 and 5.0, given that Cu^{2+} hydrolyzes at a higher pH and the acidity of general heavy metal wastewater.

Results presented in Figure 5 reveal that the copper removal rate was lower when the pH was lower and the copper removal rate was higher when the pH was higher. As a whole, as long as the PEI-T dosage was increased, the removal rate of Cu^{2+} could also reach the maximum value. The reason is that when the pH was low, the equilibrium shown in formula (2) shifted to the right. The thiol group in PEI-T mainly existed in the form of -SH, which was not easy to chelate with Cu^{2+}, and the excessively high H^+ concentration changed the Cu (II)/Cu (I) standard potential (Huang 1998), which enables H^+ to oxidize Cu^+ to Cu^{2+}. Since the stability of the complex of Cu^{2+} and PEI-T was lower than that of Cu^+ and PEI-T, the Cu^{2+} removal rate was reduced. At higher pH levels, the equilibrium shifted to the left and the molecule was dominated by the negatively charged form -S-, which easily chelates with Cu^{2+} and Cu^+, making Cu^{2+} removal easier.

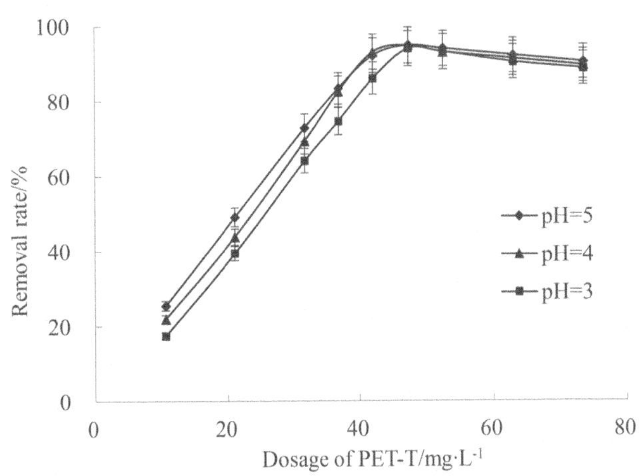

Figure 5 | Effect of pH on the removal of Cu^{2+} by PEI-T.

Flocculation effect and interaction between Cu^{2+} and turbidity

A water sample containing Cu^{2+} 25 mg • L^{-1} and turbidity of 127NTU was taken, and the dosing concentration of PEI-T for flocculation experiments was changed at pH 5. A water sample containing only Cu^{2+} but no turbidity was compared for the case with turbidity but no Cu^{2+}, as shown in Figure 6.

It can be seen from Figure 6 that in the case where Cu^{2+} and turbidity coexisted, the removal rates of Cu^{2+} and turbidity were higher than those when they existed separately, especially for turbidity removal, so the removal rate of Cu^{2+} and turbidity substances can be improved. The reason may be, on the one hand, that Cu^{2+} can neutralize the excessive negative charge of both the turbidity substances and the flocculant, reducing the repulsion between them. The flocculant can effectively play the role of 'adsorption bridging'; therefore, the turbid substances in the water can be combined into a large number of flocs. On the other hand, due to the sweeping and net-catching effects of a large number of flocs, the copper ion complex precipitates rapidly. In addition, due to the large specific surface area of the floc, a portion of the soluble copper ions will be adsorbed on the floc surface and further removed as the floc is precipitated. The effect on turbidity was large because the first-formed heavy metal chelate can fully play the role of electric neutralization and sweeping net capture of turbid substances, and strengthen flocculation.

Effects of Ca^{2+} and Mg^{2+}

Figure 7 reveals the effect of Ca^{2+} and Mg^{2+} on PEI-T flocculation and the copper removal effect at pH5. The effects of Ca^{2+} and Mg^{2+} were studied at different dosages of PEI-T. The finding presented in Figure 7 reveals that the presence of Ca^{2+} and Mg^{2+} promoted the removal of Cu^{2+}. The main reason is that the sulfur atom, which is mainly composed of -S- in the PEI-T molecule, was not easy to coordinate with the alkali metal or alkaline earth metal, is difficult to coordinate with Ca^{2+}, Mg^{2+}, and has poor stability after coordination (Song 1990). Therefore, Ca^{2+} and Mg^{2+} would not compete with Cu^{2+} for thiol ligands. In the case where -S- was chelated with Ca^{2+}, there would still be a certain amount of negatively charged groups on the polymer, and the electrostatic repulsion between them would not be conducive to flocculation and precipitation. When the presence of Ca^{2+} and Mg^{2+} in the solution increased the ion concentration, the electric double layer compressed due to the electric neutralization. Due to this, the electric double layer was reduced in thickness, reducing the zeta potential, weakening the repulsion effect and accelerating the speed of flocculation, which allowed Cu^{2+} to be removed more effectively. Earlier studies also revealed that this effect was more significant when the ion number was high (Shi & Zhang 1989), therefore Mg^{2+} and Ca^{2+} promoted the copper removal rate more than Na^{+}. The promotion effect of Mg^{2+} on Cu^{2+} removal in solution was more obvious than that of

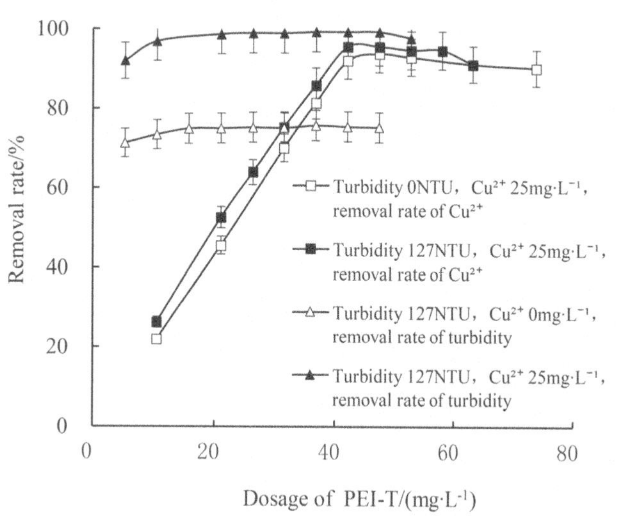

Figure 6 | Effect of Cu^{2+} and turbidity on removal by each other.

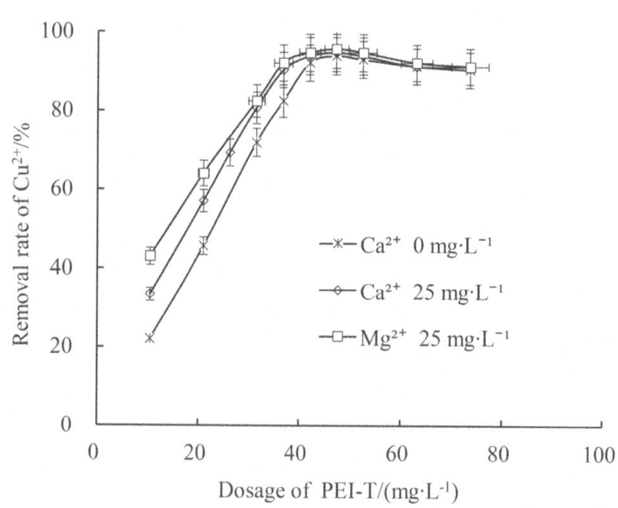

Figure 7 | The effect of Ca^{2+} and Mg^{2+} on the removal of Cu^{2+} by PEI-T.

Ca^{2+} because the effective nuclear charge number of Mg^{2+} in the solution was higher than that of Ca^{2+} (Hu & Huang 2001).

Effect of modified flocculant on mercury removal

Effect of dosage of PEI-T on Hg^{2+} removal

The PEI-T with different concentrations was added to water samples with different concentrations of Hg^{2+} (50 mg · L^{-1}, 100 mg · L^{-1}) and the removal rate of Hg^{2+} in the water after flocculation and precipitation were measured to compare the effect of Hg^{2+}removal. It is evident from Figure 8 that the removal rate of Hg^{2+} increased with the increase in the dosage, and PEI-T had the optimal dosage for water samples with different concentrations of Hg^{2+}. Excessive low or high dosage would affect the flocculation effect.

The PEI molecule underwent a complexation reaction with Hg^{2+} in the form of coordination bonds and covalent bonds to form a stable insoluble heavy metal complex with a cross-linked space network structure to precipitate PEI-T-Hg, and also had a redox reaction (Cabrera & West 1962). The reaction formula was as follows (Equation (6)).

$$2RCOCH_2SH + Hg^{2+} \rightarrow RCOCH_2S - Hg(II) - SCH_2COR + 2H^+$$

$$4RCOCH_2SH + Hg^{2+} \rightarrow RCOCH_2S - Hg(I) - SCH_2COR + RCOCH2S - SCH_2COR \quad (6)$$

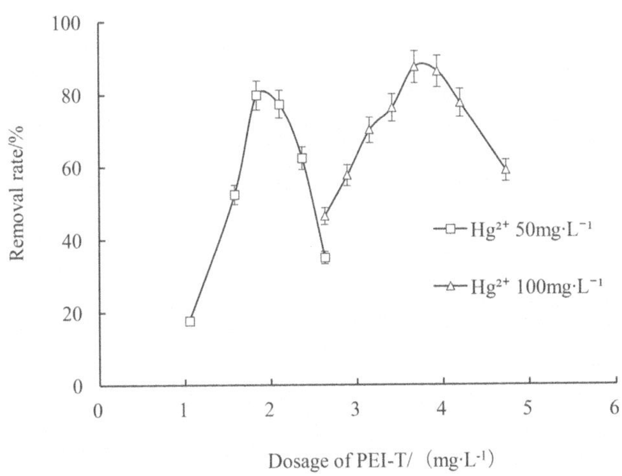

Figure 8 | Effect of PEI-T dosage on Hg^{2+} removal rate.

The removal rate of Hg^{2+} began to decrease when the dosage was higher than the optimal dosage. This trend occurred because after adding an excessive amount of PEI-T, the excess polymer's long chains protect the colloidal material and prevent particles from colliding with each other. The removal rate started to decline because the small particles were not easy to settle and the PEI-T molecules did not participate in the reaction. The -SH dissociation was negatively charged, causing electrostatic repulsion of the chelate, weakening the collisions between the particles, resulting in a decrease in the flocculation effect and a cause of the reduction in the removal rate of Hg^{2+}.

Since the sulfur atom has an empty 3D orbit and is easily polarized, it has a high affinity for heavy metal ions. As mercury belongs to the Group II elements of the periodic table, it can form a strong covalent bond with the atom that provides the electron pair. It can form a very stable complex with the mercapto group, which has a strong affinity for mercury. Therefore, the dosage of PEI-T used in the mercury removal experiment is much lower than the dose required for copper removal.

CONCLUSIONS

The synthesis of PEI-T is simple and easy. PEI-T is a highly efficient water treatment agent that can remove heavy metals and turbidity in actual wastewater due to the rapid settlement of floc particles. PEI-T has a good removal effect on copper and mercury wastewater at different concentrations. The removal efficiency of Cu^{2+} and turbidity by PEI-T was high whereas the effluent concentration was low. The pH value has a certain effect on Cu^{2+} and turbidity removal by PEI-T, but not significant. Results revealed that the turbidity removal effect was better when the pH value was low, whereas the Cu^{2+} removal effect was better at higher pH level. When PEI-T was used to remove the coexisting system of heavy metals and turbidity, the presence of heavy metals and turbidity in wastewater promoted the removal of each other. The presence of Na^+, Mg^{2+}, and Ca^{2+} in wastewater can promote the removal of heavy metals by PEI-T. The floc particles formed by the PEI-T treatment wastewater are large and settle quickly. The

PEI-T product is susceptible to oxidation and should be stored at low temperatures.

ACKNOWLEDGEMENTS

This research was co-funded by the Universities and Colleges Innovation Ability Improvement Project of the Gansu province, China (No. 2019B-193), the National Natural Science Foundation of China (No. 41967043), and the Scientific and Technological Project of Construction of the Gansu province, China (No. JK2020-30).

DATA AVAILABILITY STATEMENT

All relevant data are included in the paper or its Supplementary Information.

REFERENCES

Al-Mutairi, N. Z., Hamoda, M. F. & Al-Ghusai, L. 2004 Coagulant selection and sludge conditioning in a slaughterhouse wastewater treatment plant. *Bioresource Technology* **95** (2), 115–119.

American Public Health Association (APHA) 2005 *Standard Methods for the Examination of Water & Wastewater.* American Public Health Association, USA, Vol. 21.

Augustine, A., Tanwar, A., Tremblay, R. & Kumar, A. 2019 Flocculation processes optimization for reuse of culture medium without pH neutralization. *Algal Research* **39** (5), 101437.

Bolto, B. & Gregory, J. 2007 Organic polyelectrolytes in water treatment. *Water Research* **41** (11), 2301–2342.

Cabrera, A. M. & West, T. S. 1962 Chelating action of Thioglycolic acid with silver and some common bivalent cations. *Talanta* **9** (8), 730–733.

Chang, Q. 2011 *Water Treatment Flocculation*, 2nd ed. China Chemical Industry Press, Beijing, China (in Chinese).

Duan, M., He, Z., Wang, X., Jing, B., Xu, Z., Xiong, Y. & Fang, S. 2019 A novel interface-active cationic flocculant for the oil-water separation of oily wastewater produced from polymer flooding. *Journal of Molecular Liquid* **286**, 110868.

Fang, R., Xue, H., Wu, H. & Ye, Y. 2018 Synthesis of an efficient cationic flocculant by one-pot solution polycondensation and its flocculation performance. *Environmental Chemistry* **37** (12), 2668–2676.

Fosso-Kankeu, E., Waanders, F., van Niekerk, D., Rogers, D. & Gericke, G. 2016 Treatment of raw and processed waters from coal power plant using PACl supplemented with cationic organic polymer and bentonite. In *International Conference on Advances in Science, Engineering, Technology & Natural Resources (ICASETNR-16)*, Nov. 24–25, 2016, Parys (South Africa).

Hu, S. & Huang, Q. 2001 *Inorganic and Analytical Chemistry.* Higher Education Press, Beijing, China, pp. 15.

Huang, L. 1998 *Adsorption and Bonding Behavior of Mercaptothiadiazole Compounds in Solution Metal Composite System.* Doctoral thesis, Nanjing University, Nanjing, China, p. 115.

Li, X., Yang, B., Feng, L., Zheng, H., Zheng, G. & Wu, P. 2019 Research progress of natural polymers in wastewater treatment. *Mini Reviews Organic Chemistry* **16** (4), 335–344.

Lu, L., Pan, Z., Hao, N. & Peng, W. 2014 A novel acrylamide-free flocculant and its application for sludge dewatering. *Water Research* **57**, 304–312.

Ma, J., Wang, R., Wang, X., Zhang, H., Zhu, B., Lian, L. & Lou, D. 2019 Drinking water treatment by stepwise flocculation using polysilicate aluminum magnesium and cationic polyacrylamide. *Journal of Environmental Chemistry Engineering* **7** (3), 103049.

Meng, X. & Hu, G. 2000 *Heavy Metal-Containing Wastewater Treatment.* Chemical Industry Press, Beijing, China (in Chinese).

Namasivayam, C. & Ranganathan, K. 1998 Effect of organic ligands on the removal of Pb(II), Ni (II) and Cd (II) by waste Fe (III)/ Cr (III) hydroxide. *Water Research* **32**, 969–971.

Özverdi, A. & Erdem, M. 2006 Cu^{2+}, Cd^{2+} and Pb^{2+} adsorption from aqueous solutions by pyrite and synthetic iron sulphide. *Journal of Hazardous Material* **137**, 626–632.

Reda, A. T. & Zhang, D. X. 2019 Sorption of metal ions from aqueous solution by sulfonated calix[4]arene intercalated with layered double hydroxide. *Journal of Environmental Chemical Engineering* **7** (2), 103021.

Santos, A. S., Oliveira, L. S., Marques, A. M. T., Silva, D. C. A. & Mansur, C. R. E. 2018 Evaluation of the efficiency of polyethylenimine as flocculants in the removal of oil present in produced water. *Colloids and Surfaces A: Physicochemical & Engineering Aspects* **558**, 200–210.

Schwarzenbach, R. P., Escher, B. I., Fenner, K., Hofstetter, T. B., Johnson, C. A., von Gunten, U. & Wehrli, B. 2016 The challenge of micropollutants in aquatic systems. *Science* **313** (5790), 1072–1077. DOI: 10.1126/science.1127291.

Shi, G. & Zhang, F. 1989 *Physical Chemistry of Water Supply and Drainage.* China Construction Industry Press, Beijing, China, pp. 130–131.

Song, T. 1990 *Coordination Chemistry.* Chengdu University of Science and Technology Press, Chengdu, China, pp. 123–131.

Song, Z., Zhang, W., Gao, H. & Wang, D. 2020 Comprehensive assessment of flocculation conditioning of dredged sediment using organic polymers: dredged sediment dewaterability and release of pollutants. *Science of the Total Environment* **739**, 139884.

Vilarrasa-García, E., Cecilia, J. A., Rodríguez Aguado, E., Jiménez-Jiménez, J., Cavalcante Jr., C. L., Azevedo, D. C. S. & Rodríguez-Castellón, E. 2017 Amino-modified pillared adsorbent from water-treatment solid wastes applied to CO_2/N_2 separation. *Adsorption* **23**, 405–421.

Wang, G., Wang, Z., Chang, Q. & Xu, M. 2017 Removal and recovery of copper ions from aqueous solutions by modified polyethyleneimine. *Research in Environmental Science* **30** (6), 953–959.

Wang, H., Zhang, Y., Gao, D., Wang, G., Ma, Y., Wu, Z., Pu, H., Li, Y., Guo, Y., Wang, B., Qi, X. & Yang, J. 2019 Research on self-degradation of RGO/TiO_2-P(AM-DAC) organic-inorganic composite flocculant prepared by surface initiated polymerization and its flocculation mechanism of oil sand tailings. *European Polymer Journal* **120**, 109165.

Yang, Z., Yuan, B., Huang, X., Zhou, J., Cai, J., Yang, H., Li, A. & Cheng, R. 2012 Evaluation of the flocculation performance of carboxymethyl chitosan-graft-polyacrylamide, a novel amphoteric chemically bonded composite flocculant. *Water Research* **469** (1), 107–114.

Zhang, L., Hua, J., Zhu, W., Liu, L., Du, X., Meng, R. & Yao, J. 2018 Flocculation performance of hyperbranched polyethylenimine-grafted cellulose in wastewater treatment. *ACS Sustainable Chemical Engineering* **6** (2), 1592–1601.

First received 3 September 2020; accepted in revised form 24 October 2020. Available online 6 November 2020

Biodegradation of benzyl butyl phthalate and dibutyl phthalate by *Arthrobacter* sp. via micellar solubilization in a surfactant-aided system

Moumita Nandi, Tanushree Paul, Dipak Kumar Kanaujiya, Divya Baskaran, Kannan Pakshirajan and G. Pugazhenthi

ABSTRACT

Endocrine-disrupting chemicals (EDCs) like phthalates, mostly discharged in industrial effluents, cause detrimental effects to different life forms, and hence their removal from constituent wastewater is necessary. This study investigated the kinetics of biomass growth and biodegradation of benzyl butyl phthalate (BBP) and dibutyl phthalate (DBP) by *Arthrobacter* sp. in a surfactant-aided batch system. The effect of different surfactants on aqueous solubility of BBP and DBP was initially examined, which showed that Tween 80 resulted in maximum bioavailability and biodegradation efficiency of the phthalates by the bacterium and without inhibiting the biomass growth. Compared with BBP, DBP was found to be efficiently degraded and supported the bacterial growth within a short period of time over the entire concentration tested in the range 100–1,000 mg L^{-1}. A maximum biomass concentration of 1.819 g L^{-1} was obtained at 120 h for a DBP concentration of 600 mg L^{-1} in the presence of Tween 80, which is 5.66-fold increase in biomass concentration as compared with only DBP as the sole substrate. For evaluating the biokinetic parameters involved in DBP biodegradation, the experimental data on DBP utilization were fitted to various kinetic models as reported in the literature.

Key words | *Arthrobacter* sp., bioavailability, biodegradation, kinetic modelling, phthalates, surfactant-aided system

Moumita Nandi
Dipak Kumar Kanaujiya
Kannan Pakshirajan (corresponding author)
Department of Biosciences and Bioengineering,
Indian Institute of Technology Guwahati,
Guwahati 781039, Assam,
India
E-mail: *pakshi@iitg.ac.in*

Tanushree Paul
Kannan Pakshirajan
G. Pugazhenthi
Centre for the Environment,
Indian Institute of Technology Guwahati,
Guwahati 781039, Assam,
India

Divya Baskaran
Department of Chemical Engineering,
Sri Venkateswara College of Engineering,
Sriperumbudur, Tamil Nadu, 602117,
India

G. Pugazhenthi
Department of Chemical Engineering,
Indian Institute of Technology Guwahati,
Guwahati, Assam, 781039,
India

HIGHLIGHTS

- Biokinetics of surfactant-aided biodegradation of phthalates is reported.
- Tween 80 enhanced phthalate biodegradation by improving its bioavailability.
- *Arthrobacter* sp. showed direct uptake of micellar solubilized phthalates.
- Efficient biodegradation is achieved overcoming growth inhibition.

doi: 10.2166/ws.2020.347

GRAPHICAL ABSTRACT

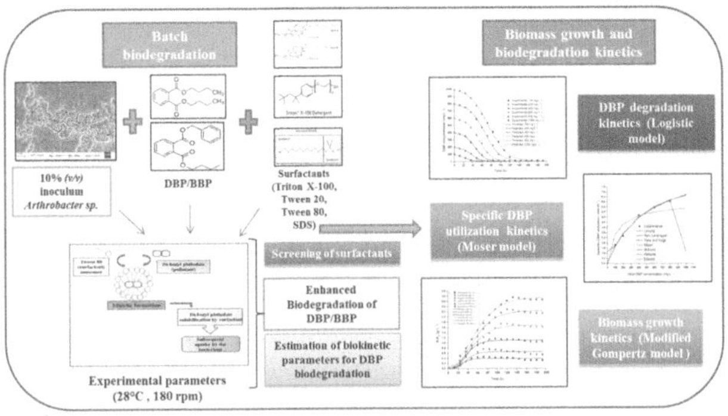

INTRODUCTION

Endocrine-disrupting chemicals (EDCs), in particular phthalates, are environmental pollutants which pose a serious threat to human beings, wildlife and the ecosystem due to their toxic, mutagenic and teratogenic nature (Lauretta *et al.* 2019; Margina *et al.* 2019; Godfray *et al.* 2019; La Merrill *et al.* 2019). Hence, their removal from contaminated systems is of recent interest. Among the different methods to treat wastewater containing EDCs, biodegradation using microorganisms is proving to be the most promising, economic and environmentally sound method (Eltoukhy *et al.* 2020; Gadupudi *et al.* 2020; Roccuzzo *et al.* 2020). However, success of this biological method is limited owing to the highly hydrophobic nature and very low bioavailability of phthalates and other EDCs to degrading microorganisms (Grinbaum *et al.* 2019; Gani & Kazmi 2020). For instance, the solubility of phthalates in water is 0.71–2.69 mg L^{-1} for benzyl butyl phthalate (BBP) and 11.2 mg L^{-1} for dibutyl phthalate (DBP) at 25 °C and hence bioavailability of these compounds is very poor. Moreover, high octanol–water partition coefficient values of these compounds indicate their high hydrophobicity which causes their accumulation in soils, sediments and even organisms (Sacan *et al.* 2005; Net *et al.* 2015; Gao & Wen 2016; Shi *et al.* 2019; Li *et al.* 2019). Hence, different strategies to improve the bioavailability and therefore biodegradation of these compounds have been proposed, which include application of surfactants (Franzetti *et al.* 2008; Paria 2008), decrease in temperature (Thomsen *et al.* 2001), solid-dispersion, particle size reduction, nanonization, cosolvency and inclusion complexation (Kumar & Singh 2016).

Among the afore-mentioned strategies, surfactants are conventionally applied to increase the aqueous solubility of these phthalates for enhancing their bioavailability and biodegradation (Aryal & Liakopoulou-Kyriakides 2013). Surfactants distinctively enhance the bioavailability of hydrophobic organic pollutants either by decreasing interfacial tension between aqueous and non-aqueous phases (Cuny *et al.* 1999) or by solubilization of hydrophobic organic compounds in the core of the micelles formed by surface-active agents (Sartoros *et al.* 2005; Hadibarata & Tachibana 2010). In the case of poorly water-soluble compounds, it has been reported that bacteria adhere to the liquid–liquid interface for lowering the interfacial tension and causing direct uptake of these pollutants, thereby enhancing their biodegradation rates (Volkering *et al.* 1998).

Among the different microorganisms reported to degrade highly recalcitrant compounds, such as phenolics, polycyclic hydrocarbons, bacteria belonging to the genus *Arthrobacter* have been shown to be highly effective (Abatenh *et al.* 2017; Guo *et al.* 2019). For instance, *Arthrobacter chlorophenolicus* A6 is reported to effectively degrade a mixture of phenolic compounds, including 4-chlorophenol, 4-nitrophenol and phenol (Unell *et al.* 2008). In addition, polycyclic aromatic hydrocarbons like phenanthrene and naphthalene were shown to be efficiently degraded by halotolerant bacteria belonging to the genus

Arthrobacter (Plotnikova *et al.* 2011). Interestingly, some studies have also reported the potential of *Arthrobacter* sp. strain YC-RL1 for utilizing bisphenol A (BPA) as the sole source of carbon for its growth and metabolism (Ren *et al.* 2016). The organism *Arthrobacter* sp. DAT1 was isolated and characterized to efficiently degrade atrazine (Wang & Xie 2012). From Table 1, biodegradation of DBP and BBP by various microorganisms is well reported in the literature (Fang *et al.* 2010; Wu *et al.* 2010; Kumar *et al.* 2017; Feng *et al.* 2018; Zhang *et al.* 2018). However, except for a study carried out by Chatterjee & Dutta (2008) on *Arthrobacter* spp. for BBP degradation, its potential to degrade other phthalates is largely unknown.

Moreover, there has been no study carried out to establish the potential of *Arthrobacter* sp. to degrade EDCs in a surfactant-aided system. Besides, details of kinetics of biodegradation and biomass growth by the organism using phthalates as the carbon source are lacking for establishing its application potential. Hence, this study was aimed at biodegradation of phthalates by *Arthrobacter* sp. using BBP and DBP as the carbon source. In order to enhance the bioavailability of these compounds, four chemical surfactants, *viz.* Tween 20, Tween 80, Triton X-100 and sodium dodecyl sulfate (SDS), were assessed as to their capability to solubilize BBP and DBP in aqueous media.

The results of biomass growth and degradation of BBP and DBP by *Arthrobacter* sp. obtained in the batch experiments conducted at different initial concentrations of the compounds and in the presence of a surfactant were compared with that obtained in experiments without any added surfactant. In order to gain further insight into the biokinetics of DBP biodegradation and biomass growth by *Arthrobacter* sp. in a surfactant-aided system the experimental data were fitted to different kinetic models reported in the literature and the biokinetic parameters involved were evaluated in the study. To our knowledge, this is the first study to show some new evidence for the biodegradation of both BBP and DBP using *Arthrobacter* sp. along with a detailed study on the biodegradation of these compounds.

MATERIALS AND METHODS

Chemicals and reagents

Benzyl butyl phthalate (>97% pure) and dibutyl phthalate (>97% pure) used in the study were obtained from TCI Chemicals Pvt. Ltd (Chennai, India). Analytical grade chemicals for preparing the bacterial growth media were obtained from Merck (Mumbai, India). Solvents (HPLC grade), including methanol, acetonitrile, and water were procured from Finar Limited (Ahmedabad, India). The surfactants Tween 20, Tween 80, Triton X-100 and SDS were purchased from either Hi-Media, Merck or Sigma-Aldrich.

Table 1 | Microorganisms reported in the literature on biodegradation of BBP and DBP

Bacterial strain	Compound	Concentration	Degradation (%)	Incubation time	References
Ochrobactrum sp. *JDC-41*	DBP	500 mg L^{-1}	87	48 hr	Wu *et al.* (2010)
Rhodococcus sp. *HS-D2*	BBP	500 mg L^{-1}	100	96 hr	Zhang *et al.* (2018)
Pleurotus ostreatus	DBP	500 mg L^{-1} 1,000 mg L^{-1}	99.6 94	312 hr 504 hr	Ahuactzin-Pérez *et al.* (2018b)
Fusarium culmorum	DBP	500 mg L^{-1} 1,000 mg L^{-1}	99 99	168 hr 228 hr	Ahuactzin-Pérez *et al.* (2018a)
Acinetobacter sp. *HS-B1* *Arthrobacter* sp. *C21*	DBP	300 mg L^{-1}	100	3 days	Wen *et al.* (2014)
Pseudomonas sp. *V21b* *Comamonas* sp. *51F*	DBP	1,000–2,000 mg L^{-1}	57 46	192 hr 192 hr	Kumar *et al.* (2017)
Bacillus megaterium YJB3	DBP	100 mg L^{-1}	82.5	5 days	Feng *et al.* (2018)
Arthrobacter sp. *WY*	BBP	1 g L^{-1}	95	44days	Chatterjee & Dutta (2008)
Acinetobacter sp. *33F*	DBP	2,000 mg L^{-1}	82.45	192 hr	Vinay & Maitra (2016)

Microorganism and culture conditions

The bacterial isolate *Arthrobacter sp.* capable of phthalate biodegradation was kindly donated by Department of Microbiology, Bose Institute, Kolkata, India. For routine growth and maintenance of the culture, Luria Bertani (LB) broth containing (g L^{-1}) peptone (10), yeast extract (5) and NaCl (5), was used. For phthalate biodegradation, minimal salt medium (MSM) containing (g L^{-1}) $MgSO_4{\cdot}7H_2O$ (0.409), $CaCl_2$ (0.0200), KH_2PO_4 (1), NH_4NO_3 (1), $Na_2HPO_4{\cdot}12H_2O$ (6), and $FeCl_3{\cdot}6H_2O$ (0.0833) and 1% (*v/v*) of each trace metal solution, i.e., 50 μL trace elements (g L^{-1}): $FeCl_3{\cdot}6H_2O$(28.33), $CaCl_2{\cdot}2H_2O$(0.8), $ZnSO_4{\cdot}7H_2O$(0.356), $CuSO_4{\cdot}5H_2O$ (0.1712), $MnSO_4{\cdot}H_2O$(0.224), $CoCl_2{\cdot}6H_2O$(1.465), H_3BO_3 (0.1) and $Na_2MoO_4{\cdot}2H_2O$ (0.3), pH 7, was used (Paul *et al.* 2019). The bacterium was cultured at 28 °C temperature and 180 rpm agitation conditions (Chatterjee & Dutta 2008).

Aqueous solubility of phthalates with surfactants

Based on the chemical nature, i.e., ionic or non-ionic, four different surfactants, namely Triton X-100, Tween 80, Tween 20 and sodium n-dodecyl sulfate (SDS), were evaluated for enhancing the aqueous solubility of BBP and DBP (Mahanty *et al.* 2007). To determine the extent of phthalate aqueous solubility in the presence of these surfactants, six different standards for each of the four surfactants were prepared from their respective stock solutions containing 5–30 times the critical micellar concentration (CMC) of the surfactants. To each of the standards, 100 mg L^{-1} of the phthalates were added in separate flasks followed by agitating the flasks overnight on an orbital shaker maintained at 180 rpm (Scigenics Biotech, Chennai, India). The resulting mixture was filtered using a 0.45-micron nylon membrane filter (Axiva Sichem Biotech, Delhi, India) to separate any insoluble compound. Soluble phthalate concentration in surfactant solutions was subsequently analyzed by high performance liquid chromatography (Shimadzu SPD-20A, Japan) as detailed later in Analytical methods.

Effect of surfactants on bacterial growth

The effect of different non-ionic surfactants (Tween 20, Tween 80, Triton X-100) and the anionic surfactant SDS on *Arthrobacter* sp. biomass growth was investigated at five different concentrations above their respective CMC values, i.e., 10–30 times of CMC of these surfactants. In these experiments, the individual surfactants served as the sole carbon and energy source. *Arthrobacter* sp. growth in the presence of surfactants was determined by measuring optical density of the culture at 600 nm using a UV-Vis spectrophotometer (Antech, California, USA). A flask without any added surfactant served as the control in this study, whereas the other flasks were incubated with different surfactants at concentrations above their respective CMC values as mentioned earlier.

Surfactant-aided biodegradation of phthalates

Based on the results of the previous experiments, biodegradation of phthalates by *Arthrobacter* sp. was evaluated by using Tween 80 as the surfactant, which yielded maximum solubility of BBP and DBP and also showed the least toxic effect on the bacterium growth as compared with the other surfactants. All experiments in this study were performed in duplicate using Erlenmeyer flasks (250 mL) with 100 mL working volume, and each of the flasks contained MSM, inoculum (10% (*v/v*)), phthalate (BBP or DBP) and the surfactant (Tween 80). The flasks were added with different initial BBP/DBP concentrations of 100, 200, 400, 600, 800, 1,000 mg L^{-1} in MSM (pH 7), and 10% (*v/v*) freshly grown bacterial cultures from the mid-log phase were used as the inoculum for this batch study. The flasks were then incubated for 192 h at 28 °C and continuous shaking of 180 rpm in an orbital incubator shaker. Control experiment was carried out using a flask containing either BBP or DBP but without any added surfactant (Tween 80) and keeping all the other parameters the same. Samples were collected at regular time intervals of 12 h and analyzed for biomass and BBP/DBP concentrations.

Biomass dry weight was expressed as cell dry weight (CDW, mg L^{-1}), and the biomass specific growth rate (μ) was calculated as per the following Equation (1):

$$\mu = \frac{1}{X}\frac{dX}{dt} \tag{1}$$

where μ is the biomass specific growth rate (h^{-1}), X is the biomass concentration ($g\ L^{-1}$) corresponding to time t (h).

The biokinetics of DBP utilization by *Arthrobacter* sp. were analyzed by fitting the experimental data to first order, logarithmic and Logistic kinetic models (Equations (2)–(4)).

First order

$$S = S_0 e^{(-Kt)} \tag{2}$$

Logarithmic

$$S = S_0 + X_0[1 - e^{(\mu_{max} t)}] \tag{3}$$

Logistic

$$S = \frac{S_0 + X_0}{1 + \frac{X_0}{S_0}\left[e^{\left(\left(\frac{\mu_{max}}{K_S}\right)(S_0+X_0)t\right)}\right]} \tag{4}$$

where S and S_O are the instantaneous and initial DBP concentrations ($g\ L^{-1}$), respectively. X_O is the initial biomass concentration ($g\ L^{-1}$), K is the first-order rate constant (h^{-1}), μ_{max} is the maximum DBP specific utilization rate (h^{-1}), and K_S is the half-saturation coefficient ($mg\ L^{-1}$). In addition, the following modified form of the Gompertz model was used to explain DBP utilization by *Arthrobacter sp.*

$$S_0 - S = S_{max} e^{\left[-e^{\left[\left(\frac{r_{s\,max} \times 2.71828}{S_{max}}\right)(\lambda - t) + 1\right]}\right]} \tag{5}$$

where r_{smax} is the maximum rate of DBP utilization ($mg\ L^{-1}\ h^{-1}$), S_{max} is the maximum utilizable DBP concentration ($mg\ L^{-1}$) and λ is lag time for DBP utilization (h).

The effect of initial DBP concentration on its specific utilization rate (q, h^{-1}) by the bacterium was calculated by using the following Equation (6), and the experimental data fitted to different bio-kinetic models reported in the literature (Equations (7)–(13)):

$$q = -\frac{1}{X}\frac{dS}{dt} \tag{6}$$

Monod

$$q = \frac{q_{max} S}{K_S + S} \tag{7}$$

Haldane

$$q = \frac{q_{max} S}{K_S + S + \frac{S^2}{K_i}} \tag{8}$$

Han–Levenspiel

$$q = \frac{q_{max} S\left[1 - \frac{S}{S_m}\right]^n}{K_S + S\left[1 - \frac{S}{S_m}\right]^m} \tag{9}$$

Edward

$$q = \frac{q_{max} S}{K_S + S + \left(\frac{S^2}{K_i}\right)\left(1 + \frac{S}{K_s}\right)} \tag{10}$$

Moser

$$q = \frac{q_{max} S^n}{K_S + S^n} \tag{11}$$

Yano and Koga

$$q = \frac{q_{max} S}{K_s + S + \left(\frac{S^2}{K_1}\right)^n} \tag{12}$$

Luong

$$q = \frac{q_{max} S}{K_s + S}\left[1 - \frac{S}{S_m}\right]^n \tag{13}$$

where q is the specific DBP utilization rate (h^{-1}), q_{max} is the maximum specific DBP utilization rate (h^{-1}), S is the DBP concentration ($mg\ L^{-1}$), S_m is the DBP concentration above which net growth ceases ($mg\ L^{-1}$), K_s is the half-saturation constant (mg L^{-1}), K_i is the inhibition constant (mg L^{-1}), K_1 is a positive constant and n and m are empirical constants.

In addition to the above, the following two bio-kinetic models (Equations (14) and (15)) were further applied to simulate the biomass growth.

Gompertz

$$X - X_0 = X_{max} e^{\left[-e^{\left[\left(\frac{r_{xmax} \times 2.71828}{X_{max}} \right)^{(\lambda - t) + 1} \right]} \right]} \tag{14}$$

Logistic

$$X - X_0 = \frac{X_{max}}{\left[1 + e^{\left[\left(\frac{4 r_{xmax}}{X_{max}} \right)^{(\lambda - t) + 2} \right]} \right]} \tag{15}$$

where X_o is the initial biomass concentration, X is the biomass concentration at time t, X_{max} is the maximum biomass concentration (g L^{-1}) and r_{xmax} is the maximum rate of biomass growth (g L^{-1} h^{-1}).

Analytical methods

Residual concentrations of BBP and DBP in the samples were quantified using a high pressure liquid chromatography (HPLC) (Shimadzu SPD-20A, Japan) fitted with C-18 column (Shim-pack GIST-HP C18) of dimensions 250 mm × 4.6 mm and a UV detector set at 220 and 254 nm (dual mode on) at an oven temperature of 30 °C. Acetonitrile and deionized water in the ratio 70:30 was used as the mobile phase at a flow rate of 0.5 mL min^{-1} and a run time of 25 min for this analysis.

The biomass growth of *Arthrobacter* sp. in this study was estimated by analyzing optical density (OD_{600}) of the culture, and for which absorbance (optical density) of the bacterial culture was measured at 600 nm using a UV-Vis spectrophotometer. Biomass determination in this study was verified based on its cell dry weight (CDW) by lyophilizing the biomass obtained after centrifugation followed by washing the cell pellet with sterile 0.8% w/v NaCl saline and weighing of dry biomass.

RESULTS AND DISCUSSION

Screening of surfactants based on enhancement of phthalate aqueous solubility and effect on *Arthrobacter* sp. biomass growth

Figure S1 (supplementary information) shows the effect of different surfactants on enhancing the aqueous solubility of BBP and DBP, which revealed the following order for both the compounds: SDS > Tween 80> Triton X-100 > Tween 20. Hence, Tween 80 and SDS proved effective as compared with the other chemical surfactants. In the literature, Aryal & Liakopoulou-Kyriakides (2013) reported increased solubilization in presence of surfactant and successive biodegradation of phenanthrene and pyrene by *Arthrobacter* strain using Tween 20 and Tween 80 as the surfactants. The authors also observed that the biodegradation of both the polycyclic aromatic hydrocarbons in the presence of the surfactants followed the first-order kinetics model. Additionally, molar solubilization ratio (MSR) of pyrene using different surfactants was studied by Mahanty *et al.* (2007), which showed that Tween 80 has a higher MSR value compared to other surfactants for enhancing the aqueous solubility of the polycyclic aromatic hydrocarbons (PAH).

The effect of non-ionic (Tween 20, Tween 80, and Triton X-100) and anionic (SDS) surfactants on the biomass growth of *Arthrobacter* sp. after 48 h of incubation was further analyzed, and the results revealed that the bacterial growth enhanced with a rise in Tween 80 (Figure S2.a) and Tween 20 (Figure S2.b) concentrations compared to that without any added surfactant in the medium. However, in the case of Triton X-100 and SDS, the bacterial growth was either reduced or totally inhibited with an increase in their concentration in the media (Figure S2.c and d). These results showed that *Arthrobacter* sp. utilized Tween 80 and Tween 20 as the sole source of carbon for its metabolism and growth. In the literature, Chatterjee & Dutta (2008) reported that biodegradation of BBP with Tween 80 caused a two-fold increase in degradation rate as compared with that in the absence of any added surfactant. SDS and Triton X-100 are known to inhibit biomass growth of *Streptococcus faecalis* ATCC 9790 by inducing lysis of cells in the exponential growth phase (Cornett & Shockman 1978).

Besides, Mahanty *et al.* (2007) reported that surfactants with a high MSR value, e.g. Tween 80 exhibits a low toxicity on microorganisms capable of degrading PAHs. However, Aryal & Liakopoulou-Kyriakides (2013) observed an increase in cell number of *Arthrobacter* strain Sphe3 with increasing Tween 80 and Tween 20 concentrations. Hence, based on the results obtained in this study, Tween 80 was further chosen for performing the batch biodegradation experiments with *Arthrobacter sp.*

Effect of Tween 80 on biomass growth by *Arthrobacter* sp. at different phthalate concentrations

Figure 1 compares the time profile of *Arthrobacter* sp. biomass growth at different initial concentrations of BBP with or without Tween 80, which reveals that the bacterial biomass (CDW) steadily increased with an increase in BBP concentration even at 1,000 mg L^{-1} in the presence of the surfactant. From Figure 1(a) and 1(b), it is observed that in the presence of Tween 80 a biomass concentration of 1.744 g L^{-1} is achieved at 120 h for 600 mg L^{-1} initial BBP concentration, which is 4.59 fold increase in biomass concentration value obtained without the surfactant addition. Similarly, for an initial DBP concentration of 600 mg L^{-1} in the presence of Tween 80, a biomass concentration (maximum) of 1.819 g L^{-1} is obtained at the end of 120 h, which is a 5.66 fold increase in biomass concentration without the surfactant (Figure 2(a) and 2(b)). These results clearly reveal that the biomass growth rate is relatively lower in the absence of surfactant, probably due to poor solubility and therefore very low bioavailability of these compounds in the media for an efficient utilization by the bacterium. All these results clearly reveal that in the presence of surfactant, the bacterium is very capable of utilizing phthalates as the sole carbon source. However, in the presence of Tween 80, *Arthrobacter* sp. showed a lag period in its growth due to BBP, which was not observed with DBP. The lag phase in growth of the bacterium due to BBP can be attributed to its greater recalcitrance compared to DBP. Hence, in order to further confirm the uptake of BBP and DBP by the bacterium for its growth, their biodegradation in the presence of Tween 80 was analyzed.

Effect of Tween 80 on phthalates biodegradation by *Arthrobacter* sp. at different initial concentrations

Time profile of BBP and DBP biodegradation at different initial concentrations in the absence or presence of Tween 80 are depicted in Figures 3 and 4, respectively. From Figures 3(a) and 4(a), it could be seen that both the compounds were almost completely degraded (98%) for an initial concentration of BBP and DBP up to 600 mg L^{-1}. At a concentration above 600 mg L^{-1} and in the absence of Tween 80, the degradation efficiency decreased due to limited solubility of the compounds in the liquid medium; the values were 30.1% and 29.2% for 1,000 mg L^{-1} of

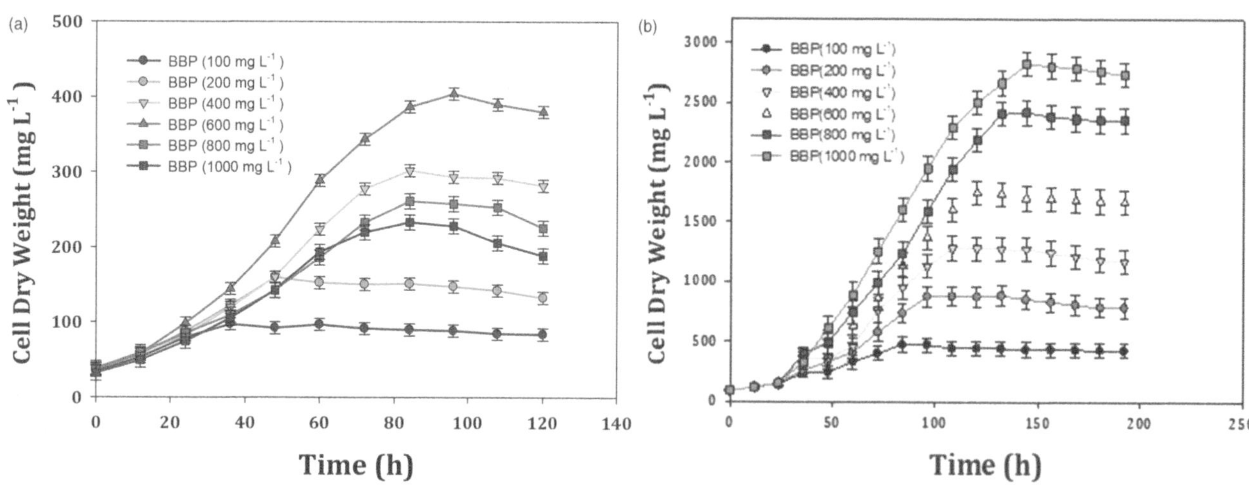

Figure 1 | Time profile of *Arthrobacter* sp. biomass growth at different initial concentrations of BBP in (a) absence and (b) presence of Tween 80.

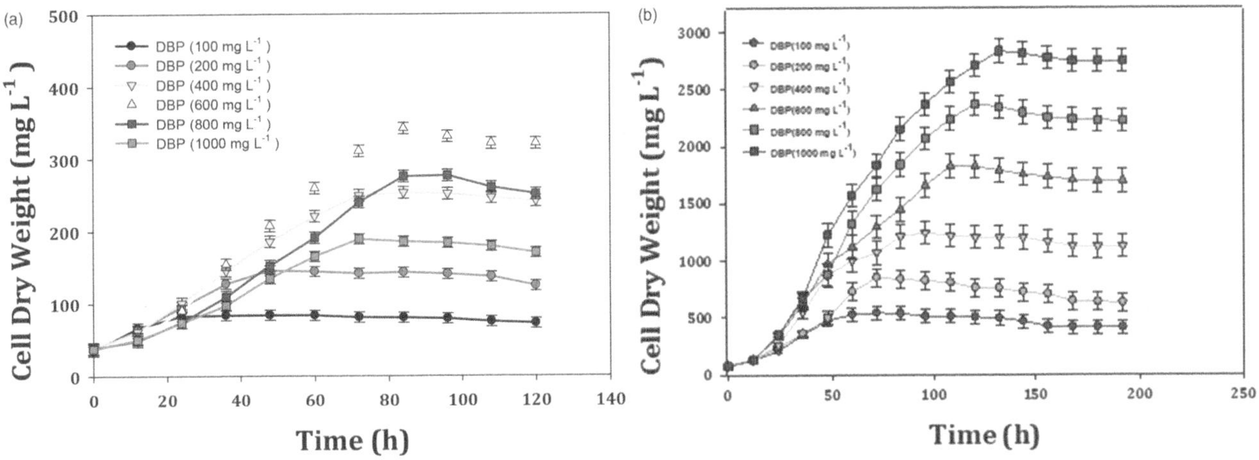

Figure 2 | Time profile of *Arthrobacter* sp. biomass growth at different initial concentrations of DBP in (a) absence and (b) presence of Tween 80.

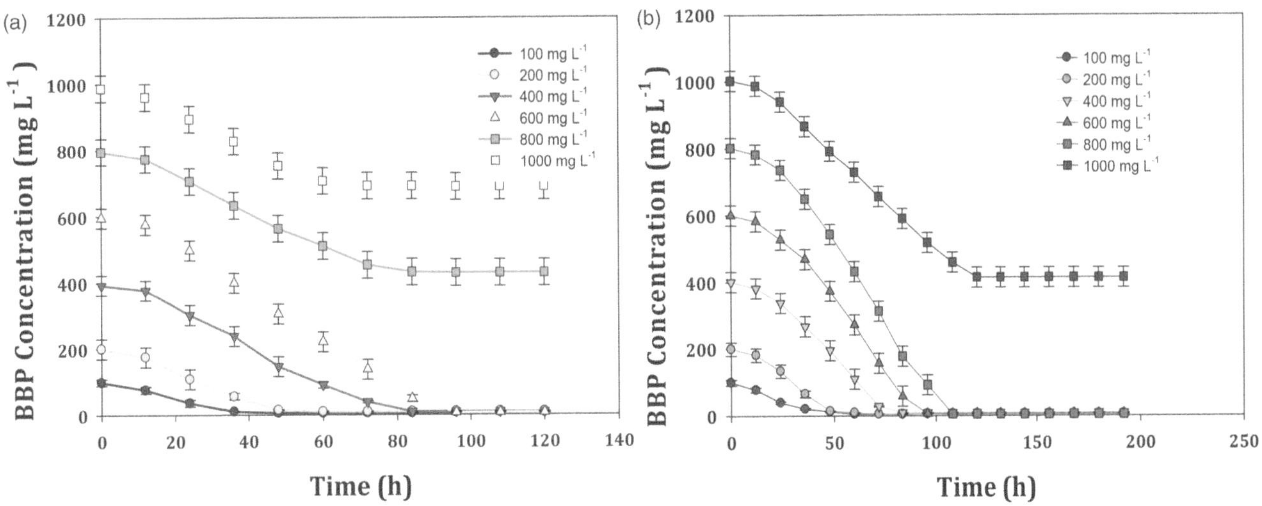

Figure 3 | Time profile of BBP biodegradation at different initial concentrations in (a) absence and (b) presence of Tween 80.

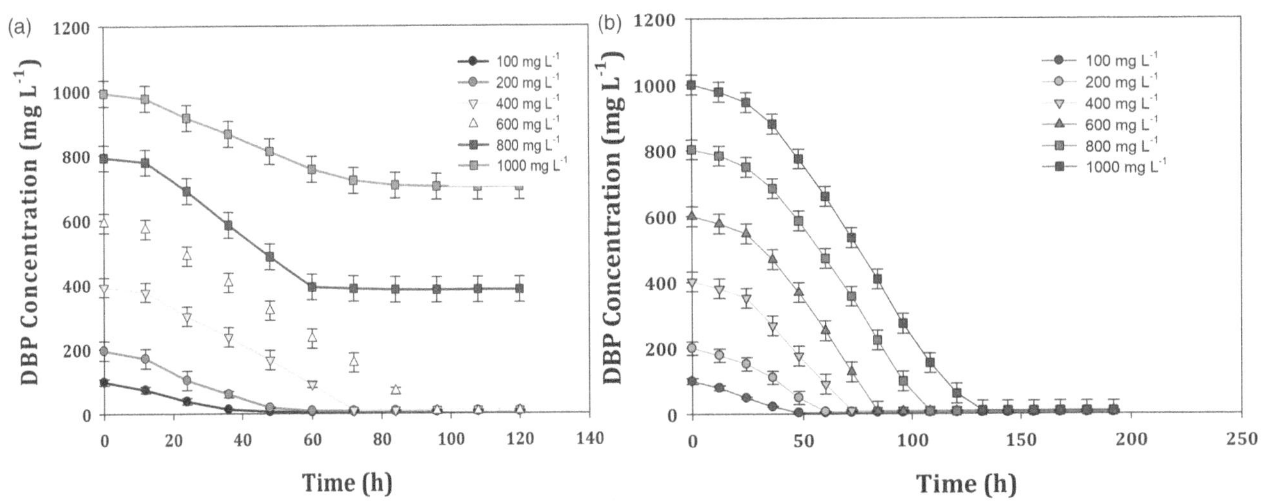

Figure 4 | Time profile of DBP biodegradation at different initial concentrations in (a) absence and (b) presence of Tween 80.

BBP and DBP, respectively. Conversely, in the presence of Tween 80, BBP and DBP degradation rapidly increased after 24 and 12 h of culture, respectively (Figures 3(b) and 4(b)). These results further demonstrate that DBP was almost completely degraded (99.1%) even up to 1,000 mg L^{-1} concentration. In the case of BBP, complete degradation was achieved only up to 800 mg L^{-1} concentration, and a degradation efficiency of 58% was obtained for 1,000 mg L^{-1} initial concentration. Hence, compared with BBP, DBP is more effectively degraded by *Arthrobacter* sp. due to its less recalcitrant structure. It is reported that owing to the presence of a single benzene ring, DBP is more easily degradable than the other phthalates (Wu *et al.* 2010; Vinay & Maitra 2016).

Surfactants are known to stimulate the uptake of micellar pollutants, resulting in their efficient biodegradation and an increased growth rate of microorganisms (Kaczorek & Olszanowski 2011). This phenomenon occurs mainly due to either phthalate and surfactant being utilized co-metabolically or direct uptake of phthalate containing surfactant micelles (Volkering *et al.* 1998). Hence, corroborating the results of biodegradation of BBP and DBP by *Arthrobacter* sp. with that of the previous experiment on biomass growth of the bacterium, it could be surmised that the latter is the mechanism of action by Tween 80 in this study.

It has been reported that DBP degradation by *Bacillus megaterium* YJB3 involves intermediates such as mono-butyl phthalate (MBP), phthalic acid (PA), dihydroxyphthalate, and protocatechuate (PCA) (Feng *et al.* 2018). Besides these compounds, mono-butyl phthalate and PA are the two transient metabolites formed during the degradation of DBP by *Enterobacter* sp. T5 (Wu *et al.* 2010). The biodegradation pathway of these two phthalates (BBP and DBP) used in this study by *Arthrobacter* sp. follows a similar pattern as already reported in various literature mentioned before. The pathway of BBP and DBP degradation was investigated by analyzing transient metabolites/intermediates at every stage of degradation using LC-MS, and the results revealed that both BBP and DBP followed the same degradation pathway (PA pathway) (Figure 5). The first degradation product of both BBP and DBP was mono-butyl phthalate, which was further broken down by

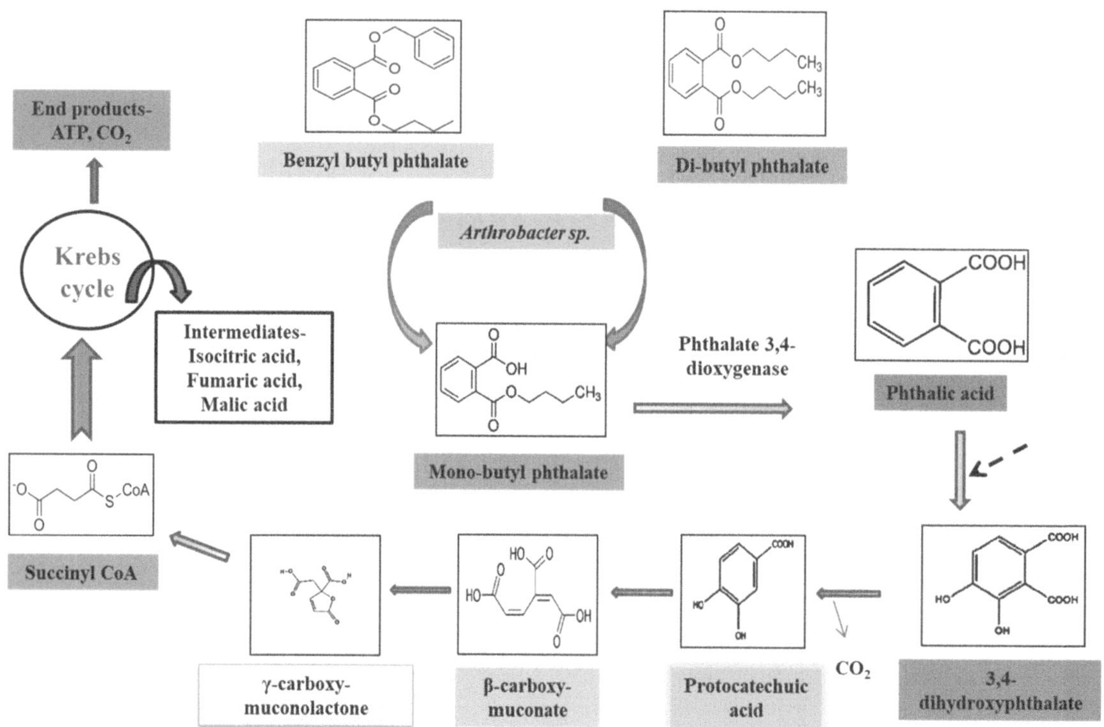

Figure 5 | An overview of degradation pathway of BBP and DBP by *Arthrobacter* sp.

phthalate 3,4-dioxygenase enzyme to form PA. During the degradation process, other pathway metabolites formed were 3,4-dihydroxyphthalate, protocatechuic acid, β-carboxy-muconate, γ-carboxy-muconolactone, followed by its successive degradation to form succinyl coenzyme A (CoA) which then enters Krebs (tricarboxylic acid) cycle via β-oxidation.

In the literature, *Pleurotus ostreatus* is reported to degrade 500 and 1,000 mg L^{-1} of DBP with efficiencies 99.6 and 94% at the end of 312 and 504 h, respectively (Ahuactzin-Pérez *et al.* 2018a). In another study, *Fusarium culmorum* was observed to degrade DBP at the same initial concentrations with 99% efficiency after an incubation period of 168 and 228 h, respectively (Ahuactzin-Pérez *et al.* 2018b). Compared with these literature reports, *Arthrobacter* sp. is found to be highly efficient in quickly degrading DBP even up to 1,000 mg L^{-1}. In order to gain further insight into DBP degradation by the bacterium, kinetics of the biodegradation process were further analyzed.

Kinetics of DBP utilization and biomass growth by *Arthrobacter* sp.

Experimental and predicted DBP utilization by *Arthrobacter* sp. obtained using different biokinetic models are shown in Figure 6, and estimated values of biokinetic parameters from these models are presented in Table 2. Among the three different models tested – First order, Logarithmic, and Logistic kinetic models, the Logistic model accurately fitted the experimental data for all the initial concentrations of DBP with a coefficient of determination (R^2) value greater than 0.98 (Table 2 and Figure 6(c)). Hence, it could be appropriately stated that the utilization of DBP by *Arthrobacter* sp. is growth-associated. The estimated maximum DBP utilization rate value of 0.0012–1.8309 h^{-1} from this growth-associated model further confirmed that *Arthrobacter* sp. is highly efficient in utilizing DBP in the presence of Tween 80. Wen *et al.* (2014) reported complete degradation of DBP as the sole substrate for an initial concentration of 300 mg L^{-1} within 3 days of culture, and kinetic analysis of the results revealed that DBP biodegradation followed first-order kinetics but only for an initial concentration less than 100 mg L^{-1}.

In order to further understand the growth-associated kinetics of DBP utilization by the bacterium, the values of specific DBP utilization rate obtained for all initial concentrations of the compound were fitted to various kinetic models as described in the literature. Kinetic variables involved were evaluated by fitting the experimental data to Luong, Han–Levenspiel, Yano and Koga, Moser, Monod, Haldane and Edward kinetic models values of the kinetic parameters are presented in Table 3. The results show that the Moser model well described the experimental data with R^2 value of 0.9998 followed by the other models. However, the Yano–Koga and Han–Levenspiel models yielded low R^2 values of less than 0.8. The values of maximum specific DBP utilization rate (q_{max}) and half-saturation coefficient (K_s) using the best-fitting Moser model were calculated to be 9.5561 h^{-1} and 535.03 mg L^{-1}, respectively. Figure 7 further confirms that the Yano–Koga and Han–Levenspiel models did not accurately fit the experimental values. Moreover, the substrate inhibition constant (K_i) value of 15,236.4 estimated from the Haldane model proved that *Arthrobacter* sp. is highly capable of utilizing a very high concentration of DBP in presence of Tween 80 without causing inhibition on its growth (Sinharoy *et al.* 2019).

The effect of different initial DBP concentration on *Arthrobacter* sp. biomass growth was further modelled using modified Gompertz and Logistic models. Figure 8 presents the experimental and the predicted *Arthrobacter* sp. biomass growth at different initial DBP concentrations; the biokinetic parameters assessed from the two models are presented in Table 4. It can be clearly observed that the modified Gompertz model is better than the Logistics model in explaining the experimental results ($R^2 > 0.97$). The predicted maximum specific growth rate of *Arthrobacter* sp. increased from 0.0161 to 0.0332 h^{-1}, with an increase in the initial DBP concentrations from 100 to 1,000 mg L^{-1}, and the predicted values of maximum biomass concentration varied in the range from 0.4011–2.769 g L^{-1} for 100–1,000 mg L^{-1} initial concentrations of DBP. The experimental values of maximum biomass concentration obtained in the range 0.540–2.829 g L^{-1} were also close to the values estimated by the modified Gompertz model. Moreover, these results confirm that an increase in concentration of DBP did not inhibit the biomass growth of *Arthrobacter* sp. which agrees with the findings reported by Dou *et al.* (2009).

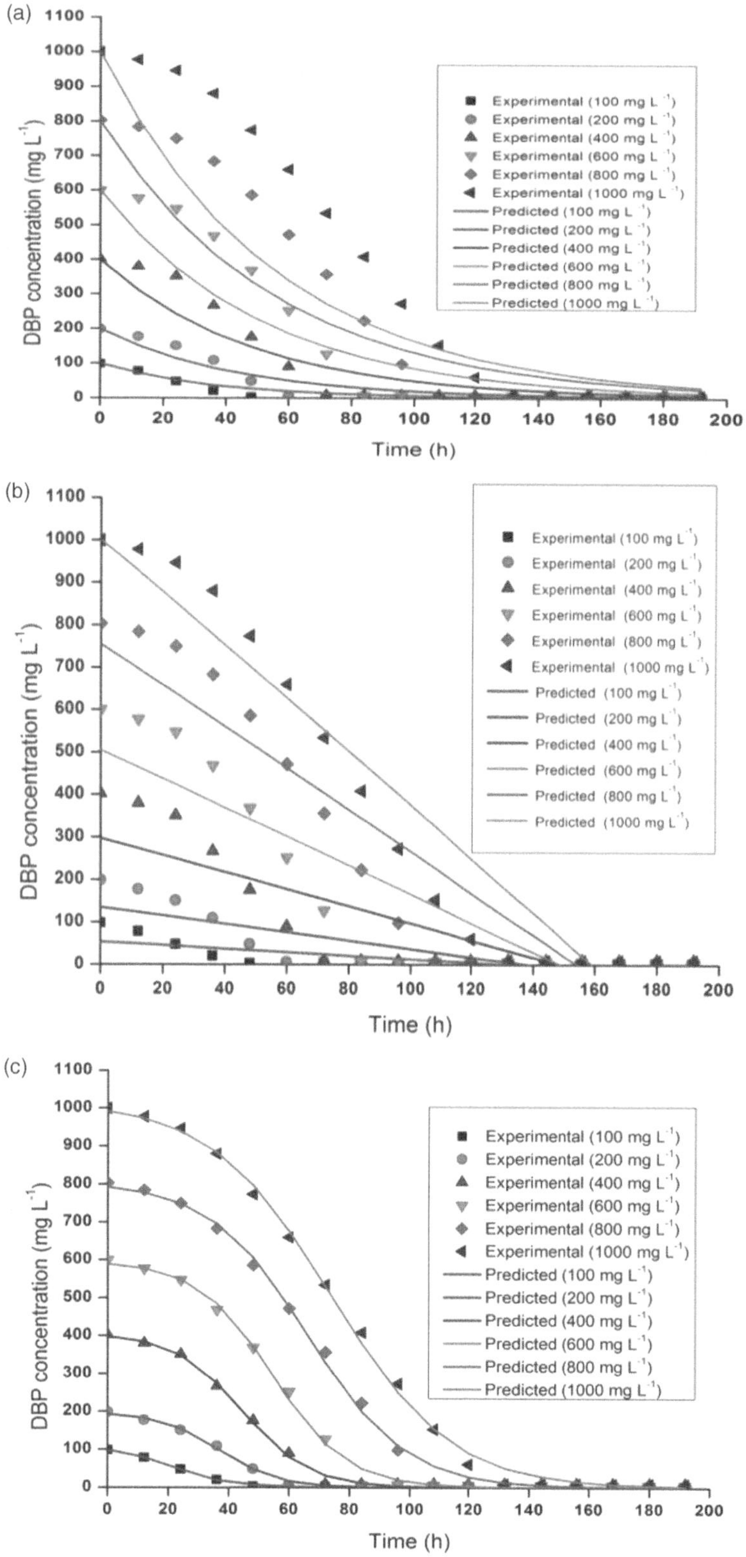

Figure 6 | Experimental and predicted values of DBP utilization due to (a) First order, (b) Logarithmic and (c) Logistic kinetic models.

Table 2 | Estimated values of bio-kinetic model parameters on DBP utilization in this study

Model	Initial DBP concentration (mg L^{-1})					
	100	**200**	**400**	**600**	**800**	**1,000**
First order						
S_0 (mg L^{-1})	106.283	225.071	470.897	718.358	971.707	1,209.646
K_1 (h^{-1})	0.0398	0.0281	0.0237	0.020	0.017	0.014
R^2	0.989	0.890	0.849	0.797	0.77	0.761
Logarithmic						
S_0 (mg L^{-1})	54.073	135.186	297.540	505.005	755.028	1,000.926
X_0 (g L^{-1})	3,578.791	9,743.567	20,232.296	45,480.901	48,270.741	32,082.66
U_{max} (h^{-1})	0.0001	0.0001	9.83986E-05	7.43076E-05	0.0001	0.0001
R^2	0.548	0.658	0.666	0.7607	0.845	0.9156
Logistic						
S_0 (mg L^{-1})	98.883	192.720	398.155	588.892	791.764	992.143
X_0 (g L^{-1})	11.347	4.264	7.632	7.333	13.524	24.621
U_{max} (h^{-1})	0.223	0.0973	0.641	0.090	0.001	1.830
K_S (mg L^{-1})	241.854	185.362	2,913.940	667.982	15.877	37,098.27
R^2	0.998	0.999	0.999	0.999	0.987	0.980

Table 3 | Estimated values of biokinetic model parameters on specific DBP utilization rate by *Arthrobacter* sp.

Model	q_{max} (h^{-1})	K_s (mg L^{-1})	K_i (mg L^{-1})	S_m (mg L^{-1})	n	m	K_p	R^2
Luong	9.838	477.34	–	3,846.5	1,497.1		–	0.997
Han–Levenspiel	11.501	596.08	–	999.6	3.492	3.971	–	0.778
Yano–Koga	5.932	108.92	–	–	−0.1705	–	999.9	0.600
Moser	9.556	535.03	–	–	1.029	–	–	0.999
Monod	9.838	477.34	–	–	–	–	–	0.999
Haldane	10.577	529.49	15,236.4	–	–	–	–	0.999
Edward	10.718	541.71	–	34,898.2	–	–		0.997

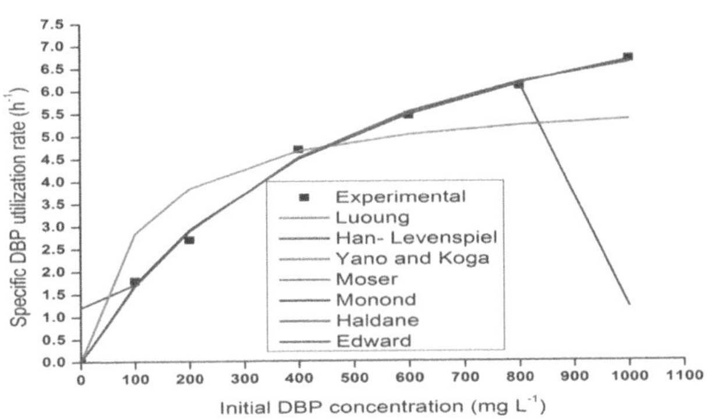

Figure 7 | Experimental and predicted values of specific DBP utilization rate at different initial DBP concentrations.

Hence, the results of kinetic analysis of biomass growth and DBP degradation by *Arthrobacter* sp. are useful for further upscaling of the present simple batch degradation system to a continuous pilot bioreactor system. Furthermore, proper monitoring and control of biokinetic parameters involved in the biodegradation process can help maintain high degradation efficiency of phthalates by the bacterium at a large scale. However, owing to their very low CMC, biosurfactants instead of the chemical surfactant Tween 80 evaluated in this study can be evaluated for enhancing bioavailability and biodegradation of EDCs by potent organisms. Moreover, identification of metabolites formed during

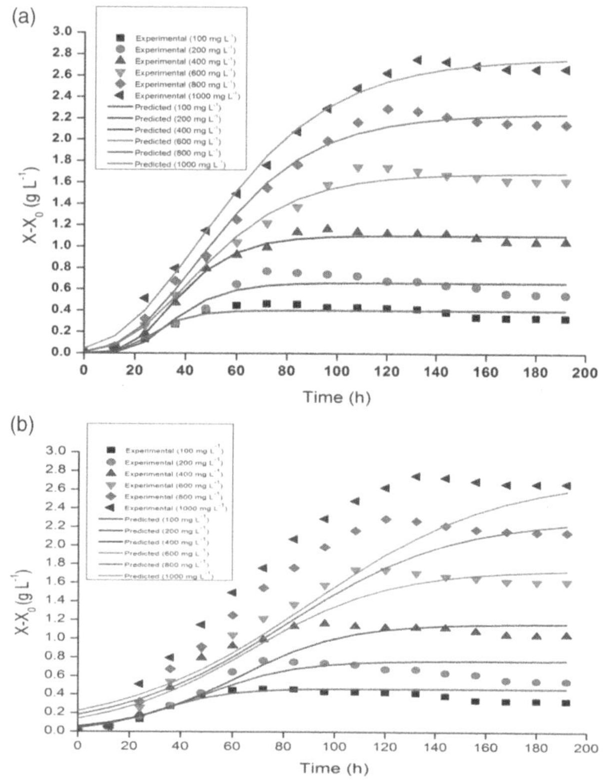

Figure 8 | Experimental and predicted *Arthrobacter* sp. biomass growth at different initial DBP concentrations using (a) modified Gompertz and (b) Logistic models.

Table 4 | Estimated values of biokinetic model parameters on *Arthrobacter* sp. biomass growth at different DBP concentrations

	Initial DBP concentration (mg L^{-1})					
Model	**100**	**200**	**400**	**600**	**800**	**1,000**
Modified Gompertz						
X_{max} (g L^{-1})	0.401	0.660	1.104	1.689	2.247	2.769
μ_{max} (h^{-1})	0.0161	0.019	0.0276	0.025	0.031	0.033
λ (h)	15.573	19.023	18.132	15.499	17.226	13.468
R^2	0.980	0.976	0.995	0.984	0.994	0.982
Logistic						
X_{max} (g L^{-1})	0.401	0.660	1.098	1.67	2.211	2.706
μ_{max} (h^{-1})	0.0141	0.0176	0.026	0.0249	0.031	0.033
λ (h)	14.901	19.727	18.860	17.001	19.770	16.306
R^2	0.926	0.947	0.964	0.939	0.958	0.947

phthalates biodegradation and details of the pathway followed by *Arthrobacter sp.* would throw more light on controlling the biodegradation process. In addition, biodegradation of phthalates in the presence of co-pollutants might be studied to establish the robustness of the process in a realistic situation.

CONCLUSIONS

Among the different surfactants evaluated in this study, Tween 80 was found to be highly effective for enhancing the bioavailability of both BBP and DBP and their biodegradation by *Arthrobacter* sp. Phthalate biodegradation and biomass growth by the bacterium in the presence of Tween 80 indicated that the phthalate uptake mechanism involved direct uptake of surfactant micelles containing the pollutant. Kinetics of DBP utilization, *Arthrobacter* sp. biomass growth and specific DBP utilization rate revealed that *Arthrobacter* sp. was highly capable and effective in degrading DBP even at very high concentrations in the presence of the surfactant without any inhibitory effect on its biomass growth. This study demonstrated an excellent potential of *Arthrobacter* sp. for the successful degradation of phthalates using a surfactant-aided system.

ACKNOWLEDGEMENTS

The financial grant by the Department of Science and Technology (DST/TM/INDO-UK/2K17/30(G) for this research work is gratefully acknowledged. The authors thank the Indian Institute of Technology, Guwahati, for providing necessary infrastructure and equipment for carrying out this research.

DATA AVAILABILITY STATEMENT

All relevant data are included in the paper or its Supplementary Information.

REFERENCES

Abatenh, E., Gizaw, B., Tsegaye, Z. & Wassie, M. 2017 The role of microorganisms in bioremediation-review. *Journal of*

Environmental Microbiology **1**, 02–09. https://doi.org/10.17352/ojeb.000007.

Ahuactzin-Pérez, M., Tlecuitl-Beristain, S., García-Dávila, J., Santacruz-Juárez, E., González-Pérez, M., Gutiérrez-Ruíz, M. C. & Sánchez, C. 2018a Mineralization of high concentrations of the endocrine disruptor dibutyl phthalate by *Fusarium culmorum*. *3 Biotech* **8** (1), 42. https://doi.org/10.1007/s13205-017-1065-2.

Ahuactzin-Pérez, M., Tlécuitl-Beristain, S., García-Dávila, J., Santacruz-Juárez, E., González-Pérez, M., Gutiérrez-Ruíz, M. C. & Sánchez, C. 2018b Kinetics and pathway of biodegradation of dibutyl phthalate by *Pleurotus ostreatus*. *Fungal Biology* **122** (10), 991–997. https://doi.org/10.1016/j.funbio.2018.07.001.

Aryal, M. & Liakopoulou-Kyriakides, M. 2013 Biodegradation and kinetics of phenanthrene and pyrene in the presence of nonionic surfactants by *Arthrobacter* strain Sphe3. *Water, Air, & Soil Pollution* **224** (2), 1426. https://doi.org/10.1007/s11270-012-1426-8.

Chatterjee, S. & Dutta, T. K. 2008 Complete degradation of butyl benzyl phthalate by a defined bacterial consortium: role of individual isolates in the assimilation pathway. *Chemosphere* **70** (5), 933–941. https://doi.org/10.1016/j.chemosphere.2007.06.058.

Cornett, J. B. & Shockman, G. D. 1978 Cellular lysis of *Streptococcus faecalis* induced with Triton X-100. *Journal of Bacteriology* **135** (1), 153–160. https://doi.org/10.1128/JB.135.1.153-160.1978.

Cuny, P., Faucet, J., Acquaviva, M., Bertrand, J.-C. & Gilewizc, M. 1999 Enhanced biodegradation of phenanthrene by marine bacterium in the presence of synthetic surfactants. *Letters in Applied Microbiology* **29**, 242–245. https://doi.org/10.1046/j.1365-2672.1999.00623.x.

Dou, J., Liu, X. & Ding, A. 2009 Anaerobic degradation of naphthalene by the mixed bacteria under nitrate reducing conditions. *Journal of Hazardous Materials* **165** (1–3), 325–331. https://doi.org/10.1016/j.jhazmat.2008.10.002.

Eltoukhy, A., Jia, Y., Nahurira, R., Abo-Kadoum, M. A., Khokhar, I., Wang, J. & Yan, Y. 2020 Biodegradation of endocrine disruptor Bisphenol A by *Pseudomonas putida* strain YC-AE1 isolated from polluted soil, Guangdong, China. *BMC Microbiology* **20** (1), 11. https://doi.org/10.1186/s12866-020-1699-9.

Fang, C. R., Yao, J., Zheng, Y. G., Jiang, C. J., Hu, L. F., Wu, Y. Y. & Shen, D. S. 2010 Dibutyl phthalate degradation by *Enterobacter* sp. T5 isolated from municipal solid waste in landfill bioreactor. *International Biodeterioration & Biodegradation* **64** (6), 442–446. https://doi.org/10.1016/j.ibiod.2010.04.010.

Feng, N. X., Yu, J., Mo, C. H., Zhao, H. M., Li, Y. W., Wu, B. X., Cai, Q. Y., Li, H., Zhou, D. M. & Wong, M. H. 2018 Biodegradation of di-n-butyl phthalate (DBP) by a novel endophytic *Bacillus megaterium* strain YJB3. *Science of the Total Environment* **616**, 117–127. https://doi.org/10.1016/j.scitotenv.2017.10.298.

Franzetti, A., Di Gennaro, P., Bestetti, G., Lasagni, M., Pitea, D. & Collina, E. 2008 Selection of surfactants for enhancing diesel hydrocarbons-contaminated media bioremediation. *Journal of Hazardous Materials* **152** (3), 1309–1316. https://doi.org/10.1016/j.jhazmat.2007.08.005.

Gadupudi, C. K., Rice, L., Xiao, L. & Kantamaneni, K. 2020 Endocrine disrupting compounds removal methods from wastewater in the United Kingdom: a review. *Preprints* **2020**, 2020110078. https://doi.org/10.3390/sci1010015.v1.

Gani, K. M. & Kazmi, A. A. 2020 Ecotoxicological risk evaluation and regulatory compliance of endocrine disruptor phthalates in a sustainable wastewater treatment scheme. *Environmental Science and Pollution Research* **27** (8), 7785–7794. https://doi.org/10.1007/s11356-019-07418-7.

Gao, D. W. & Wen, Z. D. 2016 Phthalate esters in the environment: a critical review of their occurrence, biodegradation, and removal during wastewater treatment processes. *Science of the Total Environment* **541**, 986–1001. https://doi:org/10.1016/j.scitotenv.2015.09.148.

Godfray, H. C. J., Stephens, A. E., Jepson, P. D., Jobling, S., Johnson, A. C., Matthiessen, P., Sumpter, J. P., Tyler, C. R. & McLean, A. R. 2019 A restatement of the natural science evidence base on the effects of endocrine disrupting chemicals on wildlife. *Proceedings of the Royal Society B* **286** (1897), 20182416. https://doi.org/10.1098/rspb.2018.2416.

Grinbaum, M., Camponovo, A., Desseigne, J. M., Poupault, P., Meisterman, E., Chatelet, B., Davaux, F. & Lempereur, V. 2019 Phthalates: potential sources and control measures. In: *BIO Web of Conferences* **12**, 04008. https://doi.org/10.1051/bioconf/20191204008.

Guo, X., Xie, C., Wang, L., Li, Q. & Wang, Y. 2019 Biodegradation of persistent environmental pollutants by *Arthrobacter* sp. *Environmental Science and Pollution Research* **26** (9), 8429–8443. https://doi.org/10.1007/s11356-019-04358-0.

Hadibarata, T. & Tachibana, S. 2010 Characterization of phenanthrene degradation by strain *Polyporus* sp. s133. *Journal of Environmental Sciences* **22** (1), 142–149. https://doi.org/10.1016/S1001-0742(09)60085-1.

Kaczorek, E. & Olszanowski, A. 2011 Uptake of hydrocarbon by *Pseudomonas fluorescens* (P1) and *Pseudomonas putida* (K1) strains in the presence of surfactants: a cell surface modification. *Water, Air, & Soil Pollution* **214** (1–4), 451–459. https://dx.doi.org/10.1007%2Fs11270-010-0436-7.

Kumar, S. & Singh, P. 2016 Various techniques for solubility enhancement: an overview. *The Pharma Innovation Journal* **5**, 23–28.

Kumar, V., Sharma, N. & Maitra, S. S. 2017 Comparative study on the degradation of dibutyl phthalate by two newly isolated *Pseudomonas* sp. v21b and *Comamonas* sp. 51F. *Biotechnology Reports* **15**, 1–10. https://doi.org/10.1016/j.btre.2017.04.002.

La Merrill, M. A., Vandenberg, L. N., Smith, M. T., Goodson, W., Browne, P., Patisaul, H. B., Guyton, K. Z., Kortenkamp, A., Cogliano, V. J., Woodruff, T. J. & Rieswijk, L. 2019 Consensus on the key characteristics of endocrine-disrupting chemicals as a basis for hazard identification. *Nature Reviews Endocrinology* 1–13. https://doi.org/10.1038/s41574-019-0273-8.

Lauretta, R., Sansone, A., Sansone, M., Romanelli, F. & Appetecchia, M. 2019 Endocrine disrupting chemicals: effects on endocrine glands. *Frontiers in Endocrinology* **10**, 178. https://doi.org/10.3389/fendo.2019.00178.

Li, Y., Yan, H. Q. & Yu, X. Y. 2019 Uptake and accumulation of di-n-butyl phthalate in six leafy vegetables under hydroponic conditions. *Food Production, Processing and Nutrition* **1** (1), 9. https://doi.org/10.1186/s43014-019-0009-0.

Mahanty, B., Sarma, S. J. & Pakshirajan, K. 2007 Evaluation of different surfactants for use in pyrene biodegradation by *Mycobacterium frederiksbergense*. *International Journal of Chemical Sciences* **5**, 1505–1512. http://gyan.iitg.ernet.in/handle/123456789/178.

Margina, D., Niţulescu, G. M., Ungurianu, A., Mesnage, R., Goumenou, M., Sarigiannis, D. A., Aschner, M., Spandidos, D. A., Renieri, E. A., Hernández, A. F. & Tsatsakis, A. 2019 Overview of the effects of chemical mixtures with endocrine disrupting activity in the context of real-life risk simulation (RLRS): an integrative approach. *World Academy of Sciences Journal* **1** (4), 157–164. https://doi.org/10.3892/wasj.2019.17.

Net, S., Sempéré, R., Delmont, A., Paluselli, A. & Ouddane, B. 2015 Occurrence, fate, behavior and ecotoxicological state of phthalates in different environmental matrices. *Environmental Science & Technology* **49** (7), 4019–4035. https://doi.org/10.1021/es505233b.

Paria, S. 2008 Surfactant-enhanced remediation of organic contaminated soil and water. *Advances in Colloid and Interface Science* **138** (1), 24–58. https://doi.org/10.1016/j.cis.2007.11.001.

Paul, T., Baskaran, D., Pakshirajan, K. & Pugazhenthi, G. 2019 Continuous bioreactor with cell recycle using tubular ceramic membrane for simultaneous wastewater treatment and bio-oil production by oleaginous *Rhodococcus opacus*. *Chemical Engineering Journal* **367**, 76–85. https://doi.org/10.1016/j.cej.2019.02.050.

Plotnikova, E. G., Yastrebova, O. V., Anan'ina, L. N., Dorofeeva, L. V., Lysanskaya, V. Y. & Demakov, V. A. 2011 Halotolerant bacteria of the genus *Arthrobacter* degrading polycyclic aromatic hydrocarbons. *Russian Journal of Ecology* **42** (6), 502. https://doi.org/10.1134/S1067413611060130.

Ren, L., Jia, Y., Ruth, N., Shi, Y., Wang, J., Qiao, C. & Yan, Y. 2016 Biotransformations of bisphenols mediated by a novel *Arthrobacter* sp. strain YC-RL1. *Applied Microbiology and Biotechnology* **100** (4), 1967–1976. https://doi.org/10.1007/s00253-015-7076-1.

Roccuzzo, S., Beckerman, A. P. & Trögl, J. 2020 New perspectives on the bioremediation of endocrine disrupting compounds from wastewater using algae-, bacteria- and fungi-based technologies. *International Journal of Environmental Science and Technology* 1–18. https://doi.org/10.1007/s13762-020-02691-3.

Sacan, M. T., Özkul, M. & Erdem, S. S. 2005 Physico-chemical properties of PCDD/PCDFs and phthalate esters. *SAR and QSAR in Environmental Research* **16** (5), 443–459. https://doi.org/10.1080/10659360500320602.

Sartoros, C., Yerushalmi, L., Béron, P. & Guiot, S. R. 2005 Effects of surfactant and temperature on biotransformation kinetics of anthracene and pyrene. *Chemosphere* **61** (7), 1042–1050. https://doi.org/10.1016/j.chemosphere.2005.02.061.

Shi, M., Sun, Y., Wang, Z., He, G., Quan, H. & He, H. 2019 Plastic film mulching increased the accumulation and human health risks of phthalate esters in wheat grains. *Environmental Pollution* **250**, 1–7. https://doi.org/10.1016/j.envpol.2019.03.064.

Sinharoy, A., Baskaran, D. & Pakshirajan, K. 2019 Sustainable biohydrogen production by dark fermentation using carbon monoxide as the sole carbon and energy source. *International Journal of Hydrogen Energy* **44** (26), 13114–13125. https://doi.org/10.1016/j.ijhydene.2019.03.130.

Thomsen, M., Carlsen, L. & Hvidt, S. 2001 Solubilities and surface activities of phthalates investigated by surface tension measurements. *Environmental Toxicology and Chemistry: An International Journal* **20** (1), 127–132. https://doi.org/10.1002/etc.5620200113.

Unell, M., Nordin, K., Jernberg, C., Stenström, J. & Jansson, J. K. 2008 Degradation of mixtures of phenolic compounds by *Arthrobacter chlorophenolicus* A6. *Biodegradation* **19** (4), 495–505. https://doi.org/10.1007/s10532-007-9154-2.

Vinay, K. & Maitra, S. S. 2016 Efficient degradation of dibutyl phthalate and utilization of phthalic acid esters (PAES) by *Acinetobacter* species isolated from MSW (municipal solid waste) leachate. *Leachate* **18** (4), 817–830. https://doi.org/10.30955/gnj.002028.

Volkering, F., Breure, A. M. & Rulkens, W. H. 1998 Microbiological aspects of surfactant use for biological soil remediation. *Biodegradation* **8** (6), 401–417. https://doi.org/10.1023/A:1008291130109.

Wang, Q. & Xie, S. 2012 Isolation and characterization of a high-efficiency soil atrazine-degrading *Arthrobacter* sp. strain. *International Biodeterioration & Biodegradation* **71**, 61–66. https://doi.org/10.1016/j.ibiod.2012.04.005.

Wen, Z. D., Gao, D. W. & Wu, W. M. 2014 Biodegradation and kinetic analysis of phthalates by an *Arthrobacter* strain isolated from constructed wetland soil. *Applied Microbiology and Biotechnology* **98** (10), 4683–4690. https://doi.org/10.1007/s00253-014-5568-z.

Wu, X. L., Wang, Y. Y., Liang, R. X., Dai, Q. Y. & Chao, W. L. 2010 Degradation of di-n-butyl phthalate by newly isolated *Ochrobactrum* sp. *Bulletin of Environmental Contamination and Toxicology* **85** (3), 235–237. https://doi.org/10.1007/s00128-010-0080-3.

Zhang, Y., Chen, H., Liu, J., Geng, G., Liu, D., Geng, H. & Xiong, L. 2018 Genome sequencing and biodegradation characteristics of the n-butyl benzyl phthalate degrading bacterium-*Rhodococcus* sp. HS-D2. *International Biodeterioration & Biodegradation* **128**, 56–62. https://dx.doi.org/10.1016/j.ibiod.2016.08.024.

First received 22 August 2020; accepted in revised form 19 November 2020. Available online 1 December 2020

Assessing the productivity of coriander under different irrigation depths and fertilizers applied with continuous and pulsed drip systems

Valentin Ruben Orcon Zamora, Manassés Mesquita da Silva, José Amilton Santos Júnior, Gerônimo Ferreira da Silva, Dimas Menezes and Ceres Duarte Guedes Cabral de Almeida

ABSTRACT

In this study, the authors analyzed the consequences of irrigation management strategies as an essential factor to save water and maintain high crop yields. The present research aimed at comparing coriander (*Coriandrum sativum* L.), cv. Verdão, yields under fertigation via drip irrigation with a continuous application and in pulses, with 40, 60, 80, 100 and 120% of the crop evapotranspiration (ETc) being recovered. These treatments were distributed in randomized blocks in a 2×5 factorial scheme, with three replications, in a greenhouse located at the Federal Rural University of Pernambuco, northeast Brazil (08°01′6.50″ S and 34°56′46″ W, average elevation 6.5 m). The continuous irrigation consisted of the application of water depth in a single daily event, while pulsed irrigation comprised the application of the same irrigation depth split up into five events with a 1-hour interval between pulses. Crop evapotranspiration (ETc) was determined via the water balance by lysimetric drainage. The fertilizers were applied in every watering. The biometric and productivity crop pointed that pulsed irrigation to a mean depth of 58% of the ETc corresponds to the results obtained with the application of the depth equivalent to 100% of the ETc by continuous irrigation, consequently with reduced input costs.

Key words | ARDUINO, *Coriandrum sativum* L., crop evapotranspiration, deficit irrigation, irrigation management

Valentin Ruben Orcon Zamora
Manassés Mesquita da Silva
José Amilton Santos Júnior
Gerônimo Ferreira da Silva
Agricultural Engineering Department,
Federal Rural University of Pernambuco (UFRPE), St Dom Manuel de Medeiros,
s/n, Dois Irmãos, Recife 52171-900,
PE,
Brazil

Dimas Menezes
Agronomy Department,
Federal Rural University of Pernambuco (UFRPE), St Dom Manuel de Medeiros,
s/n, Dois Irmãos, Recife 52171-900,
PE,
Brazil

Ceres Duarte Guedes Cabral de Almeida (corresponding author)
Dom Agostinho Ikas Agricultural School,
Federal Rural University of Pernambuco,
Dr Francisco Correia Avenue, 643,
São Lourenço da Mata, Pernambuco 54735-000,
Brazil
E-mail: *ceres.codai.ufrpe@gmail.com*

HIGHLIGHTS

- Drip irrigation pulsed saves 40% water compared to continuous irrigation.
- Frequency of watering and fertilization determines crop yield.
- Pulsed fertigation provides higher fresh and dry mass with deficit irrigation.

doi: 10.2166/ws.2021.008

GRAPHICAL ABSTRACT

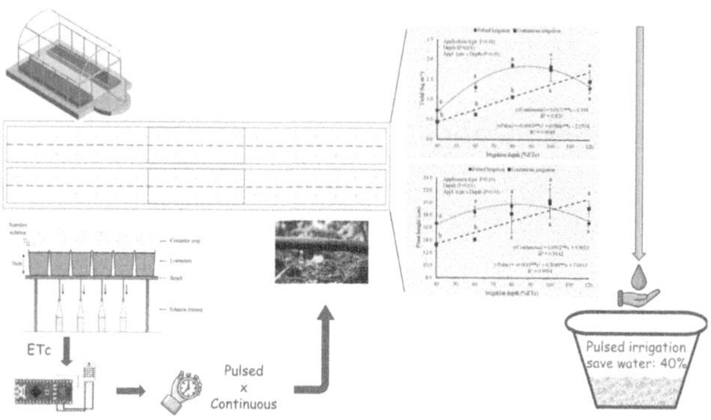

INTRODUCTION

Coriander (*Coriandrum sativum* L.) is a widely consumed herb in the northeast region of Brazil, especially in the form of green shoots, but also as ground seeds (Silva *et al.* 2012). Despite the economic and social importance of this region, coriander cultivation is still practiced in an empirical way (Cavalcante *et al.* 2016), with few studies that seek to optimize water and nutrient use.

Although India is the largest producer, consumer and exporter of spices in the world, mainly those known seed spices, as coriander (Meena *et al.* 2019; Tridge 2019), and Brazil is not among the largest world producers, the high consumption of coriander in the North and Northeast of Brazil allows the inference that the future production scenario is promising. Therefore, researches that explore the way of tillage, reduction of water consumption via irrigation and use of fertigation can subsidize the increase of the production area in a sustainable way. The information available is still incipient regarding the proper irrigation management and specifically regarding the use of pulse drip fertigation. Menezes *et al.* (2020), in similar conditions to this study, evaluated the accumulation of nutrients in coriander under continuous and pulse drip fertigation depths, and observed that the highest accumulation of K, Mn, Cu and Zn were obtained with pulse fertigation combined with water depths lower than 100% ETc. In this sense, it is possible to define the most appropriate production strategies for coriander and other leafy vegetables, rationalizing use of agricultural inputs and reducing production costs.

The deficiency of knowledge about irrigation management techniques of the growers results in a low average yield of most vegetable crops; on the other hand, an increase in yields of drip irrigated vegetables has also been reported by several researchers (Imtiyaz *et al.* 2000; Jha *et al.* 2017; Silva *et al.* 2018). Water use optimization is necessary for northeast Brazil, especially in the semiarid region, where farmers are completely dependent on irrigation for plant growth. Besides that, cost reduction in terms of water, nutrients, energy and operational costs is important for the viability of agricultural businesses (Sandri *et al.* 2014). Therefore, the development of sustainable water use practices in agriculture is essential for the rationalization of food production.

A technique that increases the water and nutrient use efficiency is localized irrigation, applied in pulses; that is, instead of applying irrigation in a single event, it is divided into several events. Eid *et al.* (2013) comment that this favors the water movement in the horizontal direction, improving the distribution of soil moisture and increasing the wet soil volume in the root zone. In this sense, some researchers have also associated it with the use of lower depths for crop evapotranspiration, the so-known 'deficit irrigation'. This statement is supported by the maintenance of soil moisture throughout the day being likely to be a suitable strategy to mitigate the harmful effects of deficit water and useful in times of limited water availability, mainly in semiarid regions (Martin *et al.* 2012; Eid *et al.* 2013; Phogat *et al.* 2013; Stallmann *et al.* 2020).

However, approaches of pulsed drip fertigation with soil water depletion are still scarce, although, in Brazil, some studies have been performed with pulse dripping on lettuce (Almeida *et al.* 2017), bean pod (Almeida *et al.* 2016) and coriander (Menezes *et al.* 2020). In Egypt, this approach has been tested in soybean and potato crops (Abdelraouf *et al.* 2012; Eid *et al.* 2013) as well the advantage of using composite conditioner with pulsed irrigation to reduce deep percolation and obtain a wide horizontal spread of wetting (Abd-elhakim 2019), while studies in Saudi Arabia have applied it to tomato crops (Elnesr *et al.* 2015).

This study evaluated the productivity of coriander (*Coriandrum sativum* L.), cv. Verdão, under five irrigation rates (40%–120% of estimated soil water depletion) and fertigated via a drip irrigation system in continuous and pulsed applications.

MATERIALS AND METHODS

The experiment was carried out between August and September 2017 in a greenhouse at the Federal Rural University of Pernambuco, Recife City, in northeast Brazil (08° 01′ 6.50″ S and 34° 56′ 46″ W, average elevation is 6.5 m above sea level). The local climate, as classified by Köppen, is As' type, tropical hot and humid (Jales *et al.* 2012). Air temperature and relative humidity were monitored via an electronic sensor. The average maximum and minimum air temperatures registered were 34.35 and 25.66 °C, respectively, with average maximum air relative humidity of 92.22% and minimum of 55.90%.

The experimental design consisted of randomized blocks in a 2 × 5 factorial scheme, with three replicates, resulting in 30 experimental plots. Two types of water application (continuous and pulsed) and five ETc replacement depths (120, 100, 80, 60 and 40%) were evaluated. The experimental plots consisted of 30 masonry beds of 1.02 m^2 (5.10 m (length) × 0.20 m (width) × 0.20 m (height)), waterproofed with polyethylene plastic film and with drainage tubes installed in the longitudinal direction of each bed. The sandy soil used in this experiment (Table 1) was characterized according to the methodologies of Teixeira *et al.* (2017).

Due to the acidic condition of the soil, acidity needed to be corrected with the application of limestone. Mineral fertilization was performed with phosphorus, P-fertilization, in planting (simple superphosphate -SSP) according to the recommendations of Cavalcanti (2008), and the other nutrients were applied daily via fertigation (Furlani 1998). The continuous irrigation consisted of the application of the calculated depth once a day; irrigation in pulses consisted

Table 1 | Chemical and physical characterization of the soil

Chemical attributes									
pH water	**Ca^{2+}**	**Mg^{2+}**	**Na^+**	**K^+**	**Al^{3+}**	**$Al + H^+$**	**P^+**	**O.C.[a]**	**O.M.[b]**
	$cmol_c\ dm^3$						$mg\ dm^{-3}$	$g\ kg^{-1}$	
5.1	2.0	1.5	0.01	0.01	0.2	4.68	2.0	5.62	9.69
Physical attributes									
Sand	**Loam**	**Clay**	**Pd[c]**	**Bd[d]**	**θfc[e]**	**θpwp[f]**			
	$g\ kg^{-1}$		$kg\ dm^{-3}$		$m^3\ m^{-3}$				
904	32	64	2.5	1.5	0.10	0.09			

[a]Organic carbon.
[b]Organic matter.
[c]Particle density.
[d]Bulk density.
[e]Field capacity.
[f]Permanent wilting point.

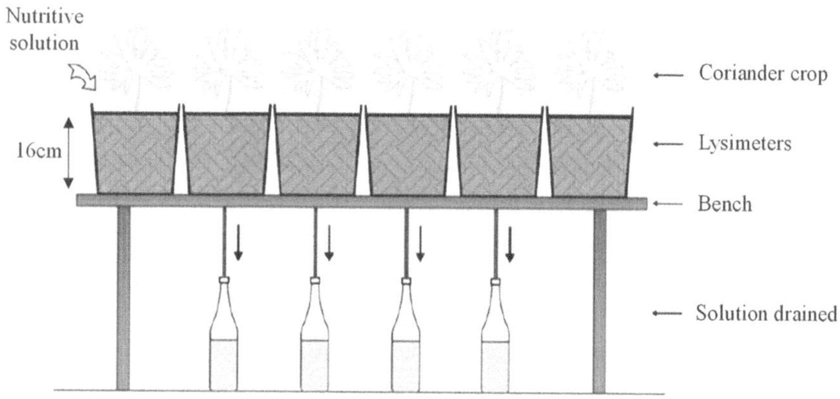

Figure 1 | Schematic diagram of non-weighting drainage lysimeter design: pot design and collecting container.

Table 2 | Irrigation depth applied per treatment

	Etc									
	120%		100%		80%		60%		40%	
Depth (mm)	**Pulse**	**Cont.**	**Pulse**	**Cont.**	**Pulse**	**Cont.**	**Pulse**	**Cont.**	**Pulse**	**Cont.**
Initial[a]	19.2	19.2	19.2	19.2	19.2	19.2	19.2	19.2	19.2	19.2
Treatment[b]	105.6	105.6	88.0	88.0	70.4	70.4	52.8	52.8	35.2	35.2
Total[c]	124.8	124.8	107.2	107.2	89.6	89.6	72.0	72.0	74.4	74.4

[a]Total depth applied in the first 10 days after sowing.
[b]Total depth applied from the tenth to the thirty-fourth day after sowing.
[c]Total depth applied during the cycle.

of dividing to the same depth into five applications a day, with an interval of 1 hour between pulses. The fertilizers were applied in every watering. The crop evapotranspiration (ETc) replacement depths were calculated supported by drainage lysimeters, installed in the experimental environment (Figure 1). The daily water consumption (ETc) was calculated by subtracting the volume of water collected as drainage during a 24-h period from the amount that was supplied by irrigation depth during the same period. Treatments started on the 10th day after sowing; until then, 100% of the ETc was applied (Table 2).

The gross irrigation depth (d gross) was the product of ETc and the efficiency of the irrigation system. The irrigation time was obtained considering the space occupied by each plant, the average flow of the dripper and the number of emitters per plant. The gross irrigation depth and the irrigation application times for each treatment were determined automatically using an electronic controller, commanded by ARDUINO. Thus, via a programming code, the relays were switched on, which turned on the solenoid valves, applying the required irrigation depth according to the treatments.

The same water quality and composition of the nutritive solution, according to Furlani (1998), have been used in all treatments; a single 1,000-L reservoir was filled with a nutritive solution with an initial electrical conductivity of 2.5 dS m^{-1}. The drip irrigation system used was composed of one drip hose (outside diameter – OD: 16 mm) per plot with non-pressure compensating (NPC) drippers spaced 0.30 m apart, with a nominal flow of 0.60 L h^{-1} and a 120-mesh disk filter. The pressurizing system consisted of one 0.5-cv centrifugal pump motor to keep the service pressure at 10 mwc (meters of water column), according to the manufacturer's recommendation.

Water distribution uniformity was evaluated twice, at the beginning and the end of the experiment, following the methodology described by Merriam & Keller (1978). The sampling flow used four drippers per lateral line: the first emitter, the emitters placed in the center of the lateral line, and the last emitter, with a collection time of twelve minutes measured in a digital chronometer. The mean flow measured in the drippers was 0.55 L h^{-1}, and the water distribution uniformity was 98.20% at the beginning and 97.09% at the end of the experiment, indicating high uniformity.

Sowing of coriander (*Coriandrum sativum* L.), Verdão cultivar, was performed directly in the experimental plots, with 20 seeds per pit, spaced at 0.10 m between rows and 0.15 m between plants, and to control border effects the useful area was 0.12 m^2. Ten days after sowing, thinning was performed, leaving six plants per pit. Throughout the crop development, weeds were controlled manually. Although there was no presence of diseases, pests, such as caterpillars and aphids, were controlled with a single dose of the contact insecticide Decis®.

At 35 days after sowing, the following variables were analyzed: (i) fresh and dry mass of shoot and root, using a precision scale (0.001 g) and after drying of the fresh material in a forced ventilation oven at 65 °C until reaching constant weight; (ii) percentage of the shoot and root dry mass (% SDM and % RDM, respectively), based on the ratio between dry and fresh mass; (iii) height of the plant, average values of the useful area and measurement from the base of the stem until the last leaf; and (v) crop productivity in terms of fresh mass production (shoot + root) per square meter.

The analysis of variance (ANOVA) was used to determine whether the means were different, using the F test. Regression analysis was used to compare the quantitative factors, and the agglomerative test of Scott Knott averages was applied with a significance threshold of 0.05 for the qualitative factors. For all analyses, we used the statistical software package SISVAR (Ferreira 2011).

RESULTS AND DISCUSSION

There was a significant effect ($p < 0.01$) of the type of application and irrigation depth on the percentage of root dry mass (% RDM). For the percentage of shoot dry mass (% SDM) the effect was significant ($p < 0.01$) only for the irrigation depth. For shoot and root fresh mass (SFM and RFM, respectively), shoot and root dry mass (SDM, RDM), plant height and yield (Y), there was a significant effect ($p < 0.01$) for the interaction between the two factors. The coefficients of variation were lower than 19%, indicating adequate uniformity in the experiment.

Under continuous irrigation, shoot fresh mass (SFM) increased linearly at a rate of 14.17 g m^{-2} to 120% of the ETc, while under pulsed irrigation the maximum yield (1,621.80 g m^{-2}) was reached with irrigation depth equivalent to 88.42% ETc. However, we highlight that above 100% of ETc, there was no difference ($p > 0.05$) in the SFM under continuous irrigation or pulse irrigation (Figure 2(a)). On the other hand, under continuous application of irrigation depth equivalent to 100% ETc, the crops produced an estimated SFM of 1,222.67 g m^{-2}, while under pulse irrigation, the same yield was obtained with depth irrigation equivalent to 58.28% ETc, which resulted in saving 41.71% of the water volume required per crop cycle, equivalent to 446.96 m^3 ha^{-1} (Figure 2(a)). In contrast, Linhares *et al.* (2015) and Aguiar *et al.* (2015) verified a maximum SFM of 645.30 g m^{-2} (6,453 kg ha^{-1}) and 1,197.80 g m^{-2} (11,978 kg ha^{-1}) for the same cultivar at 35 and 36 days after sowing, respectively.

Under continuous irrigation, root fresh mass (RFM) was increased up to 120% of the ETc, at a rate of 1.53 g m^{-2} at each 10 mm added to the fertigation depth, distinguishing ($p < 0.01$) data obtained when pulsed irrigation was used, except under 40% ETc (Figure 2(b)). When irrigation occurred via pulses, the maximum RFM was 208.75 g m^{-2} under the estimated depth at 86.46% of ETc. In another analysis, under the ETc estimated at 56.92%, the plants produced the same RFM (153.96 g m^{-2}) observed when 100% ETc under continuous irrigation was applied (Figure 2(b)).

At the SFM and RFM of the treatments under depth lower than 100% of the ETc, pulsed irrigation resulted in higher value when compared to continuous irrigation. These results point that this irrigation strategy optimizes water and nutrient use in the fertigation process, as also verified by Assouline *et al.* (2012) and Warner *et al.* (2009), who explained this result with the high nutrient availability and water absorption.

The shoot (SDM) and root dry mass (RDM) showed similar results, with an increase of 1.40 and 0.12 g m^{-2}, respectively, at each increment of 10 mm in the fertigation depth when applied continuously up to 120% of ETc. The SDM and RDM data obtained under 100% ETc (129.75

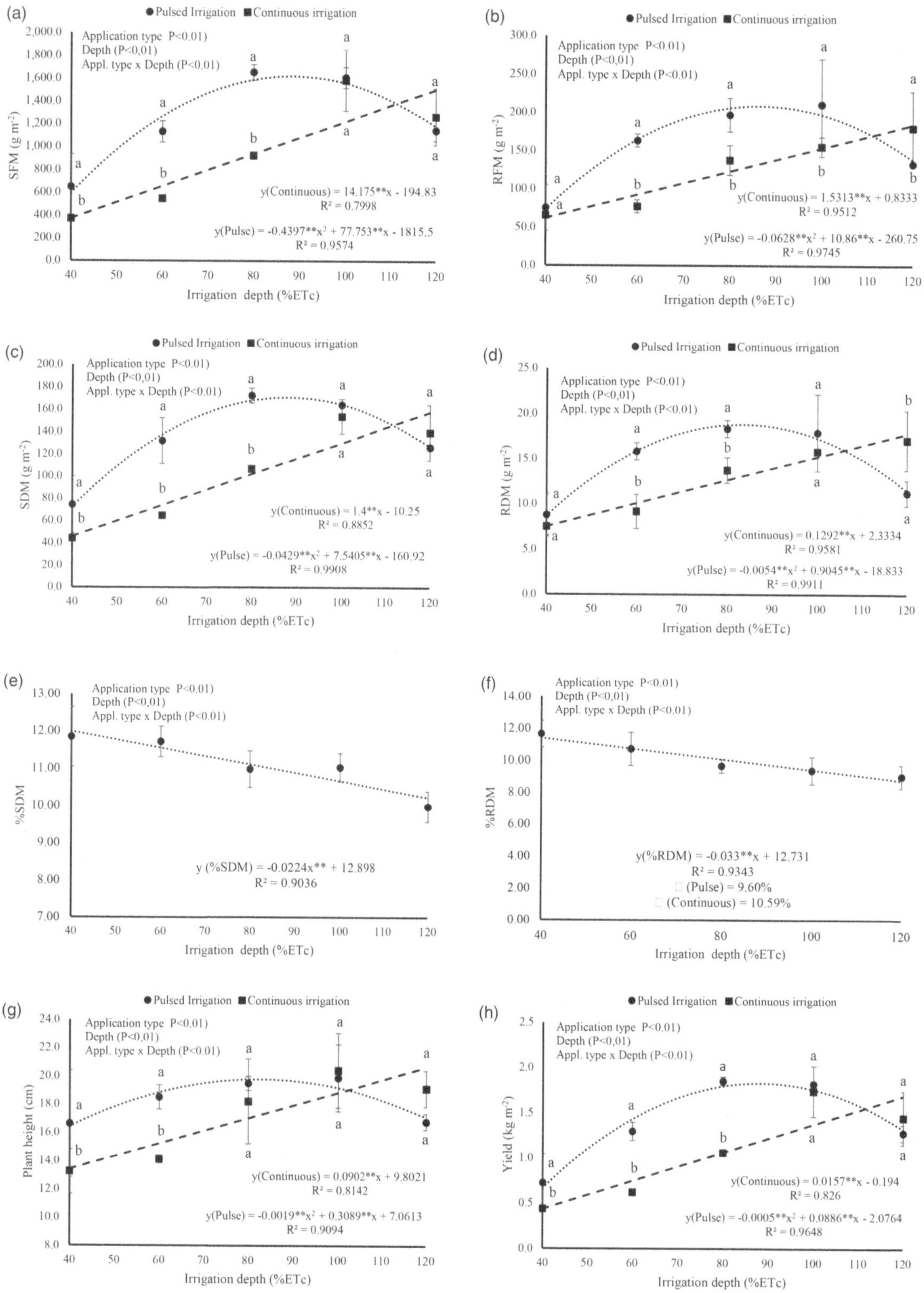

Figure 2 | Shoot and root fresh (a and b) and shoot and root dry mass (c and d), percentage of shoot dry mass (e) and root dry mass (f), the height of plants (g), and coriander cv. Verdão yield (h), under irrigation levels applied via drip irrigation system with and without pulses. Different letters indicate significant differences between the continuous irrigation and pulsed irrigation (Scott Knott test, $P < 0.05$).

and 15.25 g m^{-2}, respectively) were obtained with a depth estimated at 57.00 and 57.25% of ETc when pulsed irrigation was used. As a result, the maximum SFM and RFM yield increased by 170.42 g m^{-2} (87.88% ETc) and 19.04 g m^{-2} (83.75% ETc), respectively (Figure 2(c) and 2(d)).

Aguiar *et al.* (2015) obtained, for the cultivar Verdão (161.40 g m^{-2}), a maximum SDM value of 161.40 g m^{-2} at 36 days after sowing under organic fertilization. In the present research, the maximum SDM was 73.4% of the ETc under pulsed fertigation. Eid *et al.* (2013) suggest that the increase of the wet bulb in the root zone may favor nutrient absorption and, consequently, the increase in SDM and RDM in irrigation depths less than 100% of ETc. Therefore, the response of the plants to pulsed irrigation (SFM, RFM, SDM, and RDM) indicates the reduction of input costs, such as costs of water, nutrients, electrical energy, and maintenance, up to the depth that maximizes production; in this case, less than 100% of ETc.

The reduction of % SDM and the % RDM at the ratio of 0.17 and 0.259% for each 10-mm increase in the ETc replacement may be associated with the increase in the water content at a ratio higher than the dry mass gain, which was also favored by the greater water and nutrient availability (Figure 2(e) and 2(f)). Continuous irrigation technique increased % RDM, suggesting lower dry mass gain. Sousa *et al.* (2017), exposing sorghum to different irrigation depths, also verified a reduction in the percentage of leaves dry mass with the increase of the irrigation depth; similarly, Delazari *et al.* (2017) reported a positive increase in the % RDM of sweet potato until reaching a maximum depth, with a subsequent decrease.

Plant height was higher ($p < 0.01$) under pulsed irrigation to irrigation depth at 40 and 60% of ETc compared to continuous irrigation. However, under 100% of the ETc, plant height of 18.82 cm was reached under continuous fertigation; this same plant height was obtained with 60.84% of ETc applied via pulses (Figure 2(g)). On the other hand, the estimated depth at 81.29% of ETc, when pulse irrigation was applied, resulted in the maximum plant height (19.61 cm). Lower values were obtained by Linhares *et al.* (2015) when evaluating organic fertilization, obtaining a maximum plant height of 18.10 cm at 35 days after sowing for the same cultivar. Other authors studying the same cultivar obtained higher plant height values. For example, Aguiar *et al.* (2015) evaluated the production of coriander grown with organic compounds and obtained a plant height of 31 cm for the highest fertilization dose at 36 days after sowing. Pereira *et al.* (2015) evaluated coriander cultivar performance as a function of irrigation management in a semiarid region and obtained a maximum plant height of 40.75 cm at 40 days after sowing. Marsaro *et al.* (2014), evaluating coriander cultivar yields under greenhouse and field conditions, report a mean plant height of 29.16 cm at 43 days after sowing. Angeli *et al.* (2016), assessing the production and water use efficiency in coriander under irrigation and nitrogen fertilization, reported a plant height of 43 cm for the applied dose of 105 kg ha^{-1} of N, not indicating the variety used. Rajasekar *et al.* (2013), evaluated the influence of a growing environment (shade net house and open field), observing a mean plant height of 34 and 24 cm in the shade net and open conditions, respectively.

Comparing the results obtained in the present research with the data described above, the higher results found in the other studies are most likely due to the application of chemical fertilizers to stable soils. Besides that, under field conditions, cultivation time was up to 43 days. Another important factor to consider is that all other trials, except Marsaro *et al.* (2014), were established in the open field, where the direct solar radiation allows a better photosynthetic activity and, therefore, a greater formation of photoassimilates.

Coriander productivity was higher ($p < 0.01$) when irrigation was lower than 100% of the ETc and applied in pulses compared to continuous irrigation. As a result, the maximum productivity (1.84 kg m^{-2}) was reached under the estimated depth of 88.60% of the ETc, while under continuous irrigation with 100% of the ETc, crop productivity was 1.37 kg m^{-2}. This result could be obtained with pulsed irrigation to an estimated depth of 57.85% of the ETc (Figure 2 h).

Some authors (Linhares *et al.* 2014; Marsaro *et al.* 2014) have obtained lower yields with the same cultivar when compared to the yields observed in this study, while others (Maciel *et al.* 2012; Angeli *et al.* 2016) have obtained higher yields. Although these results were obtained under different experimental conditions and harvesting times, the maximum productivity at a depth lower than 100% of the ETc, presented here, serves as a guideline for the establishment of management strategies to optimize water and nutrient use (Assouline *et al.* 2012).

Drip pulse irrigation management can increase efficiency and reduce nutrient losses in irrigated vegetable fields. The split irrigation of the necessary daily depth is feasible as it provides good practice to save water, energy and nutrients. In the case of the coriander crop, according to our results, adopting 60% of the ETc applied in pulses with an interval of 60 minutes between two pulses is a good irrigation management strategy for greenhouse conditions.

Further investigations are, however, necessary to identify the methodology and/or protocol to define the number of pulses and time between two successive pulses, considering dripper flow, crop, soil, and local climate. Besides that, identify the possibility of using similar management with other crops, with brackish water, fertigation, and apply successive cycles with deficit irrigation and the application of a leaching depth between cycles.

CONCLUSIONS

1. Pulsed fertigation is recommended for coriander crop and provides higher shoot and root values of fresh and dry mass as well as plant height and yield values, with ETc replacement of less than 100%.
2. The reduction in input costs, such as costs for water, nutrients, electric energy, and maintenance, in the production of coriander can be obtained by adopting pulsed irrigation, because maximum production is reached with less than 100% of the ETc.
3. The biometric and productivity variables showed that pulsed irrigation to a mean depth around 60% of the ETc corresponds to the results obtained with the application of the depth equivalent to 100% of the ETc by continuous irrigation.

ACKNOWLEDGEMENT

This study was financed in part by the Coordenação de Aperfeiçoamento de Pessoal de Nível Superior – Brazil (CAPES) - Finance Code 001.

CONFLICT OF INTEREST

On behalf of all authors, the corresponding author states that there is no conflict of interest.

DATA AVAILABILITY STATEMENT

All relevant data are included in the paper or its Supplementary Information.

REFERENCES

Abd-elhakim, A. I. 2019 Soil conditioner effect on soil wetting patterns under pulsed drip irrigation system. *Misr Journal of Agricultural Engineering* **36** (2), 473–492. http://dx.doi.org/10.21608/mjae.2019.94653.

Abdelraouf, R. E., Refaie, K., Abou-Hussein, S. D. & El-Metwally, I. M. 2012 Effect of pulse irrigation on clogging emitters, application efficiency and water productivity of potato crop under organic agriculture conditions. *Australian Journal of Basic and Applied Sciences* **6** (3), 807–816. http://www.ajbasweb.com/old/ajbas/2012/March/807-816.pdf

Aguiar, A. M., Souza, J. A. E., Souza, R. F., Carvalho, C. A. S. & Ferreira, C. P. 2015 Produção de coentro (*Coriandrum sativum* L.) cultivado com composto orgânico em Irituia – Pará (*Coriander production (Coriandrum sativum L.) grown with organic compound in Irituia – Pará*). *Cadernos de Agroecologia* **10** (3), 1–5. http://revistas.aba-agroecologia.org.br/index.php/cad/article/view/19812

Almeida, W. F., Lima, L. A. & Pereira, G. M. 2017 Gotejamento por pulsos e cobertura do solo na produtividade da alface americana. *Revista Engenharia Agrícola* **35** (6), 1009–1018. http://dx.doi.org/10.1590/1809-4430-Eng.Agric.v35n6p1009-1018/2015.

Almeida, W. F., Paz, V. P. S., Jesus, A. P. C., Oliveira, A. S. & Silva, J. S. 2016 Interação entre a salinidade da água de irrigação e o gotejamento por pulsos na cultura do feijão-vagem. In: *IV Inovagri International Meeting*, Fortaleza-CE. http://dx.doi.org/10.7127/iv-inovagri-meeting-2017-res2920752

Angeli, K. P., Delazari, F. T., Nick, C., Ferreira, M. G. & Silva, D. J. H. 2016 Yield components and water use efficiency in coriander under irrigation and nitrogen fertilization. *Revista Brasileira de Engenharia Agrícola e Ambiental* **20** (5), 415–420. http://dx.doi.org/10.1590/1807-1929/agriambi.v20n5p415-420.

Assouline, S., Möller, M., Furman, A., Narkis, K. & Silber, A. 2012 Impact of water regime and growing conditions on soil-plant interactions: from single plant to field scale. *Vadose Zone Journal* **11**, 3–14. https://doi.org/10.2136/vzj2012.0006.

Cavalcante, A. R., Santos Junior, J. A., Gheyi, H. R., Dias, N. S. & Paz, V. P. S. 2016 Produção e composição mineral do coentro em sistema hidropônico de baixo custo (*Production and content of nutrients and sodium of coriander in hydroponic system of low cost*). *Irriga* **21**, 685–696. https://doi.org/10.15809/irriga.2016v21n4p685-696.

Cavalcanti, F. J. A. (coord.) 2008 Recomendações de adubação para o Estado de Pernambuco: 2ª. aproximação *(Fertilization Recommendations for the State of Pernambuco: 2nd. Approximation)*. 3.ed. rev. Recife: Instituto Agronômico de Pernambuco – IPA. 212 p.

Delazari, F. T., Ferreira, M. G., Silva, G. H., Dariva, F. D., Freitas, D. S. & Nick, C. 2017 Eficiência no uso da água e acúmulo de matéria na batata-doce em função de lâminas de irrigação (*Water use efficiency and matter accumulation in sweet potato due to different irrigation dephts*). *Irriga* **22** (1), 115–128. https://doi.org/10.15809/irriga.2017v22n1p115-128.

Eid, A. R., Bakry, B. A. & Taha, M. H. 2013 Effect of pulse drip irrigation and mulching systems on yield, quality traits and irrigation water use efficiency of soybean under sandy soil conditions. *Agricultural Sciences* **4** (1), 249–261. https://doi.org/10.4236/as.2013.45036.

Elnesr, M. N., Alazba, A. A., Zein El-Abedein, A. I. & El-Adl, M. M. 2015 Evaluating the effect of three water management techniques on tomato crop. *Plos One* **10** (6), 1–17. https://doi.org/10.1371/journal.pone.0129796.

Ferreira, D. F. 2011 Sisvar: a computer statistical analysis system. *Ciência e Agrotecnologia* **35** (6), 1039–1042. http://dx.doi.org/10.1590/S1413-70542011000600001.

Furlani, P. R. 1998 Instruções para o cultivo de hortaliças de folhas pela técnica de hidroponia NFT *(Instructions for Growing Leafy Vegetables Using the Hydroponics Technique NFT)*. Instituto Agronômico, Campinas.

Imtiyaz, M., Mgadla, N. P., Manase, S. K., Chendo, K. & Mothobi, E. O. 2000 Yield and economic return of vegetable crops under variable irrigation. *Irrigation Science* **19** (2), 87–93. https://doi.org/10.1007/s002710050005.

Jales, M. C., Feitosa, F. A. N., Koening, M. L., Bastos, R. B. & Machado, R. C. A. 2012 O Ecossistema recifal de Serrambi (nordeste do Brasil): Biomassa Fitoplanctônica e Parâmetros Hidrológicos (*The reef ecosystem of Serrambi (Northeastern of Brazil): phytoplanktonic biomass and hydrological parameters*). *Atlântica* **34** (2), 87–102. http://dx.doi.org/10.5088/atl.2012.34.2.87.

Jha, B. K., Mali, S. S., Naik, S. K. & Sengupta, T. 2017 Yield, water productivity and economics of vegetable production under drip and furrow irrigation in Eastern Plateau and Hill Region of India. *International Journal of Agricultural Science and Research* **7** (3), 43–50.

Linhares, P. C. F., Oliveira, J. D., Pereira, M. F. S., Fernandes, J. P. P. & Dantas, R. P. 2014 Espaçamento para a cultura do coentro adubado com palha de carnaúba nas condições de Mossoró-RN (*Spacing for the culture of cilantro fertilized with carnauba straw at the conditions of Mossoró-RN*). *Revista Verde de Agroecologia e Desenvolvimento Sustentável* **9** (3), 1–6. https://www.gvaa.com.br/revista/index.php/RVADS/article/view/2781

Linhares, P. C. F., Pereira, M. F. S., Moreira, J. C., Paiva, A. C. C., Assis, J. P. & Sousa, R. P. 2015 Rendimento do coentro (*Coriandrum sativum* L.) adubado com esterco bovino em diferentes doses e tempos de incorporação no solo (*Yield of the coriander [Coriandrum sativum (L.)] fertilized with manure at different doses and incorporation times in the soil*). *Revista Brasileira de Plantas Medicinais* **17** (3), 462–467. http://dx.doi.org/10.1590/1983-084X/13_006.

Maciel, G. M., Costa, C. P. & Sala, F. C. 2012 Linhagens de coentro com pendoamento tardio sob dois sistemas de plantio (*Coriander lines with later-bolting resistance under two planting systems*). *Horticultura Brasileira* **30** (4), 607–612. http://dx.doi.org/10.1590/S0102-05362012000400008.

Marsaro, R., Melo, K. D. A., Seabra Junior, S. & Borges, L. S. 2014 Produção de cultivares de coentro em diferentes telados e campo aberto (*Production of different cultivars of coriander below shading and open field*). *Revista Cultivando o Saber* **7** (4), 362–373. https://www.fag.edu.br/upload/revista/cultivando_o_saber/54eb2b5ad9ebe.pdf

Martin, J. D., Carlesso, R., Aires, N. P., Gatto, J. C., Dubou, V., Fries, H. M. & Scheibler, R. B. 2012 Irrigação deficitária para aumentar a produtividade da água na produção de silagem de milho (*Deficit irrigation in order to increase the productivity of water use in maize silage production*). *Irriga* **1** (1), 192–205. http://dx.doi.org/10.15809/irriga.2012v1n01p192.

Meena, M. D., Lal, G., Meena, S. S., Lal, S. & Chaudhary, N. 2019 Seed spices export from India: prospects and constraints. *International Journal of Seed Spices* **9** (2), 12–20.

Menezes, S. M., Silva, G. F., Silva, M. M., Morais, J. E. F., Santos Junior, J. A., Menezes, D. & Rolim, M. M. 2020 Continuous and pulse fertigation on dry matter production and nutrient accumulation in coriander. *Dyna* **87** (212), 18–25. http://doi.org/10.15446/dyna.v87n212.78569.

Merriam, J. L. & Keller, J. 1978 *Farm Irrigation System Evaluation: A Guide for Management. Agricultural & Irrigation Engineering Department*. Utah State University, Logan, 271p.

Pereira, J. C. A., Silva, S. S., Dantas, N. J., Paz, P. J. & Azevedo, C. A. V. 2015 Desempenho de cultivares de coentro em função do manejo da irrigação na região semiárida. In: *Congresso Técnico Científico da Engenharia e da Agronomia - Contecc*, Fortaleza-CE.

Phogat, V., Skewes, M. A., Mahadevan, M. & Cox, J. W. 2013 Evaluation of soil plant system response to pulsed drip irrigation of an almond tree under sustained stress conditions. *Agricultural Water Management* **118**, 1–11. https://doi.org/10.1016/j.agwat.2012.11.015.

Rajasekar, M., Arumugam, T. & Kumar, S. R. 2013 Influence of weather and growing environment on vegetable growth and yield. *Journal of Horticulture and Foresty* **5** (10), 160–167. https://doi.org/10.5897/JHF2013.0317.

Sandri, D., Pereira, J. A. P. & Vargas, R. B. 2014 Custos de produção e rentabilidade produtiva da melancia sob diferentes lâminas e sistemas de irrigação (*Production costs and profitability of watermelon under different water depths and irrigation systems*). *Irriga*, Botucatu **19** (3), 414–429. https://doi.org/10.15809/irriga.2014v19n3p414.

Silva, M. A. D., Coelho Júnior, L. F. & Santos, A. P. 2012 Vigor de sementes de coentro (*Coriandrum sativum* L.) provenientes de sistemas orgânico e convencional (*Evaluation of force of coriander (Coriandrum sativum L.) seeds from the organic and conventional systems*). *Revista Brasileira de Plantas Medicinais* **14**, 192–196. http://dx.doi.org/10.1590/S1516-05722012000500012.

Silva, V. P. R., Sousa, I. F., Tavares, A. L., Silva, T. G. F., Silva, B. B., Holanda, R. M., Brito, J. I. B., Braga, C. C., Souza, E. P. & Silva, M. T. 2018 Evapotranspiration, crop coefficient and water use efficiency of coriander grown in tropical environment. *Horticultura Brasileira*. **36**, 446–452. http://dx.doi.org/10.1590/S0102-053620180404.

Sousa, P. G. R., Viana, T. V. A., Carvalho, C. M., Azevedo, B. M., Sousa, J. P. F. & Campelo, D. H. 2017 Características agronômicas do sorgo forrageiro submetido a lâminas de irrigação e cobertura morta no semiárido (*Agronomic characteristics of the foraging sorghum submitted to the irrigation levels and mulch in semiarid regions*). *Revista Brasileira de Agricultura Irrigada* **11** (8), 2239–2248. https://doi.org/10.7127/rbai.v11n800731.

Stallmann, J., Schweiger, R., Pons, C. A. A. & Müller, C. 2020 Wheat growth, applied water use efficiency and flag leaf metabolome under continuous and pulsed deficit irrigation. *Scientific Reports* **10**, 10112–10124. https://doi.org/10.1038/s4159 8-020-66812-1.

Teixeira, P. C., Donagemma, G. K., Fontana, A. & Teixeira, W. G. 2017 Manual de métodos de análise de solo *(Manual of Soil Analysis Methods)*. 3.ed. rev. ampl. Embrapa Informação Tecnológica, Brasília, 574p.

Tridge 2019. *Market Intelligence, Coriander*. Available at: https://www.tridge.com/production

Warner, R., Hoffman, O. & Wilhoit, J. 2009 *The Effects of Pulsing Drip Irrigation on Tomato Yield and Quality in Kentucky*. Fruit and Vegetable Crop Research Report, pp. 39–40. http://www2.ca.uky.edu/agcomm/pubs/pr/pr603/pr603.pdf

First received 26 August 2020; accepted in revised form 27 December 2020. Available online 11 January 2021

Comparative analysis of select techniques and metrics for data reconciliation in smart energy distribution network

Jeyanthi Ramasamy, Sriram Devanathan and Dhanalakshmi Jayaraman

ABSTRACT

Reliability of each state of process in many chemical process industries largely relies upon water and vitality supplies. In this way, there is great necessity to have an improved and controlled smart energy distribution network (SEDN) in industries. In SEDNs, sensor information related to flow control and optimization serves as a basis for modelling of energy management systems. Therefore, it is important to ensure that sensor data are accurate and precise. However, they are affected by random noise and measurement biases, which compromise the quality of measurements. Data Reconciliation (DR) is one such approach popularly used in industries to reduce the adverse impact of random errors present in pipe flow measurements. In this study, Python-based simulations of weighted least squares (WLS) and principal component analysis (PCA) based DR techniques are implemented on the selected flow streams of SEDN, and reconciled estimates are obtained. The results show that Root Mean Square Error (RMSE) is the best performance metric since it is more sensitive to small changes in the measurement values and the reconciled estimates. Further, it is observed that PCA-DR performs better than WLS-DR in reducing the random error (and thereby achieving greater precision of measured values).

Key words | data reconciliation, performance metrics, principal component analysis (PCA), smart energy management network

Jeyanthi Ramasamy (corresponding author)
Department of Electronics and Communication Engineering, Amrita School of Engineering,
Amrita Vishwa Vidyapeetham,
Bengaluru,
India
E-mail: *r_jeyanthi@blr.amrita.edu*

Sriram Devanathan
Center of Excellence in Advanced Materials & Green Technologies, Department of Chemical Engineering and Materials Science, Amrita School of Engineering,
Amrita Vishwa Vidyapeetham,
Coimbatore,
India

Dhanalakshmi Jayaraman
Department of Chemical Engineering,
SSN college of Engineering,
Chennai,
India

HIGHLIGHTS

- Application of data reconciliation (DR) techniques to treat random errors present in flow sensor data used by water distribution networks.
- Selection of best performing metric to evaluate data reconciliation (DR) techniques.
- Analyze the performance of selected DR techniques for small and large scale networks using Python-based simulation.

INTRODUCTION

In most chemical industries, utilities such as water and energy play an important role. A sensor-based smart energy distribution network (SEDN) is required to monitor the consumption of these supplies. SEDN can aid in providing better process quality, more efficient operation, more accurate forecasting of supply and demand. SEDN operation is usually supported by Supervisory Control and Data Acquisition (SCADA) systems (Park & Jung 2014; Quevedo *et al.* 2014; Kročová 2016). Therefore, measurement of process flow variables is an essential part of this process. The precision of the measurements is very important, without which modelling and analysis can be misleading.

doi: 10.2166/ws.2020.314

However, usually measured variables are contaminated by fixed and random errors (Schönenberger 2015; Tran *et al.* 2019). These random errors creep into measurements from various sources like high frequency pickups, low resolution, and signal converters (Câmara *et al.* 2017). Data reconciliation (DR) is an approach usually applied to treat random errors present in a measured variable under a constrained process environment. The most important difference between DR and other signal processing techniques is that DR uses process constraints; that is, mass and energy balances of the process network, while the latter do not.

At times it is not possible to measure all the process variables in a process due to practical difficulty. In such situations, unmeasured variables could be estimated through soft sensors or solving DR problem (Miao *et al.* 2009; Rieger *et al.* 2010; Quevedo *et al.* 2014; Narasimhan & Bhatt 2015; Xu *et al.* 2020). DR is a simple approach and it has even become an integral part of software packages like ASPEN, VALI, VisualMesa, SimSci DATACON, and Sigmafine (Narasimhan & Bhatt 2015; Camara *et al.* 2017). In the data reconciliation landscape, the techniques include Weighted Least Squared Data Reconciliation (WLS-DR), Quasi-Weighted Least Squared Data Reconciliation (QWLS-DR), Robust-DR, and a few more recent techniques, for dealing with random errors. (Rieger *et al.* 2010; Zhang *et al.* 2010; Fuente *et al.* 2015; Lin *et al.* 2019; Xie *et al.* 2019).

Principal Component Analysis (PCA) is a multivariate data processing technique extensively used for dimension reduction where data on a large number of variables are available. It is also widely used in behavioural modelling of large water management systems, monitoring of water distribution including leakage, abnormal use of water, illegal connections, process monitoring for multi input-multi output (MIMO) processes, soft sensor modelling, data reconciliation (DR), and gross error detection (GED). In Helness *et al.* (2019), Varshith *et al.* (2019), Bhattacharyya *et al.* (2017), Fuente *et al.* (2015), Narasimhan & Bhatt (2015), Narasimhan & Shah (2008), Park & Jung (2014), the use of PCA in data reconciliation and gross error detection techniques is illustrated.

In order to prove that the DR techniques are actually accomplishing the task of improving the precision of measured data, performance metrics are needed. In Spuler *et al.* (2015), various performance metrics were explained to evaluate regression methods applied for decoding neural signals. From this, Correlation Coefficient (CC), Global deviation (GD), Signal to Noise Ratio (SNR), and Root Mean Square Error (RMSE) are identified as the most suitable metrics to evaluate DR techniques. The other performance measures are Relative Error Reduction (RER), Measurement Relative Error (MRE), and Reconciled Relative Reduction Error (RRE), which are explained in Zhang *et al.* (2010). In order to find the best metric, factors that lead to deterioration of data should be looked at, and that is discussed in the following section of this study.

In this paper, integrated water supply networks are considered whose pipe flow streams are assumed to be contaminated by gaussian noise. Two DR techniques, WLS-DR and PCA-based DR, are applied to measurements, and reconciled estimates are obtained. The selected metrics are applied to evaluate the performance of DR techniques and the best metric is then found. Further, the same is implemented and evaluated for other datasets that have different variances and serially correlated errors.

DATA RECONCILIATION (DR) TECHNIQUES

Data reconciliation is generally applied to reduce the effect of random errors present in the process variables. The reconciled estimates are obtained from the information contained in the models. The objective function of a generalised DR problem (Zhang *et al.* 2010; Korpela *et al.* 2016; Syed *et al.* 2016; Srinivasan *et al.* 2017; Xie *et al.* 2019) is formulated as:

$$\min f(M,\ \hat{M}) = \sum_{i=1}^{n}\sum_{j=1}^{N}\rho\left(\frac{m_{ij}-\hat{m}_{ij}}{\sigma_i}\right),\ \text{subject to } \mathrm{F}(\hat{M}) = 0 \tag{1}$$

where, M is the raw measurement matrix, $\hat{M}$ is the reconciled estimate of M, ρ is the cost function, m_j is the jth measurement of ith variable, $\hat{m}_{ij}$ is the jth reconciled estimate of ith variable, n is the number of process variables, N is the sample size, σ is the standard deviation of process variable and $\mathrm{F}(\hat{M})$ is the process constraint.

Optimised measurements of a system can be estimated which have lesser effect of random noise. Given below are certain prerequisites which are essential in order to apply DR to a dataset.

i. The process constraints, which consist of material and energy balance equations, should be defined.
ii. The set of measured process variables must be specified and the inaccuracies in these measurements must be specified in terms of the associated variances and covariance.

Application of DR to linear steady state processes is discussed below. Consider $a(j)$ to be the true measurement vector of a steady state system for each sample j where $j = 1,2,3,\ldots, N$. For a steady state process, equality constraints are derived as,

$$\tilde{C}a(j) = 0 \tag{2}$$

where $\tilde{C}$ is an incidence matrix of the process; $a(j)$ is the jth true measurement vector and it can be termed as $[a_1, a_2, \ldots, a_n]^T$, N is the sample size, and n is the number of process variables. The process constraint matrix is derived from the incidence matrix $\tilde{C}$.

Let the measurement of process variables be $m(j)$, then the generalised measurement model (Narasimhan & Bhatt 2015) is represented in Equation (3):

$$m(j) = a(j) + r(j) + \delta(j) \tag{3}$$

where r(j) is a random error vector and $\delta(j)$ is the measurement bias.

The measurement model shown in Equation (3) represents a realistic model since in practice, over a long period, measurements may change only due to random error. Here, the true value vector $a(j)$, variance of measurements (σ^2) and process incidence matrix ($\tilde{C}$) are considered (realistic assumption) to be fixed.

There are a few other assumptions about random errors considered.

For independently identically distributed (i.i.d.) data,

i. random errors are normally distributed, i.e. $r(j) \sim N(0, \sigma^2)$,
ii. process variables and errors are not correlated i.e., E[a(j) r(j)] = 0 and
iii. errors are not serially correlated i.e., E[r(j)r(j – 1)] = 0.

Sometimes, data are serially correlated due to process dynamics, recycling, adding controller loop, etc. Serial correlation may affect the performance of DR techniques. Hence, performance evaluation of DR techniques for identically distributed but non-independent (non-*i.i.d.*) measurements is considered here (Lin *et al.* 2019; Jeyanthi & Devanathan 2020).

Assumptions considered for non-*i.i.d* are:

i. random errors are normally distributed, i.e. $r(j) \sim N(0, \sigma^2)$,
ii. process variables and errors are not correlated i.e., E[a(j) r(j)] = 0 and
iii. errors are serially correlated i.e., E[r(j)r(j – 1)] ≠0.

In non-*i.i.d.* data, the Auto Regressive Moving Average (ARMA) model is considered for the random error and its structure is expressed as in Equation (4):

$$\begin{aligned} r(j) = {} & \varphi_1 r(j-1) + \varphi_2 r(j-2) + \cdots\cdots\cdots \\ & + \varphi_p r(j-p) + \omega(j) - \theta_1\omega(j-1) - \theta_1\omega(j-2) \\ & - \cdots\cdots\cdots - \theta_q\omega(j-q) \end{aligned} \tag{4}$$

where, φ is an auto regressive parameter, θ is a moving average parameter and ω is the white noise; p and q are the order of the model.

The weighted least squared-DR

The conventional approach to implement DR is the weighted least square data reconciliation (WLS-DR), which is used to get higher precision estimates of process variables from measurements which have noise added. The basic idea of this approach is to minimise the residuals (mass and energy imbalances resulting from measurement error) in the nodes (containers in the process where multiple flows converge) so that the estimates are much closer to true values. The generalised problem of steady state DR problem (Zhang *et al.* 2010; Fuente *et al.* 2015; Valle *et al.* 2018; Lin *et al.* 2019; Xie *et al.* 2019) is explained in the following Equation (5):

$$\min(M, A, R) = \sum_{i=1}^{n}\sum_{j=1}^{N} w_{ij}(m_{ij} - \hat{m}_{ij})^2 \tag{5}$$

subject to $f(A) = 0$ and $g(A) \leq 0$, where, M - raw measurement, A- reconciled estimate of M, w_{ij} - the jth weight of the ith measurement variable, and $f(a)$ and $g(a)$ represent the constraints of the process. The optimum value of w_{ij} will provide good reconciled estimates for m_{ij}. For a known covariance matrix (Λ), the reconciled estimates of the measurement of variables $\hat{a}(j)$ can be obtained by:

$$\hat{a}(j) = \Lambda C^T (C \Lambda C^T)^{-1} C m(j) \tag{6}$$

where, $\hat{a}$ is reconciled estimates for $m(j)$, Λ is the error covariance matrix and C is the constraint matrix.

The obtained estimates are normally distributed and satisfy the process constraints.

Principal component analysis (PCA) based data reconciliation

PCA is a statistical technique that is generally used to reduce the dimensionality of data. PCA can also be used to identify linear relationships between variables. It identifies fewer uncorrelated variables, called principal components (PCs), from a large set of data. The goal of principal components is to explain the maximum amount of variance with the fewest PCs. Narasimhan & Bhatt (2015) have described an approach for applying PCA-based DR, a recent technique to obtain reconciled estimates. For a large data set, PCA-DR proves to be effective. This can be deployed when the error covariance matrix is known. The data matrix M is transformed as in Equation (7)

$$\mathrm{M_s} = \upharpoonleft^{-1} M = (\upharpoonleft^{-1} A + \upharpoonleft^{-1} R) \tag{7}$$

where $\upharpoonleft$ is a decomposition of error covariance matrix (Λ) of the process network.

$$\Lambda = \upharpoonleft \upharpoonleft^{\mathrm{T}} \tag{8}$$

Therefore the covariance matrix of M_s can be calculated by Equation (9)

$$\Lambda_{\mathrm{M_s}} = \frac{1}{\mathrm{N}} \mathrm{M_s M_s^T} \tag{9}$$

The positive square root of the eigenvalues of $\Lambda_{\mathrm{M_s}}$ are termed as singular values of M_s. The first highest γ singular values are positive and are equal to the rank of M_s while the remaining n- γ singular values are small and used to define the dependent variables in the process. SCREE plot (Narasimhan & Bhatt 2015) can be used to select the largest eigen values (γ) $\Lambda_{\mathrm{M_s}}$.

The singular value analysis (SVA) is a technique used generally in control loop selection. The highest singular values denote the principal components (PCs) of the process variables. The PCs explain the most influential variables of the process network. The PCs of the process are extracted by applying Singular Value Decomposition (SVD) of the transformed matrix

$$\mathrm{SVD(M_s)} = U\,S\,V^T \tag{10}$$

where, the columns of U are the input singular vectors and the columns of V are the output column vectors. S is a diagonal matrix of non-zero γ singular values in order from largest to smallest. U and V are also unitary matrices such that $UU^T = I$ and $VV^T = I$.

The decomposed matrix M_s can be expressed by vector containing highest singular values γ and the vector containing smallest singular values n-γ:

$$M_s = \sqrt{N} U_{1s} S_{1s} V_{1s}^T + \sqrt{N} U_{2s} S_{2s} V_{2s}^T \tag{11}$$

The reconciled estimates of M are derived from the first part of M_s, $\sqrt{N} U_{1s} S_{1s} V_{1s}^T$ and calculated as $\hat{A} = \upharpoonleft \hat{M}_s$. The second part of $\mathrm{M_s}$ expresses the interaction among variables. So the identified constraint matrix of the process can be derived as $\hat{C} = U_{2s}^T \upharpoonleft^{-1}$. It may be noted that the PCA-DR estimates do not satisfy the original constraints, but rather satisfy the identified constraint matrix $\hat{C}$.

Performance metrics

It is important to evaluate any technique so as to ensure that it is reliable even for a large set of data. The various performance metrics must capture the error properties like bias, scaling, and other types of errors present in the measurement. They should be sensitive enough to capture these error properties and perform accurately. Among various performance metrics, Correlation Coefficient (CC), Root Mean Square Error (RMSE), Global Deviation (GD), Signal to Noise Ratio (SNR), and Relative Error Reduction (RER) are selected (Chai & Draxler 2014; Narasimhan & Bhatt 2015; Spuler *et al.* 2015; Xie *et al.* 2019). In this paper, two cases are taken for the study. The first is where only random errors (both *i.i.d* and non-*i.i.d*) are present in the measurements, and the second is where both random error and measurement bias are present in measurements.

The performance metrics are explained below.

Root mean square error (RMSE)

This is a measure of the square root of differences between estimated values and measured values. In short, it is a measure of the precision achieved by each DR technique in reducing the errors of different models for a particular dataset. It is always positive and a value of '0' would indicate a perfect fit to the data (i.e. 100% elimination of errors). RMSE is commonly used in climatology, soft sensor modelling, forecasting, and regression analysis to verify experimental results.

$$RMSE_i = \sqrt{\frac{\sum_{j=1}^{N} (\mathrm{m_{ij}} - \mathrm{a}_{ij})^2}{N}} \tag{12}$$

Global deviation (GD)

This is a measure of average squared difference between true and measured values.

$$\mathrm{GD_i} = \left(\frac{\sum_{j=1}^{N} (\mathrm{m_{ij}} - a_{ij})}{\mathrm{N}}\right)^2 \tag{13}$$

Correlation coefficient (CC)

The Pearson correlation coefficient is a statistical measure that calculates the strength and direction of the linear relationship between two variables. The value of CC is between −1 and 1. It cannot capture non-linear relationships between two variables.

$$\mathrm{CC_i} = \frac{\sum_{j=1}^{N} (\mathrm{m_{ij}} - \bar{m}_{ij})(\mathrm{a_{ij}} - \bar{a}_{ij})}{\sqrt{\sum_{j=1}^{N} (\mathrm{m_{ij}} - \bar{m}_{ij})^2 \sum_{i=1}^{N} (\mathrm{a_{ij}} - \bar{a}_{ij})^2}} \tag{14}$$

Signal to noise ratio (SNR)

With respect to statistics, SNR is defined as a measure of the ratio between variance of mean deviation from the true value of the measured variable to the variance of the measured variable:

$$\mathrm{SNR_i} = \frac{\mathrm{var}(\mathrm{m_{ij}} - a_{ij})}{\mathrm{var}(\mathrm{m_i})} \tag{15}$$

Relative error reduction (RER)

RER is another measure used to evaluate the performance of reconciliation techniques. This measure is the ratio of relative errors between raw measurement and reconciled estimates. Higher value of RER indicates a better reconciliation technique:

$$RER_i = \frac{\sum_i (MR_{ei} - RR_{ei})}{MR_{ei}} \text{ and,}$$
$$MR_{ei} = \frac{|a_{ij} - m_{ij}|}{a_{ij}},\ RR_{ei} = \frac{|a_{ij} - \hat{m}_{ij}|}{a_{ij}} \tag{16}$$

where, m is the measurements, a is the true value of the measurement, $\bar{m}$ is the mean of measurement variable I and $\hat{m}$ is the reconciled estimate. For raw measurements, RER is considered as zero.

SIMULATION

To analyse the performance of DR techniques, two benchmark process systems, a small scale recycle network and a large scale process network, have been chosen for the study. The selection of the benchmark systems (Valle *et al.* 2018; Varshith *et al.* 2019; Jeyanthi & Devanathan 2020) is based on the number of variables and interacting nodes in the process. This would lend credence to the performance evaluation of the techniques included in the study.

Example 1

A small recycle network (Xie *et al.* 2019; Jeyanthi & Devanathan 2020) shown in Figure 1 consists of five interacting nodes with seven flow variables (F1, F2, F3, F4, F5, F6, and F7). The process constraints are linear and estimated from Equation (17).

Depending on interdependencies between nodes and direction of flow, base values are assumed and true values are generated. The true value vector $(a(j))$ for all variables is derived from the base values, $b = [10\ 30\ 30\ 10\ 20\ 10\ 10]^T$, process dynamics of each flow variable with variance of 0.01, and assuming the leaks and the losses across the nodes are null. The mass balance equation at each node is defined by Equation (17),

$$\text{Input flow variable of ith node} - \text{Output flow variable of 'i' th node} = 0 \tag{17}$$

The incidence matrix for this process is shown in Equation (18)

$$C = \begin{bmatrix} 1 & -1 & 0 & 1 & 0 & 1 & 0 \\ 0 & 1 & -1 & 0 & 0 & 0 & 0 \\ 0 & 0 & 1 & -1 & -1 & 0 & 0 \\ 0 & 0 & 0 & 0 & 1 & -1 & -1 \\ -1 & 0 & 0 & 0 & 0 & 0 & 1 \end{bmatrix} \tag{18}$$

Input and output flow variables are denoted as '1' and '− 1' respectively. The process constraint matrix is derived from the incidence matrix by removing node '5', which has non-recycled variables in the network.

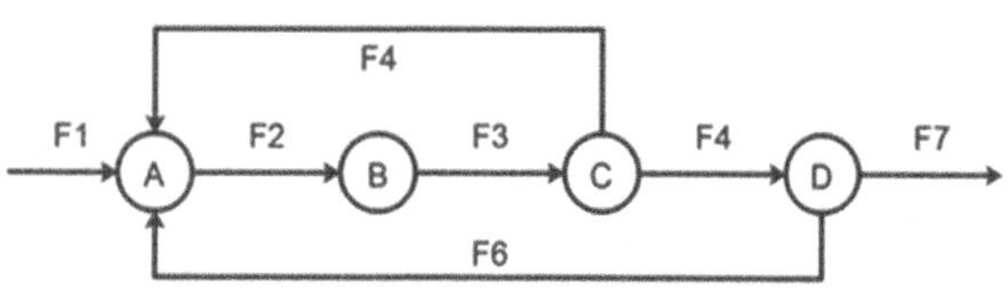

Figure 1 | Recycle network.

In order to evaluate the performance of each technique, a few flow variables with specific magnitudes are identified. The performance indices are calculated as explained in the previous section. The reconciled estimates are obtained using corresponding DR techniques as explained above, and the results are compared with raw data. The performance index calculation procedure is explained as follows:

Step 1: Obtaining raw measurement

Step 2: Applying DR technique

Step 3: Calculate reconciled estimates $(\hat{A})$

Step 4: Calculate Performance Index

The performance of the recycled network is shown in Table 1. For all variables F1 to F7, SNR has not shown much variation amongst DR techniques. CC and GD have no significant index to show the performance improvement of DR techniques. RER has prominent changes for WLS and PCA-DR techniques, but it has negative values for non-performing reconciled estimates. RMSE has significant changes in different DR techniques. Also, it has a non-negative index. From the perception, it is obvious that RMSE is the best measurement to assess the performance of DR techniques. PCA-DR is sensitive to the magnitude and performs poorly for variables having higher base values F2, F3, and F5. WLS-DR performs well for feedback variables F4 and F6, and poorly for other variables.

Example 2

The large process network (Varshith *et al.* 2019) shown in Figure 2 consists of 11 nodes representing the balance equations and 28 flow variables. The base value of each variable is referred as in Table 2. The constraint matrix for this process is calculated as in Equation (17).

Figure 3 shows the performance indices of selected performance metrics of flow variables F7, F16, F22, and F27. It is observed that in Figure 3(a), RMSE is the best metric for capturing the minute changes in noise present in the data, followed by RER and CC. SNR and GD remain constant throughout, proving to be of no valid significance in evaluating the different data sets. As in the case of variables F16 and F22 shown in Figure 3(b) and 3(c) respectively, SNR and GD remain constant here as well.

Table 1 | Performance of DR techniques for recycle network

Flow variable	DR technique	Performance Metrics SNR	CC	GD	RER	RMSE
1	Raw data	1.0107	0.0000	0.0001	0.0000	1.0065
	WLS-DR	1.0235	0.0000	0.0000	0.3288	0.6753
	PCA-DR	1.0482	0.0000	0.0003	0.5280	0.4745
2	Raw data	1.0107	0.0000	0.0001	0.0000	1.0065
	WLS-DR	1.0062	0.0000	0.0001	−0.3301	1.3390
	PCA-DR	1.0057	0.0000	0.0007	−0.3810	1.3904
3	Raw data	1.0107	0.0000	0.0001	0.0000	1.0065
	WLS-DR	1.0062	0.0000	0.0001	−0.3301	1.3390
	PCA-DR	1.0057	0.0000	0.0007	−0.3810	1.3904
4	Raw data	1.0107	0.0000	0.0001	0.0000	1.0065
	WLS-DR	1.0919	0.0000	0.0000	0.6529	0.3487
	PCA-DR	1.0482	0.0000	0.0003	0.5280	0.4745
5	Raw data	1.0107	0.0000	0.0001	0.0000	1.0065
	WLS-DR	1.0107	0.0000	0.0001	0.0000	1.0065
	PCA-DR	1.0125	0.0000	0.0000	0.0755	0.9305
6	Raw data	1.0107	0.0000	0.0001	0.0000	1.0065
	WLS-DR	1.0919	0.0000	0.0000	0.6529	0.3487
	PCA-DR	1.0482	0.0000	0.0003	0.5280	0.4745
7	Raw data	1.0107	0.0000	0.0001	0.0000	1.0065
	WLS-DR	1.0235	0.0000	0.0000	0.3288	0.6753
	PCA-DR	1.0482	0.0000	0.0003	0.5280	0.4745

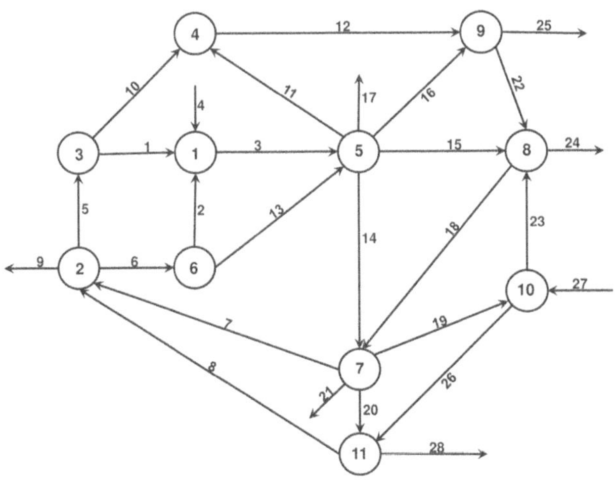

Figure 2 | Large process network.

In contrast, there is a variation in the evaluation outcome of RMSE, RER, and CC. RMSE and RER display improvement for PCA-DR data and CC does not show any improvement for WLS-DR data. The variable F27 shown in Figure 3(d) shows the variation for RMSE, RER, and CC. Since RMSE is majorly consistent and is sensitive to small changes in noise, it is being used as the main evaluation metric. RER almost complements RMSE, but is less sensitive at times.

To analyse the performance of PCA-DR and WLS-DR on the process data, flow variables F16-F1-F28-F5-F3-F18-F27 have been considered in the order of increasing base magnitude. Figure 4(a) shows the performance of the DR techniques when $\Sigma = I$. The results obtained show that WLS-DR estimates are consistent and are base magnitude independent. The estimates obtained on performing DR are maximum likelihood estimates and they are more accurate when compared to the raw process measurement. PCA-DR performs linearly as the magnitude increases. For high magnitudes, its performance is less accurate when compared to the estimates of WLS-DR. Thus, we can effectively say that the performance of PCA-DR decreases as the base magnitude of the flow variables increases.

Table 2 | Base values for flow variables

Flow Variable	1	2	3	4	5	6	7	8	9	10	11	12	13	14
Base values	10	10	30	10	20	20	30	20	10	10	10	20	10	10
Flow variable	15	16	17	18	19	20	21	22	23	24	25	26	27	28
Base values	10	5	5	40	5	5	10	20	20	10	5	30	45	15

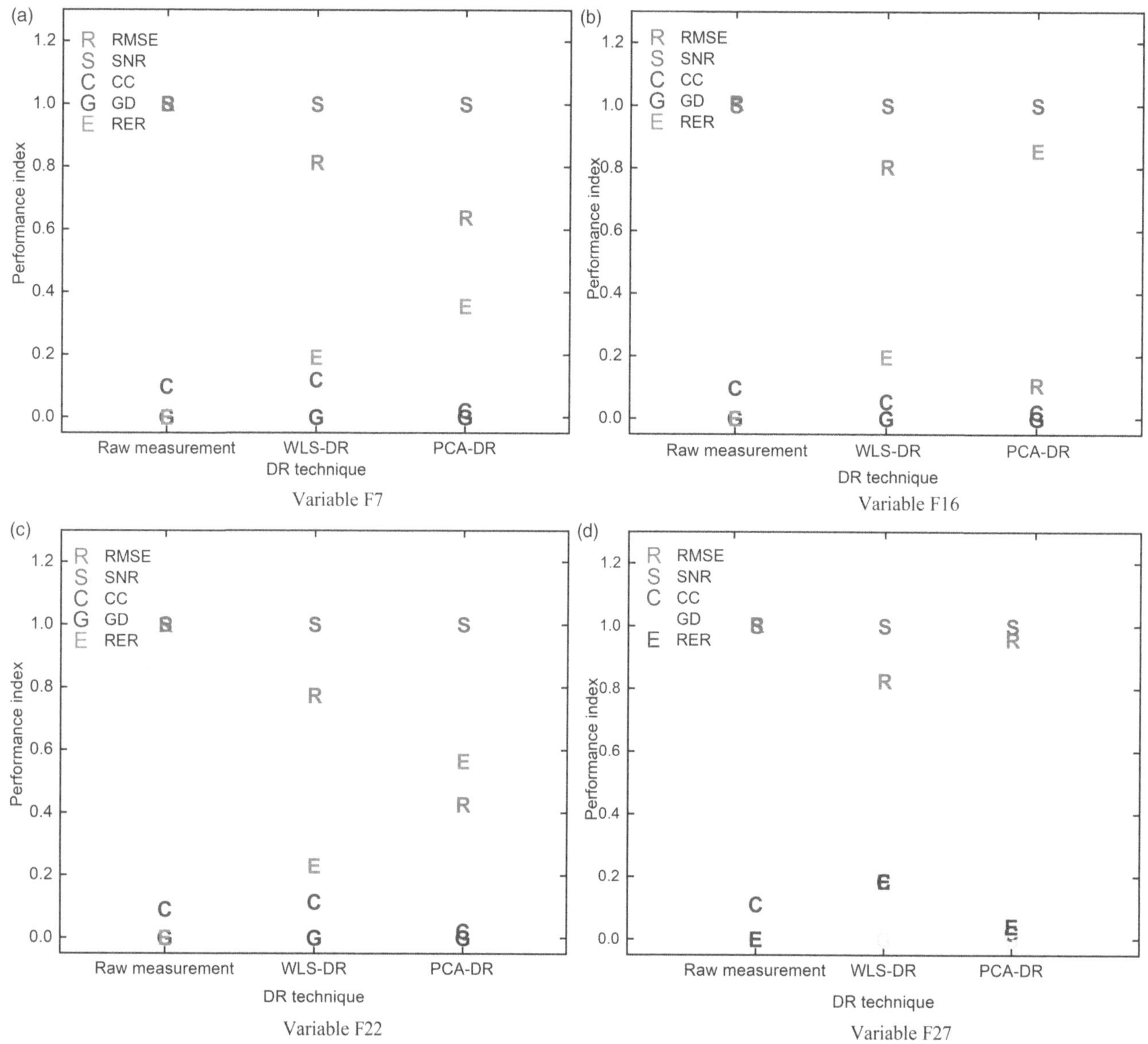

Figure 3 | Performance metrics of flow variable (*i.i.d*).

Figure 4(b) compares the performance of WLS-DR and PCA-DR for different variances in data for variable F16. It is seen that as the variance increases, the performance of WLS-DR decreases. PCA-DR performs well when compared to WLS-DR. This shows that increase in variance does not affect

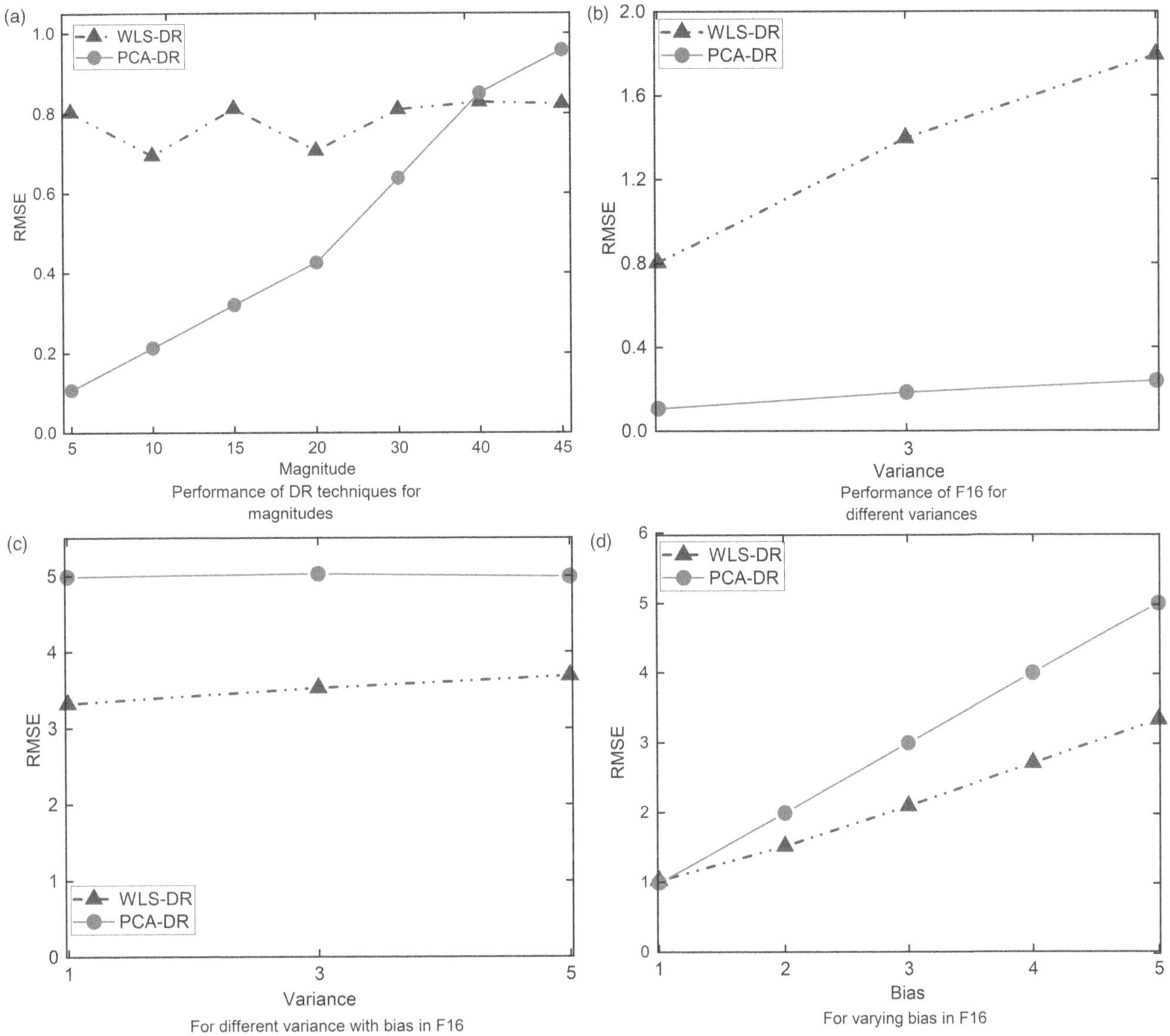

Figure 4 | Performance of DR techniques for different scenario.

the PCA-DR performance as it did for increase in magnitude.

Figure 4(c) shows the performance of DR techniques for different variance when bias (δ) of 5σ present in variable F16. As variance increases, RMSE of WLS-DR estimates also increases. PCA-DR estimates vary slightly and its RMSE remain around the gross error value. This indicates that PCA-DR can be used for detecting the presence of bias even when there are situations where variance changes.

Figure 4(d) shows that as the value of gross error increases, RMSE of PCA-DR is varying linearly with bias, and its magnitude is equal to that of the gross error. The RMSE of WLS-DR is also varying linearly with gross error, but not in a way that is equal to the magnitude of the gross error. The PCA-DR can be combined with gross error detection techniques to identify gross errors present in measurements.

Measurements are usually contaminated by random and gross errors. So, it is important to analyse the effect of gross

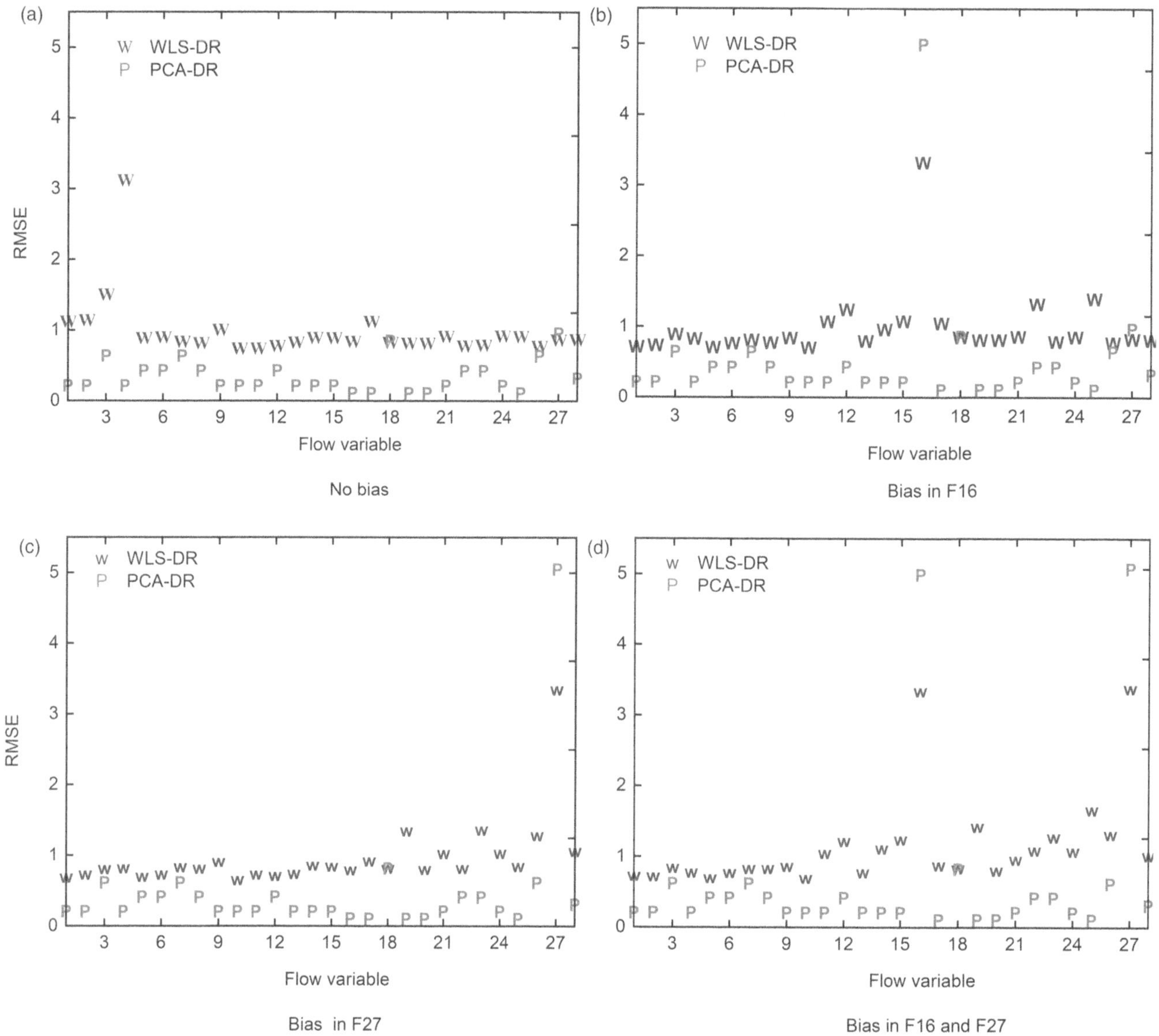

Figure 5 | Performance of DR techniques for the bias ($\delta = 5$ unit).

errors on the performance of DR techniques. Figure 5 shows two scenarios based on the presence of gross error (bias) of magnitude 5σ, in 5(a) variables F16 and F27 and 5(b) F16 and F27. The DR techniques didn't increase the accuracy of estimates. The WLS-DR distributed the gross error, and PCA-DR did not perform at all. The PCA-DR, as seen previously, depends on magnitude of variable in *i.i.d.* data; in the presence of bias, irrespective of base value, the result was the value of amount of gross error present.

To analyse the performance of the DR techniques on non-*i.i.d.* data as well, data with Auto Regressive Moving Average (ARMA) noise was simulated. The ARMA(1, 1) had φ_1 & θ_1 values as 0.4 and 0.2, while the ARMA(2, 2) had φ_1, φ_2, and θ_1, θ_2 values as 0.5, 0.4, and −0.4, 0.2 respectively. The DR techniques were then applied on this data and RMSE was calculated for evaluating the techniques. Figure 6 shows the comparison between DR techniques applied to *i.i.d.* and non-*i.i.d.* data. It is observed that PCA-DR

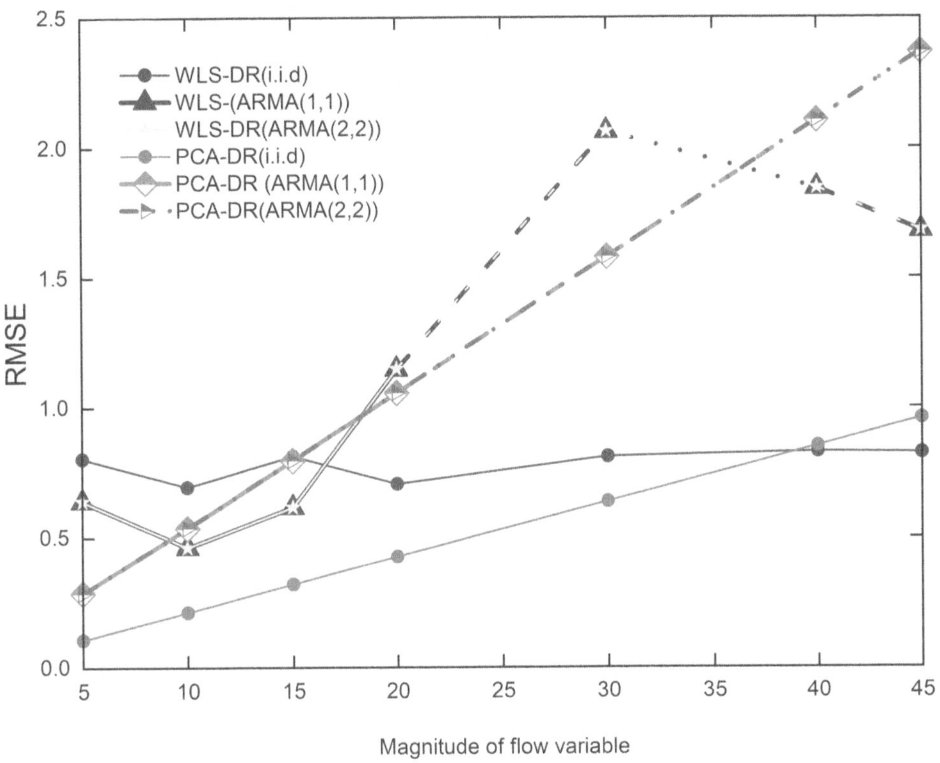

Figure 6 | Performance of DR techniques for *i.i.d.* and non-*i.i.d.* for varying magnitudes of flow variable.

estimates of non-*i.i.d.* data for both ARMA (1, 1) and ARMA (2, 2) noise are similar, and serial correlation makes the PCA-DR worsen due to erroneous variance estimation. WLS-DR when applied to the non-*i.i.d.* data shows high inconsistency in the estimates and gives inaccurate results, thus concluding that WLS-DR cannot be used for non-*i.i.d.* data. When comparing holistically with *i.i.d.* data, we can say that both PCA-DR and WLS-DR perform better when applied to *i.i.d.* data than to non-*i.i.d.* data.

CONCLUSION

The work presented here can be applied to monitoring of water inflows in urban smart water network management, waste water treatment and other areas where water supply is vital. We agree that the present work focuses on techniques that contribute significantly to augmenting the large body of knowledge in metrology. However, we would like to emphasize that these techniques find significant and direct applications in resource reconciliation in water supply networks. When it comes to water, while the specific challenges of commercial (process) use, domestic use, power use, mining use, utility use, etc. may differ, what they all have in common, when seen in the context of water supply networks, is that such networks contain the two basic constituent elements of streams (water flow) and nodes (convergence/divergence points of flows). The techniques outlined in this paper are then generally applicable for all such networks, as they employ a useful relationship constraint, namely the mass balances. Thus, it is evident that the current work has direct and important relevance to water supply, via the treatment it provides for resource reconciliation. In such networks, the reconciliation must, of necessity, address physical leaks as well as measurement biases – both of which form the core focus of data reconciliation and gross error techniques.

This work presents an approach that integrates the process of selecting a suitable data reconciliation technique with the process of selecting the best performance evaluation metric under different scenarios. The study reveals that the RMSE is the best metric (among those considered), since it is very sensitive to small changes in the measurement and estimates. SNR, GD, and CC were seen to not capture significant changes in reconciliation. RER performs in a very different way from that of RMSE, and has negative values for poorly reconciled estimates.

From the two examples discussed in this paper, PCA-DR is a good technique when compared to WLS-DR, for variables with smaller magnitudes of *i.i.d.* data. When the difference in magnitudes between process variables increases PCA-DR performs less accurately compared to WLS-DR. As in the case of changing variance, PCA-DR is better than WLS-DR. For the biased data both techniques failed in reducing the random errors. For serially correlated (non-i.i.d.) data, PCA and WLS-DR both are performing poor, due to variance change.

DATA AVAILABILITY STATEMENT

All relevant data are included in the paper or its Supplementary Information.

REFERENCES

Bhattacharyya, A., Yogi, V., Singla, S., Bhushan, M., Kelkar, M. G., Tiwari, A. P., Pramanik, M. & Belur, M. N. 2017 Adaptive, online models to detect and estimate gross error in SPNDs. In: *Proceedings of 2017 Indian Control Conference (ICC 2017)*, Guwahati, India. IEEE Control Systems Society. doi:10.1109/INDIANCC.2017.7846467.

Câmara, M. M., Soares, R. M., Feital, T., Anzai, T. K., Diehl, F. C., Thompson, P. H. & Pinto, J. C. 2017 Numerical aspects of data reconciliation in industrial applications. *Processes* **5** (56), 1–38. https://doi.org/10.3390/pr5040056.

Chai, T. & Draxler, R. 2014 Root mean square error (RMSE) or mean absolute error (MAE)? Arguments against avoiding RMSE in the literature. *Geoscientific Model Development* **7**, 1247–1250. doi:10.5194/gmd-7-1247-2014.

Fuente, M. J., Gutierrez, G., Gomez, E., Sarabia, D. & Prada, D. 2015 Gross error management in data reconciliation. In: *Proceedings of 9th International Symposium on Advanced Control of Chemical Processes*. IFAC, British Columbia, Canada. https://doi.org/10.1016/j.ifacol.2015.09.037.

Helness, H., Damman, S., Sivertsen, E. & Ugarelli, R. 2019 Principal component analysis for decision support in integrated water management. *Water Supply* **19** (8), 2256–2262. https://doi: 10.2166/ws.2019.106.

Jeyanthi, R. & Devanathan, S. 2020 Addressing higher order serial correlation in techniques for gross error detection. *Journal of Computational and Theoretical Nanoscience* **17**, 297–302. doi:10.1166/jctn. 2020.8665.

Korpela, T., Suominen, O., Majanne, Y., Laukkanen, V. & Lautala, P. 2016 Robust data reconciliation of combustion variables in multi-fuel fired industrial boilers. *Control Engineering Practice* **55**, 101–115. https://doi.org/10.1016/j.conengprac.2016.07.002.

Krōcová, S. 2016 Water supply systems and their influence on increasing operational safety in industry. *Perspectives in Science* **7**, 236–239. https://doi.org/10.1016/ j.pisc.2015.11.038.

Lin, Y., Kruger, U., Gu, F., Ball, A. & Chen, Q. 2019 Monitoring non stationary and dynamic trends for practical process fault diagnosis. *Control Engineering Practice* **84**, 139–158. https://doi.org/10.1016/j.conengprac.2018.11.020.

Miao, Y., Su, H., Gang, R. & Chu, J. 2009 Industrial process: data reconciliation and gross error detection. *Measurement+Control* **42** (7), 209–215. https://doi.org/10.1177 /002029400904200704.

Narasimhan, S. & Bhatt, N. 2015 Deconstructing principal components analysis using a data reconciliation perspective. *Computers and Chemical Engineering* **77**, 74–84. https://doi.org/10.1016/j.compchemeng.2015.03.016.

Narasimhan, S. & Shah, S. L. 2008 Model identification and error covariance matrix estimation from noisy data using PCA. *Control Engineering Practice* **16**, 146–155. https://doi.org/10.1016/j.conengprac.2007.04.006.

Park, S. & Jung, Y. S. 2014 Principal component analysis of water pipe flow data, 16th conference on water distribution system analysis, WDSA 2014. *Procedia Engineering* **89**, 395–400. doi:10.1016/j.proeng.2014.11.204.

Quevedo, J., Pascual, J., Puig, V., Saludes, J., Sarrate, R., Escobet, A., Espin, S. & Roquet, J. 2014 Flowmeter data validation and reconstruction methodology to provide the annual efficiency of a water transport network: the ATLL case study in Catalonia. *Water Supply* **14** (2), 337–346. https://doi.org/10.2166/ws.2013.203.

Rieger, L., Takacs, I., Villez, K., Siegrist, H., Lessard, P., Vanrolleghem, P. A. & Comeau, Y. 2010 Data reconciliation for wastewater treatment plant simulation studies: planning for high-quality data and typical sources of errors. *Water Environment Research* **82** (5), 426–433. https://www.jstor.org/stable/25679798.

Schönenberger, U. 2015 *Data Reconciliation and Gross Error Detection for Wastewater Treatment Processes*. Master Thesis, Process Engineering in Urban Water Management, ETH Zürich.

Spuler, M., Sanz, S. S., Birbaumer, N., Rosenstiel, W. & Murguialday, A. R. 2015 Comparing metrics to evaluate performance of regression methods for decoding of neural signals. In: *Proceedings of 37th Annual International*

Conference of EMBC'15, Milan, Italy. IEEE Engineering in Medicine and Biology Society, pp. 1083–1086. https://doi.org/10.1109/EMBC.2015.7318553.

Srinivasan, S., Billeter, J., Narasimhan, S. & Bonvin, D. 2017 Data reconciliation for chemical reaction systems using vessel extents and shape constraints. *Computers and Chemical Engineering* **101**, 44–58. https://doi.org/10.1016/j.compchemeng.2017.02.003.

Syed, M. S., Kerry, M., Dooley, K. M., Madron, F. & Knopf, C. 2016 Enhanced turbine monitoring using emissions measurements and data reconciliation. *Applied Energy* **173**, 355–365. https://doi.org/10.1016/j.apenergy.2016.04.059.

Tran, K. P., Nguyen, H. D., Tran, P. H. & Heuchenne, C. 2019 On the performance of CUSUM control charts for monitoring the coefficient of variation with measurement errors. *International Journal of Advanced Manufacturing Technology* **104** (5–8), 1903–1917. https://doi.org/10.1007/s00170-019-03987-6.

Valle, E. C., Kalid, R. A., Secchi, A. R. & Kiperstok, A. 2018 Collection of benchmark test problems for data reconciliation and gross error detection and identification. *Computers & Chemical Engineering* **111**, 134–148. https://doi.org/10.1016/j.compchemeng. 2018.01.002.

Varshith, C. R., Rishika, S. R., Ganesh, S. & Jeyanthi, R. 2017 Principal component analysis based data reconciliation for a steam metering circuit. In: *Advances in Intelligent Systems and Computing, Soft Computing and Signal Processing: Proceedings of ICSCSP 2018*. 898 (2), pp. 619–625. https://doi.org/10.1007/978-981-13-3393-4_63.

Xie, S., Yang, C., Yuan, X., Wang, X. & Xie, Y. 2019 A novel robust data reconciliation method for industrial processes. *Control Engineering Practice* **83**, 203–212. https://doi.org/10.1016/j.conengprac.2018.11.006.

Xu, Z., Ying, Z., Li, Y., He, B. & Chen, Y. 2020 Pressure prediction and abnormal working conditions detection of water supply network based on LSTM. *Water Supply* **20** (3), 963–974. https://doi.org/10.2166/ws.2020.013.

Zhang, Z., Shao, Z., Chen, X., Wang, K. & Qian, J. 2010 Quasi-weighted least squares estimator for data reconciliation. *Computers & Chemical Engineering* **34**, 154–162. https://doi.org/10.1016/j.compchemeng.2009.09.007.

First received 28 August 2020; accepted in revised form 30 October 2020. Available online 12 November 2020

Comparison of antecedent precipitation based rainfall-runoff models

Pankaj Upreti and C. S. P. Ojha

ABSTRACT

The Soil Conservation Service Curve Number (SCS-CN) method is one of the popular methods for calculating storm depth from a rainfall event. The previous research identified antecedent rainfall as a key element that controls the non-linear behaviour of the model. The original version indirectly uses five days antecedent rainfall to identify the land condition as dry, normal or wet. This leads to a sudden jump once the land condition changes. To obviate this, the present work intends to improve the performance of antecedent rainfall-based SCS-CN models. Two forms of SCS-CN model (M_1 and M_2), two recently developed P-P_5 based models (M_3 and M_4), and an alternate approach of considering P_5 in the SCS-CN model (M_5 and M_6), as proposed here, were investigated. Based on the evaluation of several error metrics, the new proposed model M_6 has performed better than other models. The performance of this model is evaluated using rainfall-runoff events of 114 watersheds located in the USA. The median value of Nash-Sutcliffe Efficiency was found as 0.78 for the M_6 model followed by M_5 (0.75), M_3 (0.73), M_4 (0.72), M_2 (0.63) and M_1 (0.61) model.

Key words | antecedent runoff condition (ARC), maximum potential retention, optimization, SCS-CN method, surface runoff estimation, US watersheds

Pankaj Upreti (corresponding author)
C. S. P. Ojha
Department of Civil Engineering,
Indian Institute of Technology,
Roorkee 247667,
India
E-mail: *pankaj_upretiiac@yahoo.com*;
pankaj.upreticot@gmail.com

Pankaj Upreti
Department of Agricultural Engineering,
GMV Rampur Maniharan,
Saharanpur 247451,
India

HIGHLIGHTS

- The study has been done on significant runoff producing events for which runoff coefficient is greater than 0.12.
- Research supports the superior performance of the proposed model in US watersheds.
- It acknowledges the applicability of five days antecedent rainfall (P_5) on runoff prediction and shows improvement in model performance under all statistical indices.

GRAPHICAL ABSTRACT

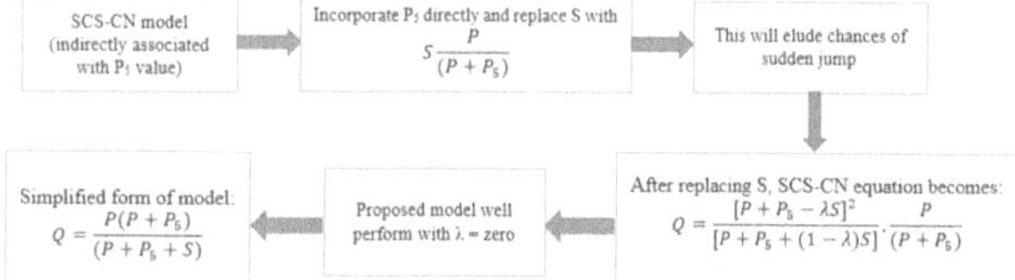

doi: 10.2166/ws.2020.315

ABBREVIATIONS

SCS-CN	Soil Conservation Service Curve Number
AMC	antecedent moisture condition
ARC	antecedent runoff condition
CN	curve number
S	potential maximum retention
NEH-4	National Engineering Handbook-Section 4
I_a	initial abstraction
λ	initial abstraction coefficient
USDA-ARS	United States Department of Agriculture-Agricultural Research Service
RMSE	root mean square error
NSE	Nash-Sutcliffe Efficiency
PBIAS	percentage bias
RGS	rank and grading system

INTRODUCTION

In surface water hydrology, the correct estimation of runoff is arguably the most important assignment for hydrologists and water scientists. This is useful in water supply, water allocation, hydrological analysis, watershed management, flood forecasting, flood protection work, integrated water management, control of non-point source pollution, etc. The runoff amount depends on rainfall intensity and its amount, wetness condition of the soil, watershed slope, various abstractions, land use, land cover, and many more factors. Hewlett *et al.* (1977) developed a rainfall-runoff relationship for a forest catchment and found negligible effects of rainfall intensity on peak discharge and runoff volume. Several studies have been conducted at event scale on various catchments of Europe to understand surface runoff generation process (Jordan 1994; Meyles *et al.* 2003; Latron *et al.* 2008). In case of events having both low-intensity rainfall and low-rainfall magnitude, the antecedent wetness state plays a major role in controlling runoff amount than rainfall intensity or its magnitude and a more precise and correct estimate can be achieved by using antecedent rainfall characteristics. If the joint dependence between rainfall event and antecedent precipitation is not taken into account properly, the runoff amount is consistently underestimated and when only one or two days' antecedent rainfall is considered, the magnitude of design flood can be as much as 30% less than actual (Pathiraja *et al.* 2012). The impact of prior rainfall of 48 hrs to 120 hrs of a rainfall event is more significant in determining the resulting runoff compared to other factors such as average and maximum rainfall intensity, event magnitude, and rainfall duration.

The relationship of intensity of storm event with antecedent moisture condition has become an increasing research subject in the context of anthropogenic climate change. To understand the role of antecedent moisture in rainfall runoff modelling, a tank experiment has been conducted based on high resolution observation in Hohai University, China, and a non-linear relationship was observed between soil moisture index and total runoff (Song & Wang 2019). If a model is calibrated without considering the impact of prior rainfall on the land surface, there will be a high likelihood that calibrated parameters will not produce the correct runoff yield. The water content in the upper soil surface may be an important factor in the rainfall-runoff relationship. A relatively lower sensitivity of antecedent rainfall on surface runoff has been found, maybe, due to the relatively dry upper surface in semiarid small watersheds located in Southeastern Arizona in the USA, contrary to other reported results (Zhang *et al.* 2011). The surface runoff is a threshold process and is governed by the runoff coefficient, which increases abruptly when a certain moisture threshold is exceeded. Similar results were found with changing values of moisture threshold by Radatz *et al.* (2013) and Zhao *et al.* (2015). A soil moisture accounting procedure was proposed by Michel *et al.* (2005) but criticized by Sahu *et al.* (2007) for not circumventing an undesirable sudden jump. Verma *et al.* (2020) conceptualized a theory of activation soil moisture (ASM) by coupling the infiltration component with the Soil Moisture Accounting (SMA) concept. The rainfall intensity, state of antecedent moisture and process of surface runoff are the three main reasons which evolve spatial and temporal non-linearity and complicate the watershed response. Most of the research shows that the rainfall-to-runoff transformation process is non-linear and the prime factor of nonlinearity is the antecedent moisture condition

(Radatz *et al.* 2013; Zhao *et al.* 2015). This antecedent moisture, along with soil characteristics, regulates the runoff process and affects the capability to store new water due to the occurrence of rainfall as well as the infiltration capacity of the soil (Merz & Blöschl 2009). It is a challenging task to evaluate the impact of antecedent rainfall on runoff at a watershed scale and it is required in the planning and management of a watershed.

There are distinct mathematical models available in hydrology with various merits, demerits and complications. The SCS-CN method is the simplest and universally accepted model for event-based rainfall-runoff modeling and integrated with various hydrological, water quality, or erosion estimation models. After a long experimental work, this method was developed for conditions prevailing in the USA. The model requires only maximum potential storage (S, mm) of the watershed in the form of a curve number (CN) to calculate runoff depth. Generally, soil type, land use and adopted practices are not much changed, but antecedent moisture condition status is changed frequently with the antecedent rainfall amount, which affects the runoff flow pattern. Antecedent moisture condition (AMC) classes were re-labelled as antecedent runoff condition (ARC) classes (NRCS 2004). The observation and monitoring of soil moisture are difficult; that is why antecedent precipitation, which can address initial soil moisture status by using five days of rainfall (Brocca *et al.* 2008), is in common use. This term is used as a predictor to decide the antecedent runoff condition, which is categorized in three levels i.e. ARC I (dry), ARC II (normal), and ARC III (wet). These levels form a discrete relation between antecedent rainfall and curve number and are responsible for an undesirable sudden jump in runoff estimation (Mishra *et al.* 2008). Generally, the CN value obtained from NEH-4 table or calculated from the P-Q dataset is CN_{II} (normal) and can be converted into CN_I (dry) or CN_{III} (wet) as per the antecedent rainfall amount. The CN model uses the antecedent five-day rainfall to categorize it into either dry, normal, or wet conditions to account for watershed initial losses (Sahu *et al.* 2007).

To obviate the error in predicting runoff calculations due to the sudden jump in curve number value, consideration of pre-storm rainfall is required in event-based runoff modelling, which minimizes the error and tries to correct the runoff value. This study has been done to evaluate the relative significance of antecedent precipitation (P_5) on the calculated runoff amount. Very few studies have been done to investigate the effect of antecedent rainfall on runoff behaviour. This is assessed using six variants of models, which are introduced here.

SCS-CN MODEL (M_1 AND M_2)

The original SCS-CN equation (USDA 1985) calculates runoff depth directly using rainfall data. This universally accepted equation is well documented, easy to understand, and useful due to its simplicity. This water balance equation, based on two assumptions, is expressed, respectively, as follows:

$$\frac{Q}{P - I_a} = \frac{F}{S} \tag{1}$$

$$I_a = \lambda S \tag{2}$$

where P and Q are rainfall and runoff depth in mm, I_a is an initial abstraction, F is cumulative filtration which is equal to P-I_a-Q and S is maximum potential retention in mm. After simplification of the above-stated assumptions, the expression is as follows:

$$Q = \frac{[P - \lambda S]^2}{[P + (1 - \lambda)S]}, \quad \textit{if } P > I_a \textit{ or } \lambda S, \textit{ otherwise } Q = 0 \tag{3}$$

λ value can vary from 0 to 1 but assumes as 0.2, while S value in mm varies from 0 to ∞. A dimensionless curve number (CN) for any watershed can be used to estimate S by using the following equation:

$$S = \frac{25400}{CN} - 254 \tag{4}$$

In addition, S can be directly obtained ($\lambda = 0.2$) using a gauged dataset of P and Q by solving Equation (3) with the quadratic equation as:

$$S = 5[(P + 2Q) - (4Q^2 + 5PQ)^{0.5}] \tag{5}$$

This Equation (5) produced a data-derived value of S (normal condition or S_2). Hawkins (1993) suggested a median CN value gives a better representation of the curve number ($\lambda = 0.2$) of any watershed using the *P-Q* dataset. Two mathematical equations proposed by Mishra *et al.* (2008) can be used to convert CN_{II} or S_{II} to CN_I (S_I) and CN_{III} (S_{III}) for dry (ARC I) or wet conditions (ARC III), respectively. These are:

$$S_I = 2.2754\, S_{II} \tag{6}$$

$$S_{III} = 0.43\, S_{II} \tag{7}$$

The conversion of S_{II} to S_I or S_{III} depends on the previous 5 days' rainfall. The value of $P_5 < 35.6$ mm, 35.6 mm $\leq P_5 \leq$ 53.3 mm, and $P_5 > 53.3$ mm assumes dry, normal, or wet conditions, respectively, for any storm event.

AJMAL & KIM MODEL (M_3 AND M_4)

Ajmal & Kim (2015) proposed two new equations and investigated runoff prediction over 15 South Korean catchments. These two equations are as follows:

$$Q = \frac{P^2}{P+S} - \frac{P+P_5}{S^2} \tag{8}$$

$$Q = \frac{P^2}{P+S} - \frac{P_5}{S} \tag{9}$$

It is to be noted that in both the above equations, Equations (8) and (9), S is the only unknown parameter. These equations are unique in nature and also exclude λ. Their performance were found to be significantly better than the SCS-CN model for the studied catchments. Equations (8) and (9) directly used P_5 value instead to decide antecedent conditions as dry, normal, or wet before rainfall and to some extent were able to overcome the sudden jump in runoff computation. To check its applicability in other watersheds, similar equations are to be used in this study by utilizing the P-P_5 dataset of 114 US watersheds.

PROPOSED MODEL (M_5 AND M_6)

The previous research suggested antecedent rainfall as a key component. The original model considers antecedent rainfall only to classify the land condition as dry, normal, or wet and ignored the exact amount of 5 days' prior rainfall (P_5) which creates an undesirable sudden jump in runoff computation. This can be explained by an example. If we assume a rainfall amount of 100 mm (which occurs in the growing season) over different land conditions with CN values as 60, 70, 80, and 90; and the antecedent five days rainfall variation from 1 to 100 mm. These CN values have been assumed, since for most of the watersheds its value falls in between 55 to 95 (Hawkins *et al.* 1985).

From Figure 1 (assuming $\lambda = 0.2$ and using Equation (3), it can be seen that the calculated values of runoff are 1.29 mm for ARC I ($P_5 < 35.6$ mm, CN = 60 or $S =$ 385.30 mm), 18.57 mm for ARC II ($P_5 = 35.6$ mm to 53.3 mm, CN = 60 or $S = 169.33$ mm), and 46.13 mm for ARC III ($P_5 > 53.3$ mm, CN = 60 or $S = 72.81$ mm). It means if five days' prior rainfall value changes the antecedent rainfall condition from ARC I to ARC II, the runoff value suddenly jumps from 1.29 mm to 18.57 mm for 100 mm rainfall. Similarly, shifting the antecedent state from ARC II to ARC III, runoff jumps from 18.57 mm to 46.13 mm. Similar jumps are illustrated in Figure 1 at CN = 70, 80, and 90 for ARC I, ARC II and ARC III. This undesirable jump is due to the indirect use of P_5 and it reduces the model efficiency.

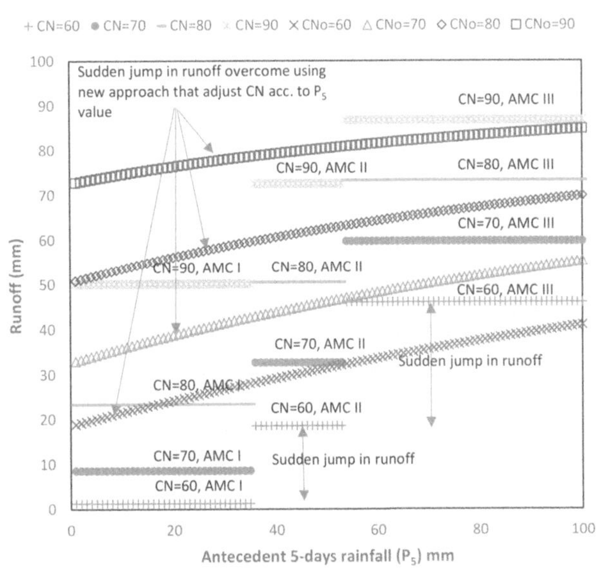

Figure 1 | Computed runoff generated from 100 mm of rainfall with different curve number (CN = 60, 70, 80, 90) and varying 5-days antecedent rainfall (P_5 = 0 to 100 mm). Here CN_0 is curve number when P_5 is equal to zero.

In the proposed model, instead of only selecting antecedent rainfall conditions (ARC), P_5 is incorporated directly as a model component. This is done by replacing S with $S\frac{P}{(P+P_5)}$ and placing it in the original SCS-CN equation with the sole objective of improving runoff prediction. In this fashion, the replacement of S will ensure the change in watershed characteristics (S) with different rainfall depth and P_5 value; and will avoid the chances of an unexpected jump in runoff calculation. The proposed model gives the following equation as:

$$Q=\frac{\left[P-\lambda S\left(\frac{P}{P+P_5}\right)\right]^2}{\left[P+(1-\lambda)S\left(\frac{P}{P+P_5}\right)\right]} \tag{10}$$

After simplification, Equation (10) yields

$$Q=\frac{[P+P_5-\lambda S]^2}{[P+P_5+(1-\lambda)S]}\cdot\frac{P}{(P+P_5)} \tag{11}$$

$\lambda=0.2$ makes Equation (11) as:

$$Q=\frac{[P+P_5-0.2S]^2}{[P+P_5+0.8S]}\cdot\frac{P}{(P+P_5)} \tag{12}$$

This expression provides variation in the S value and such replacement changes the runoff value only for those events for which P_5 is greater than zero. With increasing or decreasing P_5, the S value changes significantly. If there is no capacity to store water in the soil, whole rain will convert into runoff i.e. $S\to 0$. Similarly, if the S value tends to infinity (CN $\to 0$), all water goes for storage and there will no runoff. Thus, these two boundary conditions are satisfied using this empirical model. For improving the model performance, instead of a fixed λ value, it can also be obtained after optimization. The best λ value obtained was zero in the tested watersheds, which leads to further simplification of Equation (11) as:

$$Q=\frac{P(P+P_5)}{(P+P_5+S)} \tag{13}$$

This equation is a version of the SCS-CN equation with incorporating P_5 and represents a simplified form of model M_6. The different models using P_5 value either directly or indirectly and related equations are also summarized in Table 1.

MATERIAL AND METHODS

Study area and data selection criteria

This study has been done over 114 US watersheds having areas varying from 0.17 ha to 30,351.45 ha. Data is taken

Table 1 | Parameter description of different models

S.No.	Model	Equation to be used and description of model	Remarks
1.	M_1	According to the P-Q dataset, determine S for individual storms using Equation (5). The median value of S for the whole dataset was calculated, which represented watershed S at normal condition. The runoff depth was calculated using Equation (3) at $\lambda=0.2$. S can be converted using Equation (6) or Equation (7) based on antecedent condition as normal, dry or wet.	SCS-CN model (USDA 1985)
2.	M_2	S value was optimized using Equation (3) at $\lambda=0.2$, according to antecedent normal, dry, or wet condition which is decided based on P_5 value and using Equation (5), Equation (6), or Equation (7) respectively.	SCS-CN model (NRCS 2004)
3.	M_3	Equation (8) was used which directly consider P_5 and optimized S value	Ajmal model (Ajmal & Kim 2015)
4.	M_4	Equation (9) was used which directly consider P_5 and optimized S value	Ajmal model (Ajmal & Kim 2015)
5.	M_5	Optimized S value with $\lambda=0.2$ using Equation (12)	Proposed model
6.	M_6	Optimized S value with varying λ using Equation (11)	Proposed model

from the United States Department of Agriculture-Agricultural Research Service (USDA-ARS) database which is a collection of storm depth (P mm), runoff depth (Q mm), and antecedent five days rainfall (P_5 mm) and generated from the breakpoint rainfall-runoff dataset. The information of watershed ID with its serial number, situated location, and states have been provided in Table 2. The average storm depth (P), average runoff depth (Q) and average antecedent rainfall value (P_5) in mm unit for different watersheds are illustrated in Figure 2(a). In addition, the drainage area of different watersheds (ha) in log scale and number of available and selected events for this study are presented in Figure 2(b).

Hawkins *et al.* (1985, 2008) suggested that only larger storms ($P/S \geq 0.46$) should be used for better runoff prediction. This condition increases the chances of producing runoff by more than 90% and minimizes the biasing effect. Later, some other studies have been performed and support these criteria (Stewart *et al.* 2012; Ajmal & Kim 2015). In this study, the same criteria were used but in a different way. For doing this, the original equation of SCS-CN model can be reformed into the following form as:

$$\frac{Q}{P} = \frac{\left(\frac{P}{S} - 0.2\right)^2}{\frac{P}{S}\left(\frac{P}{S} + 0.8\right)} \tag{14}$$

Here, $P/S = 0.46$ put into Equation (14), and finds Q/P as 0.12. It mean we can replace $P/S \geq 0.46$ by $Q/P \geq 0.12$. Thus, this study excluded lower runoff producing rainfall events (runoff coefficient value is less than 0.12) and sorted only those events for which $C > 0.12$. After adopting this criterion out of 28,849 events of 114 US watersheds, 11,784 events were selected for this study.

Parameter estimation

To obtain the best possible results with different models, optimization has been carried out by minimizing the sum of the square difference between observed and computed runoff employing Microsoft Excel (Solver) and using Equation (15). The main intention of optimization is to obtain a realistic and conceptually unique value of model parameter (Lal *et al.* 2015).

$$\sum_{i}^{n} (Q_{oi} - Q_{ci})^2 \Rightarrow Minimum \tag{15}$$

Model M_1 used a fixed value of S (using Equation (3)) while the rest of the models obtained their S value through the optimization process. The initial estimate of S in the M_2 model was taken as 125 mm and allowed to vary in between 1 and 500, while in other models S was allowed to vary in the range of 1 to 1,000 with an initial estimate of 250 mm. Model M_6 optimized both the S and λ value. In the M_6 model, the initial estimate of λ was assumed as 0.05 and permitted to vary between 0 and 1. The statistical summary of S and λ value for different models are presented in Table 3.

Performance evaluation

The statistical indices-based assessment has been done to analyze the performance of different models. The root mean square error (RMSE), Nash-Sutcliffe Efficiency (NSE), percent bias (PBIAS), and n(t) criteria have been used on 114 US watersheds to evaluate the performance of different models. The mathematical equation of these criteria are respectively given below:

$$RMSE = \left[\frac{1}{n}\sum_{i=1}^{n} (Q_o - Q_c)_i^2\right]^{1/2} \tag{16}$$

$$NSE = 1 - \left[\frac{\sum_{i=1}^{n} (Q_o - Q_c)_i^2}{\sum_{i=1}^{n} (Q_o - \overline{Q_O})_i^2}\right] \tag{17}$$

$$PBIAS = \left[\frac{\sum_{i=1}^{n} (Q_o - Q_c)_i}{\sum_{i=1}^{n} (Q_o)_i}\right] \tag{18}$$

$$n(t) = \frac{SD_{OR}}{RMSE} - 1 \tag{19}$$

where Q_o, Q_c and SD_{OR} are observed runoff, computed runoff and variation in observed runoff given by standard deviation, respectively and i is an integer varying from 1 to N. The *n(t)* value depicts the number of times the cumulative variation in mean observation is more than the mean error.

Table 2 | Different watersheds ID's with S.No., location, state and goodness of fit of different models based on NSE (Nash-Sutcliffe Efficiency) and RGS (Rank and grading system)

				Model rank based on NSE						Obtained score based on rank					
S. No.	**WS ID**	**Location**	**State**	**M_1**	**M_2**	**M_3**	**M_4**	**M_5**	**M_6**	**M_1**	**M_2**	**M_3**	**M_4**	**M_5**	**M_6**
1	9004	Americus	Georgia	Neg	0.26	0.69	0.67	0.76	0.88	1	2	4	3	5	6
2	13007	Blacksburg	Virginia	Neg	Neg	0.08	0.08	Neg	Neg	1	2	6	5	3	4
3	13008	Blacksburg	Virginia	Neg	0.62	0.89	0.89	0.79	0.92	1	2	5	4	3	6
4	13009	Blacksburg	Virginia	0.42	0.44	0.67	0.67	0.53	0.65	1	2	6	5	3	4
5	13010	Blacksburg	Virginia	Neg	Neg	0.57	0.56	0.21	0.59	1	2	5	4	3	6
6	13012	Blacksburg	Virginia	0.13	0.13	0.58	0.57	0.41	0.61	1	2	5	4	3	6
7	13014	Blacksburg	Virginia	0.83	0.87	0.95	0.95	0.92	0.94	1	2	6	5	3	4
8	13015	Blacksburg	Virginia	Neg	Neg	0.89	0.89	0.66	0.90	1	2	5	4	3	6
9	17001	Edwardsville	Illinois	0.75	0.75	0.79	0.78	0.80	0.82	1	2	4	3	5	6
10	17002	Edwardsville	Illinois	0.72	0.72	0.77	0.76	0.79	0.80	1	2	4	3	5	6
11	17003	Edwardsville	Illinois	0.72	0.72	0.72	0.71	0.78	0.80	2	4	3	1	5	6
12	17004	Edwardsville	Illinois	0.70	0.70	0.81	0.80	0.80	0.84	1	2	5	3	4	6
13	19001	Lafayette	Indiana	Neg	Neg	0.07	Neg	0.15	0.39	1	2	4	3	5	6
14	19004	Lafayette	Indiana	0.55	0.57	0.58	0.56	0.63	0.68	1	3	4	2	5	6
15	19005	Lafayette	Indiana	0.81	0.84	0.84	0.83	0.86	0.87	1	4	3	2	5	6
16	19006	Lafayette	Indiana	0.29	0.29	0.51	0.45	0.43	0.58	1	2	5	4	3	6
17	19007	Lafayette	Indiana	0.50	0.57	0.65	0.64	0.65	0.73	1	2	5	3	4	6
18	19008	Lafayette	Indiana	0.78	0.79	0.81	0.78	0.84	0.84	2	3	4	1	5	6
19	19009	Lafayette	Indiana	0.77	0.83	0.78	0.76	0.87	0.88	2	4	3	1	5	6
20	19010	Lafayette	Indiana	0.85	0.85	0.83	0.80	0.90	0.90	3	4	2	1	5	6
21	19011	Lafayette	Indiana	0.81	0.85	0.89	0.87	0.91	0.91	1	2	4	3	5	6
22	22003	Fourmile CK	Iowa	0.40	0.46	0.70	0.68	0.68	0.70	1	2	5	4	3	6
23	22004	Fourmile CK	Iowa	0.43	0.47	0.69	0.68	0.68	0.69	1	2	5	4	3	6
24	25001	Kingdom City	Missouri	0.78	0.78	0.81	0.80	0.81	0.82	1	2	4	3	5	6
25	26001	Coshocton	Ohio	0.19	0.34	0.70	0.69	0.66	0.71	1	2	5	4	3	6
26	26010	Coshocton	Ohio	0.45	0.46	0.68	0.66	0.62	0.70	1	2	5	4	3	6
27	26013	Coshocton	Ohio	Neg	Neg	0.27	0.26	0.01	0.27	1	2	6	4	3	5
28	26014	Coshocton	Ohio	0.64	0.65	0.70	0.69	0.70	0.75	1	2	4	3	5	6
29	26016	Coshocton	Ohio	0.19	0.35	0.65	0.64	0.53	0.63	1	2	6	5	3	4
30	26018	Coshocton	Ohio	0.59	0.69	0.74	0.72	0.75	0.82	1	2	4	3	5	6
31	26030	Coshocton	Ohio	0.55	0.57	0.70	0.69	0.64	0.71	1	2	5	4	3	6
32	26031	Coshocton	Ohio	Neg	Neg	0.52	0.49	0.24	0.48	1	2	6	5	3	4
33	26035	Coshocton	Ohio	0.35	0.35	0.72	0.72	0.55	0.74	1	2	5	4	3	6
34	26036	Coshocton	Ohio	0.56	0.61	0.71	0.71	0.66	0.68	1	2	6	5	3	4
35	26863	Coshocton	Ohio	0.84	0.84	0.88	0.87	0.87	0.89	1	2	5	3	4	6
36	33002	Bentonville	Arkansas	Neg	Neg	0.36	0.33	Neg	0.32	1	2	6	5	3	4
37	33005	Bentonville	Arkansas	0.92	0.93	0.96	0.96	0.95	0.97	1	2	5	4	3	6
38	33006	Bentonville	Arkansas	0.76	0.80	0.82	0.82	0.82	0.85	1	2	4	3	5	6

(continued)

Table 2 | continued

S. No.	WS ID	Location	State	Model rank based on NSE M$_1$	M$_2$	M$_3$	M$_4$	M$_5$	M$_6$	Obtained score based on rank M$_1$	M$_2$	M$_3$	M$_4$	M$_5$	M$_6$
39	34002	Cherokee	Oklahoma	0.68	0.68	0.70	0.68	0.73	0.75	1	2	4	3	5	6
40	34006	Cherokee	Oklahoma	0.48	0.48	0.59	0.57	0.60	0.64	1	2	4	3	5	6
41	34007	Cherokee	Oklahoma	0.55	0.57	0.61	0.59	0.66	0.69	1	2	4	3	5	6
42	34008	Cherokee	Oklahoma	0.50	0.50	0.58	0.56	0.61	0.65	1	2	4	3	5	6
43	35001	Guthrie	Oklahoma	0.66	0.66	0.78	0.77	0.78	0.79	1	2	5	3	4	6
44	35002	Guthrie	Oklahoma	Neg	Neg	0.76	0.75	0.55	0.67	1	2	6	5	3	4
45	35003	Guthrie	Oklahoma	0.68	0.68	0.79	0.78	0.80	0.81	1	2	4	3	5	6
46	35008	Guthrie	Oklahoma	0.19	0.41	0.67	0.66	0.67	0.74	1	2	4	3	5	6
47	35010	Guthrie	Oklahoma	0.30	0.40	0.66	0.64	0.68	0.74	1	2	4	3	5	6
48	35011	Guthrie	Oklahoma	Neg	Neg	Neg	Neg	Neg	Neg	1	2	5	4	3	6
49	37001	Stillwater	Oklahoma	0.84	0.88	0.89	0.88	0.89	0.89	1	2	5	3	4	6
50	37002	Stillwater	Oklahoma	0.72	0.74	0.77	0.75	0.78	0.80	1	2	4	3	5	6
51	42008	Riesel	Texas	0.68	0.73	0.69	0.68	0.75	0.75	2	4	3	1	5	6
52	42010	Riesel	Texas	0.70	0.74	0.81	0.80	0.80	0.83	1	2	5	4	3	6
53	42012	Riesel	Texas	0.64	0.67	0.73	0.72	0.73	0.75	1	2	4	3	5	6
54	42013	Riesel	Texas	0.27	0.54	0.71	0.70	0.74	0.75	1	2	4	3	5	6
55	42014	Riesel	Texas	0.55	0.59	0.71	0.70	0.74	0.77	1	2	4	3	5	6
56	42015	Riesel	Texas	0.66	0.69	0.77	0.76	0.76	0.78	1	2	5	3	4	6
57	42016	Riesel	Texas	0.63	0.63	0.69	0.68	0.72	0.76	1	2	4	3	5	6
58	42017	Riesel	Texas	0.72	0.74	0.78	0.77	0.79	0.80	1	2	4	3	5	6
59	42037	Riesel	Texas	0.80	0.82	0.83	0.82	0.87	0.87	1	3	4	2	5	6
60	42038	Riesel	Texas	0.75	0.75	0.71	0.69	0.78	0.78	3	4	2	1	5	6
61	42039	Riesel	Texas	0.75	0.78	0.75	0.74	0.83	0.83	3	4	2	1	5	6
62	42040	Riesel	Texas	0.77	0.80	0.77	0.75	0.85	0.85	2	4	3	1	5	6
63	44001	Hastings	Nebraska	0.61	0.61	0.74	0.72	0.74	0.77	1	2	5	3	4	6
64	44002	Hastings	Nebraska	0.71	0.71	0.79	0.78	0.83	0.84	1	2	4	3	5	6
65	44003	Hastings	Nebraska	0.73	0.73	0.76	0.74	0.80	0.86	1	2	4	3	5	6
66	44004	Hastings	Nebraska	0.67	0.67	0.79	0.78	0.79	0.85	1	2	5	3	4	6
67	44005	Hastings	Nebraska	0.58	0.59	0.62	0.60	0.66	0.71	1	2	4	3	5	6
68	44006	Hastings	Nebraska	0.57	0.59	0.50	0.48	0.60	0.60	3	4	2	1	5	6
69	44007	Hastings	Nebraska	0.83	0.83	0.83	0.80	0.85	0.85	3	4	2	1	5	6
70	44008	Hastings	Nebraska	0.81	0.81	0.85	0.83	0.85	0.85	1	2	4	3	5	6
71	44009	Hastings	Nebraska	0.81	0.82	0.87	0.85	0.87	0.88	1	2	4	3	5	6
72	44010	Hastings	Nebraska	0.77	0.77	0.81	0.78	0.84	0.85	1	2	4	3	5	6
73	44011	Hastings	Nebraska	0.72	0.72	0.75	0.73	0.78	0.80	1	2	4	3	5	6
74	44012	Hastings	Nebraska	0.75	0.76	0.82	0.80	0.81	0.83	1	2	5	3	4	6
75	44013	Hastings	Nebraska	0.59	0.62	0.75	0.72	0.76	0.82	1	2	4	3	5	6
76	44014	Hastings	Nebraska	0.56	0.60	0.65	0.61	0.71	0.75	1	2	4	3	5	6

(continued)

Table 2 | continued

				Model rank based on NSE						Obtained score based on rank					
S. No.	**WS ID**	**Location**	**State**	**M_1**	**M_2**	**M_3**	**M_4**	**M_5**	**M_6**	**M_1**	**M_2**	**M_3**	**M_4**	**M_5**	**M_6**
77	44015	Hastings	Nebraska	0.72	0.72	0.80	0.78	0.81	0.82	1	2	4	3	5	6
78	44016	Hastings	Nebraska	0.48	0.50	0.63	0.61	0.63	0.67	1	2	5	3	4	6
79	44017	Hastings	Nebraska	0.66	0.68	0.78	0.76	0.83	0.85	1	2	4	3	5	6
80	44018	Hastings	Nebraska	0.66	0.68	0.79	0.77	0.83	0.85	1	2	4	3	5	6
81	44019	Hastings	Nebraska	0.52	0.55	0.70	0.67	0.71	0.76	1	2	4	3	5	6
82	44020	Hastings	Nebraska	0.72	0.72	0.79	0.76	0.81	0.83	1	2	4	3	5	6
83	44021	Hastings	Nebraska	0.68	0.69	0.78	0.75	0.80	0.81	1	2	4	3	5	6
84	44022	Hastings	Nebraska	0.31	0.34	0.70	0.70	0.60	0.67	1	2	6	5	3	4
85	44023	Hastings	Nebraska	0.75	0.75	0.81	0.79	0.82	0.83	1	2	4	3	5	6
86	44024	Hastings	Nebraska	0.05	0.23	0.42	0.38	0.46	0.56	1	2	4	3	5	6
87	44025	Hastings	Nebraska	0.64	0.66	0.74	0.72	0.76	0.78	1	2	4	3	5	6
88	44026	Hastings	Nebraska	0.63	0.63	0.78	0.76	0.79	0.80	1	2	4	3	5	6
89	44027	Hastings	Nebraska	0.53	0.56	0.69	0.66	0.74	0.76	1	2	4	3	5	6
90	44028	Hastings	Nebraska	0.66	0.66	0.78	0.76	0.79	0.80	1	2	4	3	5	6
91	45001	Safford	Arizona	0.06	0.16	0.43	0.40	0.43	0.52	1	2	4	3	5	6
92	45002	Safford	Arizona	0.77	0.79	0.82	0.81	0.78	0.80	1	3	6	5	2	4
93	47002	Albuquerque	New Maxico	0.25	0.28	0.50	0.47	Neg	0.48	2	3	6	4	1	5
94	47003	Albuquerque	New Maxico	Neg	Neg	Neg	Neg	Neg	Neg	2	3	5	4	1	6
95	61001	Monticello	Illinois	Neg	Neg	0.71	0.70	0.62	0.74	1	2	5	4	3	6
96	61002	Monticello	Illinois	Neg	Neg	0.67	0.66	0.53	0.67	1	2	5	4	3	6
97	61003	Monticello	Illinois	Neg	0.42	0.66	0.63	0.69	0.75	1	2	4	3	5	6
98	61004	Monticello	Illinois	Neg	0.22	0.52	0.49	0.44	0.61	1	2	5	4	3	6
99	62001	Oxford	Mississippi	0.56	0.58	0.75	0.75	0.76	0.82	1	2	4	3	5	6
100	62002	Oxford	Mississippi	0.81	0.82	0.82	0.81	0.85	0.85	2	4	3	1	5	6
101	62005	Oxford	Mississippi	0.60	0.60	0.82	0.82	0.84	0.87	1	2	4	3	5	6
102	62007	Oxford	Mississippi	0.71	0.77	0.85	0.85	0.83	0.85	1	2	6	4	3	5
103	62010	Oxford	Mississippi	0.73	0.74	0.76	0.76	0.79	0.81	1	2	4	3	5	6
104	62011	Oxford	Mississippi	Neg	0.14	0.32	0.30	0.33	0.55	1	2	4	3	5	6
105	62012	Oxford	Mississippi	0.24	0.27	0.50	0.49	0.46	0.55	1	2	5	4	3	6
106	62014	Oxford	Mississippi	0.73	0.74	0.73	0.70	0.79	0.80	3	4	2	1	5	6
107	62018	Oxford	Mississippi	0.52	0.63	0.79	0.79	0.74	0.79	1	2	6	5	3	4
108	63103	Tombstone	Arizona	0.65	0.91	0.95	0.95	0.89	0.92	1	3	6	5	2	4
109	63104	Tombstone	Arizona	0.78	0.83	0.88	0.87	0.83	0.86	1	2	6	5	3	4
110	63106	Tombstone	Arizona	0.47	0.55	0.72	0.70	Neg	0.63	2	3	6	5	1	4
111	63112	Tombstone	Arizona	0.63	0.64	0.63	0.57	0.75	0.77	3	4	2	1	5	6
112	76001	Santa Rita	Arizona	0.37	0.53	0.70	0.68	0.50	0.65	1	3	6	5	2	4
113	76002	Santa Rita	Arizona	Neg	0.40	0.64	0.64	Neg	0.51	2	3	6	5	1	4
114	76003	Santa Rita	Arizona	0.39	0.42	0.59	0.57	0.48	0.60	1	2	5	4	3	6

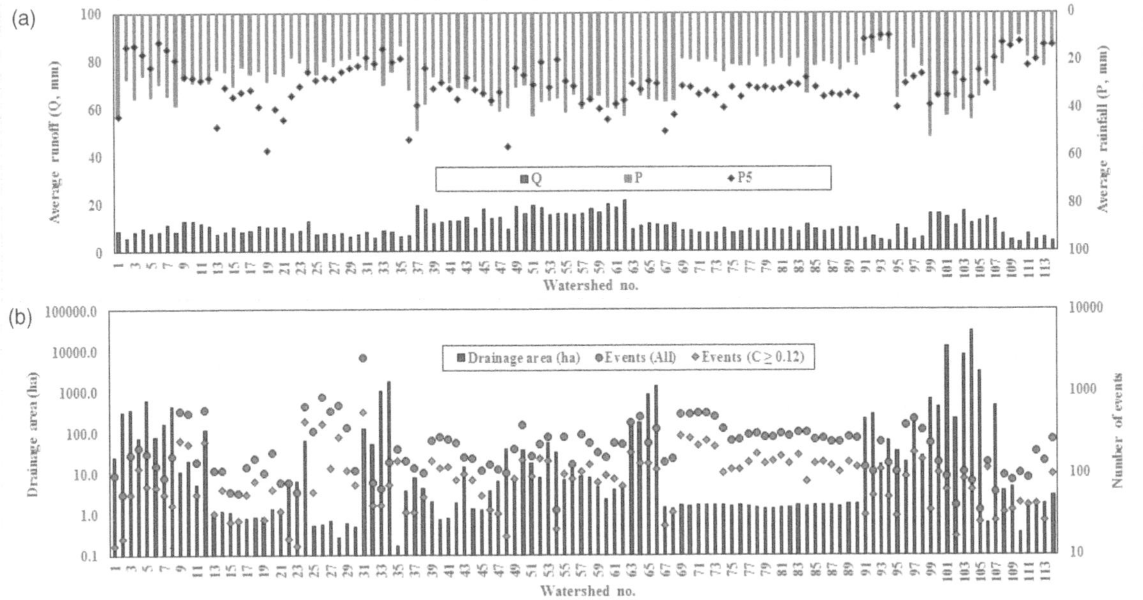

Figure 2 | (a) Average values of rainfall (P), runoff (Q) and antecedent five days rainfall (P_5); (b) number of available events, number of selected events (runoff coefficient $C \geq 0.12$) and drainage area of the studied 114 US watersheds.

Table 3 | Parameter statistics of different models over 114 US watersheds (S_{med} of M_1 model obtained from storm-event data and rest parameters for M_2 to M_6 model obtained by optimization)

Statistic	M_1 S_{med}, mm	M_2 S, mm	M_3 S, mm	M_4 S, mm	M_5 S, mm	M_6 λ	M_6 S, mm
Minimum	8.96	6.76	26.37	24.67	25.73	0.00	35.07
Maximum	58.41	203.04	339.01	338.81	394.93	0.53	580.50
Mean	25.36	33.04	95.94	95.30	80.68	0.02	129.08
Median	23.33	22.70	74.09	73.37	59.21	0.00	98.50
SD	11.36	33.19	62.74	62.75	58.86	0.08	94.28
SE	1.06	3.11	5.88	5.88	5.51	0.01	8.83
CI	1.75	5.11	9.66	9.67	9.07	0.01	14.52
Q1	16.46	18.08	57.34	56.94	50.07	0.00	76.54
Q3	31.98	33.72	115.64	114.82	86.65	0.00	149.11
UB	27.11	38.15	105.60	104.97	89.75	0.03	143.61
LB	23.61	27.93	86.27	85.63	71.61	0.01	114.56

The higher $n(t)$ value reflects the appropriateness of the model for efficient runoff computation.

To select the best model for a particular watershed, NSE based rank and grading system (RGS) was applied on different models and ranks (I to VI) were assigned according to their NSE value (Verma *et al.* 2018). Their awarded score was added to select the best model among all. The percentage improvement achieved by the best model over rest models can be addressed by r^2 criteria. The mathematical expression of r^2 is as follows:

$$r^2 = \frac{(NSE_b - NSE_r)}{1 - NSE_e} \tag{20}$$

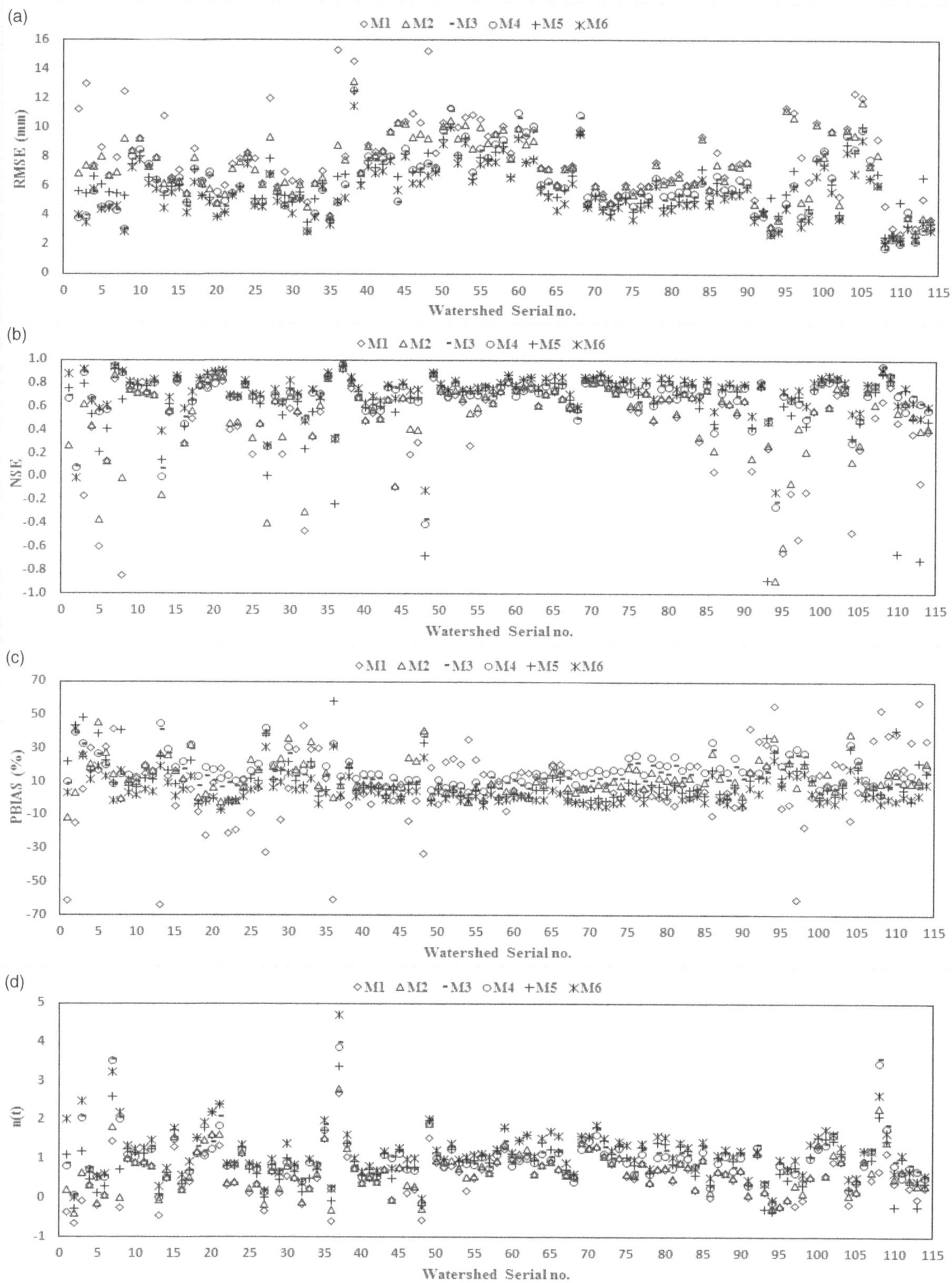

Figure 3 | Variation in (a) Root Mean Square Error (RMSE, mm); (b) Nash-Sutcliffe Efficiency (NSE); (c) PBIAS (%) and (d) n(t) values with watershed serial no. using different models.

where NSE_b and NSE_r are the efficiencies of best model and referenced model. More than 10% indicates a significant improvement (Senbeta *et al.* 1999).

RESULTS AND DISCUSSION

The performance of all 114 US watersheds resulting from applications of each of the models has been compared using RMSE, NSE, PBIAS, n(t), rank and grading system, and by visual assessment. The performance of all models for individual watersheds was compared by using a scatter plot (Figure 3) and the collective performance of models can be seen in a Box and Whisker plot for all statistical indices (Figure 4). The best model suggestion was based on total score obtained using rank and grading system (RGS) for all 114 US watersheds. The value of *S* was optimized for all six models which are presented in Table 3. The mean and median value of *S* was found to be higher for the M_6 model and lowest for the M_1 model. The higher *S* value also increases the standard deviation (SD) and standard error (SE) value of the M_6 model.

The RMSE values for all models are presented in Figure 3(a). The lower the RMSE, the better the performance of the model, and vice versa. It can be seen that the existing SCS-CN model shows a worse performance than the other models. The median value of RMSE for M_1, M_2, M_3, M_4, M_5, and M_6 models were found to be 7.54, 7.23, 5.70, 5.88, 5.85 and 5.32 mm, respectively, which is lowest for the M_6 model. Its mean value was also lowest (5.53 mm) for the M_6 model. When we compared all models based on RMSE value, out of 114 watersheds the M_6 model performs best with minimum RMSE values in 95 watersheds. Next is the model M_3, which had the lowest RMSE values in the remaining 19 watersheds. From the RMSE value, it can be inferred that the M_6 model performs well followed by M_3, M_5, M_4, M_2, and M_1. The higher RMSE values exhibit the worst performance and are found mostly in the M_1 and M_2 models. Figure 4(a) exhibits the overall RMSE performance of all models by using a Box and Whisker plot. Noteworthy, the

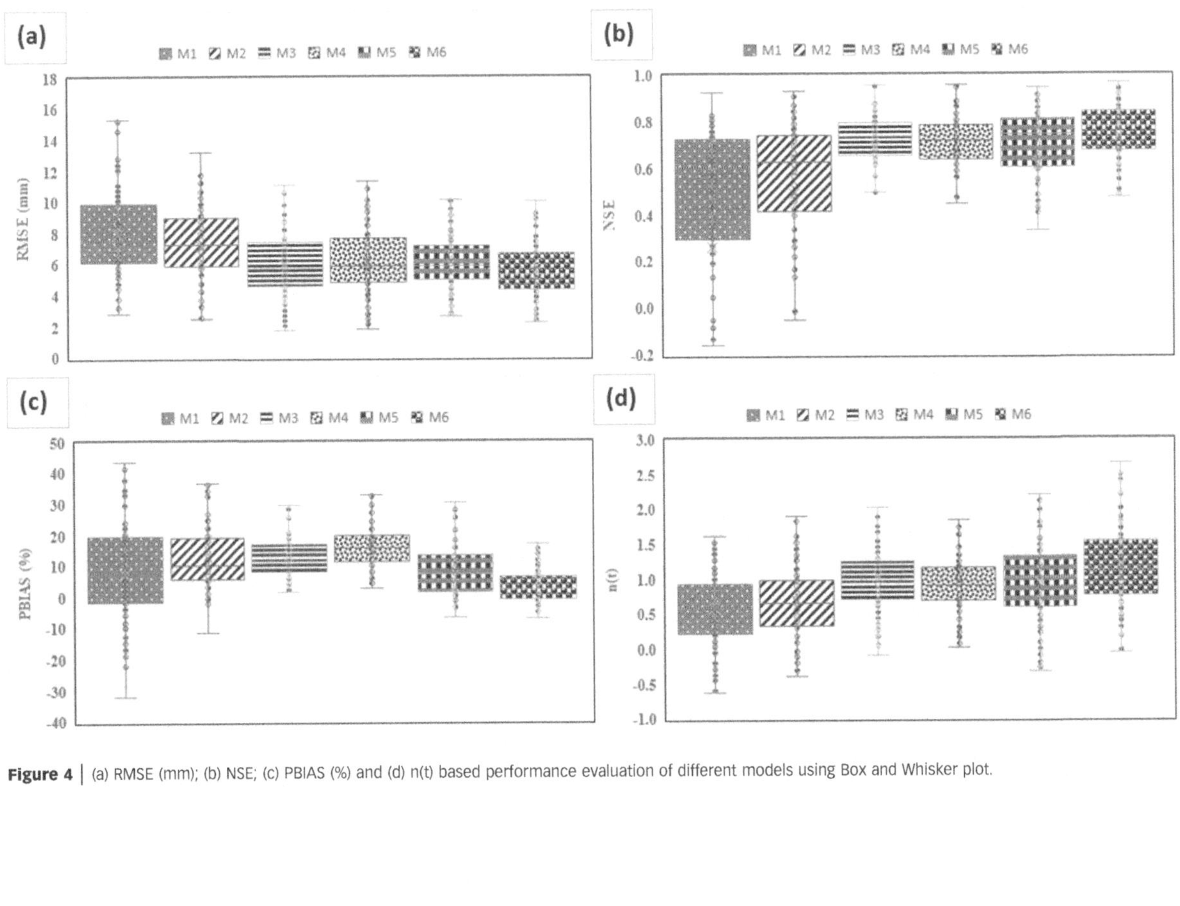

Figure 4 | (a) RMSE (mm); (b) NSE; (c) PBIAS (%) and (d) n(t) based performance evaluation of different models using Box and Whisker plot.

M_3 and M_4 models follow the same trend and the RMSE value of the M_3 model is better than the M_4 model.

The NSE value indicates the model efficiency and a high value means a better model. The NSE value was found to be better in the M_6 model for most of the watersheds (Figure 3(b)). NSE values fall in a good range ($0.80 \leq NSE < 0.90$) in 41 watersheds followed by 30, 25, 19, 15, and 11 for M_5, M_3, M_4, M_2, and M_1, respectively. The cumulative median value for all watersheds was found to be highest for the M_6 model as 0.78 followed by M_5 (0.75) > M_3 (0.73) > M_4 (0.72) > M_2 (0.63) > M_1 (0.61). The mean value was very low, as 0.62 for the M_5 model, since some of the watersheds' NSE values were found to be very low. In some cases it was negative, which indicates model prediction was worse than the mean observed value. From Figure 4(b), it is evident that model M_6 performed superiorly, with higher NSE than other models and model M_3 and M_4 depict nearly the same performance.

PBIAS quantifies a model tendency either to overestimate or underestimate. A PBIAS value of zero indicates perfect fit, its negative value indicates overestimation and vice versa. Models M_3 and M_4 underestimated (positive PBIAS) the runoff in all 114 watersheds and the mean PBIAS value was also as high as 14.13% and 17.12% respectively compared to the other models. Models M_1, M_2, and M_5 underestimated the runoff in 80, 107, and 101 watersheds and overestimated (negative PBIAS) in 34, 7, and 13 watersheds respectively. The mean and median PBIAS were both lowest and nearer to zero for the M_6 model with 4.77% and 2.62% respectively. The PBIAS results indicated that the M_6 model performance was very good in 91 watersheds and unsatisfactory in 4 watersheds only. By using the M_6 model, most of the watersheds fall either in the very good or good range (Figure 3(c)). The PBIAS statistical index prefers the M_6 model because it gives a more consistent and better PBIAS value (Figure 4(c)). The overall statistics of RMSE, NSE and PBIAS values for all models are presented in Table 4. The overall SD and SE values of RMSE, NSE, and PBIAS for 114 watersheds were found to be less for the M_6 model. For the M_5 model, SD and SE values are quite high because in some watersheds the performance of the model was very poor, which increases dispersion from its mean value, lowers the mean and maximizes their range.

Like the previous three indices, the $n(t)$ value was also found to be best in most of the watersheds using the M_6

Table 4 | Statistics of performance indicators for various models over 114 US watersheds

Statistic	M_1 RMSE	M_2 RMSE	M_3 RMSE	M_4 RMSE	M_5 RMSE	M_6 RMSE	M_1 NSE	M_2 NSE	M_3 NSE	M_4 NSE	M_5 NSE	M_6 NSE	M_1 PBIAS	M_2 PBIAS	M_3 PBIAS	M_4 PBIAS	M_5 PBIAS	M_6 PBIAS
Minimum	2.81	2.46	1.77	1.82	2.61	2.23	−6.98	−1.96	−0.36	−0.40	−1.20	−0.13	−63.72	−11.44	1.48	2.78	−6.67	−7.01
Maximum	15.35	13.20	12.49	12.58	12.48	11.46	0.92	0.93	0.96	0.96	0.95	0.97	57.65	45.88	40.80	45.07	58.52	41.35
Mean	7.92	7.22	6.01	6.18	6.11	5.53	0.26	0.49	0.69	0.67	0.62	0.73	7.49	12.65	14.13	17.12	10.44	4.77
Median	7.54	7.23	5.70	5.88	5.85	5.32	0.61	0.63	0.73	0.72	0.75	0.78	7.53	10.34	12.19	15.82	7.17	2.62
SD	2.61	2.14	2.06	2.09	1.74	1.82	1.14	0.45	0.20	0.21	0.40	0.18	21.55	10.44	8.39	8.41	12.07	8.32
SE	0.24	0.20	0.19	0.20	0.16	0.17	0.11	0.04	0.02	0.02	0.04	0.02	2.02	0.98	0.79	0.79	1.13	0.78
CI	0.40	0.33	0.32	0.32	0.27	0.28	0.17	0.07	0.03	0.03	0.06	0.03	3.32	1.61	1.29	1.30	1.86	1.28
Q1	6.10	5.93	4.62	4.84	4.97	4.35	0.30	0.42	0.65	0.64	0.61	0.68	−1.46	5.88	8.38	11.33	1.83	−0.73
Q3	9.84	8.96	7.43	7.55	7.19	6.70	0.72	0.74	0.79	0.78	0.81	0.84	18.84	18.77	16.96	19.95	12.74	6.42
UB	8.32	7.55	6.32	6.50	6.38	5.81	0.44	0.56	0.72	0.70	0.68	0.76	10.81	14.26	15.42	18.41	12.30	6.05
LB	7.52	6.89	5.69	5.85	5.84	5.25	0.09	0.42	0.66	0.64	0.55	0.70	4.18	11.04	12.83	15.82	8.58	3.49

Table 5 | Comparative performance and identification of the best model using different statistical indicators after studying 114 US watersheds

Performance Index	Range	Rating	M_1	M_2	M_3	M_4	M_5	M_6
No. of watersheds with '–'ve NSE	NSE < 0	Unsatisfactory	18	12	**2**	3	7	3
NSE performance criteria set by Ritter & Muñoz-Carpena (2013)	NSE < 0.65	Unsatisfactory	65	60	27	31	34	**24**
	0.65 ≤ NSE < 0.80	Satisfactory	37	37	59	61	46	42
	0.80 ≤ NSE < 0.90	Good	11	15	25	19	30	41
	NSE ≥ 0.90	Very good	1	2	3	3	4	**7**
NSE Performance criteria set by Moriasi *et al.* (2007)	NSE ≤ 0.50	Unsatisfactory	42	37	11	15	19	**8**
	0.50 < NSE ≤ 0.65	Satisfactory	23	23	16	16	15	16
	0.65 < NSE ≤ 0.75	Good	27	29	35	35	24	23
	NSE ≥ 0.75	Very good	22	25	52	48	56	**67**
PBIAS performance criteria set by Moriasi *et al.* (2007)	PBIAS ≥ 25	Unsatisfactory	25	13	16	18	12	**4**
	15 ≤ PBIAS < 25	Satisfactory	20	28	23	44	16	9
	10 ≤ PBIAS < 15	Good	17	19	36	32	13	10
	PBIAS < 10	Very good	52	54	39	20	73	**91**
n(t) performance criteria set by Ritter & Muñoz-Carpena (2013)	*n(t)* < 0.7	Unsatisfactory	64	61	27	30	34	**24**
	0.7 ≤ n(t) < 1.2	Satisfactory	38	36	55	59	40	36
	1.2 ≤ n(t) < 2.2	Good	11	15	29	22	36	47
	n(t) > 2.2	Very good	1	2	3	3	4	**7**

model (Figure 3(d)). Similar to the NSE, the cumulative mean of the $n(t)$ value was found to be high for the M_6 model at 1.19, followed by 1.04 for model M_3. It was lowest for the M_1 model at just 0.58 followed by the M_2 model at 0.70. Table 5 summarises the guidelines for classifying the model performances according to their range as unsatisfactory, satisfactory, good and very good. This table suggests the usefulness and better performance of the M_6 model compared to other models. Model M_6 found exemplary performance in maximum cases, while poor performance was found comparatively in fewer watersheds.

The superiority of the M_6 model over the rest of the models was judged by making a cumulative frequency distribution curve for percentage improvement (r^2) criteria (Figure 5). The M_6 model performed better than the M_1, M_2, and M_5 models for all watersheds while for the M_3 and M_4 models, only 17 and 14% watersheds respectively performed better than M_6. To visually assess the performance

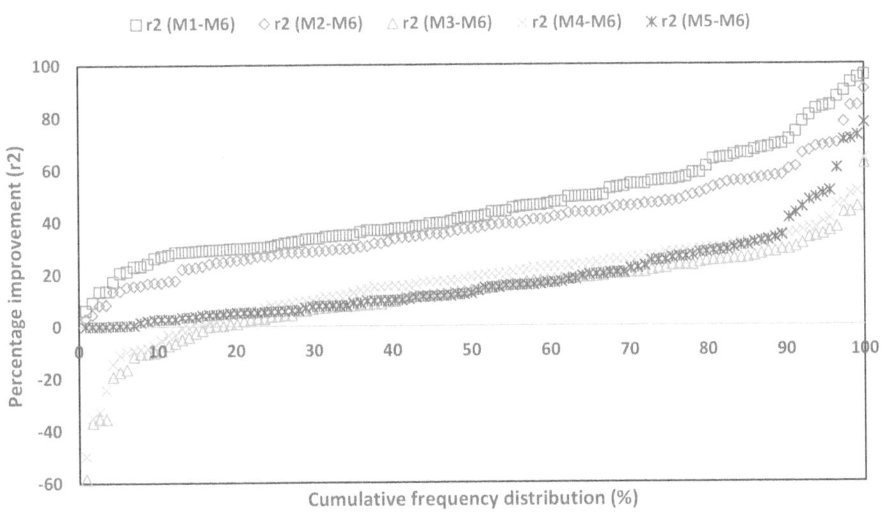

Figure 5 | Cumulative frequency distribution for percentage improvement of M_6 model over other models using r^2 criteria.

of different models a scatter plot (Figure 6) has been drawn between observed and computed runoff for WS ID- 9004, and 33005. The visual assessment indicates almost equal performance of M_6 and M_5 models for WS ID 9004 and much better performance of the M_6 model than the rest of the models for WS ID 33005.

A NSE-based rank and grading based system (RGS) was adopted to select the preference of the best model. A rank was assigned to each model for the individual watershed. A high NSE value has been given to the best rank as 1 and credit 6 point score in its account. From Table 2, for WS ID 9004, the NSE value was highest for the M_6 model (Rank I, Score 6) followed by M_5 (Rank II, Score 5), M_4 (Rank III, Score 4), M_3 (Rank IV, Score 3), M_2 (Rank V, Score 2), and M_1 (Rank VI, Score 1) as 0.88, 0.76, 0.67, 0.69, 0.26 and a negative NSE respectively. Model M_6 gained 6, 5, 4, 3, 2, and 1 score in 95, 3, 16, 0, 0, and 0 watersheds respectively. The cumulative sum of the score was highest for the M_6 model at 649 ($6\times95+5\times3+4\times16+3\times0+2\times0+1\times0=649$) followed by 504 for M_3, 471 for M_5, 368 for M_4, 264 for M_2 and 138 for the M_1 model (Figure 7). Therefore, on the basis of RGS, the M_6 model can be also rated as the best model. After the M_6 model, the next better performing models were found to be M_3, M_5, M_4, M_2, and M_1 respectively.

The results indicate that the indirect approach to categorize runoff conditions on the basis of antecedent five days rainfall as dry, normal, or wet does not reflect the best performing potential. Better is to incorporate P_5 directly in the runoff model and this certainly helps in improved model performance. Using trial and error methods, Ajmal and Kim proposed two models for South Korean catchments that directly consider the P_5 value in model formulation. Both model performances were found to be better than the SCS-CN model and increase the model efficiency of US watersheds. In this study, the SCS-CN model was modified by incorporating P_5 value and provides a more rational basis to M_5 and M_6 models. Despite being empirical in nature, it is needless to reiterate that the new model proposed in this work shows the promise of better performance using several error metrics.

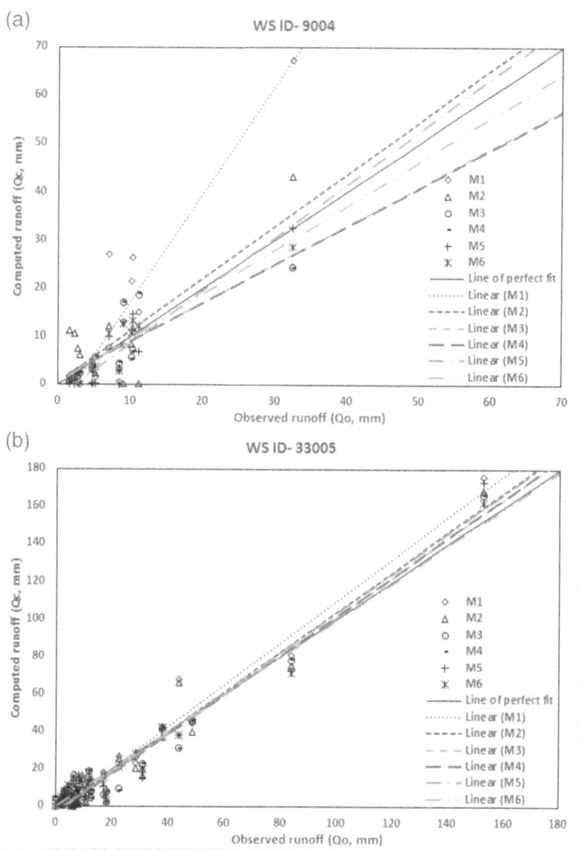

Figure 6 | Plot of observed runoff versus computed runoff to visually assess the performance of different models in the watershed (a) WS-9004 and (b) WS-33005.

CONCLUSION

Based on this study, the following can be inferred:

1. The proposed study obviates the problem of the sudden jump and improves the performance of the model. The proposed model changes the maximum potential retention value (S, mm) due to the antecedent 5 days' rainfall values.

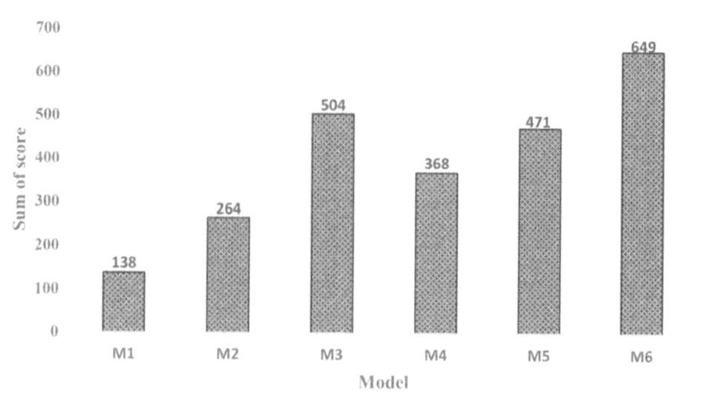

Figure 7 | RGS based sum score resulted by all the models over 114 US watersheds.

2. All statistical criteria were found to be better for the proposed model with less RMSE, high NSE, better PBIAS, high $n(t)$ value, and better overall rank and grading based score.
3. Between M_5 and M_6 models, M_5 performance was poor due to the fixed λ value. The M_5 model indicates that a fixed λ value of 0.2 does not fit well and some other value should be used. Model M_6 includes variation in λ value and it increases the efficiency of the model. For 97 watersheds, the λ value was found to be zero. The median value of parameter λ was found to also be zero. Therefore, model M_6 can be used with λ value as zero (Equation (11)) simplifying the M_5 model.

ACKNOWLEDGEMENT

A special thanks to Prof. R.H. Hawkins, Emeritus Professor, University of Arizona for providing rainfall-runoff dataset of US watersheds and Prof. C.S.P. Ojha for supervising and giving useful suggestions to complete this research work.

DATA AVAILABILITY STATEMENT

All relevant data are included in the paper or its Supplementary Information.

REFERENCES

Ajmal, M. & Kim, T. W. 2015 Quantifying excess stormwater using SCS-CN-based rainfall runoff models and different curve number determination methods. *Journal of Irrigation and Drainage Engineering* **141** (3), 04014058.

Brocca, L., Melone, F. & Moramarco, T. 2008 On the estimation of antecedent wetness conditions in rainfall–runoff modelling. *Hydrological Processes* **22** (5), 629–642.

Hawkins, R. H. 1993 Asymptotic determination of runoff curve numbers from data. *Journal of Irrigation and Drainage Engineering* **119** (2), 334–345.

Hawkins, R. H., Hjelmfelt Jr, A. T. & Zevenbergen, A. W. 1985 Runoff probability, storm depth, and curve numbers. *Journal of Irrigation and Drainage Engineering* **111** (4), 330–340.

Hawkins, R. H., Ward, T. J., Woodward, D. E. & Van Mullem, J. A. 2008 *Curve Number Hydrology: State of the Practice*. American Society of Civil Engineers, Reston, VA, 106pp.

Hewlett, J. D., Fortson, J. C. & Cunningham, G. B. 1977 The effect of rainfall intensity on storm flow and peak discharge from forest land. *Water Resources Research* **13** (2), 259–266.

Jordan, J. P. 1994 Spatial and temporal variability of stormflow generation processes on a Swiss catchment. *Journal of Hydrology* **153** (1), 357–382.

Lal, M., Mishra, S. K. & Pandey, A. 2015 Physical verification of the effect of land features and antecedent moisture on runoff curve number. *Catena* **133**, 318–327.

Latron, J., Soler, M., Llorens, P. & Gallart, F. 2008 Spatial and temporal variability of the hydrological response in a small Mediterranean research catchment (Vallcebre, Eastern Pyrenees). *Hydrological Processes* **22** (6), 775–787.

Merz, R. & Blöschl, G. 2009 A regional analysis of event runoff coefficients with respect to climate and catchment characteristics in Austria. *Water Resources Research* **45** (1), W01405.

Meyles, E., Williams, A., Ternan, L. & Dowd, J. 2003 Runoff generation in relation to soil moisture patterns in a small Dartmoor catchment, Southwest England. *Hydrological Processes* **17** (2), 251–264.

Michel, C., Andréassian, V. & Perrin, C. 2005 Soil conservation service curve number method: how to mend a wrong soil moisture accounting procedure? *Water Resources Research* **41** (2), W02011.

Mishra, S. K., Jain, M. K., Suresh Babu, P., Venugopal, K. & Kaliappan, S. 2008 Comparison of AMC-dependent CN-conversion formulae. *Water Resources Management* **22** (10), 1409–1420.

Moriasi, D. N., Arnold, J. G., Van Liew, M. W., Bingner, R. L., Harmel, R. D. & Veith, T. L. 2007 Model evaluation guidelines for systematic quantification of accuracy in watershed simulations. *Transactions of the ASABE* **50** (3), 885–900.

NRCS 2004 Chapter 10 *Estimation of Direct Runoff From Storm Rainfall*. In: Part 630 Hydrology National Engineering Handbook. Natural Resource Conservation Service, U.S.D.A., Washington, DC.

Pathiraja, S., Westra, S. & Sharma, A. 2012 Why continuous simulation? The role of antecedent moisture in design flood estimation. *Water Resources Research* **48**, 6.

Radatz, T. F., Thompson, A. M. & Madison, F. W. 2013 Soil moisture and rainfall intensity thresholds for runoff generation in southwestern Wisconsin agricultural watersheds. *Hydrological Processes* **27** (25), 3521–3534.

Ritter, A. & Muñoz-Carpena, R. 2013 Performance evaluation of hydrological models: statistical significance for reducing subjectivity in goodness-of-fit assessments. *Journal of Hydrology* **480**, 33–45.

Sahu, R. K., Mishra, S. K., Eldho, T. I. & Jain, M. K. 2007 An advanced soil moisture accounting procedure for SCS curve number method. *Hydrological Processes* **21** (21), 2872–2881.

Senbeta, D. A., Shamseldin, A. Y. & O'Connor, K. M. 1999 Modification of the probability-distributed interacting storage capacity model. *Journal of Hydrology* **224** (3–4), 149–168.

Song, S. & Wang, W. 2019 Impacts of antecedent soil moisture on the rainfall-runoff transformation process based on high-resolution observations in soil tank experiments. *Water* **11** (2), 296.

Stewart, D., Canfield, E. & Hawkins, R. 2012 Curve number determination methods and uncertainty in hydrologic soil groups from semiarid watershed data. *Journal of Hydrologic Engineering* **17**, 1180–1187.

USDA 1985 *National Engineering Handbook, Supplement A, Section 4*. US Department of Agriculture, Washington, DC.

Verma, S., Singh, P. K., Mishra, S. K., Jain, S. K., Berndtsson, R., Singh, A. & Verma, R. K. 2018 Simplified SMA-inspired 1-parameter SCS-CN model for runoff estimation. *Arabian Journal of Geosciences* **11** (15), 420.

Verma, S., Singh, P. K., Mishra, S. K., Singh, V. P., Singh, V. & Singh, A. 2020 Activation soil moisture accounting (ASMA) for runoff estimation using soil conservation service curve number (SCS-CN) method. *Journal of Hydrology* **589**, 125114.

Zhang, Y., Wei, H. & Nearing, M. A. 2011 Effects of antecedent soil moisture on runoff modeling in small semiarid watersheds of southeastern Arizona. *Hydrology and Earth System Sciences* **15** (10), 3171–3179.

Zhao, N., Yu, F., Li, C., Zhang, L., Liu, J., Wenbin, M. & Wang, H. 2015 Soil moisture dynamics and effects on runoff generation at small hillslope scale. *Journal of Hydrologic Engineering* **20** (7), 05014024.

First received 31 August 2020; accepted in revised form 30 October 2020. Available online 11 November 2020

An inexact multi-stage interval-parameter partial information programming model for water resources management under uncertainties

Hongguang Chen and Zhongjun Wang

ABSTRACT

The urban water shortage crisis around the world is increasing. In this study, an inexact multi-stage interval-parameter partial information programming model (IMIPM) is proposed for urban water resources planning and management under uncertainties. Optimization techniques of two-stage stochastic programming (TSP), interval-parameter programming (IPP), linear partial information theory (LPI) and multistage stochastic programming (MSP) are combined into one general framework. IMIPM is used to tackle uncertainties like interval numbers, water inflow probabilities expressed as linear partial information, dynamic features in a long planning time and joint probabilities in water resources management. It is applied to Harbin where the manager needs to allocate water from multi-water sources to multi-water users during multi-planning time periods. Four water flow probability scenarios are obtained, which are associated with uncertainties of urban rainfall information. The results show that the dynamics features and uncertainties of system parameters (such as water allocation targets and shortage) are considered in this model by generating a set of representative scenarios within a multistage context. The results also imply that IMIPM can truly reflect the actual urban water resources management situation, and provide managers with decision-making space and technical support to promote the sustainable development of economics and the ecological environment in cities.

Key words | inexact multistage stochastic programming, linear partial information theory, uncertainties, urban water resources management

Hongguang Chen (corresponding author)
School of Water Conservancy and Civil Engineering,
Northeast Agricultural University,
Harbin 150030,
China
E-mail: *chg218@126.com*

Zhongjun Wang
College of Engineering,
Northeast Agricultural University,
Harbin 150030,
China

HIGHLIGHTS

- Combine techniques of interval-parameter method, multistage stochastic method, linear partial information theory, and two-stage stochastic programming into one framework.
- Authors apply it in a real-world water resources problem.
- An interactive algorithm is introduced to deal with this model.
- The results can support managers in having multiple decision alternatives.
- It can apply in water or other resource problems.

doi: 10.2166/ws.2020.265

GRAPHICAL ABSTRACT

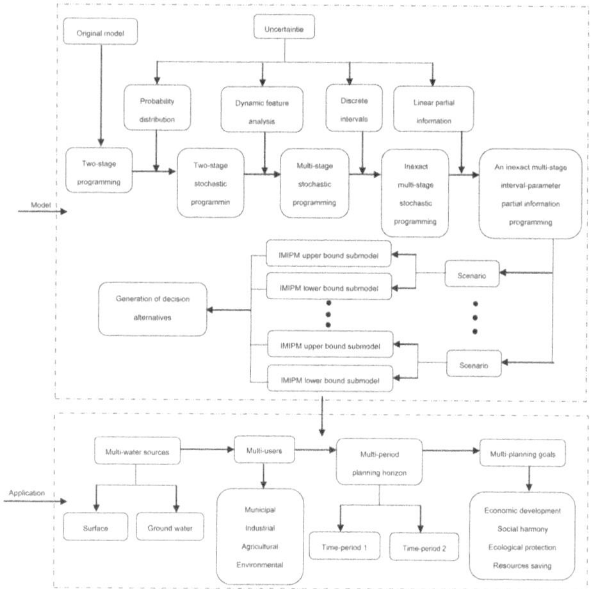

INTRODUCTION

The world is increasingly considering water shortages. Because of urbanization, climate change, water pollution and so on, many countries are facing water shortage, which is a major challenge to economic and social progress (Yannopoulos *et al.* 2019; WWAP 2019). Cities are particularly vulnerable to the impact of water scarcity, which is directly affected by rapid urban growth and reduced freshwater supply (Ray & Shaw 2018). By 2050, 66% of the world's population will live in cities (54% in 2014), increasing the water scarcity crisis (Jensen & Wu 2018). Water problems are common throughout China. About two-thirds of Chinese cities face significant water shortages, and groundwater in some areas is over-utilized (Yang *et al.* 2019). The sustainability of water resources is a prerequisite for sustainable urban development (Wang *et al.* 2019). Driven by population growth, socio-economic development, and ever-changing consumption patterns, the world's water demand continues to increase (WWAP 2016). Several important aspects of water shortage need to be emphasized. Firstly, the available water varies greatly from season to season. Therefore, the amount of available water throughout the year is different from the amount that water users can rely on when there is only one season (Water Resources Group 2008). Furthermore, groundwater is being drained. The cultivating, drinking, industrial and environmental uses mean that make water resources are extricated almost every single day (Wada *et al.* 2010). Thirdly, most countries have enough water to meet the development needs of population growth and maintain rivers needed to protect natural environment. The problem is that we have done a terrible job in managing water resources (Water Resources Group 2008).

In urban water resources management, lots of framework parameters and their interrelationships may bring uncertainty and complexity. The complexity of various framework components and economic penalties for violation of guarantee objectives may complicate this uncertainty. In addition, due to the diversity of water supply and demand in time and space, the plan required for water distribution may undergo greater changes. Therefore, feasible optimization strategies must be developed to support urban water management (Li *et al.* 2008a, 2008b; Fan *et al.* 2011).

In the past few decades, many optimization techniques have been used to solve these water management problems, including fuzzy, stochastic, chance-constrained and interval programming methods (abbreviated as FMP, SMP, CCP and IMP) (Charnes *et al.* 1971; Huang 1996; Wang 1997; Jairaj & Vedula 2000; Ben & Masri 2005; Guo & Huang 2009; Li *et al.* 2010). Among them, many researchers have tried to solve these uncertainties through two-stage and multi-stage stochastic programming methods (TSP and MSP). TSP has powerful functions for problems that need to be analyzed for policy scenarios and whose coefficients on the right-side are random but have a known probability distribution (Li *et al.* 2006). In TSP, a decision is first made before the value of the random variable is known. Then the 'penalty' can be minimized from a second-stage decision after a random event occurs and its value is known. A penalty can be generated if promised water (decided in the first stage) is not reached (Huang & Loucks 2000). Extensive research has been conducted on the TSP method. For example, Schultz *et al.* (1996) studied two-stage stochastic integer programming. Barbarosoğlu & Arda (2004) proposed a two-stage probabilistic stochastic programming model to coordinate the transportation of critical first-aid items during emergency response to disaster areas. Huang & Loucks (2000) proposed an inaccurate two-stage stochastic programming (ITSP) model. It overcomes uncertainties where parameter information cannot be expressed as a probability density function (PDF). Recently, Maqsood *et al.* (2005) developed the interval parameter fuzzy two-stage stochastic planning (IFTSP) method, which integrates the ITSP model and fuzzy programming technique for water resources management under uncertainties. In fact, many practical large-scale problems involve a series of decisions that interact with periodic results over time. The above TSP method cannot effectively represent multiple dynamic changes over time, especially when there is a sequential interaction.

As an extension of dynamic stochastic optimization techniques, various multi-stage stochastic programming methods have been developed to address dynamic features in water resources management. MSP extends TSP by allowing management measures to be taken based on updated information at each time-stage (Pereira & Pinto 1991). Recently, inexact multi-stage stochastic programming (IMSP) has been developed to manage water resources under uncertainty. Li & Huang (2007) developed inexact multistage stochastic quadratic programming (IMQP) to support water resources management in Heshui River Basin, China.

These above-mentioned methods still cannot resolve the problem of when information is only known partially. In previous inexact optimization programming techniques, probability distribution information was defined as certain values. However, for many practical problems, the probability distribution may not be clearly determined. This means that probability information can only be partially measured (Fan *et al.* 2011). To solve such problems, Ben & Masri (2005) used the α-cut technique to transform the problem into a stochastic program that contains linear partial information on probability distribution. Nevertheless, previous studies had not considered using the integrated method of incorporating interval programming, multi-stage stochastic programming, and linear partial information theory into one general optimization system to deal with complex water resource management problems.

Therefore, as an extension of previous studies, the purpose of this research is to establish an inexact multi-stage interval-parameter partial information programming model (IMIPM) and apply it to urban water resources management. This is the first attempt in which interval-parameter programming (IPP), multi-stage stochastic programming (MSP), and linear partial information (LPI) theory are integrated into one general framework system. It can support the process of water resources management and planning under uncertainties and randomness. To prove the applicability of the developed model, we take the actual water resources management in Harbin, China, as an example. The results of modeling will help water resource managers to choose the ideal water resource allocation plan under various system conditions.

METHODOLOGY

Model development

It is assumed that the water resource manager is responsible for allocating water from different water resources to multiple water users during multiple time stages. The manager can describe the problem as expanding the normal income of the water allocation system in the city. On the one hand, if the guaranteed amount of water is conveyed, at that point it will produce benefits to the nearby economy; if this is not the case, the water will have to come from other more expensive sources or users will need to diminish advancement plans, which will bring about financial punishments on the local economy.

On the other hand, the water flow in each time-stage is usually expressed as a random variable, and the water distribution plan has dynamic characteristics, which demands proper decisions to be made at discrete time points and discrete probability levels of uncertain variables. Therefore, the problem can be expressed as an MSP model. Figure 1 is a T-stage decision tree, which represents the dynamic characteristics of the MSP model. Nodes represent decisions in related stages, and arcs are used to implement random variables. The number of stages is equal to one plus the number of nodes over the planning horizon (Li *et al.* 2008a; Zhou *et al.* 2013; Liu *et al.* 2015). Furthermore, these uncertain parameters can be estimated as discrete intervals, so the problem can be refined into an interval multi-stage stochastic programming model (Li *et al.* 2008b). Thus we have:

$$\max f^{\pm} = \sum_{i=1}^{I}\sum_{j=1}^{J}\sum_{t=1}^{T} NB_{ijt}^{\pm} K_{ijt}^{\pm} - \sum_{t=1}^{T}\sum_{i=1}^{I}\sum_{j=1}^{J}\sum_{h=1}^{N} p_{ht} C_{ijt}^{\pm} S_{ijht}^{\pm} \quad (1a)$$

subject to:

constraints of available water amount

$$\sum_{i=1}^{I}\sum_{j=1}^{J} (K_{ijt}^{\pm} - S_{ijht}^{\pm}) \le q_{iht}^{\pm} + \varepsilon_{(t-1)}^{\pm} \quad \forall i, j, h, t \quad (1b)$$

constraints of surplus water amount

$$\varepsilon_{h(t-1)}^{\pm} = q_{ih(t-1)}^{\pm} - \sum_{i=1}^{I}\sum_{j=1}^{J} (K_{ij(t-1)}^{\pm} - S_{ijh(t-1)}^{\pm}) + \varepsilon_{h(t-2)}^{\pm} \quad \forall i, j, h, t \quad (1c)$$

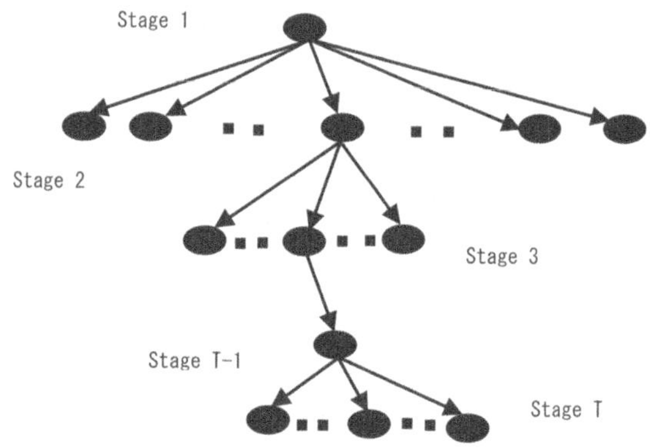

Figure 1 | Structure of the decision tree.

constraints of maximum water demand

$$\sum_{i=1}^{I} K_{ijt}^{\pm} \le K_{jt\max} \quad \forall i, j, t \quad (1d)$$

constraints of minimum water demand

$$\sum_{i=1}^{I} (K_{ijt}^{\pm} - S_{ijht}^{\pm}) \ge K_{jt\min} \quad \forall i, j, h, t \quad (1e)$$

constraints of water pipeline transportation capacity

$$\sum_{i=1}^{I} (K_{ijt}^{\pm} - S_{ijht}^{\pm}) \le e_{it}^{\pm} \quad \forall i, j, h, t \quad (1f)$$

constraints of water allocation amounts

$$K_{ijt\max} \ge K_{ijt}^{\pm} \ge S_{ijht}^{\pm} \ge 0 \quad \forall i, j, h, t \quad (1g)$$

constraints of the probability of water flow

$$p_{ht} \ge 0 \quad \forall h, t \quad (1h)$$

$$\sum_{h=1}^{N} p_{ht} = 1 \quad \forall t \quad (1i)$$

where $f^{\pm}$, $NB_{ijt}^{\pm}$, $K_{ijt}^{\pm}$, $C_{ijt}^{\pm}$, $S_{ijht}^{\pm}$, $q_{iht}^{\pm}$ and $\varepsilon_t^{\pm}$ are interval parameters/variables; + and --- mean the upper and lower bound of interval parameters/variables, separately; $f^{\pm}$ is the total water allocation system benefit over the planning horizon, (10^8¥); $NB^{\pm}_{ijt}$ is the system benefit of allocating water from water source i to user j per unit during time period t, (¥/m^3); i is water resource, $i = 1, 2, \ldots, I$; j is water user, $j = 1, 2, \ldots, J$; t is the planning time period, $t = 1, 2, \ldots, T$; $K^{\pm}_{ijt}$ is the optimized water-allocation target from water source i to user j per unit during period t, (10^8 m^3); $C^{\pm}_{ijt}$ is the economic punishment of the system from water source i to user j per unit of water not delivered during period t, $C^{\pm}_{ijt} \geq NB^{\pm}_{ijt}$, (¥/m^3); $q^{\pm}_{iht}$ is the total water availability of water source i during period t when the water flow probability is p_{ht}; p_{ht} is probability when the water flow level is h; 'L', 'M', 'H' means the water flow level is low, medium, and high, respectively. Each water flow level would correspond to a probability level p_{ht} in each time stage/period (e.g. L-L-L-...-L), with $p_{ht} \geq 0$ and $\sum_{h=1}^{N} p_{ht} = 1$, $\forall t$ (Liu *et al.* 2015; Wu *et al.* 2015). $S^{\pm}_{ijht}$ is the amount of water deficit due to the water-allocation target $K^{\pm}_{ijt}$ not being met when the water availability of i is $q^{\pm}_{iht}$ during period t, (10^8 m^3).

Model (1) can effectively solve the dynamic characteristics of water resources management and uncertainties expressed in interval numbers. However, model (1) only works when the water inflow probabilities p_{ht} are known as certain values. In real-world problems, the probability p_{ht} is usually measured and estimated by the number of random events (water flow level, such as 'low', 'medium', 'high') in a certain period of time, which causes a problem that different water users, such as industrial or environmental users may get their own data collection from different time periods. Therefore, in this case, they may generate different PDFs for future water flow levels. This means that the water flow probability distribution cannot be estimated as a certain value. Consequently, the uncertainties in the probability distribution of water flow level can be resolved as follows (Kofler 2001): (1) the probability p_h is between α and β, where α and β are fixed numbers (i.e., $\alpha \leq p_h \leq \beta$); (2) the probability of p_h plus p_i equal to γ (i.e., $p_j + p_i = \gamma$). This leads to the following set (Ben & Masri 2005):

$$\pi = \Big\{ P{:}(p_1, p_2, p_3, \ldots, p_N) \in \Re^N s.t. HP \leq b, \quad \sum_{h=1}^{N} p_h = 1, \quad p_h \geq 0, \quad h = 1, 2, \ldots, N \Big\} \tag{2}$$

where $H = (h_{ih})$ is $s \times N$ fixed matrices and $b = (b_i)$ is $s \times 1$ fixed matrices. Equation (2) can express the probability distribution information of water flow level as linear partial information (LPI) (Kofler & Zweifel 1991). It can better tackle uncertainties expressed as partial probability distributions. Thus, we have the inexact multi-stage interval-parameter partial information programming model (IMIPM). It is capable of analyzing various policy scenarios over a multiple time-stage context. The model is as follows:

$$\max f^{\pm} = \sum_{i=1}^{I}\sum_{j=1}^{J}\sum_{t=1}^{T} NB^{\pm}_{ijt} K^{\pm}_{ijt} - \sum_{t=1}^{T}\sum_{i=1}^{I}\sum_{j=1}^{J}\sum_{h=1}^{N} p_{ht} C^{\pm}_{ijt} S^{\pm}_{ijht} \tag{3a}$$

subject to:

$$\sum_{i=1}^{I}\sum_{j=1}^{J} (K^{\pm}_{ijt} - S^{\pm}_{ijht}) \leq q^{\pm}_{iht} + \varepsilon^{\pm}_{(t-1)} \quad \forall i, j, h, t \tag{3b}$$

$$\varepsilon^{\pm}_{h(t-1)} = q^{\pm}_{ih(t-1)} - \sum_{i=1}^{I}\sum_{j=1}^{J} (K^{\pm}_{ij(t-1)} - S^{\pm}_{ijh(t-1)}) + \varepsilon^{\pm}_{h(t-2)} \quad \forall i, j, h, t \tag{3c}$$

$$\sum_{i=1}^{I} K^{\pm}_{ijt} \leq K_{jt\max} \quad \forall i, j, t \tag{3d}$$

$$\sum_{i=1}^{I} (K^{\pm}_{ijt} - S^{\pm}_{ijht}) \geq K_{jt\min} \quad \forall i, j, h, t \tag{3e}$$

$$\sum_{i=1}^{I} (K^{\pm}_{ijt} - S^{\pm}_{ijht}) \leq e^{\pm}_{it} \quad \forall i, j, h, t \tag{3f}$$

$$K_{ijt\max} \geq K^{\pm}_{ijt} \geq S^{\pm}_{ijht} \geq 0 \quad \forall i, j, h, t \tag{3g}$$

$$p_{ht} \geq 0 \quad \forall h, t \tag{3h}$$

$$\sum_{h=1}^{N} p_{ht} = 1 \quad \forall t \tag{3i}$$

$$Hp \leq b \tag{3j}$$

Solution method

In the above formulas, formulas (3h)–(3j) are constraints that can express partial information with a probability distribution. To solve model (3), firstly, select $(n - 1)$ inequalities from constraint (3j) and express them as equations. Then associate them with constraint (3i) and generate n linear equations. Finally, by tackling the formulated system of linear equations, we can obtain an extreme point of the linear partial information (LPI) proposed by constraints (3h) to (3j). After considering all possible systems of linear equations from (3j) and (3i) (constraint (3i) will always be involved), then we can obtain all the extreme points of LPI. Assume $P^{l*} = (p_1^{l*}, \ldots, p_n^{l*})$ $(l = 1, \ldots, r)$ are all the extreme points that constraints (3h) to (3j) can generate (Kofler 2001; Ben & Masri 2005; Fan *et al.* 2011). Then for one P^l, model (3) can be changed to an inexact multi-stage interval-parameter programming model (IMSP) presented as follows:

$$\max f^{l\pm} = \sum_{i=1}^{I}\sum_{j=1}^{J}\sum_{t=1}^{T} NB_{ijt}^{\pm} K_{ijt}^{\pm} - \sum_{t=1}^{T}\sum_{i=1}^{I}\sum_{j=1}^{J}\sum_{h=1}^{N} p_{ht}^{l*} C_{ijt}^{\pm} S_{ijht}^{l\pm} \tag{4a}$$

subject to:

$$\sum_{i=1}^{I}\sum_{j=1}^{J} (K_{ijt}^{\pm} - S_{ijht}^{l\pm}) \le q_{iht}^{\pm} + \varepsilon_{(t-1)}^{\pm} \quad \forall i, j, h, t \tag{4b}$$

$$\varepsilon_{h(t-1)}^{\pm} = q_{ih(t-1)}^{\pm} - \sum_{i=1}^{I}\sum_{j=1}^{J} (K_{ij(t-1)}^{\pm} - S_{ijh(t-1)}^{l\pm}) + \varepsilon_{h(t-2)}^{\pm} \quad \forall i, j, h, t \tag{4c}$$

$$\sum_{i=1}^{I} K_{ijt}^{\pm} \le K_{jt\max} \quad \forall i, j, t \tag{4d}$$

$$\sum_{i=1}^{I} (K_{ijt}^{\pm} - S_{ijht}^{l\pm}) \ge K_{jt\min} \quad \forall i, j, h, t \tag{4e}$$

$$\sum_{i=1}^{I} (K_{ijt}^{\pm} - S_{ijht}^{l\pm}) \le e_{it}^{\pm} \quad \forall i, j, h, t \tag{4f}$$

$$K_{ijt\max} \ge K_{ijt}^{\pm} \ge S_{ijht}^{l\pm} \ge 0 \quad \forall i, j, h, t \tag{4g}$$

Model (4) can be converted into two sets of deterministic submodels, which correspond to the upper and lower bounds of the objective function value. This transformation process is based on the following interactive algorithm according to Huang & Loucks (2000):

Step 1: In model (4), water demand targets ($K_{ijt}^{\pm}$) are first-stage decision variables, which need to be known before the water flow levels (random variables) are defined. In this study, the values of $K_{ijt}^{\pm}$ are identified by having y_{ijt} be decision variables and this could relate to maximizing system benefit. Let $K_{ijt}^{\pm} = K_{ijt}^{-} + \Delta K_{ijt} y_{ijt}$, where $\Delta K_{ijt} = K_{ijt}^{+} - K_{ijt}^{-}$ and $y_{ijt} \in [0, 1]$. Thus, when $K_{ijt}^{\pm}$ approaches upper bounds ($y_{ijt} = 1$), if the promised water is fulfilled, it will generate high system profits; however, if the promised water is not provided, a high penalty will need to be paid. Conversely, when $K_{ijt}^{\pm}$ reaches lower bounds ($y_{ijt} = 0$), the profit of the system may be lower, but the risk of violating the promised goal is also very low, so there will be lower penalties.

Step 2: Since the requirement is the maximum benefit of the water resources system, the upper bound submodel 1 for f^{l+} should be solved first. Then we have:

$$\max f^{l+} = \sum_{i=1}^{I}\sum_{j=1}^{J}\sum_{t=1}^{T} NB_{ijt}^{+} (K_{ijt}^{-} + y_{ijt}^{l} \Delta K_{ijt}) - \sum_{t=1}^{T}\sum_{i=1}^{I}\sum_{j=1}^{J}\sum_{h=1}^{N} p_{ht}^{l*} C_{ijt}^{-} S_{ijht}^{l-} \tag{5a}$$

subject to:

$$\sum_{i=1}^{I}\sum_{j=1}^{J} (y_{ijt}^{l} \Delta K_{ijt} - S_{ijht}^{l-}) \le q_{iht}^{+} + \varepsilon_{(t-1)}^{+} - \sum_{i=1}^{I}\sum_{j=1}^{J} K_{ijt}^{-} \quad \forall i, j, h, t \tag{5b}$$

$$\varepsilon_{h(t-1)}^{+} = q_{ih(t-1)}^{+} - \sum_{i=1}^{I}\sum_{j=1}^{J} (K_{ij(t-1)}^{-} + y_{ij(t-1)}^{l} \Delta K_{ij(t-1)} - S_{ijh(t-1)}^{l-}) + \varepsilon_{h(t-2)}^{+} \quad \forall i, j, h, t \tag{5c}$$

$$\sum_{i=1}^{I} (y_{ijt}^{l} \Delta K_{ijt}) \le K_{jt\max} - \sum_{i=1}^{I} K_{ijt}^{-} \quad \forall i, j, t \tag{5d}$$

$$\sum_{i=1}^{I} (y_{ijt}^{l} \Delta K_{ijt} - S_{ijht}^{l-}) \ge K_{jt\min} - \sum_{i=1}^{I} K_{ijt}^{-} \quad \forall i, j, h, t \tag{5e}$$

$$\sum_{i=1}^{I}(y^{l}_{ijt}\Delta K_{ijt}-S^{l-}_{ijht})\le e^{+}_{it}-\sum_{i=1}^{I}K^{-}_{ijt}\quad \forall i,j,h,t \tag{5f}$$

$$K_{ijt\max}\ge K^{\pm}_{ijt}\ge S^{l-}_{ijht}\ge 0\qquad \forall i,j,h,t \tag{5g}$$

Step 3: The solution for model (5) provides the extreme upper bounds of system benefit. Let y^{l}_{ijtopt}, $\max f^{l+}$ and $S^{l-}_{ijhtopt}$ be solutions of submodel 1. Then the optimized water-allocation targets are $K^{\pm}_{ijtopt}=K^{-}_{ijt}+y^{l}_{ijtopt}\Delta K_{ijt}$ and we should add the $S^{l+}_{ijht}\ge S^{l-}_{ijhtopt}$ constraint to submodel 2.

Step 4: Then submodel 2 for f^{l-}which corresponds to the lower bound of the system benefit can be solved as follows:

$$\max f^{l-}=\sum_{i=1}^{I}\sum_{j=1}^{J}\sum_{t=1}^{T}NB^{-}_{ijt}(K^{-}_{ijt}+y^{l}_{ijtopt}\Delta K_{ijt})-\sum_{t=1}^{T}\sum_{i=1}^{I}\sum_{j=1}^{J}\sum_{h=1}^{N}p^{l*}_{ht}C^{+}_{ijt}S^{l+}_{ijht} \tag{6a}$$

subject to:

$$\sum_{i=1}^{I}\sum_{j=1}^{J}(y^{l}_{ijtopt}\Delta K_{ijt}-S^{l+}_{ijht})\le q^{-}_{iht}+\varepsilon^{-}_{(t-1)}-\sum_{i=1}^{I}\sum_{j=1}^{J}K^{-}_{ijt}\quad \forall i,j,h,t \tag{6b}$$

$$\varepsilon^{-}_{h(t-1)}=q^{-}_{ih(t-1)}-\sum_{i=1}^{I}\sum_{j=1}^{J}(K^{-}_{ij(t-1)}+y^{l}_{ij(t-1)opt}\Delta K_{ij(t-1)}-S^{l+}_{ijh(t-1)})+\varepsilon^{-}_{h(t-2)}\quad \forall i,j,h,t \tag{6c}$$

$$\sum_{i=1}^{I}(y^{l}_{ijtopt}\Delta K_{ijt})\le K_{jt\max}-\sum_{i=1}^{I}K^{-}_{ijt}\quad \forall i,j,t \tag{6d}$$

$$\sum_{i=1}^{I}(y^{l}_{ijtopt}\Delta K_{ijt}-S^{l+}_{ijht})\ge K_{jt\min}-\sum_{i=1}^{I}K^{-}_{ijt}\quad \forall i,j,h,t \tag{6e}$$

$$\sum_{i=1}^{I}(y^{l}_{ijtopt}\Delta K_{ijt}-S^{l+}_{ijht})\le e^{-}_{it}-\sum_{i=1}^{I}K^{-}_{ijt}\quad \forall i,j,h,t \tag{6f}$$

$$K_{ijt\max}\ge K^{\pm}_{ijt}\ge S^{l+}_{ijht}\ge S^{l-}_{ijht}\ge 0\qquad \forall i,j,h,t \tag{6g}$$

From models (5) and (6), we get the solutions for model (4):

$$f^{l\pm}_{opt}=[f^{l-}_{opt},f^{l+}_{opt}] \tag{7a}$$

$$S^{l\pm}_{ijhtopt}=[S^{l-}_{ijhtopt},S^{l+}_{ijhtopt}] \tag{7b}$$

And the optimal water allocation under this scenario can be obtained:

$$F^{l\pm}_{ijhtopt}=T_{ijhtopt}-S^{l\pm}_{ijhtopt},\ \forall i,j,h,t \tag{7c}$$

Repeat models (5) and (6) for $l=1,2,\ldots,r$, and various optimal solution intervals of model (3) can be obtained under all extreme points of partial probability for the water flow level. Therefore, the optimal solution intervals for model (3) are as follows:

$$S^{\pm}_{ijhtopt}=(S^{1\pm}_{ijhtopt},S^{2\pm}_{ijhtopt},\ldots,S^{r\pm}_{ijhtopt}) \tag{8a}$$

$$y_{ijtopt}=(y^{1}_{ijtopt},y^{2}_{ijtopt},\ldots,y^{r}_{ijtopt}) \tag{8b}$$

$$F^{\pm}_{ijhtopt}=(A^{1\pm}_{ijhtopt},A^{2\pm}_{ijhtopt},\ldots,A^{r\pm}_{ijhtopt}) \tag{8c}$$

$$f^{\pm}=(f^{1\pm}_{opt},f^{2\pm}_{opt},\ldots,f^{r\pm}_{opt}) \tag{8d}$$

According to the principle proposed by Kofler & Zweifel (1991), the final optimal objective function for model (3) can be presented as:

$$\max f^{\pm}_{opt}=[\min(\max f^{l-}),\ \max(\max f^{l+})] \tag{9}$$

CASE STUDY

Harbin is located between 125°42′–130°10′ east longitude and 44°04′–46°40′ north latitude. It is a transportation, political, and financial center in the northeast of China. Harbin is very short of water resources, with average annual rainfall of 538.80 mm, and annual per capita water resources share of only 1,350.00 m^3, accounting for about 1/2 of the national per capita level. The urban per capita share for water is only 218.00 m^3. It is one of the cities with severe water shortage in northern China. In this case, it is particularly important to optimize the allocation of limited water resources.

The developed method is applied to water resources management in Harbin, aiming at allocating water from multiple water sources to multiple water users over multiple time periods, and therefore achieving multiple planning goals. There are four users, these being municipal, industrial, agricultural and environmental users. Agriculture uses 70% of the world's available fresh water (Li & Qian 2018). The total area of Harbin is 53×10^3 km^2, and the area of arable land is about 2.00×10^6 hm^2. Thus, the agricultural water consumption is huge in Harbin. The crops are ripe once a year, and in the summer half-year (May–October), warm-temperature crops are grown. In the winter half-year (November–April of the following year), because the temperature is below 0 °C, no crops can be planted. Therefore, the planning time periods are divided into two parts, the first is the winter half-year and the second is the summer half-year. Figure 2 shows the general framework and application of the IMIPM method, which is based on IPP, MSP and LPI techniques. Table 1 is a summary of the related economic parameters in this water distribution system.

In order to solve the uncertainties in water flow level, this paper selects three time intervals, which are 1987–1996, 1997–2006, 2007–2016, and divides water inflow level into low, medium and high levels, which mean rainfall is 270–440 mm, 440–600 mm, and above 600 mm, respectively. The hydrological characteristic of Harbin is that the annual rainfall mainly occurs in the summer half-year, and the winter half-year has extremely low temperature and much less rainfall. The proportion of rainfall in the winter half-year to the annual rainfall is small, so for Harbin, it is feasible to take one year as a hydrological cycle. Therefore, for every ten years, the frequency of low, medium, and high water inflow level is estimated as the probability of water inflow level of Harbin. Figure 3 shows the water inflow level frequency of Harbin. Thus, in this study case, the general partial distribution constraints (3h) to (3j) can be expressed as follows: $0.10 \le p_1 \le 0.20$, $0.50 \le p_2 \le 0.70$, $0.20 \le p_3 \le 0.30$, $p_1 + p_2 + p_3 = 1$.

Table 2 presents the results of extreme points obtained by the above constraints. According to the probabilities of three water inflow levels fluctuating within three different intervals, only four combinations of these three probabilities will satisfy the constraints, which is represented by 3D graphics in Figure 4. Table 3 shows the total available water with different water inflow levels during two planning time periods.

Table 4 shows the water allocation targets from multi-water source to multi-users during multi-time periods. Table 5 presents the water conveyance capacity of the pipeline, the minimum water demand and the maximum water demand for users during each time-stage. To handle the dynamic uncertainties in this problem, a decision tree based on the two time-stages is generated (shown in Figure 5). Each node is branched with three nodes for the following planning period when the probability of the current water inflow (q_{ht}) is p_{ht}.

RESULTS ANALYSIS AND DISCUSSION

Tables 6–10 show the results generated through IMIPM. Most non-zero decision variables and solutions are interval numbers. In other words, relevant decisions must be sensitive to uncertain modeling inputs. For example, the interval solutions of $S_{ijht}^{l\pm}$ reflect variations of system conditions caused by uncertain inputs of $NB_{ijt}^{l\pm}$, $K_{ijt}^{l\pm}$, $C_{ijt}^{l\pm}$, $q_{iht}^{l\pm}$, $\varepsilon_{(t-1)}^{l\pm}$, and $e_{it}^{l\pm}$.

As shown in Table 6, for scenario A, when water inflow level is low (0.1) and $q_{iht}^{A\pm}$ reaches its upper bound in period 1, only the agricultural users have about $[4.54,3.51]\times10^8$ m^3 of water shortage. When $q_{iht}^{A\pm}$ reaches its lower bound, every user has a certain degree of water shortage. When water inflow level is medium and high (0.6, 0.3, respectively) in period 1, water quantity is increasing. There is no water deficit for municipal and industrial users. Ecological users have a small amount of surface and ground water shortage, $[0,0.11]\times10^8$ m^3 and $[0,0.05]\times10^8$ m^3 respectively.

As shown in Table 7, for scenario A, under the worst situation (joint probability = 1%), when the water inflow levels are low during the whole planning horizon, the total allocated water for the four users would be $F_{ijh2opt}^{A\pm}=[23.28,28.64]\times10^8 m^3$ but $S_{ijh2}^{A\pm}=[12.01,17.37]\times10^8 m^3$,

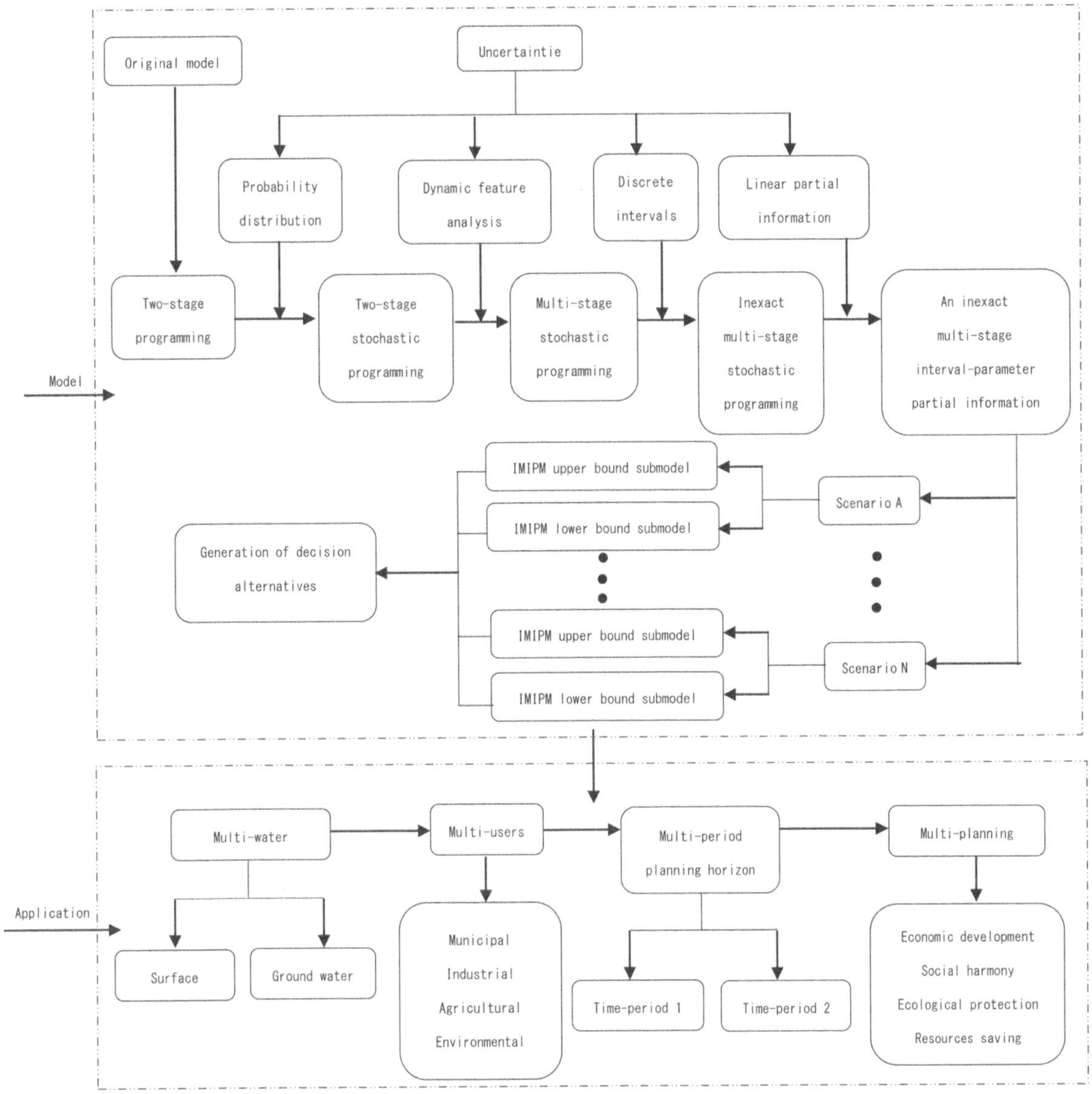

Figure 2 | The framework and application of the IMIPM.

Table 1 | Net benefit and penalty (¥/m^{-3})

User	Benefit of per unit water (¥/m^{-3})		Penalty of per unit water (¥/m^{-3})
	Surface water	Ground water	
1	[2.40,3.20]	[1.70,2.50]	[4.50,4.90]
2	[3.30,4.20]	[2.80,3.60]	[4.30,4.80]
3	[1.60,2.80]	[0.90,2.10]	[2.10,3.00]
4	[3.10,3.60]	[2.40,2.90]	[4.30,5.40]

indicating a serious shortage. There is a big chance that water inflow levels are both mediums in two periods compared with other situations, with the probability being 36%. In this situation, the surface water shortage for agricultural and environmental users would be $[3.31,4.85]\times 10^8\,m^3$, $[0,0.30]\times 10^8\,m^3$ respectively. There is no other ground water shortage except $[4.32,6.43]\times 10^8\,m^3$ for the agricultural user. The total allocated water would be

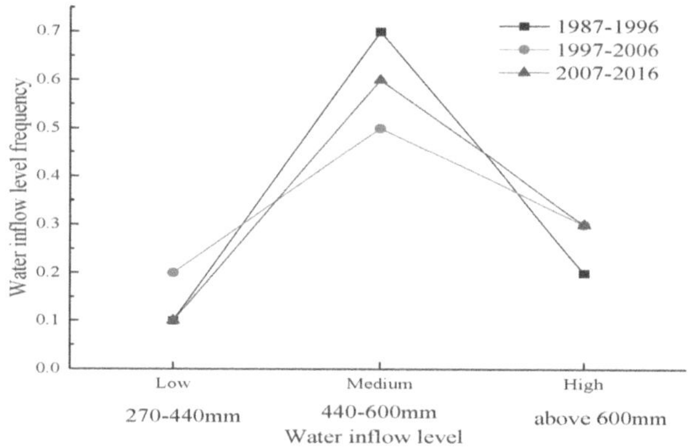

Figure 3 | water flow level frequency of Harbin.

Table 2 | The extreme points for partial probability distribution

Flow level	Scenario A	Scenario B	Scenario C	Scenario D
Low (L)	0.10	0.10	0.20	0.20
Medium (M)	0.60	0.70	0.60	0.50
High (H)	0.30	0.20	0.20	0.30

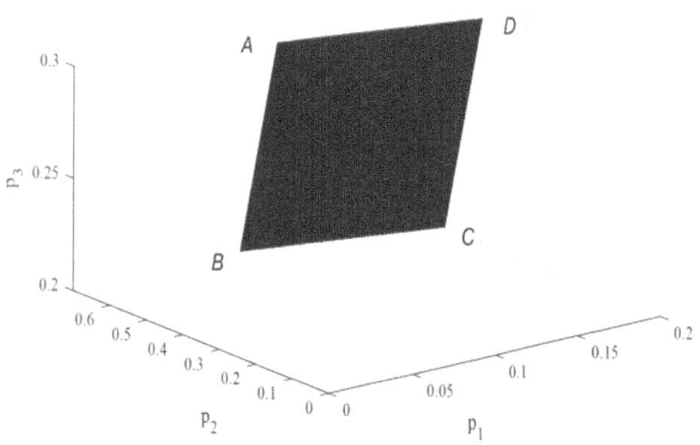

Figure 4 | Partial probability distribution of water inflow level.

$F^{A\pm}_{ijh2\text{opt}} = [29.07, 33.02] \times 10^8\text{m}^3$ but $S^{A\pm}_{ijh2} = [7.63, 11.58] \times 10^8\text{m}^3$, indicating a significant shortage. Under the best situation (joint probability is 9%), when the water inflow levels are high during the whole planning horizon, the total allocated water would be $F^{A\pm}_{ijh2\text{opt}} = [30.21, 37.22] \times 10^8\text{m}^3$, but $S^{A\pm}_{ijh2} = [3.43, 10.44] \times 10^8\text{m}^3$, indicating that the water demands of the four users would generally be satisfied. The objective function value under scenario A is $f^{A\pm}_{\text{opt}} =$ ¥ $[76.04, 154.62] \times 10^8$. It suggests that the system benefit would alter correspondingly from f^{A-}_{opt} to f^{A+}_{opt} as decision variables vary within their lower and upper bounds.

For scenario B, the water allocation scheme is the same as Scenario A's in period 1 as shown in Table 6. This means that, during period 1, if the water inflow probability of low flow level is 0.10, the changes of medium and high water flow level probability have no effect on the water distribution scheme. When the water inflow level is M-M during two periods, the joint probability is 49% (Table 8), and at the same time, the optimized water allocation to the four users is reduced to $[25.34, 30.66] \times 10^8\text{m}^3$ compared with scenario A. The probability of M-M water inflow level in scenario B is higher than it is in scenario A, however, the probability of M-H water inflow level in scenario B is lower than it is in scenario A. This indicates that greater water quantity would be available in scenario A than in scenario B during period 2 when the water inflow level is medium in period 1. The objective function value under scenario B is $f^{B\pm}_{\text{opt}} =$ ¥ $[68.54, 160.17] \times 10^8$.

For scenario C, values of decision variable y_{ij1} are the same as those in scenario A, B and D during period 1 (Table 6). However, there are different water allocation schemes in each scenario since they have different water inflow probabilities of low, medium, and high levels. As shown in Table 9, when the water inflow level is high in period 1, the water shortage for the four users in scenario C ($S^{C\pm}_{ijh1} = [0.95, 4.47] \times 10^8\text{m}^3$) is much higher than it is in scenario A and B ($S^{A\pm}_{ijh1} = S^{B\pm}_{ijh1} = [0, 0.47] \times 10^8\text{m}^3$). That is because the water inflow probability of low level is high while the water inflow probability of medium and high levels are low compared with scenario A and B. Compared with scenario A and B, optimized water allocation from surface and ground water to agricultural users is smaller in scenario C during period 2. This means that the available water in scenario C is less than in scenario A and B. The objective function value under scenario C is $f^{C\pm}_{\text{opt}} =$ ¥ $[61.05, 145.45] \times 10^8$.

The water distribution plans are similar under the four scenarios during period 1. The reason is that in period 1, the urban water demand and available water are both low, and especially the agricultural users' water demand is drastically reduced, so that in order to firstly meet the basic water demand of various users there is no big difference in water allocation scheme during period 1. For scenario D,

Table 3 | Available water in different flow levels during in two periods

Flow level	Probability distribution	Available water (10^8 m^3) t = 1 Surface water	t = 1 Ground water	t = 2 Surface water	t = 2 Ground water
Low (L)	[0.10,0.20]	[7.27,9.12]	[3.68,3.95]	[15.45,19.37]	[7.83,8.40]
Medium (M)	[0.50,0.70]	[9.36,10.26]	[4.31,4.90]	[19.90,21.74]	[9.17,10.41]
High (H)	[0.20,0.30]	[11.37,13.90]	[5.00,5.38]	[24.15,29.50]	[10.62,11.30]

Table 4 | Water allocation targets for users

User j	Water allocation targets (10^8 m^3) t = 1 Surface water	t = 1 Ground water	t = 2 Surface water	t = 2 Ground water
1	[0.56,0.87]	[0.99,1.59]	[0.77,1.21]	[1.36,2.20]
2	[1.02,2.00]	[1.04,1.94]	[1.41,2.76]	[1.44,2.68]
3	[6.02,8.22]	[0.03,0.19]	[15.49,21.12]	[7.68,10.44]
4	[0.03,0.19]	[0.06,0.11]	[0.07,0.49]	[0.16,0.28]

when the water flow level is L-L and M-L (the joint probabilities are 4% and 10% respectively, shown in Table 10), that is, when the available water is low or moderate during period 1, as long as the available water is low during period 2, the optimal water allocation of surface water to industrial users is always $[0.83, 2.27] \times 10^8 m^3$. It can be seen that in scenario D, there is no significant increase in $F^{D\pm}_{ijh2opt}$ even though the probability of H-H (10%) increases. The reason may be that 10% is still a small probability event to happen. Therefore, the model chose a conservative water

Table 5 | Water conveyance capacity of pipeline and minimum, maximum water demand for users

User j	Water conveyance capacity of pipeline (10^8 m^3) t = 1	t = 2	Minimum, maximum water distribution (10^8 m^3) t = 1 Minimum	t = 1 Maximum	t = 2 Minimum	t = 2 Maximum
1	[3.06,4.78]	[3.74,5.84]	1.44	3.15	1.76	3.85
2	[4.66,7.50]	[5.69,9.17]	1.97	5.40	2.41	6.60
3	[7.92,12.58]	[21.42,34.01]	6.97	13.50	18.86	36.50
4	[0.14,0.25]	[0.38,0.68]	0.08	0.81	0.21	2.19

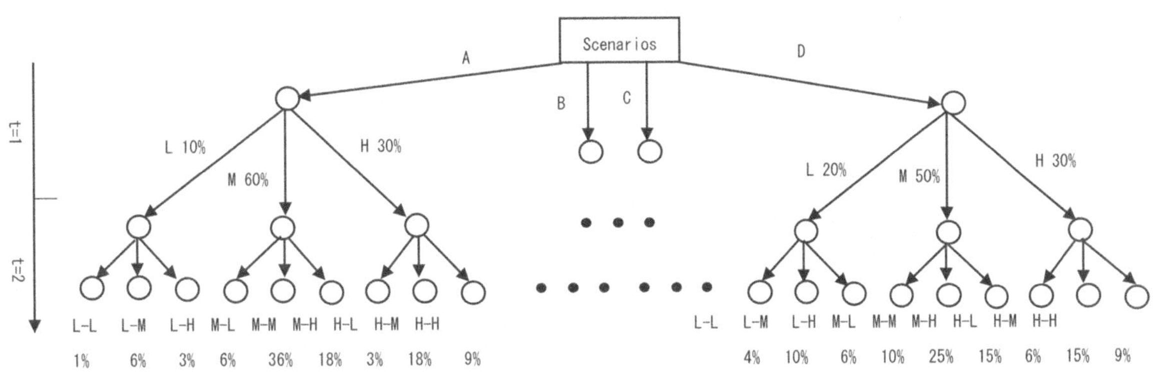

Figure 5 | Multi-stage decision tree based on the two planning time periods.

Table 6 | Water distribution scheme during $t = 1$.

User j	Water inflow level h	Scenario A: Probability (%)	Scenario A: Optimized water target $K_{ij1}^{A\pm}$ Surface water	Scenario A: Optimized water target $K_{ij1}^{A\pm}$ Ground water	Scenario A: Water deficit $S_{ijh1}^{A\pm}$ Surface water	Scenario A: Water deficit $S_{ijh1}^{A\pm}$ Ground water	Scenario A: Optimized water allocation $F_{ijh1opt}^{A\pm}$ Surface water	Scenario A: Optimized water allocation $F_{ijh1opt}^{A\pm}$ Ground water	Scenario B: Probability (%)	Scenario B: Optimized water target $K_{ij1}^{B\pm}$ Surface water	Scenario B: Optimized water target $K_{ij1}^{B\pm}$ Ground water	Scenario B: Optimized water deficit $S_{ijh1}^{B\pm}$ Surface water	Scenario B: Optimized water deficit $S_{ijh1}^{B\pm}$ Ground water	Scenario B: Optimized water allocation $F_{ijh1opt}^{B\pm}$ Surface water	Scenario B: Optimized water allocation $F_{ijh1opt}^{B\pm}$ Ground water
1	L	10%	0.87	1.37	[0,0.19]	[0,0.18]	[0.68,0.87]	1.19	10%	0.87	1.37	[0,0.19]	[0,0.18]	[0.68,0.87]	1.19
2	L	10%	2.00	1.52	[0,1.55]	0	[0.45,2.00]	1.52	10%	2.00	1.52	[0,1.55]	0	[0.45,2.00]	1.52
3	L	10%	8.22	4.06	2.16	[2.38,3.15]	6.06	[0.91,1.68]	10%	8.22	4.06	2.16	[2.38,3.15]	6.06	[0.91,1.68]
4	L	10%	0.19	0.06	[0,0.11]	0	[0.08,0.19]	0.06	10%	0.19	0.06	[0,0.11]	0	[0.08,0.19]	0.06
1	M	60%	0.87	1.37	0	0	0.87	1.37	70%	0.87	1.37	0	0	0.87	1.37
2	M	60%	2.00	1.52	0	0	2.00	1.52	70%	2.00	1.52	0	0	2.00	1.52
3	M	60%	8.22	4.06	[1.02,1.81]	[1.43,2.70]	[6.41,7.20]	[1.36,2.63]	70%	8.22	4.06	[1.02,1.81]	[1.43,2.70]	[6.41,7.20]	[1.36,2.63]
4	M	60%	0.19	0.06	[0,0.11]	0	[0.08,0.19]	0.06	70%	0.19	0.06	[0,0.11]	0	[0.08,0.19]	0.06
1	H	30%	0.87	1.37	0	0	0.87	1.37	20%	0.87	1.37	0	0	0.87	1.37
2	H	30%	2.00	1.52	0	0	2.00	1.52	20%	2.00	1.52	0	0	2.00	1.52
3	H	30%	8.22	4.06	[0,0.30]	[0,0.06]	[7.92,8.22]	[0,3.11]	20%	8.22	4.06	[0,0.30]	[0,0.06]	[7.92,8.22]	[0,3.11]
4	H	30%	0.19	0.06	[0,0.05]	0	[0.14,0.19]	0.06	20%	0.19	0.06	[0,0.05]	[0,0.06]	[0.14,0.19]	[0,0.06]
		$y_{111opt} = 1$ $y_{121opt} = 1$ $y_{131opt} = 1$ $y_{141opt} = 1$ $y_{211opt} = 0.63$ $y_{221opt} = 0.53$ $y_{231opt} = 1$ $y_{241opt} = 0$							$y_{111opt} = 1$ $y_{121opt} = 1$ $y_{131opt} = 1$ $y_{141opt} = 1$ $y_{211opt} = 0.63$ $y_{221opt} = 0.53$ $y_{231opt} = 1$ $y_{241opt} = 0$						

(continued)

Table 6 | continued

User	Water inflow level	Scenario C							Scenario D						
		Probability (%)	Optimized water target $K_{ij1}^{C\pm}$		Optimized water deficit $S_{ijh1}^{C\pm}$		Optimized water allocation $F_{ijh1opt}^{C\pm}$		Probability (%)	Optimized water target $K_{ij1}^{D\pm}$		Optimized water deficit $S_{ijh1}^{D\pm}$		Optimized water allocation $F_{ijh1opt}^{D\pm}$	
			Surface water	Ground water	Surface water	Ground water	Surface water	Ground water		Surface water	Ground water	Surface water	Ground water	Surface water	Ground water
1	L	20%	0.87	1.37	[0,0.37]	0	[0.50,0.87]	1.37	20%	0.87	1.37	[0,0.37]	0	[0.5,0.87]	1.37
2	L	20%	2.00	1.52	[0,1.37]	[0,0.18]	[0.63,2.00]	[1.34,1.52]	20%	2.00	1.52	[0,1.37]	[0,0.18]	[0.63,2.00]	[1.34,1.52]
3	L	20%	8.22	4.06	2.16	[2.38,3.15]	6.06	[0.91,1.68]	20%	8.22	4.06	2.16	[2.38,3.15]	6.06	[0.91,1.68]
4	L	20%	0.19	0.06	[0,0.11]	0	[0.08,0.19]	0.06	20%	0.19	0.06	[0,0.11]	0	[0.08,0.19]	0.06
1	M	60%	0.87	1.37	0	0	0.87	1.37	50%	0.87	1.37	0	0	0.87	1.37
2	M	60%	2.00	1.52	0	[0,0.15]	2.00	[1.37,1.52]	50%	2.00	1.52	0	0	2.00	1.52
3	M	60%	8.22	4.06	[1.02,1.87]	[1.43,2.49]	[6.35,7.20]	[1.57,2.63]	50%	8.22	4.06	[1.02,1.81]	[1.43,2.70]	[6.41,7.20]	[1.36,2.63]
4	M	60%	0.19	0.06	[0,0.05]	[0,0.06]	[0.14,0.19]	[0,0.06]	50%	0.19	0.06	[0,0.11]	0	[0.08,0.19]	0.06
1	H	20%	0.87	1.37	0	0	0.87	1.37	30%	0.87	1.37	0	0	0.87	1.37
2	H	20%	2.00	1.52	0	0	2.00	1.52	30%	2.00	1.52	0	0	2.00	1.52
3	H	20%	8.22	4.06	[0,0.30]	[0.95,4.06]	[7.92,8.22]	[0,3.11]	30%	8.22	4.06	[0,0.30]	[0.95,4.06]	[7.92,8.22]	[0,3.11]
4	H	20%	0.19	0.06	[0,0.05]	[0,0.06]	[0.14,0.19]	[0,0.06]	30%	0.19	0.06	[0,0.05]	[0,0.06]	[0.14,0.19]	[0,0.06]
		$y_{111opt} = 1\ y_{121opt} = 1\ y_{131opt} = 1\ y_{141opt} = 1\ y_{211opt} = 0.63\ y_{221opt} = 0.53\ y_{231opt} = 1\ y_{241opt} = 0$							$y_{111opt} = 1\ y_{121opt} = 1\ y_{131opt} = 1\ y_{141opt} = 1\ y_{211opt} = 0.63\ y_{221opt} = 0.53\ y_{231opt} = 1\ y_{241opt} = 0$						

Table 7 | Water distribution scheme of scenario A during $t = 2$

User	Water inflow level	Probability (%)	Associated water flow	Associated probability (%)	Optimized water target $K^{A\pm}_{ij2}$		Optimized water deficit $S^{A\pm}_{ijh2}$		Optimized water allocation $F^{A\pm}_{ijh2opt}$	
					Surface water	Ground water	Surface water	Ground water	Surface water	Ground water
1	L	10%	L-L	1%	0.77	2.20	[0,0.77]	[0,0.44]	[0,0.77]	[1.76,2.20]
2	L	10%	L-L	1%	2.76	2.68	[0,2.76]	[0,0.27]	[0,2.76]	[2.41,2.68]
3	L	10%	L-L	1%	21.12	10.44	5.68	[6.33,7.02]	15.44	[3.42,4.11]
4	L	10%	L-L	1%	0.40	0.28	[0,0.39]	[0,0.04]	[0.01,0.40]	[0.24,0.28]
1	M	60%	L-M	6%	0.77	2.20	0	0	0.77	2.20
2	M	60%	L-M	6%	2.76	2.68	0	0	2.76	2.68
3	M	60%	L-M	6%	21.12	10.44	[3.31,4.85]	[4.32,6.43]	[16.27,17.81]	[4.01,6.12]
4	M	60%	L-M	6%	0.40	0.28	[0,0.30]	0	[0.10,0.40]	0.28
1	H	30%	L-H	3%	0.77	2.20	0	0	0.77	2.20
2	H	30%	L-H	3%	2.76	2.68	0	0	2.76	2.68
3	H	30%	L-H	3%	21.12	10.44	[0,5.16]	[3.43,4.98]	[15.96,21.12]	[5.46,7.01]
4	H	30%	L-H	3%	0.40	0.28	[0,0.30]	0	[0.10,0.40]	0.28
1	L	10%	M-L	6%	0.77	2.20	[0,0.77]	[0,0.44]	[0,0.77]	[1.76,2.20]
2	L	10%	M-L	6%	2.76	2.68	[0,2.76]	[0,0.27]	[0,2.76]	[2.41,2.68]
3	L	10%	M-L	6%	21.12	10.44	5.68	[6.33,7.02]	15.44	[3.42,4.11]
4	L	10%	M-L	6%	0.40	0.28	[0,0.39]	[0,0.04]	[0.01,0.40]	[0.24,0.28]
1	M	60%	M-M	36%	0.77	2.20	0	0	0.77	2.20
2	M	60%	M-M	36%	2.76	2.68	0	0	2.76	2.68
3	M	60%	M-M	36%	21.12	10.44	[3.31,4.85]	[4.32,6.43]	[16.27,17.81]	[4.01,6.12]
4	M	60%	M-M	36%	0.40	0.28	[0,0.30]	0	[0.10,0.40]	0.28
1	H	30%	M-H	18%	0.77	2.20	0	0	0.77	2.20
2	H	30%	M-H	18%	2.76	2.68	0	0	2.76	2.68
3	H	30%	M-H	18%	21.12	10.44	[0,5.16]	[3.43,4.98]	[15.96,21.12]	[5.46,7.01]
4	H	30%	M-H	18%	0.40	0.28	[0,0.30]	0	[0.10,0.40]	0.28
1	L	10%	H-L	3%	0.77	2.20	[0,0.77]	[0,0.44]	[0,0.77]	[1.76,2.20]
2	L	10%	H-L	3%	2.76	2.68	[0,2.76]	[0,0.27]	[0,2.76]	[2.41,2.68]
3	L	10%	H-L	3%	21.12	10.44	5.68	[6.33,7.02]	15.44	[3.42,4.11]
4	L	10%	H-L	3%	0.40	0.28	[0,0.39]	[0,0.04]	[0.01,0.40]	[0.24,0.28]
1	M	60%	H-M	18%	0.77	2.20	0	0	0.77	2.20
2	M	60%	H-M	18%	2.76	2.68	0	0	2.76	2.68
3	M	60%	H-M	18%	21.12	10.44	[3.31,4.85]	[4.32,6.43]	[16.27,17.81]	[4.01,6.12]
4	M	60%	H-M	18%	0.40	0.28	[0,0.30]	0	[0.10,0.40]	0.28
1	H	30%	H-H	9%	0.77	2.20	0	0	0.77	2.20
2	H	30%	H-H	9%	2.76	2.68	0	0	2.76	2.68
3	H	30%	H-H	9%	21.12	10.44	[0,5.16]	[3.43,4.98]	[15.96,21.12]	[5.46,7.01]
4	H	30%	H-H	9%	0.40	0.28	[0,0.30]	0	[0.10,0.40]	0.28

$y_{112opt} = 0$ $y_{122opt} = 1$ $y_{132opt} = 1$ $y_{142opt} = 0.79$ $y_{212opt} = 1$ $y_{222opt} = 1$ $y_{232opt} = 1$ $y_{242opt} = 1$

Total net benefit (¥10^8) : $f^{A\pm}_{opt} = [76.04, 154.62]$

Table 8 | Water distribution scheme of scenario B during $t = 2$

User	Water inflow level	Probability (%)	Associated water flow	Associated probability (%)	Optimized water target $K_{ij2}^{B\pm}$		Optimized water deficit $S_{ijh2}^{B\pm}$		Optimized water allocation $F_{ijh2opt}^{B\pm}$	
					Surface water	Ground water	Surface water	Ground water	Surface water	Ground water
1	L	10%	L-L	1%	0.77	2.20	0	[0,1.35]	0.77	[0.85,2.20]
2	L	10%	L-L	1%	2.76	2.68	[0,2.11]	[0,1.45]	[0.60,2.76]	[1.23,2.68]
3	L	10%	L-L	1%	18.76	10.44	5.68	6.33	13.08	4.11
4	L	10%	L-L	1%	0.40	0.28	[0,0.18]	0	[0.22,0.40]	0.28
1	M	70%	L-M	7%	0.77	2.20	0	0	0.77	2.20
2	M	70%	L-M	7%	2.76	2.68	[0,0.62]	0	[2.14,2.76]	2.68
3	M	70%	L-M	7%	18.76	10.44	[3.31,4.42]	[4.32,7.61]	[14.34,15.45]	[2.83,6.12]
4	M	70%	L-M	7%	0.40	0.28	[0,0.3]	0	[0.10,0.40]	0.28
1	H	20%	L-H	2%	0.77	2.20	0	0	0.77	2.20
2	H	20%	L-H	2%	2.76	2.68	0	0	2.76	2.68
3	H	20%	L-H	2%	18.76	10.44	[0,1.62]	[3.43,6.62]	[17.41,18.76]	[4.28,7.01]
4	H	20%	L-H	2%	0.40	0.28	[0,0.30]	0	[0.10,0.40]	0.28
1	L	10%	M-L	7%	0.77	2.20	0	[0,1.35]	0.77	[0.85,2.20]
2	L	10%	M-L	7%	2.76	2.68	[0,2.11]	[0,1.45]	[0.60,2.76]	[1.23,2.68]
3	L	10%	M-L	7%	18.76	10.44	5.68	[6.33,6.89]	13.08	[3.55,4.11]
4	L	10%	M-L	7%	0.40	0.28	[0,0.18]	0	[0.22,0.40]	0.28
1	M	70%	M-M	49%	0.77	2.20	0	0	0.77	2.20
2	M	70%	M-M	49%	2.76	2.68	[0,0.62]	0	[2.14,2.76]	2.68
3	M	70%	M-M	49%	18.76	10.44	[3.31,4.42]	[4.32,7.61]	[14.34,15.45]	[2.83,6.12]
4	M	70%	M-M	49%	0.40	0.28	[0,0.30]	0	[0.10,0.40]	0.28
1	H	20%	M-H	14%	0.77	2.20	0	0	0.77	2.20
2	H	20%	M-H	14%	2.76	2.68	0	0	2.76	2.68
3	H	20%	M-H	14%	18.76	10.44	[0,1.62]	[3.43,6.62]	[17.41,18.76]	[4.28,7.01]
4	H	20%	M-H	14%	0.40	0.28	[0,0.30]	0	[0.10,0.40]	0.28
1	L	10%	H-L	3%	0.77	2.20	[0,0.77]	[0,0.39]	[0,0.77]	[1.81,2.20]
2	L	10%	H-L	3%	2.76	2.68	[0,1.15]	[0,2.41]	[1.61,2.76]	[0.27,2.68]
3	L	10%	H-L	3%	18.76	10.44	4.19	[6.33,6.89]	14.57	[3.35,4.11]
4	L	10%	H-L	3%	0.40	0.28	[0,0.18]	0	[0.22,0.40]	0.28
1	M	70%	H-M	21%	0.77	2.20	0	0	0.77	2.20
2	M	70%	H-M	21%	2.76	2.68	[0,0.62]	0	[2.14,2.76]	2.68
3	M	70%	H-M	21%	18.76	10.44	[3.31,4.73]	[4.32,7.61]	[14.03,15.45]	[2.83,6.12]
4	M	70%	H-M	21%	0.40	0.28	[0,0.30]	0	[0.10,0.40]	0.28
1	H	20%	H-H	6%	0.77	2.20	0	0	0.77	2.20
2	H	20%	H-H	6%	2.76	2.68	0	0	2.76	2.68
3	H	20%	H-H	6%	18.76	10.44	[0,1.62]	[3.43,6.16]	[17.14,18.76]	[4.28,7.01]
4	H	20%	H-H	6%	0.40	0.28	[0,0.30]	0	[0.10,0.40]	0.28

$y_{112opt} = 0\ y_{122opt} = 1\ y_{132opt} = 0.58\ y_{142opt} = 0.79\ y_{212opt} = 1\ y_{222opt} = 1\ y_{232opt} = 1\ y_{242opt} = 1$

Total net benefit (¥10^8) : $f_{opt}^{B\pm} = [68.54, 160.17]$

Table 9 | Water distribution scheme of scenario C during $t = 2$

User	Water inflow level	Probability (%)	Associated water flow	Associated probability (%)	Optimized water target $K_{ij2}^{C\pm}$		Optimized water deficit $S_{ijh2}^{C\pm}$		Optimized water allocation $F_{ijh2opt}^{C\pm}$	
					Surface water	Ground water	Surface water	Ground water	Surface water	Ground water
1	L	20%	L-L	4%	0.77	2.20	0	0	0.77	2.20
2	L	20%	L-L	4%	2.76	2.68	0	0	2.76	2.68
3	L	20%	L-L	4%	21.12	10.44	[5.68,10.16]	[6.33,7.80]	[10.96,15.44]	[2.64,4.11]
4	L	20%	L-L	4%	0.40	0.28	[0,0.30]	0	[0.10,0.40]	0.28
1	M	60%	L-M	12%	0.77	2.20	0	0	0.77	2.20
2	M	60%	L-M	12%	2.76	2.68	[0,2.17]	[0,1.30]	[0.60,2.76]	[1.38,2.68]
3	M	60%	L-M	12%	21.12	10.44	[3.31,7.83]	[4.32,5.30]	[13.29,17.81]	[5.14,6.12]
4	M	60%	L-M	12%	0.40	0.28	[0,0.30]	0	[0.10,0.40]	0.28
1	H	20%	L-H	4%	0.77	2.20	0	0	0.77	2.20
2	H	20%	L-H	4%	2.76	2.68	0	[0,2.41]	2.76	[0.27,2.68]
3	H	20%	L-H	4%	21.12	10.44	[0,8.39]	[3.43,4.15]	[12.73,21.12]	[6.29,7.01]
4	H	20%	L-H	4%	0.40	0.28	[0,0.30]	0	[0.10,0.40]	0.28
1	L	20%	M-L	12%	0.77	2.20	0	0	0.77	2.20
2	L	20%	M-L	12%	2.76	2.68	0	0.15	2.76	1.53
3	L	20%	M-L	12%	21.12	10.44	[5.68,10.16]	[6.33,7.80]	[10.96,15.44]	[2.64,4.11]
4	L	20%	M-L	12%	0.40	0.28	[0,0.30]	0	[0.10,0.40]	0.28
1	M	60%	M-M	36%	0.77	2.20	0	0	0.77	2.20
2	M	60%	M-M	36%	2.76	2.68	0	0	2.76	2.68
3	M	60%	M-M	36%	21.12	10.44	[3.31,4.17]	[4.32,7.61]	[16.95,17.81]	[2.83,6.12]
4	M	60%	M-M	36%	0.40	0.28	[0,0.30]	0	[0.10,0.40]	0.28
1	H	20%	M-H	12%	0.77	2.20	0	0	0.77	2.20
2	H	20%	M-H	12%	2.76	2.68	0	0	2.76	2.68
3	H	20%	M-H	12%	21.12	10.44	[0,3.98]	[3.43,6.16]	[17.14,21.12]	[4.28,7.01]
4	H	20%	M-H	12%	0.40	0.28	[0,0.30]	0	[0.10,0.40]	0.28
1	L	20%	H-L	4%	0.77	2.20	0	0	0.77	2.20
2	L	20%	H-L	4%	2.76	2.68	0	0	2.76	2.68
3	L	20%	H-L	4%	21.12	10.44	[5.68,10.16]	[6.33,7.80]	[10.96,15.44]	[2.64,4.11]
4	L	20%	H-L	4%	0.40	0.28	[0,0.30]	0	[0.10,0.40]	0.28
1	M	60%	H-M	12%	0.77	2.20	0	0	0.77	2.20
2	M	60%	H-M	12%	2.76	2.68	0	0	2.76	2.68
3	M	60%	H-M	12%	21.12	10.44	[3.31,4.17]	[4.32,7.61]	[16.95,17.81]	[2.83,6.12]
4	M	60%	H-M	12%	0.40	0.28	[0,0.30]	0	[0.10,0.40]	0.28
1	H	20%	H-H	4%	0.77	2.20	0	0	0.77	2.20
2	H	20%	H-H	4%	2.76	2.68	0	0	2.76	2.68
3	H	20%	H-H	4%	21.12	10.44	[0,3.98]	[3.43,6.16]	[17.14,21.12]	[4.28,7.01]
4	H	20%	H-H	4%	0.40	0.28	[0,0.30]	0	[0.10,0.40]	0.28

$y_{112opt} = 0$ $y_{122opt} = 1$ $y_{132opt} = 1$ $y_{142opt} = 0.79$ $y_{212opt} = 1$ $y_{222opt} = 1$ $y_{232opt} = 1$ $y_{242opt} = 1$

Total net benefit (¥10^8) : $f_{opt}^{C\pm} = [61.05,\ 145.45]$

Table 10 | Water distribution scheme of scenario D during $t = 2$

User	Water inflow level	Probability (%)	Associated water flow	Associated probability (%)	Optimized water target $K_{ij2}^{D\pm}$ Surface water	Ground water	Optimized water deficit $S_{ijh2}^{D\pm}$ Surface water	Ground water	Optimized water allocation $F_{ijh2opt}^{D\pm}$ Surface water	Ground water
1	L	20%	L-L	4%	0.77	2.20	[0,0.77]	[0,0.39]	[0,0.77]	[1.81,2.20]
2	L	20%	L-L	4%	2.27	2.68	[0,1.44]	[0,1.52]	[0.83,2.27]	[1.16,2.68]
3	L	20%	L-L	4%	21.12	10.44	5.68	6.33	15.44	4.11
4	L	20%	L-L	4%	0.40	0.28	[0,0.19]	[0,0.28]	[0.21,0.40]	[0,0.28]
1	M	50%	L-M	10%	0.77	2.20	[0,0.02]	0	[0.75,0.77]	2.20
2	M	50%	L-M	10%	2.27	2.68	0	0	2.27	2.68
3	M	50%	L-M	10%	21.12	10.44	[3.31,5.09]	[4.32,7.61]	[16.03,17.81]	[2.83,6.12]
4	M	50%	L-M	10%	0.40	0.28	[0,0.30]	0	[0.10,0.40]	0.28
1	H	30%	L-H	6%	0.77	2.20	0	0	0.77	2.20
2	H	30%	L-H	6%	2.27	2.68	0	0	2.76	2.68
3	H	30%	L-H	6%	21.12	10.44	[0,4.26]	[3.43,5.88]	[16.86,21.12]	[4.56,7.01]
4	H	30%	L-H	6%	0.40	0.28	[0,0.02]	[0,0.28]	[0.38,0.40]	[0,0.28]
1	L	20%	M-L	10%	0.77	2.20	[0,0.77]	[0,0.39]	[0,0.77]	[1.81,2.20]
2	L	20%	M-L	10%	2.27	2.68	[0,1.44]	[0,1.52]	[0.83,2.27]	[1.16,2.68]
3	L	20%	M-L	10%	21.12	10.44	5.68	6.33	15.44	4.11
4	L	20%	M-L	10%	0.40	0.28	[0,0.19]	[0,0.28]	[0.21,0.40]	[0,0.28]
1	M	50%	M-M	25%	0.77	2.20	0	0	0.77	2.20
2	M	50%	M-M	25%	2.27	2.68	[0,0.02]	0	[2.25,2.27]	2.68
3	M	50%	M-M	25%	21.12	10.44	[3.31,5.09]	[4.32,7.61]	[16.03,17.81]	[2.83,6.12]
4	M	50%	M-M	25%	0.40	0.28	[0,0.30]	0	[0.10,0.40]	0.28
1	H	30%	M-H	15%	0.77	2.20	0	0	0.77	2.20
2	H	30%	M-H	15%	2.27	2.68	0	0	2.27	2.68
3	H	30%	M-H	15%	21.12	10.44	[0,4.26]	[3.43,5.88]	[16.86,21.12]	[4.56,7.01]
4	H	30%	M-H	15%	0.40	0.28	[0,0.02]	[0,0.28]	[0.38,0.40]	[0,0.28]
1	L	20%	H-L	6%	0.77	2.20	[0,0.77]	[0,0.39]	[0,0.77]	[1.81,2.20]
2	L	20%	H-L	6%	2.27	2.68	[0,1.23]	[0,1.73]	[1.04,2.27]	[0.95,2.68]
3	L	20%	H-L	6%	21.12	10.44	5.68	6.33	15.44	4.11
4	L	20%	H-L	6%	0.40	0.28	[0,0.40]	[0,0.07]	[0,0.40]	[0.21,0.28]
1	M	50%	H-M	15%	0.77	2.20	0	0	0.77	2.20
2	M	50%	H-M	15%	2.27	2.68	0	[0,0.02]	2.27	[2.66,2.68]
3	M	50%	H-M	15%	21.12	10.44	[3.31,5.39]	[4.32,7.31]	[15.73,17.81]	[3.13,6.12]
4	M	50%	H-M	15%	0.40	0.28	[0,0.02]	[0,0.28]	[0.38,0.40]	[0,0.28]
1	H	30%	H-H	9%	0.77	2.20	0	0	0.77	2.20
2	H	30%	H-H	9%	2.27	2.68	0	0	2.27	2.68
3	H	30%	H-H	9%	21.12	10.44	[0,3.98]	[3.43,6.16]	[17.14,21.12]	[4.28,7.01]
4	H	30%	H-H	9%	0.40	0.28	[0,0.30]	0	[0.10,0.40]	0.28

$y_{112opt} = 0$ $y_{122opt} = 0.64$ $y_{132opt} = 1$ $y_{142opt} = 0.79$ $y_{212opt} = 1$ $y_{222opt} = 1$ $y_{232opt} = 1$ $y_{242opt} = 1$

Total net benefit (¥10^8) : $f_{opt}^{D\pm} = [63.83, 146.98]$

allocation plan. The objective function value under scenario D is $f_{\text{opt}}^{D\pm} =$ ¥ $[63.83,\ 146.98] \times 10^8$.

According to the results, water allocations are sensitive to $NB_{ijt}^{\pm}$ and $C_{ijt}^{\pm}$. Municipal and industrial uses should be ensured because they provide the highest benefit ($NB_{ijt}^{\pm}$) when their water demand is met, whereas they are subject to the highest penalty ($C_{ijt}^{\pm}$) if the promised water is not achieved. In comparison with that, agricultural and environmental uses correspond to lower benefits ($NB_{ijt}^{\pm}$) and penalties ($C_{ijt}^{\pm}$). Consequently, the allocation water to agricultural users should be diminished firstly in the case of inadequate water supplies, followed by that to environment users when necessary. We can get the final objective function value of $f_{\text{opt}}^{\pm} =$ ¥ $[61.05, 160.17] \times 10^8$ from Tables 6–10. The possible reasons that the range of $f_{\text{opt}}^{\pm}$ varies greatly are as follows: (a) the water resources parameter interval values change too much in this area; (b) although the penalties of per unit water shortage for agricultural users are low, the large amount of water shortage leads to excessive penalties. In all of the four scenarios, there are always users having a certain degree of water shortage. It can be seen that the contradiction between supply and demand of water resources in Harbin is outstanding. It is of great significance to study how to improve the efficiency of water resources utilization (such as optimizing industrial structure).

CONCLUSION

In this study, an inexact multi-stage interval-parameter partial information programming model (IMIPM) has been proposed for urban water resources planning and management under various uncertainties. The model is a first attempt to combine optimization techniques of interval-parameter programming (IPP), multistage stochastic programming (MSP), linear partial information (LPI) theory, and two-stage stochastic programming into one framework. Compared with the traditional models, the uncertain parameters and their interactions in an urban water resources management system could be analyzed better. In addition, an interactive algorithm is introduced to deal with this model. Firstly, it brings out IMSP models according to every extreme point of the LPI; then each IMSP model is solved via two deterministic submodels (Huang & Loucks 2000). Finally, we could get the interval solutions for supporting managers in having multiple decision alternatives.

The developed model has been applied to a real-world case study to verify the rationality and practicality of this method. The system parameters are expressed as interval numbers. By tackling partial probability distributions through the LPI method, we have four flow level probability scenarios. The winter half-year and summer half-year can reflect time dynamics characters. Therefore, the final Harbin water resources water distribution system benefit obtained through the IMIPM model is $f_{\text{opt}}^{\pm} =$ ¥ $[61.05, 160.17] \times 10^8$. It suggests that the system benefit would alter correspondingly from f_{opt}^{-} to f_{opt}^{+} when the manager chooses different water allocation schemes.

Even though this model can consider multiple uncertainties in the water resources management system, there are still many problems that need to be overcome in future studies. For example, it is necessary to take into account the water resource losses generated in the process of water resources extraction, transmission and distribution to increase the data accuracy of the model parameters.

ACKNOWLEDGEMENT

The authors would like to express sincere thanks to the Water Conservancy Research Office of Northeast Agricultural University.

DATA AVAILABILITY STATEMENT

All relevant data are included in the paper or its Supplementary Information.

REFERENCES

Barbarosoğlu, G. & Arda, Y. 2004 A two-stage stochastic programming framework for transportation planning in disaster response. *Journal of the Operational Research Society* **55** (1), 43–53.

Ben Abdelaziz, F. & Masri, H. 2005 Stochastic programming with fuzzy linear partial information on probability distribution. *European Journal of Operational Research* **162** (3), 619–629.

Charnes, A., Cooper, W. W. & Kirby, M. J. L. 1971 Chance-constrained programming: an extension of statistical method.

In: *Optimizing Methods in Statistics* (J. S. Rustagi, ed.), Academic Press, New York, USA, pp. 391–402.

Fan, Y. R., Huang, G. H., Guo, P. & Yang, A. L. 2011 Inexact two-stage stochastic partial programming: application to water resources management under uncertainty. *Stochastic Environmental Research and Risk Assessment* **26** (2), 281–293.

Guo, P. & Huang, G. H. 2009 Inexact fuzzy-stochastic mixed-integer programming approach for long-term planning of waste management – part A: methodology. *Journal of Environmental Management* **91** (2), 461–470.

Huang, G. H. 1996 IPWM: an interval parameter water quality management model. *Engineering Optimization* **26** (2), 79–103.

Huang, G. H. & Loucks, D. P. 2000 An inexact two-stage stochastic programming model for water resources management under uncertainty. *Civil Engineering and Environmental Systems* **17** (2), 95–118.

Jairaj, P. G. & Vedula, S. 2000 Multireservoir system optimization using fuzzy mathematical programming. *Water Resources Management* **14** (6), 457–472.

Jensen, O. & Wu, H. 2018 Urban water security indicators: development and pilot. *Environmental Science & Policy* **83**, 33–45.

Kofler, E. 2001 Linear partial information with applications. *Fuzzy Sets and Systems* **118** (1), 167–177.

Kofler, E. & Zweifel, P. 1991 Convolution of fuzzy distributions in decision-making. *Statistical Papers* **32** (1), 123–136.

Li, Y. P. & Huang, G. H. 2007 Inexact multistage stochastic quadratic programming method for planning water resources systems under uncertainty. *Environmental Engineering Science* **24** (10), 1361–1378.

Li, P. & Qian, H. 2018 Water resources research to support a sustainable China. *International Journal of Water Resources Development* **34** (3), 327–336.

Li, Y. P., Huang, G. H. & Baetz, B. W. 2006 Environmental management under uncertainty – an internal-parameter two-stage chance-constrained mixed integer linear programming method. *Environmental Engineering Science* **23** (5), 761–779.

Li, Y. P., Huang, G. H., Nie, S. L. & Liu, L. 2008a Inexact multistage stochastic integer programming for water resources management under uncertainty. *Journal of Environmental Management* **88** (1), 93–107.

Li, Y. P., Huang, G. H., Yang, Z. F. & Nie, S. L. 2008b IFMP: Interval-fuzzy multistage programming for water resources management under uncertainty. *Resources, Conservation and Recycling* **52** (5), 800–812.

Li, Y. P., Huang, G. H. & Nie, S. L. 2010 Planning water resources management systems using a fuzzy-boundary interval-stochastic programming method. *Advances in Water Resources* **33** (9), 1105–1117.

Liu, X. M., Huang, G. H., Wang, S. & Fan, Y. R. 2015 Water resources management under uncertainty: factorial multi-stage stochastic program with chance constraints. *Stochastic Environmental Research and Risk Assessment* **30** (3), 945–957.

Maqsood, I., Huang, G. H. & Scott Yeomans, J. 2005 An interval-parameter fuzzy two-stage stochastic program for water resources management under uncertainty. *European Journal of Operational Research* **167** (1), 208–225.

Pereira, M. V. F. & Pinto, L. M. V. G. 1991 Multi-stage stochastic optimization applied to energy planning. *Mathematical Programming* **52** (1–3), 359–375.

Ray, B. & Shaw, R. 2018 Implications of water insecurity and future prospects in Asian cities. In: *Urban Drought: Emerging Water Challenges in Asia* (B. Ray & R. Shaw, eds), Springer, Singapore, pp. 413–427.

Schultz, R., Stougie, L. & van der Vlerk, M. H. 1996 Two-stage stochastic integer programming: a survey. *Statistica Neerlandica* **50** (3), 404–416.

Wada, Y., van Beek, L. P. H., van Kempen, C. M., Reckman, J. W. T. M., Vasak, S. & Bierkens, M. F. P. 2010 Global depletion of groundwater resources. *Geophysical Research Letters* **37**, L20402.

Wang, D. 1997 An inexact approach for linear programming problems with fuzzy objective and resources. *Fuzzy Sets and Systems* **89** (1), 61–68.

Wang, Q., Li, S. & Li, R. 2019 Evaluating water resource sustainability in Beijing, China: combining PSR model and matter-element extension method. *Journal of Cleaner Production* **206**, 171–179.

Water Resources Group 2008 *Charting Our Water Future: Full Report*. Available from: https://www.mckinsey.com/~/media/mckinsey/dotcom/client_service/sustainability/pdfs/charting%20our%20water%20future/charting_our_water_future_full_report_.ashx.

Wu, C. B., Huang, G. H., Li, W., Xie, Y. L. & Xu, Y. 2015 Multistage stochastic inexact chance-constraint programming for an integrated biomass-municipal solid waste power supply management under uncertainty. *Renewable and Sustainable Energy Reviews* **41**, 1244–1254.

WWAP (UNESCO World Water Assessment Programme) 2016 *The United Nations World Water Development Report 2016: Water and Jobs*. UNESCO, Paris, France.

WWAP (UNESCO World Water Assessment Programme) 2019 *The United Nations World Water Development Report 2019: Leaving No One Behind*. UNESCO, Paris, France.

Yang, Z., Song, J., Cheng, D., Xia, J., Li, Q. & Ahamad, M. I. 2019 Comprehensive evaluation and scenario simulation for the water resources carrying capacity in Xi'an city, China. *Journal of Environmental Management* **230**, 221–233.

Yannopoulos, S., Giannopoulou, I. & Kaiafa-Saropoulou, M. 2019 Investigation of the current situation and prospects for the development of rainwater harvesting as a tool to confront water scarcity worldwide. *Water* **11** (10), 2168.

Zhou, Y., Huang, G. H. & Yang, B. 2013 Water resources management under multi-parameter interactions: a factorial multi-stage stochastic programming approach. *Omega* **41** (3), 559–573.

First received 17 August 2020; accepted in revised form 1 October 2020. Available online 16 October 2020

A method for classifying interrelation between sectoral regulatory laws and the 'water-energy-agriculture nexus concept' in Brazil

Cássia Juliana Fernandes Torres, Camilla Hellen Peixoto de Lima, Andrea Sousa Fontes, Daniel Veras Ribeiro, Ícaro Thiago Andrade Moreira IWA and Yvonilde Dantas Pinto Medeiros

ABSTRACT

We propose a method to classify the level of interrelation between the water, energy and agriculture sectoral regulatory laws and selected main characteristics of the nexus concept. This method was created to be applied with sectoral regulatory laws so a study case was conducted in Brazil. The results show that all analysed legislations have low to medium interrelation with the criteria intrinsic to the nexus concept; the policies that stood out the most, regarding the number of criteria present in their scopes, were the national irrigation policy and the desertification policy; and in order to operationalize the nexus concept in management models in the Brazilian context, adjustments and a restructuring of regulatory laws are necessary. The proposed method was able to investigate the interrelation between regulatory legislations with the characteristics of the nexus concept as well as their strengths and weaknesses.

Key words | methodological proposal, policy analysis, water-energy-agriculture nexus

HIGHLIGHTS

- The method allows the classification of the level of interrelation between sectoral regulatory laws and the main characteristics present in the nexus concept in an innovative approach.
- A simplified structure was applied based on criteria and scales that can be implemented for other spatial scales and countries.

Cássia Juliana Fernandes Torres (corresponding author)
Ícaro Thiago Andrade Moreira IWA
Yvonilde Dantas Pinto Medeiros
Department of Environmental Sanitary Engineering, Polytechnic School,
Federal University of Bahia (UFBA),
Salvador,
Brazil
E-mail: *torres.cjf@gmail.com*

Camilla Hellen Peixoto de Lima
Civil Engineering Program, Alberto Luiz Coimbra Institute for Postgraduate Studies and Research in Engineering (COPPE),
Federal University of Rio de Janeiro (UFRJ),
Rio di Janeiro,
Brazil

Andrea Sousa Fontes
Associate Professor Federal University Recôncavo of Bahia,
Cruz das Almas,
Brazil

Daniel Veras Ribeiro
Department of Materials Science and Technology, Polytechnic School,
Federal University of Bahia (UFBA),
Salvador,
Brazil

INTRODUCTION

Historically, institutional arrangements and sectorial regulations have been commonly organized independently, pushing each sector to develop its strategies, policies, plans, and actions (Eftelioglu *et al.* 2016; Embid & Martín 2018; Huckleberry & Potts 2019).

Developing and implementing sectoral policies independently and without taking into account the trade-offs and impacts between the multiple sectors involved can lead to risk and inconsistent approaches (De Strasser *et al.* 2016). Risks are inherent in systems where different sub-sectors share the same resources (Gallagher *et al.* 2016). Consequently, they may also face similar uncertainties associated with them. Inconsistencies often refer to the fact that policies are still being formulated individually,

doi: 10.2166/ws.2021.036

even recognizing the interrelations and interdependencies between related sectors and sub-sectors (Bhaduri *et al.* 2015).

Over time, global environmental changes, including climate change, have required greater understanding and consideration of the links and interdependencies between sectors such as water, energy, and food production (Leck *et al.* 2015). The idea of individuality came to be confronted with the need for integrated studies to answer increasingly complex and multidisciplinary problems (Hoff 2011; WEF 2011; Eftelioglu *et al.* 2016).

The recognition attributed to interrelation between the water, energy and agriculture sectors was initially discussed at the World Economic Forum in 2008 (Smajgl *et al.* 2016; Embid & Martin 2017), gaining international prominence through the Conference 'The Water-Energy and Food Security Nexus – Solutions for the Green Economy' in 2011 (Hoff 2011; WEF 2011; Liu *et al.* 2017). Interrelation in this case can be defined as the reciprocal relationship between sectors on a bilateral basis. Since the conference, the term 'water-energy-food nexus' (systems organized in networks) has been used in a variety of contexts to advance the understanding of their interrelations.

For Huckleberry & Potts (2019), the nexus concept has emerged as a powerful approach to address the social and environmental challenges created by land-use and climate change. Hoff (2011) sees the nexus concept as an analytical approach to derive strategic resource management solutions with an emphasis on integrated system efficiency rather than isolated sector productivity. There is no consensus on the nexus concept, it is interpreted as processes that integrates ideas and actions from different stakeholders present in different sectors in pursuit of a common goal to achieve sustainable development (Endo *et al.* 2017).

Because the nexus concept is perceived as integrated management, this terminology is often associated with integrated water resources management (Abdullaev & Rakhmatullaev 2016; Roidt & Avellán 2019), integrated environmental policy management (Venghaus & Hake 2018), and more recently integrated solid waste management (Garcia *et al.* 2019; Roidt & Avellán 2019). Due to the aforementioned concepts' previous existence, some authors question the nexus concept's innovation in conceptual terms and its lack of coherence in more practical terms (Biggs *et al.* 2015; Cairns & Krzywoszynska 2016; Wichelns 2017; Venghaus & Hake 2018).

Despite these criticisms, there is one feature that strongly defines the nexus concept and makes it an innovative approach, which is the shift from a single sector-centric perspective or a single resource to a multi-centred and multi-level perspective (Bazilian *et al.* 2011; Bréthaut *et al.* 2019). Dealing with water-energy-food nexus interrelations in a sustainable way has become one of humanity's great global challenges (Endo *et al.* 2015). Besides that, existing integrating concept applications have shown that the transformation required to implement these concepts requires a framework that explicitly explains interdependencies not only between resources but also between their policies (Venghaus & Hake 2018). The analysis of the coherence of sectoral policies and institutional arrangements through the nexus concept represents the essential aspects of achieving their operationalization.

This paper was based on knowledge gaps and challenges presented by Bazilian *et al.* (2011), Bhaduri *et al.* (2015), Liu *et al.* (2017), Larcom & Gevelt (2017), and Embid & Martín (2018) regarding the investigation of institutional and legal aspects involving the nexus concept. The gaps and challenges were: (I) institutional organizations are not yet sufficiently structured to support the nexus dialogue (Bazilian *et al.* 2011); (II) the lack of innovative methodologies and decision support tools to deal with complex interrelations between water, energy, and agriculture, minimizing investment risks and maximizing economic returns (Bhaduri *et al.* 2015; Liu *et al.* 2017; Mercure *et al.* 2019); (III) the need to evaluate whether existing sectoral regulations are appropriate to the nexus concept requirements (Larcom & Gevelt 2017; Embid & Martín 2018); and (IV) in some cases does not require a radical overhaul of the regulatory frameworks and governance system, as some regulations may be structured to address the implications of nexus analysis, requiring merely revisions and better articulation (Larcom & Gevelt 2017).

Furthermore, to our knowledge, no study addressing the nexus concept has developed a method for classifying the level of integration of sectoral policies concerning the main features surrounding the concept. Research directed at these aspects tends to help identify the limitations and challenges surrounding the implementation of the nexus concept in management models. By recognizing this knowledge gap and challenges, this article's key questions are:

(I) What is the level of interrelation between sectoral regulatory laws and the main characteristics present in the nexus concept? and
(II) What are the main strengths and weaknesses of sectoral regulatory laws upon the nexus concept?

To investigate the key questions raised, this study aims to propose a method to classify the level of interrelation between the water, energy and agriculture sectoral regulatory laws and selected main characteristics of the nexus concept.

PROPOSED METHOD

This section presents the method proposal for the classification and evaluation of regulatory legislation in the water, energy and agriculture sectors in relation to the main characteristics involving the nexus concept. The method was based on an adaptation of the organizational structure of multicriteria analysis. Multicriteria analysis structures a decision problem in terms of several possible alternatives and assesses each of them under various criteria at the same time (Hajkowicz & Collins 2007).

Although multicriteria analysis is widely applied into the context of decision-making, which does not match the proposal of the present study, previous experience from the main authors (Srdjevi *et al.* 2004; Medeiros *et al.* 2014; Torres *et al.* 2016) of this paper with this method inspired them to create an adaptation that fit this methodological proposal. The main aspect used from multicriteria analysis was the classification phase of alternatives according to multiple criteria.

Therefore, the method was structured in three stages as shown in Figure 1: (I) Step I: Definition of the criteria (Main characteristics of the composition of the nexus concept); (II) Step II: Definition of alternatives (Regulatory Legislation); (III) Step III: Evaluation and Classification of the interrelation between criteria and alternatives.

In order to test the proposed method minimum requirements were designed for each step of the application and a study case was conducted. The country selected for this application was Brazil. The country characterization and the minimum requirements are going to be described in more detail in the following sections.

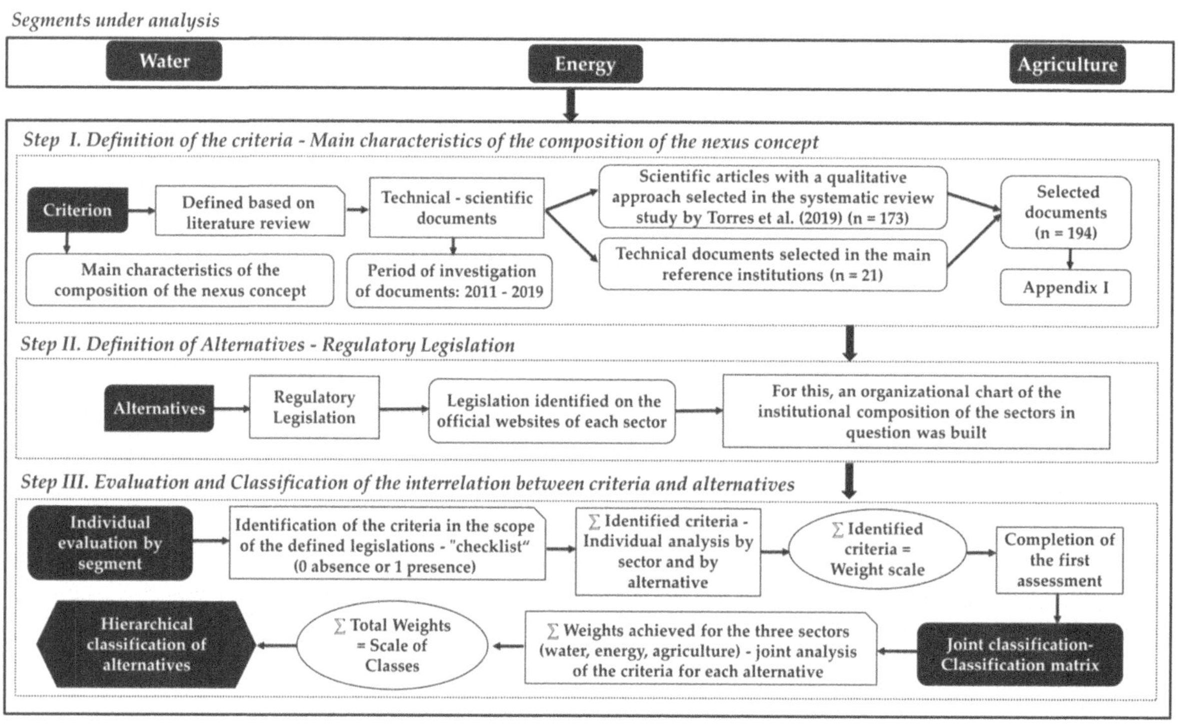

Figure 1 | Method steps.

Characterization of the case study: Brazil

Brazil is a country with continental dimensions (8,510,295.914 km^2), located in South America. The country stands out in the agricultural sector, being one of the main suppliers of raw materials to the world economy. In terms of energy, the country also excels in hydraulic generation and bioenergy production, it has a diversified matrix with a high share of renewables.

Despite the advances mentioned, in recent years, Brazil has been going through great uncertainties due to the water crises that have advanced in several regions, in addition to the consecutive political and economic conflicts at the national level with repercussions in the whole society.

Water crises and, its conflicts were a motivation to the first nexus concept studies in the country back in 2015. These studies were developed in order to propose solutions to the water crisis between two brazlian largest metropolises, Rio de Janeiro and São Paulo (Giatti *et al.* 2016). Although the nexus concept studies advanced in the country over the years (e.g. FAPESP 2015; Vilanova & Balestieri 2015; Giatti *et al.* 2016; INPE 2016; CNPq 2017; Arroyo 2018; Semertzidis *et al.* 2018; Castillo *et al.* 2019; Deveza, 2019; Mercure *et al.* 2019), none of the studies surveyed, so far, contemplates an investigation of the interrelation between regulatory legislation with the characteristics of the nexus concept.

In view of this, the development of the proposed procedure in Brazil is considered pertinent in order to identify the strengths and weaknesses of sectoral regulatory legislation aiming at greater integration between them.

Administrative political organization in Brazil

The Federal Constitution of 1988 organizes political and administrative Brazil in entities of the Federation formed by the Union, States (26), Federal District and municipalities (5,570), all endowed with autonomy in legislative, governmental, tax, and administrative powers (Brazil 1988). In other words, according to the Constitution, all the entities of the Federation can legislate, organize and manage their territory, and the Union must intervene only in situations of disorder that interfere with National integrity. Besides, it is important to mention that despite the autonomy of the Federated Entities, their legislative competences follow a hierarchy of power: Federation – State/Federal District – Municipality.

In the water sector, the Federal Constitution defines as the domain of the Union 'lakes, rivers and any body of water in lands that belong to its domain, or that bathe more than one State, serve as limits with other countries, or extend to foreign territory or from it, as well as marginal lands and river beaches' (Brazil 1988).

The energy sector is segmented between public and private companies in the attributions of generation, commercialization, transmission, and operation of energy. The Federal Constitution defines potential hydroelectric energy and mineral resources as federal assets, so the exploitation or use of these resources can only be carried out with authorization or concession from the federal government (Brazil 1988).

The agricultural sector, the Federal Constitution, played a fundamental role in its development by promoting the country's agricultural policy. This sector has undergone several cycles of transformation, from its rudimentary beginning and of little expression in the country to large mechanized agricultural complexes (agribusiness) of significant relevance in Brazil's trade balance.

Definition of criteria – main composition characteristics of the nexus concept (step I)

To identify the main characteristics that involve the nexus concept, a total of 194 technical – scientific documents were selected in the period from 2011 to 2019, as shown in Appendix I. These documents were surveys from two main sources: Torres *et al.* (2019) papers database; and technical documents from reference institutions such as, the United Nations Food and Agriculture Organization (FAO), the International Institute for Sustainable Development, the Economic Commission for Latin America and the Caribbean (ECLAC), the Institute of Stockholm Environment (SEI) and International Energy Agency (IEA).

Torres *et al.* (2019) conducted a literature review on the nexus concept, its database contains 304 papers, among those 173 qualitative scope and 131 quantitative scope. To develop the presented criteria for this study only articles of a qualitative scope were used from this database (173 papers), disregarding the articles classified as quantitative due to their greater discussion directed towards the application of the nexus concept and not merely their theoretical approach. Besides, the database additional 21 technical documents that were surveyed from the aforementioned institutions.

194 technical – scientific documents (173 papers and 21 technical documents) were used to propose 18 criteria for each sector that represent the main characteristics of the nexus concept for this study. In total there were 18 criteria (6 for water, 6 for energy and 6 for agriculture sectors).

Definition of alternatives – regulatory legislation (step II)

The alternatives are represented by the federal regulatory laws for the water, energy, and agriculture sectors. This step requires the exclusive usage of legislation of a regulatory nature. To identify these laws in Brazil, an online survey was carried out on the official websites of the ministries: for water resources, Agência Nacional de Águas (National Water Agency) <https://www.ana.gov.br/> , for energy, Ministério de Minas e Energia (Ministry of Mines and Energy) <http://www.mme.gov.br/> , and agriculture, Ministério da Agricultura, Pecuária e Abastecimento (Ministry of Agriculture, Livestock and Supply) <https://www.gov.br/agricultura/pt-br>. Additionally, government technical documents present in the Federal Government's official online diary were analyzed. Official acts of the executive, legislative and judicial public administration are published in the Diário Oficial da União (Official Gazette).

The application of the method in regulatory legislation of national scale was chosen because the Federal laws have precedence, that is, priority over State and Municipal laws according to the Federal Constitution of 1988 (Section 2.1.1). Therefore, the first changes that must be made in legislative terms must be at the Federal level, so as to guide the changes at lower levels.

Figure 2 contains an organizational chart of the institutional composition of each sector under analysis in Brazil. It can be observed that each sector is inserted in a governmental Ministry, and aligned to each Ministry are the specific secretariats, departments, councils, and bodies.

Evaluation and classification of the interrelation between criteria and alternatives (step III)

The evaluation step consisted of identifying whether the criteria are included in the scope of the alternatives (legislation). To identify the criteria in the scope of the legislation, a matrix must be filled out as a minimum requirement to move forward. The matrix contains two numerical codes, 1 to indicate presence of the criterion and 0 its absence, the outcome of this first matrix is a checklist.

After this process, the sum of all the criteria present in the alternatives analysed by the type of segment (water, energy, and agriculture) is performed. For example, in Legislation 1, x criteria of the water segment, x criteria of the energy segment and x criteria of the agriculture segment were found. The results of the sums for each segment corresponding to a weight scale as shown in Table 1.

The weight scale was built considering a total of six criteria per segment, totalling 18 criteria. There is no requirement for the number of criteria that one can use on its studies when applying this method, but if the criteria number changes a new scale may have to be adapted.

With the filling of the matrix by the weights shown in Table 1 (classification by segment) the joint classification is performed. This classification is performed considering the comparison between the sum of the weights of all the criteria fulfilled in the matrix with the classes determined in Table 2. The classes were defined symmetrically, both with the same amplitude, totaling six classes due to the maximum limit of the class being 18 (corresponding to the total number of criteria).

RESULTS AND DISCUSSION

Main composition characteristics of the nexus concept

Based on the 194 technical – scientific documents, eighteen (18) criteria were established as the main characteristics of the nexus concept for this study, with six representing the water, energy and agriculture segment (Table 3). Overall the criteria were based on two aspects: the occurrence of the terms in the analyzed technical – scientific documents and their level of relevance (essential aspects of the nexus concept). For this study, all the selected criteria have the same relevance in the evaluations of regulatory legislation, therefore no weight system was used to differentiate their relevance. All criteria are used to evaluate and classify all alternatives individually.

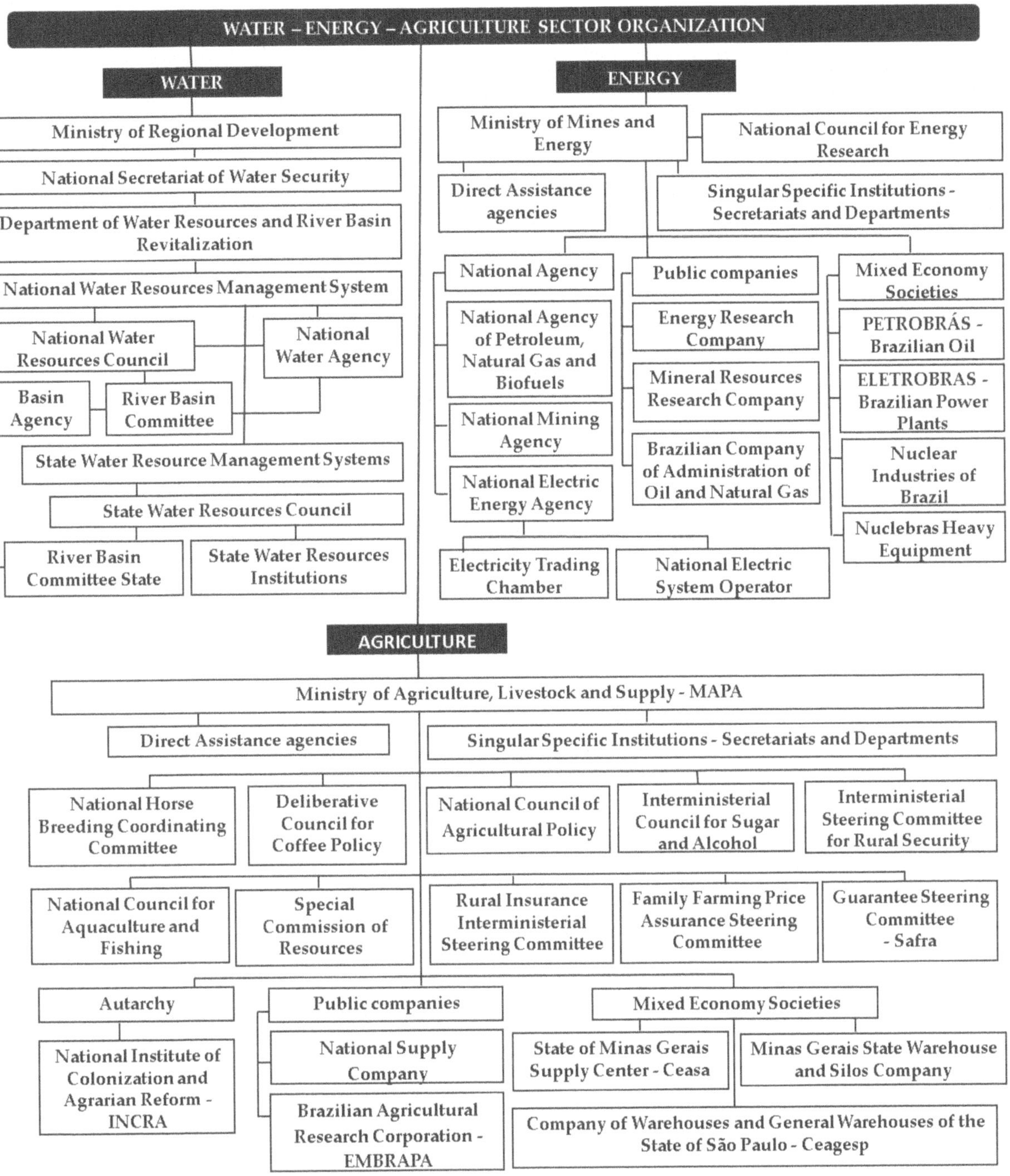

Figure 2 | Institutional organization of the Water-Energy-Agriculture sectors in Brazil.

Identification and characterization of the main regulatory laws in Brazil

In Brazil, ten federal regulatory laws were identified, as shown in Table 4. It can be seen that the laws were formulated in different time periods. Some of these laws have been updated over time, while other laws remain with the same scope.

The water sector in Brazil is governed by the National Water Resources Policy (Law n° 9433/1997), which brings water management as decentralized, participatory

Table 1 | Weight scale for individual assessment by segment

Category	Numerical Scale *	Description	Colour Scale
Very high	6	Six criteria were met	
High	5	Five criteria were met	
Intermediate	4	Four criteria were met	
Average	3	Three criteria were met	
Low	2	Two criteria were met	
Very low	1	Only one criterion was met	
It has no interrelation	0	No criteria met	

[a]Each segment has a maximum of six criteria as shown in the table.

Table 2 | Rating scale – joint classification

Description	Classes	Colour Scale
Excellent interrelation	16 -18	
High interrelation	13 -15	
Intermediate interrelation	10 - 12	
Average Interrelation	7 -9	
Low interrelation	4 - 6	
Very low interrelation	1 - 3	

and integrated (Brazil 1997). The sector has recently undergone changes in view of the update of the National Basic Sanitation Policy (Law n° 11,445/2007 updated by Law n° 14,026/2020). In its updated version, the National Water Agency (ANA) is responsible for establishing the reference standards for the regulation of public basic sanitation services (Article 1/Law n° 14,026/2020). This tends to approximate the relationship between water resources services and sanitation services in the country. In addition to this change, the new legal framework opens up the participation of private companies in public tenders for sanitation services, promoting greater competition between public and private companies, ending the preference of state-owned companies (Brazil 2020). For this sector, three legislations were selected for analysis.

Table 3 | Analysis criteria considering water, energy and agriculture segments

Water	Energy	Agriculture
Water security **(C1)**	Energy security **(C7)**	Food security **(C13)**
Conservation and rational use (efficient) of water **(C2)**	Conservation and rational use of energy **(C8)**	Encourage the use of unconventional water sources, promoting reuse **(C14)**
Integrated relationship between surface water and groundwater **(C3)**	Mitigation of greenhouse gas emissions **(C9)**	The incentive to increase alternative energy sources (bioenergy, wind and solar) in agricultural systems **(C15)**
Water resources without dissociation between quantity and quality **(C4)**	The incentive to increase renewable energy sources in the energy matrix **(C10)**	The incentive for more sustainable production by reducing greenhouse gas emissions **(C16)**
Encourage the generation and transfer of technologies related to the efficient use of water (technological efficiency associated with water efficiency) **(C5)**	Foster the generation and transfer of technologies related to energy efficiency (technology efficiency associated with energy efficiency) **(C11)**	Foster the generation and transfer of technologies related to resource efficiency (technological efficiency associated with agricultural efficiency) **(C17)**
Integration with the National Water Resources Policy and the National environmental policy[a] **(C6)**	Integration with the National Energy Policy and the National Environmental Policy[a] **(C12)**	Integration with the Agricultural Policy and the National Environmental Policy[a] **(C18)**

[a]The National Water Resources, Energy and Agricultural Policies should always be compared along with the National Environmental Policy against other laws, the only exception is when they are compared with themselves.

Table 4 | Main federal legislation present in the water, energy and agriculture sectors in Brazil

Sectors	Federal regulatory legislation	Initials
Water	Law n° 9,433/1997 – National Water Resources Policy	L1
	Law n° 11,445/2007 – National Policy to Sanitation. Amended by Law n°. 14,026/2020	L2
	Law n° 13,153/2015 – National Policy to Combat Desertification and Mitigate the Effects of Drought	L3
Energy	Law n° 9,478/1997 – National Energy Policy. Amended by Law n° 12,490/201	L4
	Law n° 10,295/2001 – National Policy for Conservation and Rational Use of Energy	L5
	Law n° 13,576/2017 – National Biofuels Policy	L6
	Law n° 11,097/2005 – Introduction of biodiesel in the energy matrix. Amended by Law n° 13,033 of 24/2014.	L7
Agricultural	Law n° 8,171/1991 – National Agricultural Policy	L8
	Law n° 12,787/2013 – National Irrigation Policy: Amended by Law n° 13,702/ 2018	L9
	Law n° 12,188/2010 – Institutes the National Policy of Technical Assistance and Rural Extension for Family Agriculture and Agrarian Reform	L10

As for the energy sector, a major breakthrough was achieved with the National Biofuels Policy (Law n° 13,576/2017) proposed to contribute to meet the country's commitments to the Paris Agreement under the United Nations Framework Convention on Climate Change (Brazil 2017). The sector in the country is governed by Law n° 9,478/1997, which defines energy policy as 'a set of programs, plans, projects and legislation to develop the energy sector' (Brazil 1997). For this sector, four legislations were selected for analysis.

In the case of the agricultural sector, three legislations were selected for investigation. In Brazil, the sector is governed by Law n° 8,171/1991. Similar to the energy sector, agricultural policy is defined as a set of programs, plans, projects and legislation to develop and coordinate the agricultural sector (Brazil 1991).

Case study application

Based on the defined criteria and the selected legislation for each sector, the application of the proposed method was conducted. Table 5 contains the filling matrix with the results of the checklist carried out in the documents of the sectoral regulatory legislation. In this case, the weight scale presented in Table 1 was applied, which corresponds to the individual assessment of alternatives through the criteria for each segment.

Overall, the results show low interrelation between the analyzed laws against the criteria defined to represent the main characteristics present in the nexus concept. Some of the facts that may have influenced this result could be related to the following:

i. Hierarchical structure of Brazilian laws: The fact that in Brazil, the water, energy and agriculture sectors were structured in a segmented manner, as presented by the Federal Constitution of 1988 (Brazil 1988). This fact was also discussed by Benites-Lazaro *et al.* (2020), which affirms that land, energy and water are managed in isolation by separate and disconnected institutional entities. Within the scope of the analyzed legislation, it was possible to notice the lack of integration between them. Each legislation tends to focus only on its sector and does not cover direct communication with other laws.

The lack of integration between the laws and the low number of criteria shows that, in order to operationalize the proposed nexus concept, regulatory laws must be restructured, and along with them their institutional arrangements. This restructuring can be carried out through changes in current laws, it may not always require the creation of new laws. For example, the national sanitation legal framework was recently updated instead of being replaced by a new law. Its update included changes in its scope and the restructuring of its related institutions. Through these proposed changes, there is a chance that actions of the water resources sector will become closer to the actions of the sanitation sector, which are both interconnected by the same regulatory agency.

ii. Climate variability decreasing environmental resources (e.g. water security problems): for example, another

Table 5 | Result of individual evaluation by segment

Sector	Criteria	Water			Energy				Agriculture		
		L1	L2	L3	L4	L5	L6	L7	L8	L9	L10
Water	C1	0	0	1	0	0	0	0	0	0	0
	C2	1	1	1	0	0	0	0	1	1	1
	C3	0	0	0	0	0	0	0	0	0	0
	C4	1	0	0	0	0	0	0	0	1	0
	C5	0	1	1	0	0	0	0	0	0	0
	C6	1	1	1	0	0	0	0	0	1	0
∑ Water criteria		**3**	**3**	**4**	**0**	**0**	**0**	**0**	**1**	**3**	**1**
Weight scale for individual assessment by Segment *		Average	Average	Intermediate	It has no interrelation	It has no interrelation	It has no interrelation	It has no interrelation	Very low	Average	Very low
Energy	C7	0	0	1	0	1	0	0	0	0	0
	C8	0	1	0	1	1	0	1	0	0	0
	C9	0	0	0	1	0	1	0	0	0	0
	C10	0	0	0	1	0	1	1	0	1	0
	C11	0	0	0	1	0	1	0	0	0	0
	C12	0	0	0	0	1	0	1	0	1	0
∑Energy criteria		**0**	**1**	**1**	**4**	**3**	**3**	**3**	**0**	**2**	**0**
Weight scale for individual assessment by Segment *		It has no interrelation	Very low	Very low	Intermediate	Average	Average	Average	It has no interrelation	Low	It has no interrelation
Agriculture	C13	0	0	1	0	0	0	0	0	0	1
	C14	0	0	1	0	0	0	0	0	0	0
	C15	0	0	0	0	0	1	0	0	1	0
	C16	0	0	1	0	0	0	0	0	1	1
	C17	0	0	0	0	0	0	0	1	1	1
	C18	0	0	0	0	0	0	0	0	1	0
∑ Agriculture criteria		**0**	**0**	**3**	**0**	**0**	**1**	**0**	**1**	**4**	**3**
Weight scale for individual assessment by Segment *		It has no interrelation	It has no interrelation	Average	It has no interrelation	It has no interrelation	Very low	It has no interrelation	Very low	Intermediate	Average

[a]Weight scale defined in Table 1. L1: National Water Resources Policy; L2: National Basic Sanitation Policy; L3: National Policy to Combat Desertification and Mitigate the Effects of Drought; L4: National Energy Policy; L5: National Policy for Conservation and Rational Use of Energy; L6: National Biofuels Policy; L7: Introduction of biodiesel in the energy matrix; L8: Agricultural Policy; L9: National Irrigation Policy; L10: National Policy for Technical Assistance and Rural Extension for Family Farming and Agrarian Reform.

factor that has been occurring in Brazil that has a direct impact on sectoral interrelations refers to water scarcity. The country experienced a severe water scarcity crisis; among those, one hit its two biggest cities (Rio de Janeiro and São Paulo) in the period from 2014 to 2016 (Giatti *et al.* 2016; Soriano *et al.* 2016; Mercure *et al.* 2019). This crisis highlighted for the government and population the dependence of the energy and agriculture on the water sector. This crisis challenged stakeholders regarding the real causes of these problems and its solutions. Besides all these problems, in this period and region, there was also a conflict between the energy and water sectors because the electricity policy encouraged the use of water for generation of energy, affecting the drinking water availability (Mercure *et al.* 2019).

Another example was mentioned by Benites-Lazaro *et al.* (2020). The authors highlight that Brazilian stakeholders reacted to certain events (e.g. expansion of sugarcane and water scarcity), yet changes in public policies have failed to consider the interdependence among various sectors. The authors believe that the nexus concept can aid policy-making because it identifies the interdependencies among such linked systems and mitigates the emerging risks to resource security. The bioethanol sector (Law n° 13,576/2017) is used as an example of a possible pilot for the implementation of a nexus concept with a high chance of success in Brazil.

It should be noted that water is a link that connects all other sectors that make up a nation's economy and survival. Therefore, the lack of this resource brings the need for reflection on the current management model. In this scenario, legislation starts to be validated, discussed and reformulated, new regulations and government programs are structured, and a closer relationship between sectors that have joint demands and share risks tends to occur to contain conflicts.

Complementing the assessment of legislation by segment, Table 6 shows the results found from the final classification of regulatory legislation. As can be seen, the policies evaluated presented similar classifications, from low to very low interrelations with the criteria for characterizing the concept of nexus, except for the Irrigation Policy (L9), which obtained the highest score, being classified as average interrelationship. No analysed legislation scored above nine criteria. In other words, no policy was rated as high or excellent interrelation.

In a hierarchical manner, the laws of greatest interrelation with the characteristics of the nexus concept for the least interrelation are:

(I) Irrigation Policy (L9);
(II) National Policy to Combat Desertification and Mitigate the Effects of Drought (L3);

Table 6 | Result of the final classification of regulatory legislation

Sector	Water			Energy				Agriculture		
	L1	L2	L3	L4	L5	L6	L7	L8	L9	L10
∑ Criteria - Water	3	3	4	0	0	0	0	1	3	1
∑ Criteria - Energy	0	1	1	4	3	3	3	0	2	0
∑ Criteria - Agriculture	0	0	3	0	0	1	0	1	4	3
∑ Total	**3**	**4**	**8**	**4**	**3**	**4**	**3**	**2**	**9**	**4**
Rating scale - Joint classification*	Very low interrelation	Low interrelation	Average Interrelation	Low interrelation	Very low interrelation	Low interrelation	Very low interrelation	Very low interrelation	Average Interrelation	Low interrelation

[a]Class scale defined in Table 2.

(III) National Policy to Sanitation (L2), National Energy Policy (L4), National Biofuels Policy (L6) and National Policy of Technical Assistance and Rural Extension for Family Agriculture and Agrarian Reform (L10), all with the same score;
(IV) National Water Resources Policy (L1), National Policy for Conservation and Rational Use of Energy (L5) and Introduction of biodiesel in the energy matrix (L7), all with the same score;
(V) Agricultural Policy (L8), with the lowest score.

The agricultural policy (L8) stood out for having the lowest score: only two criteria were present in their scope, followed by the Water Policy (P1), National policy for the conservation and rational use of energy (L5) and the policy Introduction of biodiesel in the energy matrix (L7), with three criteria met, all of them belonging to their segment.

As for the integration between regulatory laws, it was possible to notice that the irrigation law, referring to the agricultural sector, was the only one that presented integration with more than two regulatory legislations, energy, water resources, sanitation, and environmental law.

For L1 (National Water Resources Policy), no associations were identified with energy policy and agricultural policy. It is important to note that agricultural activity and energy generation, in the water resources policy, are considered as users of water, as well as mining activities, industry, among others. Therefore, any mention of these activities is described in the context of 'users', as can be seen in its article 3, item IV: 'General guidelines for action for the implementation of the National Water Resources Policy are the articulation of water resources planning with that of user sectors and with regional, state and national planning' (Brazil 1997).

By making a comparative analysis of water resources policies and other sectoral regulatory policies, it was possible to notice that for the water sector, the desertification policy (L3) obtained a greater correlation with the proposed criteria than the water policy itself (L1) and sanitation (L2). As in the agricultural sector, the irrigation law (L9) has greater integration between the criteria under analysis than the agricultural policy itself (L8). In the case of the energy sector, the energy policy (L4) and the biofuel policy (L6) obtained the same number of criteria met in their documents.

It is important to highlight that the results presented in this section are associated exclusively with the validation of the proposed method. Its application consists of investigating sectoral regulatory laws regarding their level of interrelation with criteria that represent the main characteristics of the nexus concept. The proposed method cannot be classified as a nexus concept application itself. What it can do is aid in the identification of strengths and weaknesses present in the scope of regulatory laws against the principle within the nexus concept.

In this sense, the method proposed in the present study brings an investigation of the level of interrelation of the regulatory laws that coordinate the water, energy and agriculture sectors in the country, as well as evaluating criteria of great relevance for the security of its resources and that are recommended in the nexus concept.

The application of the proposed method made it possible to identify the main weaknesses and strengths present in the scope of each law (Appendix II). Despite the results indicated, it is important to mention that the application of the method does not directly refer to the operationalization of the nexus concept, but rather a necessary and important step to achieve this goal.

The big issue in integrating regulatory legislation and inserting factors that involve systems efficiency, refers to the fact that, in practical terms, the activities carried out in the water/sanitation, agriculture and energy sectors also tend to be integrated because they are governed by these laws. Thus, it is hoped that the decision-making processes, investments and actions are developed together, whenever they have an intersection point between them. In this sense, programs and actions implemented in one sector automatically refer to gains for other correlated sectors as long as the trade-offs and risks inherent in the process are accounted for. This implies better targeting of investments with reduction of waste of public funds, maximization of synergies, strengthening of governance systems and internalization of social and environmental impacts (Rasul & Sharma 2016; Pardoe *et al.* 2017; Stephan *et al.* 2018; Albrecht *et al.* 2018). For this to happen, one of the first steps is to assess the current scope of legislation, which was proposed and made in the present study.

FUTURE TRENDS

Once the method is applied into the Federal Regulatory Laws, the next step should be to study a smaller scale; for example, state and municipal laws of water, energy and agriculture. Studying small scales could aid the investigation of the level of interrelationship between the legislations of other spatial scales and to identify their points of interconnectivity with the proposed criteria, pointing out their strengths and weaknesses. The results to be achieved with this investigation must be compared with the results identified in the present study to have a greater coverage of the country's legal situation regarding the aspects investigated. In Brazil, it is recommended to carry out the analysis by state and regions (Northeast, North, Southeast, South, Midwest), considering its territorial dimension.

Another extension that must be carried out is to compose an investigation of public policies in Brazil. The proposed method could be applied to analyze government programs and project documents in the country's water, energy and agriculture sectors. Appendix III contains a list of the main public policies for the development of these sectors that could aid in this next phase. This investigation could identify possible overlaps between the programs and their actions and their level of interrelation.

The proposed method can be easily modified and also could be applied in different countries and scales. Some of the suggestions for adaptation include: identification and changes if necessary of the criteria to define the nexus concept to another study or scales (if the criteria number change the scale of weights would have to be adapted too); and more sectors could be added to the analysis besides water, energy and agriculture (e.g. economy and the environment could be added).

CONCLUSION

The nexus concept still represents a paradigm for public management of natural resources. Traditional models of natural resource management no longer respond to today's complex social and economic problems, so new strategies need to be considered. The nexus concept could be part of this new strategy. In this paper, a new method was proposed for the classification of sectoral regulatory laws based on the corresponding main characteristics of the nexus concept for this study. With the results achieved, the following can be concluded:

- All analysed legislations have low to medium interrelation with the criteria that represent the main characteristics of the nexus concept for this study. The policies that stood out the most, regarding the number of criteria present in their scopes, were the national irrigation policy and the policy to combat desertification.
- The analyzed laws must contain points of interconnectivity (water reuse and unconventional water sources in agricultural and energy policy; use of renewable energy sources in agricultural and water policy, among others) and its guidelines must advocate efficient and safe resources.
- Older laws need to be reviewed and updated in order to achieve higher interrelation between themselves and be more efficient and synergistic.
- The strengths identified in the laws must be strengthened with interconnected public policies aiming at greater efficiency in actions and the weaknesses must be carefully investigated so that they are eradicated.
- Current regulatory laws in the agricultural sector in Brazil need to consider water, energy, and technological efficiency as a prerequisite for releasing investment in sector projects and programs.
- The Brazilian energy sector needs to be tightly integrated with water and agricultural legislations, especially when referring to the biofuels and biodiesel policies that have a strong association with the agricultural sector. Some examples that characterize advances in the more intense approximation between the sectors under study in Brazil arose from the need to solve problems present in situations of a scarcity of a certain natural resource in common between them. In other words, because they share the water, energy, and agricultural risks and insecurities, they have tended, due to circumstances, to a greater recognition of the existence of interrelations and interdependencies between

them, which has fostered some joint actions. This represents management in a crisis; however, the nexus proposal is to make this interconnected management continuous throughout the process.

In order to achieve an efficient and synergistic intersectoral and shared management, changes in the medium to long term must be initiated by its institutional and legislative base. The proposed method represents an important factor in the change process; however, its individual application alone is not sufficient to achieve more ambitious goals in terms of operationalizing the main characteristics of the nexus concept proposed for this study. It should be noted that investigations about the practical effectiveness of the regulatory laws analyzed as well as their interference in society were not considered in this study. For this, it is recommended to structure research with this objective through the application of questionnaires, interviews, development of a focus group, or other methods that help to gather information about the applicability of these laws.

ACKNOWLEDGEMENTS

We, the authors, wish to thank the Interdisciplinary Center for Energy and Environment (CIEnAm) from the Federal University of Bahia, Brazil, for all the technical support and thank the Coordination for the Improvement of Higher Education Personnel (CAPES) for sponsoring financially this research.

CONFLICTS OF INTEREST

The authors declare no conflict of interest.

DATA AVAILABILITY STATEMENT

All relevant data are included in the paper or its Supplementary Information.

REFERENCES

Abdullaev, I. & Rakhmatullaev, S. 2016 Setting up the agenda for water reforms in Central Asia: does the nexus approach help? *Environmental Earth Sciences* **75**, 870.

Albrecht, T. R., Crootof, A. & Scott, C. A. 2018 The water-energy-food nexus: a systematic review of methods for nexus assessment. *Environmental Research Letters* **13**, 1–27.

Arroyo EMV. 2018 *Incorporação do Nexo Energia-água em um Modelo de Otimização da Expansão do Sistema Energético Brasileiro. (Incorporation of the Energy-Water Nexus in A Model to Optimize the Expansion of the Brazilian Energy System).* Tese (Doutorado em Planejamento Energético), Universidade Federal do Rio de Janeiro, Rio de Janeiro, Brazil.

Bazilian, M., Rogner, H., Howells, M., Hermann, S., Arent, D., Gielen, D., Steduto, P., Mueller, A., Komor, P., Tol, R. S. J. & Yumkella, K. K. 2011 Considering the energy, water and food nexus: towards an integrated modelling approach. *Energy Policy* **39**, 7896–7906. https://doi.org/10.1016/j.enpol.2011.09.039.

Benites-Lazaro, L. L., Giatti, L. L., Sousa Junior, W. C. & Giarolla, A. 2020 Land-water-food nexus of biofuels: discourse and policy debates in Brazil. *Environmental Development* **33**, 100491. https://doi.org/https://doi.org/10.1016/j.envdev.2019.100491.

Bhaduri, A., Ringler, C., Dombrowsk, I., Mohtar, R. & Scheumann, W. 2015 Sustainability in the water–energy–food nexus. *Water International* **40**, 723–732. doi:10.1080/02508060.2015.1096110.

Biggs, E. M., Bruce, E., Boruff, B., Duncan, J. M. A., Horsley, J., Pauli, N., Mcneill, K., Neef, A., Ogtrop, V., Curnow, J., Haworth, B., Duce, S. & Imanari, Y. 2015 Sustainable development and the water–energy–food nexus: a perspective on livelihoods. *Environmental Science & Policy* **54**, 389–397.

Brazil – Federal government. 1988 *Constituição Da República Federativa Do Brasil De 1988 (Constitution Of The Federative Republic Of Brazil 1988).* Brasília, 5 de outubro de 1988.

Brazil – Federal government. 1991 Lei n° 8.171, de 17 de janeiro de 1991 Dispõe sobre a política agrícola (Law 8,171, of January 17, 1991. Provides for agricultural policy). DOU de 18.1.1991.

Brazil – Federal government. 1997 Lei n° 9433 de 08 de janeiro de 1997. Institui a Política Nacional de Recursos Hídricos (Law n°. 9433 of January 8, 1997. Institutes the National Water Resources Policy). Publicada no D.O.U. em 09 de janeiro de 1997.

Brazil – Federal government. Lei n° 13.576 de 26 de dezembro de 2017. Dispõe sobre a Política Nacional de Biocombustíveis – RenovaBio (Law n°. 13,576 of December 26, 2017. Provides for the National Biofuel Policy – RenovaBio). DOU de 27.12.2017.

Brazil – Federal government. Lei n° 14.026, de 15 de julho de 2020. Lei de Saneameto Básico – Atualizada. (Law n°. 14,026 of July 15, 2020. Basic Sanitation Law – Updated). DOU de 16.7.2020.

Bréthaut, C., Gallagher, L., Dalton, J. & Allouche, J. 2019 Power dynamics and integration in the water-energy-food nexus: learning lessons for transdisciplinary research in Cambodia. *Environmental Science & Policy* **94**, 153–162. https://doi.org/10.1016/j.envsci.2019.01.010.

Cairns, R. & Krzywoszynska, A. 2016 Anatomy of a buzzword: the emergence of 'the water-energy-food nexus' in UK natural resource debates. *Environmental Science & Policy* **64**, 164–170. https://doi.org/10.1016/j.envsci.2016.07.007.

Castillo, R. M., Feng, K., Sun, L., Guilhoto, J., Pfister, S., Miralles-Wilhelm, F. & Hubacek, K. 2019 The land-water nexus of biofuel production in Brazil: analysis of synergies and trade-offs using a multiregional input-output model. *Journal of Cleaner Production* **214**, 52–61. doi:10.1016/j.jclepro.2018.12.264.

CNPq – Conselho Nacional de Desenvolvimento Científico e Tecnológico. 2017 Chamada MCTIC/CNPq N° 19/2017 – NEXUS I: Pesquisa e Desenvolvimento em Ações Integradas e Sustentáveis para a Garantia da Segurança Hídrica, Energética e Alimentar nos Biomas Caatinga e Cerrado. Ministério da Ciência, Tecnologia, Inovações e Comunicações – MCTIC.

Deveza, A. C. P. 2019 O nexo água-energia-alimento na operação de sistemas hidrotérmicos: estudo de caso da Bacia do rio São Francisco. Dissertação (Mestrado em Planejamento Energético) – Universidade Federal do Rio de Janeiro, Rio de Janeiro, Brazil.

De Strasser, L., Lipponen, A., Howells, M., Stec, S. & Bréthaut, C. 2016 A methodology to assess the water energy food ecosystems nexus in Transboundary River Basins. *Water* **8** (59). https://doi.org/10.3390/w8020059

Eftelioglu, E., Jiang, Z., Ali, R. & Shekhar, S. 2016 Spatial computing perspective on food energy and water nexus. *Journal of Environmental Studies and Sciences* **6**, 62–76.

Embid, A. & Martín, L. 2017 El Nexo entre el agua, la energia y la alimentación en América Latina y el Caribe Planificación, marco normativo e identificación de interconexiones prioritárias. *Comisión Económica para América Latina y el Caribe (CEPAL)*/Serie Recursos Naturales e Infraestructura **179**, 1–71.

Embid, A. & Martín, L. 2018 Lineamientos de políticas públicas: Un mejor manejo de las inter-relaciones del Nexo entre el agua, la energia y la alimentación (Public policy guidelines: The best management of the interrelationships of the Nexus between water, energy and food). In: *Comisión Económica para América Latina y el Caribe (CEPAL)*/Serie Recursos Naturales e Infraestructura, n. 189.

Endo, A., Burnett, K., Orencio, P. M., Kumazawa, T., Wada, C., Ishii, A., Tsurita, I. & Taniguchi, M. 2015 Methods of the water-energy-food nexus. *Water* **7**, 5806–5830. doi:10.3390/w7105806.

Endo, A., Tsurita, I., Burnett, K. & Orencio, P. M. 2017 A review of the current state of research on the water,energy, and food nexus. *Journal of Hydrology: Regional Studies* **11**, 20–30. https://doi.org/10.1016/j.ejrh.2015.11.010.

FAPESP – Fundação de Amparo à Pesquisa do Estado de São Paulo. 2015 *Chamada de Propostas de Pesquisa: Mudanças climáticas e suas relações com Energia, Água e Agricultura (Call for Research Proposals: Climate Change and its Relationship with Energy, Water and Agriculture)*. Programa FAPESP de Pesquisa sobre Mudanças Climáticas Globais (PFPMCG), São Paulo, Brazil.

Gallagher, L., Dalton, J., Bré Thaut, C., Allan, T., Bellfield, H., Crilly, D., Cross, K., Gyawali, D., Klein, D., Laine, S., Leflaive, X., Li, L., Lipponen, A., Matthews, N., Orr, S., Pittock, J., Ringler, C., Smith, M., Tickner, D., Von Schlippenbach, U. & Vuille, F. 2016 The critical role of risk in setting directions for water, food and energy policy and research. *Current Opinion in Environmental Sustainability* **23**, 12–16. https://doi.org/10.1016/j.cosust.2016.10.002.

Garcia, D. J., Lovett, B. M. & You, F. 2019 Considering agricultural wastes and ecosystem services in food-energy-water-waste nexus system design. *Journal of Cleaner Production* **228**, 941–955. https://doi.org/10.1016/j.jclepro.2019.04.314.

Giatti, L. L., Jacobi, P. R., Favaro, A. K. M. & Empinotti, V. L. 2016 O nexo água, energia e alimentos no contexto da Metrópole Paulista (The water, energy and food nexus in the context of the metropolis of São Paulo). *Estudos Avançados* **30** (88), 1–12.

Hajkowicz, S. & Collins, K. 2007 A review of multiple criteria analysis for water resource planning and management. *Water Resources Management* **21**, 1553–1566.

Hoff, H. 2011 Understanding the Nexus. Background Paper for the Bonn 2011. In *Conference: The Water, Energy and Food Security Nexus*. Stockholm Environment Institute – SEI, Stockholm, Sweden.

Huckleberry, J. K. & Potts, M. D. 2019 Constraints to implementing the food-energy-water nexus concept: governance in the lower Colorado River Basin. *Environmental Science and Policy* **92**, 289–298. https://doi.org/10.1016/j.envsci.2018.11.027.

INPE – Instituto Nacional de Pesquisas Espaciais. Projeto de pesquisa da Rede Brasileira de Pesquisas sobre Mudanças Climáticas Globais (Rede Clima). 2016 *Mapeamento da demanda de água, energia e alimento na bacia hidrográfica do rio São Francisco diante cenários de mudanças climáticas para definir soluções sustentáveis*. INPE, São Paulo, Brazil.

Larcom, S. & Gevelt, T. V. 2017 Regulating the water-energy-food nexus: interdependencies, transaction costs and procedural justice. *Environmental Science & Policy* **72**, 55–64. https://doi.org/10.1016/j.envsci.2017.03.003.

Leck, H., Conway, D., Bradshaw, M. & Rees, J. 2015 Tracing the water–energy–food nexus: description, theory and practice. *Geography Compass* **9/8**, 445–460. https://doi.org/10.1111/gec3.12222.

Liu, J., Yang, H., Cudennec, C., Gain, A. K., Hoff, H., Lawford, R., Qi, J., De Strasser, L., Yillia, P. T. & Zheng, C. 2017 Challenges in operationalizing the water–energy–food nexus. *Hydrological Sciences Journal* **62**, 1714–1720. https://doi.org/10.1080/02626667.2017.1353695.

Medeiros, Y. D. P., Torres, C. J. F., Ceccato, L., Damasceno, A., Pessôa, Z. B. & Fontes, A. S. 2014 *Participatory Decision-Making Methodology for Water Quality Management in the Brazilian Water Basin. Sustainable Watershed Management*, 1st edn. CRC Press, Salvador, Brazil, pp. 33–41.

Mercure, J.-F., Paim, M. A., Bocquillon, P., Lindner, S., Salas, P., Martinelli, P., Berchin, I. I., Andrade Guerra, J. D., Derani, C., Albuquerque Junior, C. L., De Ribeiro, J. M. P., Knobloch, F., Pollitt, H., Edwards, N. R., Holden, P. B., Foley, A., Schaphoff, S., Faraco, R. A. & Vinuales, J. E. 2019 System complexity and policy integration challenges: the Brazilian energy- water-food nexus. *Renewable and Sustainable Energy Reviews* **105**, 230–243. https://doi.org/10.1016/j.rser.2019.01.045.

Pardoe, J., Conway, D., Namaganda, E., Vincent, K., Dougill, A. J. & Kashaigili, J. J. 2017 Climate change and the water-energy-food nexus: insights from policy and practice in Tanzania. *Climate Policy* 1–17. doi:10.1080/14693062.2017.1386082.

Rasul, G. & Sharma, B. 2016 The nexus approach to water–energy–food security: an option for adaptation to climate change. *Climate Policy* **16** (6), 682–702. doi:10.1080/14693062.2015.1029865.

Roidt, M. & Avellán, T. 2019 Learning from integrated management approaches to implement the Nexus. *Journal of Environmental Management* **237**, 609–616. https://doi.org/10.1016/j.jenvman.2019.02.106.

Semertzidis, T., Spataru, C. & Bleischwitz, R. 2018 The nexus: estimation of water consumption for hydropower in Brazil. *Journal of Sustainable Development of Energy. Water and Environment Systems* **7** (1), 122–138.

Smajgl, A., Ward, J. & Pluschke, L. 2016 The water–food–energy nexus – Realising a new paradigm. *Journal of Hydrology* **533**, 533–540. doi:10.1016/j.jhydrol.2015.12.033.

Soriano, E., Londe, L. d. R., Gregorio, L. T. D., Coutinho, M. P. & Santos, L. B. L. 2016 Crise hídrica em São Paulo sob o ponto de vista dos desastres (Water crisis in São Paulo from the point of view of disasters). *Ambiente & Sociedade* **19** (1), 21–42. https://doi.org/10.1590/1809-4422asoc150120r1v1912016.

Srdjevi, B. M., Medeiros, Y. D. P. & Faria, A. S. 2004 An objective multi-criteria evaluation of water management scenarios. *Journal of Water Resources Planning and Management* **1**, 65–84.

Stephan, R. M., Mohtar, R. H., Daher, B., Irujo, A. E., Hillers, A., Ganter, J. C., Karlberg, L., Martin, L., Nairiz, S., Rodriguez, D. J. & Sarni, W. 2018 Water–energy–food nexus: a platform for implementing the Sustainable Development Goals. *Water International* **43**, 1–8.

Torres, C. J. F., Medeiros, Y. D. P. & Freitas, I. M. D. P. 2016 Training watershed committee members to aid on the decision-making process for the execution program of the framework of water bodies. *RBRH* **21**, 314–327. https://doi.org/10.21168/rbrh.v21n2.p314-327.

Torres, C. J. F., Lima, C. d., Goodwin, B. d. A., Aguiar Junior, T. d., Fontes, A. S., Ribeiro, D. V., Silva, R. d. & Medeiros, Y. D. P. 2019 A literature review to propose a systematic procedure to develop 'Nexus thinking' considering the water–energy–food nexus. *Sustainability* **11**, 7205. doi:10.3390/su11247205.

Venghaus, S. & Hake, J.-F. 2018 Nexus thinking in current EU policies – The interdependencies among food, energy and water resources. *Environmental Science & Policy* **90**, 183–192. https://doi.org/10.1016/j.envsci.2017.12.014.

Vilanova, M. R. N. & Balestieri, J. A. P. 2015 Exploring the water-energy nexus in Brazil: the electricity use for water supply. *Energy* **85**, 415–432. https://doi.org/10.1016/j.energy.2015.03.083.

WEF – World Economic Forum. 2011 Water security: the water-food-energy-climate nexus. In: *Conference: The Water, Energy and Food Security Nexus*. Island Press, Washington.

Wichelns, D. 2017 The water-energy-food nexus: is the increasing attention warranted, from either a research or policy perspective? *Environmental Science & Policy* **69**, 113–123.

First received 31 August 2020; accepted in revised form 27 January 2021. Available online 8 February 2021

IWA Publishing's authorised EU representative for General Product Safety Regulations is Diane D'Arras, 15 rue Duret, 75116 Paris, France, e-mail: safety@iwap.co.uk.

Printed and bound by CPI Group (UK) Ltd, Croydon, CR0 4YY
12/05/2026
02108874-0005